Ergonomics and Health in Modern Offices

Ergonomics and Health in Modern Offices

Edited by

Etienne Grandjean

Institut für Hygiene und Arbeitsphysiologie,
Zurich, Switzerland

Proceedings of the International Scientific Conference on Ergonomic and Health Aspects in Modern Offices, held in Turin, Italy on 7–9 November 1983 under the auspices of the International Ergonomics Association, the Permanent Commission and International Association of Occupational Health, the University of Turin, the Societa Italiana di Medicina del Lavoro e di Igiene Industriale and the Regione Piemonte

Taylor & Francis
London and Philadelphia
1984

UK Taylor & Francis Ltd, 4 John St, London WC1N 2ET
USA Taylor & Francis Inc., 242 Cherry St, Philadelphia, PA 19106–1906

British Library Cataloguing in Publication Data

International Scientific Conference on
Ergonomic and Health Aspects in Modern
Offices *(1983 : Turin)*
Ergonomics and health in modern offices.
1. Video display terminals—Human factors
I. Title II. Grandjean, E.
651.8′ 443 TK7882.16

ISBN 0-85066-270-2

Library of Congress Cataloging in Publication Data

International Scientific Conference on Ergonomic and
Health Aspects in Modern Offices (1983: Turin, Italy)
Ergonomics and health in modern offices.

"Proceedings of the International Scientific Conference on Ergonomic and Health Aspects in Modern Offices, held in Turin, Italy on 7–9 November 1983 under the auspices of the International Ergonomics Association, the Permanent Commission and the International Association of Occupational Health, the University of Turin, the Societa Italiana di Medicina del Lavoro e di Igiene Industriale, and the Regione Piemonte."
Bibliography: p.
Includes index.
1. Offices—Hygienic aspects—Congresses. 2. White collar workers—Diseases and hygiene—Congresses. 3. Occupational diseases—Congresses. 4. Human engineering—Congresses. I. Grandjean, E. (Etienne) II. International Ergonomics Association. III. Title.
RC965.03158 1983 616.9'803 84-8470
ISBN 0-85066-270-2

Typeset by Red Lion Setters, London WC1
Printed in Great Britain by Taylor & Francis (Printers) Ltd, Basingstoke, Hants.

Contents

Preface

Most of the papers presented at the Turin Conference reveal, in different ways, that office life is becoming more technical and more complicated and that several office jobs are today more strenuous than they were in the first half of this century. Greater investments in offices are associated with an increased need for more productivity and these conditions have led, among other changes, to large, open offices. Such offices pose new environmental problems and there is an increasing use of office machines, especially computer terminals. The result is the transformation of the former 'paper office' into the present 'electronic office'.

At the traditional office desk an employee has a great variety of physical activities and a large space for various body postures and movements: he/she may look for some documents, read some texts, exchange information with colleagues, type for a short while and carry out many other activities during the course of the working day. Ergonomic shortcomings are unlikely to cause annoyance or physical discomfort.

The situation is entirely different for an operator working continuously at a VDT for several hours or for a whole day. Such an operator is tied to a man–machine system: movements are restricted, attention is directed to the screen or source documents and the hands are linked to the keyboard. These operators are more vulnerable to ergonomic shortcomings, to constrained postures, to unsuitable lighting conditions and to uncomfortable furniture. They are more sensitive to visual loads and to unnatural body postures. The recognition of these changing circumstances was the main reason for organizing an International Scientific Congress on Ergonomics and Health in Modern Offices. The aim of the Congress and of course also of the present Proceedings is to analyse objectively the effects of the new office conditions and to lay the foundations for improvements of working life in offices.

Zurich, 12 January 1984 *Etienne Grandjean*

Acknowledgement

The International Scientific Conference on Ergonomics and Health in Modern Offices (Turin, 7–9 November 1983) and the publication of the present Proceedings were generously supported by the following sponsors:

Burroughs Corporation (Plymouth, USA)
Camera di Commercio Industria Artigianto e Agricoltura (Turin, Italy)
Cassa di Risparmio di Torino (Turin, Italy)
COMAU (Turin, Italy)
Giroflex Entwicklungs A.G. (Koblenz, FR Germany)
IBM World Trade Corporation (White Plains, USA)
ILTE (Moncalieri, Italy)
Istituto Bancario San Paolo di Torino (Turin, Italy)
Olivetti (Ivrea, Italy)
Regione Piemonte (Turin, Italy)
SEAT (Turin, Italy)
Siemens (FR Germany)
Siemens Data (Milan, Italy)
SIP (Rome, Italy)
Standard Telephon and Radio A.G. (Zurich, CH)
Tandberg Data (Norway)
Toro Assocurazioni (Turin, Italy)
Università degli Studi di Torino (Turin, Italy)

THE AMBIENT ENVIRONMENT IN OFFICES

Toxic Agents Emitted from Office Machines and Materials (Introductory paper)

GIOVANNI SCANSETTI

Institute of Occupational Health, Turin University, Turin, Italy

1. *Introduction*

Office work mainly deals with handling, transmitting and storing information. For such a purpose, paper is the means currently used, although its replacement by magnetic and electronic devices is expected and partly in progress. Therefore we may look forward to tomorrow's office without paper. In time the silicon machine will replace the carbon one.

With regard to today's office only, I will consider some points of possible toxicological interest. First I will deal with the few simple instruments currently used and then the new automated systems for treating information. Finally some components of office environment will be considered.

2. *Paper, Carbon Paper*

Paper dermatitis has been reported to represent 0·8% of all occupational dermatitis (Meneghini *et al.*, 1963; Adams, 1983) though paper is extensively used in modern society.

Carbon paper can cause allergic dermatitis: this applies less to the so-called smudging type than to the non-staining one; the sensitizing substance is tricresyl phosphate (TCP), reported by Hjort (1964) to be present at a concentration of 30% in the film emulsion of the carbon paper used by a man who had a positive patch test to that particular type of carbon paper, to TCP, and to triphenyl phosphate through cross-sensitization. TCP, the *ortho* isomer of which is so toxic that the amount is kept as low as possible, is widely used as a plasticizer in the most common plastics but not frequently in the manufacture of carbon paper.

Some four other cases of sensitization to carbon paper were identified among 40 000 eczema patients at the Finsen Institute, Copenhagen: one of them had positive skin reaction to oleyl alcohol, whereas details of an allergen identified in the other three cases were not given (Hjorth, 1964).

Calnan and Connor (1972) described a case of allergic contact dermatitis due to nigrosine base (C.I. 42535B, solvent black 7), a tricyclic azine dye derived from phenazine, used to obtain extremely good black copies in high-speed computer printers.

Calnan and Fregert (personal communication to Jordan, 1975) have also documented allergy to methyl violet in carbon paper, and Jordan (1975) described a case of bilateral dermatitis in a young female secretary who had a strongly positive patch test to two types of typewriter correcting paper. The implicated chemical was probably identified in a modified phenol formaldehyde maleic anhydride resin, already found responsible for allergic dermatitis in the use of a commercial marking pen ink (Mailbach, personal communication to Jordan, 1975).

3. *Carbonless Copy Paper*

In 1954 the National Cash Register Company (Dayton, Oh.) introduced its 'no carbon required' paper (NCR): this is a carbonless copy paper (CCP) or pressure-sensitive paper working by mechanical or chemical transfer. In the former the undersurface of the top sheet is coated with a carbon-like film and the coloured ink is simply transferred by pressure to the top surface of the underlying sheet. In the chemical type the undersurface of the top sheet (called 'coated back') holds some colour-forming chemicals, while the reacting material is place on the top surface of the underlying sheet (called 'coated front'). The number of copies can be increased by coating the second and subsequent sheets on the undersurface with a colour-forming emulsion, and the top of the underlying sheets with reacting material. Moreover some areas of the top surface of the second and following sheets may have a special coating of desensitizing ink (D-ink), which is applied by printers (Dodds and Butler, 1981), to prevent the dye from being transferred (Menné *et al.*, 1981).

Colour forming chemicals mainly derive from triphenyl- and triaryl-methane are encapsulated in a colourless state. They include crystal violet (methyl violet) and malachite green lactones, of the triphenyl-methane group, and benzoyl leucomethylene blue and fluoran derivatives.

The solvents used to disperse them must have low volatility, in order to prevent drying of the capsule content (Göthe, 1981); from 1954 (the date of the original paper) until 1970 the main solvent used was a chloro-biphenyl (Aroclor), thereafter it was replaced by hydrogenated terphenyls (Weaver *et al.*, 1979), diaryl-ethanes, alkylnaphthalenes, cyclohexane, dibutyl-phthalate, often diluted with odourless kerosene for economic reasons.

Gelatin and gum arabic were used in the first commercially valuable microencapsulation process developed by NCR, the reaction forming viscous liquid microdroplets of polymer coacervate. The capsule wall may be hardened, e.g. by the addition of formaldehyde. Using two gelatins of different isoelectric point, hardening may be accomplished by adding glutaraldehyde (Sparks, 1981). Carboxymethylcellulose and synthetic polymers are now frequently used in combination, the latter including polyesters, polyamides and polyurethanes: encapsulation proceeds through a chemical method of interfacial polymerization.

The emulsion of the capsule is spread, dried and held adherent to the undersurface of the paper with water-soluble starch.

It is worth noting that four men working in a factory producing this emulsion were reported to be suffering from hand eczema; they were positively patch-tested with 1, 2-benzisothiazolin-3-one contained in Proxel, ICI, a gelatin preservative in the capsule; in two cases this sensitization was only an aggravating factor of a pre-existing eczema (Cronin, 1980).

The reactive material on the top surface of the copy sheet contains a montmorillonite clay, alkaline on the surface but acid inside, or an alternative coreacting system material, spread, dried and adhered by a styrene–butadiene latex. Attapulgite clay, the chief ingredient of fuller's earth, was used for this purpose, but recently it has been replaced by other materials in NCR paper; at the present time research is in process to produce a chemically processed bentonite, kaolin or sepiolite clay, or their combination, for use in duplicating paper (Grim, 1979).

Sepiolite, and perhaps attapulgite, occur in fibrous form, very familiar to chrysotile fibrils (Leineweber, 1980; Bignon *et al.*, 1980); mesotheliomas have been induced through peritoneal injection of attapulgite fibres in rats (Pott *et al.*, 1976), and lung fibrosis has been observed in a worker after inhalation of fibrous attapulgite (Sors *et al.*, 1979).

The complaints related to the use of CCP have frequently an epidemic characteristic, involving people engaged in various office functions such as writing, carrying out surveys on recently written documents, or collating papers. Also there is often a positive correlation between the number of forms used and discomfort; the rupture of the capsules seems important, since generally only used paper causes the complaints. These concern the skin of the exposed parts of the body, namely face and hands, consisting of dryness, itching, rashes, burning or prickling sensations and swelling, and, in some cases, of active eczema. Eyes are also involved with itching red and swollen eyelids, soreness, photophoby and injection of conjunctiva. Mucous membranes symptoms include dryness and/or burning of lips, tongue and throat, soreness in the throat and chest, sneezing, stuffed nose, and hoarseness. Symptoms such as headache, drowsiness and thirst also occur.

Patch tests with samples of papers and colour formers have always been negative. The only exception is the case described by Marks (1981), of a young woman with allergic contact dermatitis from CCP who was positive to Michler's hydrol (4, 4′ -bis(dimethylamino)benzhydrol) a component of the colour former molecule, the *para*-toluene-sulphinate of Michler's hydrol, a colourless dye salt.

Prick and photo-patch tests with the papers themselves and the different substances of the papers usually gave negative results. Prick tests with some papers which were positive in a group of cases, have been related to a non-immunological histamine liberation (Menné *et al.*, 1981).

Generally speaking it seems that the companies marketing CCP have received health complaints on a much greater scale than previously to introducing these papers 20 years ago.

Altogether the reactions to CCP look irritative in nature rather than allergic; tracing the causative agent(s) is more difficult. The first claims arose in 1972 when Masuda *et al.* demonstrated that the solvents of colour former in the capsules were frequently polychlorinated biphenyls

(PCBs): at that time the amount detected in Japanese brands of CCP was 22–64 mg/g of paper for gelatin coated sheet, and 200–280 p.p.m. for the uncoated ones. PCBs, quickly absorbed through the skin (Voes and Beems, 1971), were the cause of Yusho, the epidemic illness from oil-treated rice which exploded in Japan in the late 1960s affecting more than a thousand people. This caused the suspension of PCBs production in that country in 1972.

Because of the tendency of these products to accumulate and persist in the environment and biota, and owing to their toxic effect, production was also stopped in USA in 1976, and their use was generally restricted to closed systems.

Even if since the early 1970s PCBs have no longer been used in dissipative systems as CCPs are, it is possible that some older CCPs stocks, containing PCBs, were still in use in the mid-1970s, e.g. in Italy, as pointed out by conflicting analytical results (Belliardo *et al.*, 1979; Benvenuti *et al.*, 1979; Sampaolo *et al.*, 1980).

Moreover, the picture of CCP complaints does not agree with PCBs type of biological activity; other hypotheses were related once more to unspecified solvents (Calnan, 1979; Menné *et al.*, 1981), also because of the similarity of the symptoms to those noticed among users of photocopying machines with wet toners (Jensen and Roed-Petersen, 1979); or to the clay of the receiving surface (Magnusson, 1974). In the experience of Magnusson (1974) the use of a formaldehydic resin as adhesive for clay stopped it from being airborne, and the complaints disappeared. Such an observation is partly in contrast with the results, and subsequent interpretation, of sampling and analysing air drawn through columns containing cut-up and crumpled forms: a significant release of formaldehyde from copy paper, especially if coming from unopened new packages, was demonstrated. Formaldehyde was thus suspected of causing the irritation, although not being the only source of it (Gockel *et al.*, 1981; Schumacher, 1981). It is worth recalling that contact lens wearers are greatly sensitive to very low concentrations of formaldehyde (Steinberg, 1982).

Of course, we cannot forget that most of the complaints started from used CCP; however a formaldehyde allergy was suspected at the beginning of the 1970s by the Swedish National Board of Occupational Safety and Health as a possible cause of discomfort due to the use of NCR papers (Lidblom, 1981). More recently Swedish authors (Nörback, 1981) found that there was a significant relationship between the occurrence of complaints and the presence of D-ink on the form. Two out of seven of these inks turned out to be primary irritants, predominantly of the skin, while symptoms related to NCR papers not treated with D-ink predominantly concern mucous membranes. Suspicions have been directed to the amines contained in the irritant D-ink (Löfström, 1981). A linkage between a particular D-ink and severe skin and eye irritation was already present in an outbreak in Belgium in 1975, reported by Dodds and Butler (1981) with particular emphasis on one of its ingredients, namely 1-hydroxyethyl-2-oleyl-imidazoline.

As itching is one of the main symptoms associated with NCR paper exposure, the itchy sensation was studied quantitatively with an 'itch test' on patients with this symptom when handling copy paper who turned out to be more sensitive to an itch-producing agent than controls matched for sex and age (Jeansson, 1981).

Mucous membranes symptoms turned out to be significantly related with papers containing mono-isopropyl-biphenyl (MIPB) in the capsule as solvent (Nörbäck, 1981).

This is the most important of alkylated biphenyls, and is primarily used in the production of NCR paper (Weaver *et al.*, 1979). An abnormally large amount of MIPB impurities (biphenyl, methylbiphenyl and diisopropylbiphenyl) were found in specimens of Norwegian CCP associated with the only hitherto known epidemic in that country (Levy and Hanoa, 1980). In MIPB containing papers related to complaints both in Norway and Sweden, a specific unprecised colour former was also present.

4. *Rubber Sensitization*

Rubber fingerstalls were used by office staff as an aid to counting bank notes in Swedish Post Giro Offices (Eriksson and Östlund, 1968). British (Kirton and Wilkinson, 1972) and Danish post-sorters (Roed-Petersen *et al.*, 1977) as well as Copenhagen Telephone Company workers used them when handling no-carbon-required paper (Menné *et al.*, 1981).

This has caused outbreaks of contact band dermatitis localized on the dorsum and sides of distal phalanx: the patients were two-day patch-tested both with the rubber and its individual compounds. Swedish and British workers experienced sensitivity to rubber accelerators of similar structure, namely *N*-cyclohexyl-2-benzothiazole-sulphenamide (CBS) (Eriksson and Östlund, 1968) and dibenzthiazyl-disulphide (Kirton and Wilkinson, 1972).

The 51 Danish cases had a positive reaction to *N*-isopropyl-*N'*-phenyl-*p*-phenylendiamine (IPPD), a rubber antioxidant. Following the BRMA Code of Practice (1978) IPPD is a potent skin sensitizer; rubber which contains this compound may also cause skin sensitization.

5. *Diazocopying*

Diazocopying technique (or dye line or diazo process) utilizes a diazo-sensitized paper, into which the image of a transparent original is projected by a bright fluorescent light passing through it. Then the paper is developed (coloured) by either ammonia gas or a liquid chemical giving a positive blue-print copy. The process is used in particular for plan printing.

Sensitivity to diazonium salts (chloride; acid sulphate) was described in industrial settings: occupational asthma in men engaged in the production of the reactive dyes (Armeli, 1968), or in the first stages of the manufacture of the sensitized paper, i.e. the weighing and mixing of powders including diazonium chloride (Graham *et al.*, 1981). Contact eczema in man involved in coating rolls of paper with the diazo solution is also reported: patch-tests with the copy paper ('Amonax') and a 10% concentration of the solution were strongly positive (Harman *et al.*, 1968).

Diazocopying may induce desquamative or degenerative dermatitis, caused by ammonia impairing fingertip skin whilst handling copies (Gertler and Laubstein, 1963) and also severe cases of contact eczema starting from the hands and elbows and spreading in some cases to the whole body. The sensitivity was traced to *p*-diethylamino-azobenzene chloride, zinc chloride double salt, the active substance of the paper, as well as to other '*para*' compounds through cross-sensitization (Gertler and Laubstein, 1963). Sensitivity to diazo paper, to the aforementioned substance, and to one of its intermediates, *N,N*-diethyl-*p*-perylene-diamine, was also present in a woman engaged in diazocopying and suffering from bronchial asthma which lasted 10 years (Gertler and Laubstein, 1963).

Itching dermatitis to the hands (dorsa and palms) and face with subsequent photosensitivity was recently observed (van der Leun *et al.*, 1977; Nurse, 1980). Undeveloped and developed papers gave rise to positive patch tests; in different situations sensitivity to sunlight, indoor light, and UV-A and UV-B, was demonstrated by photo-patch tests. Thiourea was the substance involved in the manufacture of the paper which turned out to be the cause of the severe reactions to light persisting several years after the contact ceased.

Thiourea, which is also a rubber additive, and, in the past, an antithyroid drug, is used in almost all kinds of these photocopying papers as an antiyellowing agent to prevent discoloration of the paper after the breakdown of diazocompounds (Gertler and Laubstein, 1963; IARC, 1974).

Diazo-type reproducing equipment should have mechanical local exhaust ventilation to reduce the ammonia released by the machine, by the paper discharged after printing and when refilling reservoirs (Utidjian, 1976). Mainly in diazocopying discomfort and annoyance have been reported from a short exposure to concentrations of ammonia as low as 20 p.p.m. (Mangold, 1971), a little less than the present (1982) TLV-TWA (25 p.p.m.): however general experience confirms that the health of inured workers is not adversely affected by exposure to ammonia up to a TWA of 100 p.p.m., with excursions to 150 or higher levels (Ferguson *et al.*, 1977).

6. *Photocopying*

In the 1960s and 1970s office copying underwent an extremely rapid expansion, following the introduction of dry-copying which initially concerned coated paper copiers, and plain paper copiers after 1975. Both are known as indirect electrophotographic or transfer xerographic methods.

In transfer xerography the preliminary step of the process consists of producing, on the surface of a photoreceptor, a deposition of gas ions produced by a corotron. This is a corona discharge device which, operating at a voltage of 6–8 kV, ionizes the surrounding air (Wolf and Weigel, 1979). The transfer of the developed image from photoreceptor to the paper is also accomplished by charging the back of the paper by a corotron. Air breakdown by corona discharge is the first of the only two methods of any commercial importance in producing ozone, the second being UV-C irradiation of air or oxygen (Nebel, 1981). In corona discharge technology an electron propelled to a high velocity and containing energy of 6–7 eV can dissociate oxygen molecules into two atoms, which react rapidly with molecular oxygen to form ozone. This fact is also known to have been a side-effect of early corona discharge air ionizers (Hedge, 1982). Thus there is the possibility of ozone production during the activity of corona discharge devices of electrophotocopiers (Greenberg, 1965), and eventually this fact can lead to increases in indoor concentration of ozone during the activity of photocopying machines (WHO, 1978). Ozone was the only substance found by NIOSH to exceed, near a photocopying machine, the concentration existing in the rest of a studio (*O.H.a.S. News*, 1982).

For indoor settings with no identified inside sources, levels of ozone are typically 40–70% of outside concentrations (Allen *et al.*, 1978), because of destructive reactions which occur on most surfaces (WHO, 1978), the decay occurring faster on organic ones. In a room with many furnishings the half-life is in the order of minutes (Mc Intyre, 1980). Also smoking easily destroys ozone in indoor settings through the production of elevated nitric oxide concentrations (Schuck and Stephens, 1969).

Monitoring of a copier with a maximum voltage of 11 kV under normal working conditions (door open) showed at equilibrium an ozone concentration at the operators breathing zone of 0·068 p.p.m.; ozone emissions varied from less than 1 μg to 54 μg/copy (Selway *et al.*, 1980). An ozone hazard may exist when operating in badly ventilated areas, and in the summer, around midday (Bouhuys, 1974), in areas open to the outside air when outside concentrations are greater than 0·05 p.p.m. (Allen *et al.*, 1978).

Servicing turned out to be very important in reducing ozone emissions, but the return to preservicing conditions was very rapid, indicating that the procedures in use do not produce long-lasting results (Selway *et al.*, 1980). The increasing emission of ozone with time may possibly be related with interferences between residual toner on the photoreceptor, or on the corotron itself, and the deposition of electrons and charged particles which, remaining airborne, contribute more heavily to ozone production.

The pulmonary function adaptation phenomenon which occurs with repeated ozone exposure has been well documented, e.g. at a level of 0·4 p.p.m. (four times the current TLV-TWA), and consists of decreases of FVC and FEV_1 in the first 2–3 days, returning to the baseline by the fourth and fifth day (Kulle *et al.*, 1982).

Near the TLV the major health concern is eye irritation, since, on inhalation, ozone is largely removed in the nose (Buohuys, 1974); this was documented in the 1950s among female office workers who did not complain of symptoms until the TLV was reached and experienced painful ocular sensations above it (Richardson and Middleton, 1958). On the contrary asthmatic people did not differ from control subjects using forced expiratory measures, lung volume or single breath nitrogen indices after repeated exposure to ozone concentrations approximating 0·2 p.p.m. (Linn *et al.*, 1978). ACGIH (1980) recommends that exposure to the TLV of 0·1 p.p.m. should not be prolonged because of the possibility of premature ageing in a manner similar to that of other radiomimetic agents.

Toners are the thermoplastic pigmented powders which give the printed image of an original in transfer xerography. Seven out of eleven toners used in photocopiers turned out to be mutagenic in the Ames *Salmonella* assay; the same behaviour was displayed by the extracts of copies printed on plain paper (Löfroth *et al.*, 1980).

The range of the mutagenic response corresponded to a variation between 40 and 4000 revertants per mg of toner. The mutagenic activity was mainly present in the fraction of extract eluted by benzene (aromatic fraction). The further separation of this fraction was found to elute close to pyrene. The testing for mutagenicity of the smaller fractions showed that the mutagenicity peaks coincided with samples of 1,6- and 1,8-dinitropyrene.

The mutagenic response was influenced only by toner formulation, more precisely by the type of carbon black used (Rosenkranz *et al.*, 1980). Competitive adsorption interferences caused the apparent paradox of the lack of mutagenic power from individual ingredients in the toner, including carbon black. Since the manufacture of the accounted carbon blacks involved a nitration–oxidation step, the appearance of nitrated PAH appeared plausible. Analysis of health records of Xerox employees, exposed to xerographic toners since the late 1950s, reveals no signs of health effects; the medical surveillance of these workers will be continued by the Corporation.

Since, in the meaning of carbon black producers, this process is highly refined and controlled, the conditions could be adjusted to reduce toluene-extractable nitropyrenes by a factor of 50 to 200. This modified carbon black has been used since 1980 by the various Xerox toner manufacturing plants.

Whereas selenium was the first photoreceptor used by Xerox for transfer electrophotography, organic photoreceptors were introduced in higher speed copier-duplicators firstly by IBM, using a roll with a coating of 1:1 molar complex of polyvinylcarbazole and trinitrofluorenone (TNF) (Kuchera, 1982). In 1980 the latter caused concern similar to the case described for dinitropyrene-containing toners, as TNF has been recognized since 1962 as a rat carcinogen, and also resulted in being a potent direct-acting mutagen in bacterial and mammalian cell culture assays (von Burg, 1981). There is also evidence that the compound is absorbed through the skin, of which it appears to be a potent sensitizer. Following IBM conclusions, a content of TNF up to 2·6 μg may be found on hard copy processed by their machines Copier 1° and 11° and 3800 laser printer. However, such conclusions show that when TNF is attached to copy paper, it is physically bound and therefore can not be a source of exposure to users. No exposure to hazardous levels of TNF was demonstrated in the workplace air of a computer room (Eckardt and Wilcox, 1982); testing of mutagenic activity in the urine of possibly exposed workers also showed negative results (Crebelli *et al.*, 1982).

7. *Video Display Terminals*

There has recently been great concern about potential health hazards associated with electromagnetic radiations emitted by video display terminals (VDTs), mainly because users must operate in close proximity to them.

In 1977 two men employed at the *New York Times* as copy editors, and aged 35 and 29 respectively, were reported to suffer from bilateral cataracts; the duration of exposure to VDTs ranged from 4 months to 1 year. Thereafter (1980) other cases were reported at the *Baltimore Sun*.

Four out of seven pregnancies resulted in children with different birth defects within a period of about one year (1980) among 125 VDT operators at the *Toronto Star*.

Electromagnetic radiation may be emitted from VDTs: low energy X-rays can be generated in the cathode ray tube (CRT) when the electron beam strikes materials within it. Some UV and IR radiations are emitted, besides those in the visible spectrum, by the screen face depending on the phosphors used. The electronic circuitry which controls the electron beam may also generate radio frequency (RF) radiations, and in a few cases computer circuitry of some VDTs may produce microwave (MW) radiations (Weiss, 1983). Consequently measurements over the whole electromagnetic spectrum are performed in the workplace as well as in the laboratory. Laboratory evaluation may lead to a detailed characterization of actual emissions, but is very expensive and time-consuming; field measurement is a quicker method to test a large number of units, with the disadvantage of some loss in sensitivity and quality analysis (Muc, 1981). X-ray emissions at the face of the screen were either not detectable (Murray *et al.*, 1981; Weiss, 1983), by chance with the only exception of cases with faulty high power voltage supply (Weiss and Petersen, 1979), or not distinguishable from background level (Moss *et al.*, 1977). Moreover, ageing of a VDT unit would lead to a decreased X-ray emission.

UV emissions ranged from non-detectable values to 0·65 $\mu W/cm^2$ (Murray *et al.*, 1981) or less (Moss *et al.*, 1977; Weiss and Petersen, 1979).

RF and MW radiations were lower than 0·05 mW/cm^2 (Muc, 1981) or not detectable (Moss *et al.*, 1977; Murray *et al.*, 1981).

Therefore VDT emission turns out to be even lower than the most stringent occupational exposure standards, being frequently below the detection limits of the instrumentation used for measurements. The conclusions of the two NIOSH reports (Moss *et al.*, 1977; Murray *et al.*, 1981) state that properly functioning VDTs pose no threat from radiation to operators at or near the terminals.

More recently, levels of PCBs 50–80 times those recorded outside were measured in an office using VDTs (Digernes and Astrup, 1982); it has been hypothesized that skin rashes and itching reported among VDT operators (Linden and Rolfsen, 1981) might be linked with leakage of PCBs from some terminal components. The possible sources of the leakage were the capacitors and transformers in the unit, but their chemical content was unknown. This hypothesis relied mainly on qualitative differences of indoor versus outdoor PCBs levels, since quantitatively it is usual (McLeod, 1981) to find a gap quite similar to that reported by the Norwegian survey.

8. *Fluorescent Light*

Since 1960 skin melanoma mortality rates have almost doubled, and incidence rates have increased even more rapidly (Anonymous, 1981): this trend is not homogeneous by anatomic site, the rates for the head and neck, unlike other skin cancers, rising more slowly than those for the trunk and limbs (Lee, 1982).

Among others, this behaviour has supported the hypothesis that lentigo maligna, which is similar to non-melanoma skin cancers in environmental relations, may be a pathological entity different from malignant melanoma (McGovern *et al.*, 1980).

In Australia (Holman *et al.*, 1980), in Britain (Lee and Strickland, 1980) and in the State of Washington, USA (Milham, 1976) melanoma rates are high among professional and office workers and low in outdoor workers. More recently in Britain an especially high rate of lesions on the unexposed parts of the body has been observed (Beral and Robinson, 1981).

Beral *et al.* (1982) performed a case-control study in New South Wales, Australia, primarily concerned with exploring the relationship between melanoma and oral contraceptive use. White female patients aged 18–54 years attending the melanoma clinic at Sydney, and two controls for each case matched into five-year age groups, were also asked about exposure to sunlight and to fluorescent light while at home or at work.

The reporting of any exposure to fluorescent light at work was associated with a two-fold increase in melanoma risk: the RR grew with increasing duration of exposure, and was stronger for office workers (RR = 2·6) than for other women working indoors (RR = 1·8).

The cumulative proportion of cases with lesions to the trunk was similar to that of other studies (e.g. Canton of Vaud, Switzerland). However, among women exposed to fluorescent light there was a relative excess in comparison with those working outdoors, and those who had never been exposed indoors.

Various items related to recreational exposure to sunlight showed no consistent relation to melanoma risk, instead the RR tended to be lower in women who had apparently been more heavily exposed to sunlight.

The possibility that the increase of melanoma risk by fluorescent light at work should reflect the still unexplained relationship between higher socioeconomic status and the risk of melanoma (Stern, 1982) could be ruled out by the absence of a link between duration of exposure to fluorescent light, women's education attainment or husbands' social class (Beral and Evans, 1982).

As in other case-control studies using questionnaires the explained aim of the study, i.e. 'to understand why some women get skin diseases and others don't', may have aroused the level of interest of melanoma patients for questions possibly linked with their illness, such as asking whether fluorescent light was used in indoor jobs or not (Brown, 1982).

An inconsistency pointed out by the authors themselves is the lack of association between the presence of fluorescent light at home and a rise in melanoma risk: this fact would have implied a discussion on quality and quantity of fluorescent light possibly used in two so different ambients of work and life, impossible to perform because of the lack of detailed information.

As far as the localization of the lesions is concerned, predominantly to the unexposed parts of the body, the type of clothing worn might be permeable to ultraviolet light (Daniels, 1975; WHO, 1979; Molineux, 1981), as fair-skinned people are well aware, and as demonstrated, e.g. in Queensland, Australia (Beardmore, 1972).

Sun-tanning may, to some extent, protect melanocytes against the carcinogenic effects of ultraviolet light (McGovern, 1977). Therefore a person who works indoors and occasionally outdoors might be more susceptible, particularly on those parts of the body normally covered by clothing (Anonymous, 1981).

Skin photosensitivity from white fluorescent lighting has been demonstrated, especially to the 365 nm line of the long wave UV emission; furthermore patients with solar urticaria were sensitive to fluorescent lighting only on skin which was as a rule clothed, showing a higher tolerance on their face and hands (Brown *et al.*, 1969).

The near-UV region of the spectrum emitted by fluorescent lamps has been implicated for the mutagenic effects from these sources. There is an indication that the magnitude of the effects is not related to the total irradiance but rather to the relative intensities at different near-UV wavelengths and, possibly, to their interactions (Jacobsen and Krell, 1979): the use of a sheet of clear plastic diffuser material, commercially available for use in fluorescent light fixtures and impairing 99% of the transmission below 388 nm, eliminated the mutagenic effect.

Likewise, malignant transformation in mouse embryo cell culture was induced by fluorescent light with a dose-dependent relationship: the use of covers on petri dishes containing the cultures resulted in no detectable malignant transformation (Kennedy *et al.*, 1980). These results suggested that fluorescent light could contribute on a small scale to human skin carcinogenesis.

This question raised by Diethelm (1970) in Switzerland, who described exposure peculiarities in five cases of squamous and basal cells skin carcinoma, noticeable because of directionality and closeness to a fluorescent light source.

Owing to experimental and clinical observations a carcinogenic potential of UV-A can no longer be denied, even if it is much lower than for UV-B (NAS, 1981; Barrière, 1982).

A measurable amount of energy is emitted by all fluorescent lamps in the UV-A region (Peters, 1976; NAS, 1981): the irradiance range is 70–110 $\mu W/m^2$ (Jewess, 1981), which is only a very small fraction of that of the sun (Rigel *et al.*, 1983) and two orders of magnitude under the standard of 10 W/m^2.

Small but significant intensities of UV-B may also be emitted by some tubes in common use in the region around 290 nm, that is at a wavelength where solar irradiance at ground level approaches zero: this fact, or the implied differences in the ratio of short to longer UV wavelength in comparison to sunlight, could give a biologically plausible basis for the association found by Beral *et al.* (Maxwell and Elwood, 1983).

Another causal factor has been proposed (Jensen, 1982) for the possible interpretation of the aforementioned observations of Beral *et al.* (1982): the exposure to PCBs. These chemicals are used in ballast capacitors for fluorescent lighting at home or in the office (WHO, 1976). They may be emitted during burn-out, whereas emission from normally operating fluorescent light units is well below the established TLV; moreover in recent years thermal protective switches have been incorporated by manufacturers in ballasts to prevent overheating and burnout (Staiff *et al.*, 1974). However, indoor air in commercial, industrial or residential buildings contains levels of PCBs which are at least of one order of magnitude higher than outdoor levels (McLeod, 1981).

These observations are worth noting since an apparent excess of malignant melanoma has been reported in workers exposed to Aroclor 1254, and probably to other chemicals, in the Mobil Oil Refinery at Paulsboro, N.J. (Bahn *et al.*, 1976, 1977; Fishbein, 1979).

9. *Air Conditioners*

Many industries require a highly controlled level of humidity. Printing, stationery, manufacturing of textiles, these are but a few of such localities. This control is achieved with humidifiers using evaporation, cool-mist atomizing or steaming (van Assendelft *et al.*, 1979). 'Humidifier fever' was first described in 1959 by Pestalozzi. Some other industries as well as the home environment were thereafter involved (Burke *et al.*, 1977).

Offices, and operating theatres, with increasing use of air conditioning systems have also given rise to outbreaks of 'fever' (Banaszak *et al.*, 1970; Cockcroft *et al.*, 1981). A relative humidity between 50 and 60% should be maintained, in these systems, to ensure staff comfort. Traditional wet humidifiers obtain this level by using a collection pond. This can provide a breeding ground for bacteria (Ratcliffe, 1977) and eventually a complete food-chain of micro-organisms (Anonymous, 1978). Consequently, through the advent of modern ventilation techniques, ambients with no obvious dust burden, such as offices, are included among the environments producing an organic dust disease (MRC Symposium, 1977).

However, the term 'humidifier fever' cannot be attached to all observed cases. The existence of two different syndromes may be suggested, both arising in non-atopic individuals. Extrinsic allergic alveolitis (hypersensitivity pneumonia), a disease similar to farmers' lung, has been predominantly described in both the USA and Switzerland. A true humidifier fever, as seen in the United Kingdom, is more commonly known as Monday sickness or Monday night fever.

The acute, subacute and chronic forms of extrinsic allergic alveolitis caused by humidifiers are identical to the cases related to other causes. There are systemic and pulmonary findings, such as coughing, dyspnoea, malaise, fever and rales. Radiographic abnormalities ranging from just detectable changes to widespread coarse opacities, and lung function changes consisting of a reduction in VC, TLC, compliance and gas transfer were also observed. The latter two parameters may persist at a reduced level even after chest radiographs have returned to normal, or between relapses (Parkes, 1982). In some cases patient's blood contains precipitins to some genus of bacteria of the family Micromonosporaceae, especially *Thermoactinomyces vulgaris*. These have been isolated as the main pathogen in tests performed on warm water, contained in air conditioners currently used in America and to which the illness has been attributed.

Humidifier fever is an acute, influenza-like illness with patterns similar to grade C1 and 2 byssinosis. Symptoms occur mainly on Monday or in general on resumption of work, decreasing during the rest of the working week. They have a clear asthmatic feature and consist of malaise, fever, coughing and dyspnoea usually disappearing in 24 hours without any radiographic change (Parkes, 1982). Thermophilic Actinomycetes were not implicated, but antibodies against protozoa (*Negreria gruberi*; Acanthamoeba) were found in affected workers when present in humidifier system (Edwards *et al.*, 1976). Precipitins to extracts of *Bacillus subtilis* were also discovered in another group of affected people (Parrott and Blyth, 1980).

A common feature of these syndromes is the extremely low atmospheric content of material sufficient to cause an outbreak in susceptible individuals, unlike typical acute or subacute extrinsic allergic alveolitis. The timecourse, on the contrary, is very similar showing a delay of 4–6 hours between the exposure to the inhalable spores and cysts and the onset of symptoms. Later these disappear then reappear according to seasonal humidifier use. It may also be that pyrexial episodes of humidifier fever represent the initial step of events resulting in a full attack of extrinsic allergic alveolitis.

Replacing a wet system humidifier with one of steam injection would eliminate the need to recirculate water. The improvement of input filtering is also another remedial action. Removal of susceptible individuals or of contaminated appliances from the ambient may also lead to the disappearance of the symptoms. Nonetheless, those affected by chronic forms of extrinsic allergic alveolitis due to humidifiers may progress to irreversible pulmonary fibrosis.

10. *Conclusions*

In conclusion, this review shows that some toxicological problems may be of concern for office workers.

According to the relative frequency they may be listed as follows:

1. Prevalent subjective symptoms related to skin contact with chemical substances, frequently unidentified, as causal agents. Since such chemicals are embedded in other harmless solid materials, which complete disguise them (e.g. CCPs), the risk may be undetected.
2. The possibility of some electromagnetic emission even at frequencies outside the visible spectrum (e.g. fluorescent light; VDTs).
3. Some effects of exposure to known skin and/or mucous membranes irritants (e.g. ammonia; ozone).
4. Occasional reports of the presence, within handled materials, of chemicals or impurities for which a carcinogenic potential has been experimentally demonstrated.

A common feature of these problems is the office workers' unawareness of the possibility that the office environment, traditionally regarded as safe, might involve some toxic risks. From this fear may arise when beginning subjective troubles or when becoming aware of dangerous properties of some material in use: the end result is an epidemic charactertic of such troubles.

In general office machines are updated and the list of materials in use includes few formulations. Moreover machines have few critical parts, which involve chemical exposure. Therefore it is amazing that such chemicals are not screened in depth before their introduction into the office environment.

References

Adams, R.M., 1983, *Occupational Skin Diseases* (New York: Grune and Stratton), p.422.

Allen, R.J., Wadden, R.A. and Ross, E.D., 1978, Characterization of potential indoor sources of ozone, *American Industrial Hygiene Association Journal*, **39**, 466–471.

American Conference of Governmental Industrial Hygienists, 1980, *Documentation of the Threshold Limit Values* (Cincinnati: ACGIH), pp.316–317.

Andanson, J., Raulot-Lapointe, H. and Moulanier, M., 1980, Papiers autocopiants—Problèmes d'intolérance: irritation ou allergie, *Archives des Maladies Professionnelles*, **41**, 168–169.

Anonymous, 1978, Humidifier fever: a disease to look for, *British Medical Journal*, 1164–1165.
Anonymous, 1981, The aetiology of melanoma, *The Lancet*, **1**, 253–255.
Armeli, G., 1968, Asma bronchiale da sali di diazonio, *La Medicina del Lavoro*, **59**, 463–466.
van Assendelft, A., Forsen, K.O., Keskinen, H. and Alanko, K., 1979, Humidifier-associated extrinsic allergic alveolitis, *Scandinavian Journal of Working Environment and Health*, **5**, 35–41.
Bahn, A.K., Rosenwaike, I., Herrmann, N., Grover, P., Stellman, J. and O'Leary, K., 1976, Melanoma after exposure to PCBs, *New England Journal of Medicine*, **295**, 450.
Bahn, A.K., Grover, P., Rosenwaike, I., O'Leary, K. and Stellman, J., 1977, PCBs and melanoma, *New England Journal of Medicine*, **396**, 108.
Banaszak, E.F., Thiede, W.H. and Fink, J.N., 1970, Hypersensitivity pneumonitis due to contamination of an air conditioner, *New England Journal of Medicine*, **283**, 271–276.
Barrière, H. 1982, Soleil, cancers cutanés et activités professionnelles, *Archives des Maladies Professionnelles*, **43**, 679–681.
Beardmore, G.C., 1972, The epidemiology of malignant melanoma in Australia, in *Melanoma and Skin Cancer*, edited by W.H. McCarthy (Sidney: Blight), pp.39–64.
Belliardo, F., Nano, G.M., Pavan, I. and Scansetti, G., 1979, Sulla presenza di policlorobifenili in carte autocopianti, *La Medicina del Lavoro*, **70**, 391–397.
Benvenuti, F., Lepore, L., Maggio, M. and Salerno, A., 1979, Risultati di un'indagine sul contenuto di policlorobifenili in carte autocopianti, *Securitas*, **64**, 75–80.
Beral, V. and Robinson, N., 1981, The relationship of malignant melanoma, basal and squamous skin cancers to indoor and outdoor work, *British Journal of Cancer*, **44**, 886–891.
Beral, V., Evans, S., Shaw, H. and Milton, G., 1982, Malignant melanoma and exposure to fluorescent lighting at work, *The Lancet*, **2**, 290–293.
Beral, V. and Evans, S., 1981, reply to Stern, R.S. (ibidem).
Bignon, J., Sebastien, P., Gaudichet, A. and Jaurand, M.C., 1980, Biological effects of attapulgite, in *Symposium on Biological Effects of Mineral Fibres*, edited by J.C. Wagner (Lyon: IARC), pp.163–181.
Bouhuys, A., 1974, *Breathing* (New York: Grune and Stratton), pp.393–395.
British Rubber Manufacturers Association, 1978, *Code of Practice* (Birmingham: BRMA).
Brown, A.P., 1982, Melanoma and fluorescent light, *The Lancet*, **2**, 1398.
Brown, S., Lane, P.R., and Magnus, I.A., 1969, Skin photosensitivity from fluorescent lighting, *British Journal of Dermatology*, **81** 420–428.
Burke, J.W., Carrington, C.B., Strauss, R., Rink, J.N. and Gaensler, E.A., 1977, Allergic alveolitis caused by home humidifiers, *Journal of the American Medical Association*, **238**, 2705–2708.
Cakir, A., Hart, D.J. and Stewart, T.F.M., 1980, *Visual Display Terminals* (Chichester: Wiley), p.34.
Calnan, D.C., 1979, Carbon and carbonless copy paper, *Acta Dermatovenereologica*, **59**, suppl. 85, 27–32.
Calnan, C.D. and Connor, B.N., 1972, Carbon paper dermatitis due to nigrosine, *Berufsdermatosen*, **20**, 248–254.
Cockcroft, A., Edwards, J., Bevan, C., Campbell, I., Collins, G., Houston, K., Jenkins, D., Latham S., Saunders, M., and Trotman, D., 1981, An investigation of operating theatre staff exposed to humidifier fever antigen, *British Journal of Industrial Medicine*, **38**, 144–151.
Conde-Salazar, L., Romero, L., Guimaraens, D., 1982, Allergic contact dermatitis from diazo paper, *Contact Dermatitis*, **8**, 210–211.
Crebelli, R., Falcone, E., Aquilina, G., Carere, A. and Zito, R., 1982, Monitoring of the urinary mutagenicity in men exposed to low doses of trinitrofluorenone (TNF), *12th Annual Meeting of the European Environmental Mutagen Society*, June 20–24, Dipoli, Espoo, Finland.
Creux, S., Raulot-Lapointe, H., Castelain, P.Y. and Mathias, A., 1982, Les dermatoses dues aux papiers utilisés par les duplicateurs our reproducteurs de documents, *Archives des Maladies Professionnelles*, **43**, 409–410.
Cronin, E., 1980, *Contact Dermatitis* (Edinburgh: Churchill Livingstone), pp.853–854.
Daniels, F., Jr, 1975, Sunlight, in *Cancer Epidemiology and Prevention, Current Concepts*, edited by D. Schottenfeld, (Springfield: Thomas), pp.126–152.

Diethelm, R., 1970, Hautkarzinom durch Fluoreszenzlampen?, *Schweizerische Medizinische Wochenschrift*, **100**, 1159–1160.

Digernes, V. and Astrup, E.G., 1982, Are data screen terminals a source of increased PCB-concentrations in the working atmosphere?, *International Archives of Occupational and Environmental Health*, **49**, 193–197.

Dodds, W.J. and Butler, P.E.B., 1981, Carbonless copy paper, *Contact Dermatitis*, **7**, 218–219.

Eckardt, J.G. and Wilcox, C., 1982, Sampling and determination of airborne 2,4,7-trinitro-9-fluorenone in the computer room environment, June 16–18, Provo, Utah.

Edwards, J.H., Griffiths, A.J. and Mullins, J., 1976, Protozoa as sources of antigen in 'humidifier fever', *Nature*, **264**, 438–439.

Eriksson G. and Östlund, E., 1968, Rubber bank notes counters as the cause of eczema among employees at the Swedish Post Giro Office, *Acta Dermatovenereologica*, **48**, 212–214.

Ferguson, W.S., Koch, W.C., Webster, L.B. and Gould, J.R., 1977, Physiological response and adaptation to ammonia, *Journal of Occupational Medicine*, **19**, 319–326.

Fishbein, L., 1979, *Potential Industrial Carcinogens and Mutagens* (Amsterdam: Elsevier), pp.292–293.

Gertler, H. and Laubstein, H., 1963, Ueber berufsbedingte Erkrankungen bei Lichtpausern, *Berufsdermatosen*, **11**, 125–140.

Gockel, D., Horstman, S.W. and Scott, C.M. 1981, Formaldehyde emissions from carbonless copy paper forms, *American Industrial Hygiene Association Journal*, **42**, 474–476.

Göthe, C.J. 1981, in 'Carbonless copy paper and health effects', A Symposium held at the Southern Hospital, Stockholm, Sweden, *Opuscula Medica*, Suppl. 56, p.2.

Graham, V., Coe, M.J.S. and Davies, R.J., 1981, Occupational asthma after exposure to a diazonium salt, *Thorax*, **36**, 950–951.

Greenberg, L., 1965, Health hazards in the modern office, *Journal of the American Society of Safety Engineers*, **10**, 15–17.

Grim, R.E., 1979, Clays, in *Kirk–Othmer Encyclopedia of Chemical Technology*, 3rd ed, vol. 6 (New York: Wiley), pp.190–223.

Harman, R.R.M. and Sarkany, I., 1960, Study in contact dermatitis. XI. Copy paper dermatitis, *Transactions of the St John's Hospital Dermatological Society*, **44**, 37–42.

Hedge, A., 1982, Discussion of the paper 'Air ionization: an evaluation of its physiological and psychological effects, by A. Hedge and E. Eleftherakis, *Annals of Occupational Hygiene*, **25**, 409–419.

Hjorth, N., 1964, Contact dermatitis from cellulose acetate film, *Berufsdermatosen*, **12**, 86–100.

Holman, C.D.J., Mulroney, C.D. and Armstrong, B.K., 1980, Epidemiology of pre-invasive and invasive malignant melanoma in Australia, *International Journal of Cancer*, **25**, 317–323.

IARC, 1974, *Monographs on the Evaluation of the Carcinogenic Risk of Chemicals to Humans, 7, Some Antithyroid and Related Substances, Nitrofurans and Industrial Chemicals* (Lyon: IARC), pp.98–99.

IARC, 1978, *Monographs on the Evaluation of the Carcinogenic Risk of Chemicals to Humans, 18, Polychlorinated Biphenyls and Polychlorinated Terphenyls* (Lyon: IARC), pp.54–60.

Jacobsen, E. and Krell, K., 19797, Ultraviolet regions implicated in toxic and mutagenic effects of broad spectrum radiations from fluorescent lamps on L5178Y Lymphoma cells, *Mutation Research*, **62**, 533–538.

Jeansson, I., 1981, in 'Carbonless copy paper and health effects', A Symposium held at the Southern Hospital, Stockholm, Sweden, *Opuscula Medica*, suppl. 56, pp.5, 9, 11.

Jensen, A.A., 1982, Melanoma, fluorescent light and polychlorinated biphenyls, *The Lancet*, **2**, 935.

Jensen, M. and Roed-Petersen, J., 1979, Itching erythema among post office workers caused by a photocopying machine with wet toner, *Contact Dermatitis*, **5**, 389–391.

Jewess, B.W., 1981, Ultraviolet content of lamp in common use, *Society of Photo-optical Instrumentations Engineering. Proceedings*, **262**, 55–61.

Jordan, W.P., 1975, Contact dermatitis from typewriter correction paper, *Cutis*, **15**, 594–595.

Kennedy, A.R., Ritter, M.A. and Little, J.B., 1980, Fluorescent light induces malignant transformation in mouse embryo cell culture, *Science*, **207**, 1209–1211.

Kirton, V. and Wilkinson, D.S., 1972, Rubber band dermatitis in Post Office sorters, *Contact Dermatitis Newsletter*, **11**, 257–260.

Kleinmann, G.D. and Horstman, S.W., 1982, Health complaints attributed to the use of carbonless copy paper (A preliminary report), *American Industrial Hygiene Association Journal*, **43**, 432–435.

Kuchera, T.J., 1982, Reprography, in *Kirk–Othmer Encyclopedia of Chemical Technology*, 3rd ed, vol. 20 (New York: Wiley), pp.128–179.

Kulle, T.J., Sauder, L.R., Kerr, D.H., Farrell, B.P., Bermal, M.S. and Smith, D.M., 1982, Duration of pulmonary function adaptation to ozone in humans, *American Industrial Hygiene Association Journal*, **43**, 832–837.

Lee, J.A.H., 1982, Melanoma and exposure to sunlight, *Epidemiologic Reviews*, **4**, 110–136.

Lee, J.A.H. and Strickland, D., 1980, Malignant melanoma: social status and outdoor work, *British Journal of Cancer*, **41**, 757–763.

Leineweber, J.P., 1980, Dust chemistry and physics: mineral and vitreous fibres, in *Symposium on Biological Effects of Mineral Fibres*, edited by J.C. Wagner (Lyon: IARC), 881–900.

Levy, F. and Hanoa, R., 1980, cited by Nörback, D. and Göthe, C.J. (1981).

Lidblom, A., 1981, in 'Carbonless copy papers and health effects', A Symposium held at the Southern Hospital, Stockholm, Sweden, *Opuscula Medica*, suppl. 56, p.3.

Lindén, V. and Rolfsen, S., 1981, Video computer terminals and occupational dermatitis, *Scandinavian Journal of Work, Environment and Health*, **7**, 62–64.

Linn, W.S., Buckley, R.D., Spier, C.E., Blessey, R.L., Jones, M.P., Fischer, D.A. and Hackney, J.D., 1978, Health effects of ozone exposure in asthmatics, *American Review of Respiratory Disease*, **117**, 835–843.

Löfroth, G., Hefner, E., Alfheim, I. and Moeller, M., 1980, Mutagenic activities in photocopies, *Science*, **209**, 1037–1039.

Löfström, A., 1981, in 'Carbonless copy paper and health effects', A Symposium held at the Southern Hospital, Stockholm, Sweden, *Opuscula Medica*, suppl. 56, p.3.

Mangold, C.A., 1971, *Investigation of Occupational Exposure to Ammonia*, Record of Industrial Hygiene Division Investigation, Puget Sound Naval Shipyard, 29 Nov 1971.

Marks, J.G., 1981, Allergic contact dermatitis from carbonless copy paper, *Journal of the American Medical Association*, **245**, 2331–2332.

Masuda, Y., Kagawa, R. and Kuratsune, M., 1972, Polychlorinated biphenyls in carbonless copying paper, *Nature,* **237**, 41–42.

Maxwell, K.J. and Elwood, J.M., 1983, UV radiation from fluorescent lights, *The Lancet*, **2**, 579.

McGovern, V.J., 1977, Epidemiological aspects of melanoma: a review, *Pathology*, **9**, 233–241.

McGovern, V.J., Shaw, H.M., Milton, G.W. and Farago, G.A., 1980, Is malignant melanoma in a Hutchinsons's melanotic freckle a separate disease entity?, *Histopathology*, **4**, 235–242.

McIntyre, D.A., 1980, *Indoor Climate* (London: Applied Science Publishers), pp.264–265.

McLeod, K.E., 1981, Polychlorinated biphenyls in indoor air, *Environmental Science and Technology*, **15**, 926–928.

Meneghini, C.L., Rantuccio, F. and Riboldi, A., 1963, Klinisch-allergologische beobachtungen bei berüflichen ekzematösen Kontakt-dermatosen, 2. Mitteilung, *Berufsdermatosen*, **11**, 280–293.

Menné, T., Asnaes, G. and Hjorth, N., 1981, Skin and mucous membranes problems from 'no carbon required' paper, *Contact Dermatitis*, **7**, 72–76.

Milham, S., 1976, *Occupational Mortality in Washington State* (Washington: USDHEW).

Molineux, M.K. 1981, The physical environment, in *Occupational Health Practice*, 2nd ed., edited by R.S.F. Schilling (London: Butterworths), pp.403–452.

Moss, C.E., Murray, W.E., Parr, W.H., Messite, J. and Karchey, G.J., 1977, *A Report of Electromagnetic Radiation Surveys of Video Display Terminals* (Cincinnati: DHEW).

MRC, 1977, MRC symposium on 'humidifier fever', *Thorax*, **32**, 653–663.

Muc, A.M., 1981, Video display terminals: a radiation hazard?, in *Proceedings 14th Annual Meeting Human Factors Association of Canada*, edited by M.M. Matthews (University of Guelph, Canada).

Murray, W.E., Moss, C.E., Parr, W.H. and Cox, C. 1981, A radiation and industrial hygiene survey of video display terminal operations, *Human Factors*, **23**, 413–420.

National Academy of Sciences, 1981, *Indoor Pollutants* (Washington: National Academy Press), p.222.

Nebel, C., 1981, Ozone, in *Kirk–Othmer Encyclopedia of Chemical Technology*, 3rd ed., vol. 16 (New York: Wiley), pp.683–716.

Nörback, D., 1981, in 'Carbonless copy paper and health effects', A Symposium held at the Southern Hospital, Stockholm, Sweden, *Opuscula Medica*, suppl. 56, pp.8, 10, 12.

Nörback, D. and Göthe, C.J., 1981, in 'Carbonless copy paper and health effects', A Symposium held at the Southern Hospital, Stockholm, Sweden, *Opuscula Medica*, suppl. 56, p.15.

Nurse, D.S., 1980, Sensitivity ot thiourea in plan printing paper, *Contact Dermatitis*, **3**, 153–154.

Parkes, W.R., 1982, *Occupational Lung Disorders*, 2nd ed. (London: Butterworths), pp.384–385.

Parrott, W.F. and Blyth, W., 1980, Another causal factor in the production of humidifier fever, *Journal of the Society for Ocupational Medicine*, **30**, 63–68.

Pestalozzi, C., 1959, Febrile Gruppenerkrankungen in einer Modellschreinerei durch Inhalation von mit Schimmelpilzen kontaminiertem Befeuchterwasser ('Befeuchterfieber'), *Schweizerische Medizinische Wochenschrift*, **89**, 710–713.

Peters, T., 1976, *Arbeitswissenschaft für die Büropraxis*, 2nd ed. (Ludwigshafen: F. Kiehl).

Ratcliffe, S., 1977, in 'MRC symposium on humidifier fever', Engineering aspects, *Thorax*, **32**, 654.

Richardson, N.A. and Middleton, W.C., 1958, cited in WHO (1978).

Rigel, D.S., Friedman, R.J., Levenstein, M. and Greenwald, D.I., 1983, Malignant melanoma and exposure to fluorescent lighting at work, *The Lancet*, **1**, 704.

Roed-Petersen, J., Hjorth, N., Jordan W.P. and Bourlas, M., 1977, Post-sorters rubber fingerstall dermatitis, *Contact Dermatitis*, **3**, 143–147.

Rosenkranz, H.S., McCoy, E.C., Sanders, D.R., Butler, M., Kiriazides, D.K. and Mermelstein, R., 1980, Nitropyrenes: Isolation, identification and reduction of mutagenic impurities in carbon black and toners, *Science*, **209**, 1039–1043.

Sampaolo, A., Gramiccioni, L., Boniforti, L., Rufini, L., Esposito, G., Arena, C. and Binetti, R., 1980, Carte autocopianti: Ricerca de policlorobifenile (PCB), benzofurani e diossine, *Rassegna Chimica*, **4**, 183–188.

Schuck, E.A. and Stephens, E.R., 1969, Oxides of nitrogen, *Advances in Environmental Sciences*, **1**, 73–118.

Selway, M.D., Allen, R.J. and Wadden, R.A. 1980, Ozone production from photocopying machines, *American Industrial Hygiene Association Journal*, **41**, 455–459.

Sors, H., Gaudichet, A., Sebastien, P., Bignon J. and Even, P., 1979, Lung fibrosis after inhalation of fibrous attapulgite, *Thorax*, **34**, 695.

Sparks, R.E., 1981, Microencapsulation, in *Kirk–Othmer Encyclopedia of Chemical Technology*, 3rd ed., vol. 15 (New York: Wiley), pp.470–493.

Staiff, D.C., Quinby, G.E., Spencer, D.L. and Starr, H.G., Jr, 1974, Polychlorinated biphenyl emissions from fluorescent lamp ballasts, *Bulletin of Environmental Contamination and Toxicology*, **12**, 455–463.

Steinberg, M., 1982, ACGIH TLVs and the sensitive worker, in *Protection of the Sensitive Individual*, edited by W.D. Kelley (Cincinnati: ACGIH), pp.77–81.

Stern, R.S., 1982, Malignant melanoma and exposure to fluorescent lighting at work, *The Lancet*, **2**, 1277.

Utidjian, H.M.D., 1976, Criteria documents: Recommendations for an ammonia standard, *Journal of Occupational Medicine*, **18**, 200–205.

van der Leun, J.C., de Kreek, E.J., Deenstra-van Leeuwen, M. and van Weelden, H., 1977, Photosensitivity owing to thiourea, *Archives of Dermatology*, **113**, 1611.

Voes, J.B. and Beems, R.B., 1971, Dermal toxicity studies of technical polychlorinated byphenyls and fractions thereof in rabbits, *Toxicology and Applied Pharmacology*, **19**, 617–633.

von Burg, R., 1981, Toxicology updates: trinitrofluorenone, *Journal of Applied Toxicology*, **1**, 50.

Weaver, W.C., Simmons, P.B., and Thomson, Q.E., 1979, Diphenyl and terphenyls, in *Kirk–Othmer Encyclopedia of Chemical Technology*, 3rd ed., vol. 7 (New York: Wiley), pp.782–793.

Weiss, M.M., 1983, The video display terminals: is there a radiation hazard?, *Journal of Occupational Medicine*, **25**, 98–100.

Weiss, M.M. and Petersen, R.C., 1979, Electromagnetic radiation emitted from video computer terminals, *American Industrial Hygiene Association Journal*, **40**, 300–309.

WHO, 1976, *Environmental Health Criteria, 2, Polychlorinated Biphenyls and Terphenyls* (Geneva: WHO), pp.27–45.

WHO, 1978, *Environmental Health Criteria, 7, Photochemical Oxidants* (Geneva: WHO), pp.28–85.

WHO, 1979, *Environmental Health Criteria, 14, Ultraviolet Radiation* (Geneva: WHO), p.93.

Wikström, K., 1969, Allergic contact dermatitis caused by paper, *Acta Dermatovenereologica*, 49, 547–551.

Wolf, N.E. and Weigel, J.W., 1979, Electrophotography, in *Kirk–Othmer Encyclopedia of Chemical Technology*, 3rd ed., vol. 8 (New York: Wiley), pp.794–826.

Indoor Air Quality in Offices

HANS-URS WANNER
Swiss Federal Institute of Technology, Department of Hygiene and Ergonomics, CH-8092 Zurich, Switzerland

Abstract

Pollutants of room air have become important for the following reasons: on the one hand the fresh air supply has been considerably reduced in order to save energy, thus entailing a poorer air quality, on the other hand new building materials are used which may pollute the room air.

Possible sources of pollutants in the room air are tobacco smoke, building and furniture materials (e.g. particle boards), insulations and office machines. Health risks can be caused particularly by pollutants being continuously emitted by the above mentioned materials. To prevent such risks the emission of pollutants must be reduced. In order to obtain a perfect air supply, a regular control of the processing plant of an air conditioning equipment is necessary; critical points are the moistening and the cooling plants where the micro-organisms can propagate.

To eliminate the environmental pollution caused by human being and by their activities, a minimal supply of fresh air is required, the quantity of which depends on the duration and the purpose for which the room is being used. The fresh air supply should be calculated in such a way that the carbon dioxide content does not exceed 0·15%. To reach this target, 12–15 m^3 of fresh air/person/hour are sufficient; when smoking or physical activities are going on, larger quantities of fresh air are necessary.

1. *Introduction*

Interviews with office workers about the indoor climate repeatedly show that complaints about bad air quality are very frequent—especially in air-conditioned rooms. There has also been an increase in complaints about non-specific 'building illness' symptoms, such as eye and throat irritations, headache and fatigue. What are these complaints due to? Is it the air quality—or perhaps simply the working conditions in modern office buildings?

To begin with, we must keep in mind that—in contrast to the optimal conditions for thermal comfort—there has been little investigation of the influence of air quality on health and comfort. Air quality seemed not to be regarded as a problem, as long as it could be guaranteed by supplying a sufficient amount of fresh air. In the last few years, however, the importance of air quality in research has suddenly increased. This is due to the following reasons: on one hand, in order to conserve heat, air infiltration and ventilation have to be minimized; this in turn entails a general deterioration of air quality. Another factor is the use of modern building materials which emit different types of pollutant.

2. *Sources of Indoor Air Pollutants*

Pollutants of indoor air originate from the outdoor air and from different sources in the room itself (Figure 1). The following presentation of the most important pollutants and their effects on health is based for the most part on the reports of a WHO Working Group (World Health Organisation, 1979 and 1982), a symposium on indoor climate in Copenhagen (Indoor Climate, 1979), the symposium on indoor air quality in Amherst (Indoor Air Pollution, 1982) and some recent studies.

FIGURE 1. *Sources of indoor air pollution. The contribution of the outdoor air depends upon the location of the building. The pollution load caused by individual activities depends on the occupancy and the use of the room. Building materials could emit contaminants during a long period of time.*

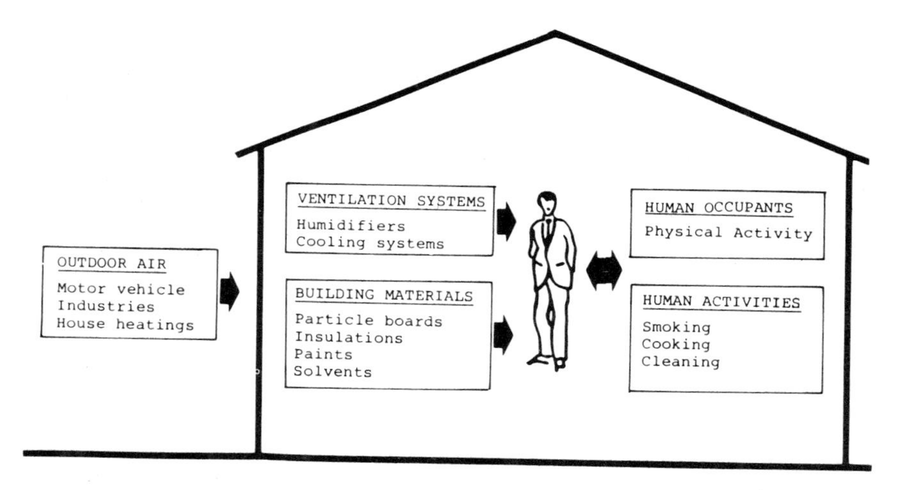

2.1. **Outdoor Air**

Pollutants of outdoor air contaminate indoor air especially in buildings which are located near streets with heavy traffic density or near industrial areas. Present studies on the relationship between outdoor and indoor air pollution show that pollutants in ambient air affect the indoor air quality to different degrees (Yocom, 1982). As a rule the ratio indoor to outdoor is higher for non-reactive gases (carbon monoxide for example) than for reactive gases such as sulphur dioxide, nitrogen oxides and respirable particulates. These lower ratios could be due to the fact that reactive gases and particulates can be absorbed by walls and furniture. In air-conditioned rooms, the amount of outdoor air pollution can be additionally reduced by means of filters. Air washers can also be effective in the elimination of water-soluble pollutants such as sulphur dioxide.

2.2. Ventilation Systems

The air can also be contaminated by air-conditioning systems. Important sources of possible pollutants are: humidifiers, cooling systems and air ducts. Water in humidifiers as well as all wet surfaces and cooling systems provide, with small amounts of organic pollutants, an ideal environment for the growth of micro-organisms (bacteria and fungi) which can then spread in the ventilated rooms (Wanner, 1980).

Bacteria which can pollute the air in the above manner (for example Gram negative germs) are usually harmless for a healthy person, but can lead to serious complications for those suffering from respiratory diseases. Fungi, which also grow in air-washers, can cause hypersensitivity lung disease and allergic alveolitis. In the last few years we have heard a lot of Legionnaire's disease. This is an infection of the lungs which is caused by a certain type of Gram negative bacterium. Such germs were observed in contaminated humidifiers and cooling systems. There were descriptions of cases of illness which clearly point to a transmission by air.

2.3. Human Occupants

The human occupant itself affects the quality within a room by continuously emitting humidity, carbon dioxide, particles, micro-organisms and perspiration. An increase in the concentration of the different pollutants depends directly on the occupancy of the room as well as the activities of each occupant, and the metabolic rate of each.

The commonly used parameters to measure man-caused pollution of indoor air are carbon dioxide and body odours (Figure 2).

The evaluation of the instantaneous odour situation can be affected by sensory perception, i.e. through subjective odour intensity assessment carried out by test persons (Huber *et al.*, 1983). As a rule the recommended level which maintains acceptable indoor air quality is a carbon dioxide concentration of 0·15%.

2.4. Human Activities

Among all pollutants, the one which affects indoor air most directly is tobacco smoke. Depending on the smoking intensity of cigarettes, cigars and pipes, and the size and degree

FIGURE 2. *The carbon dioxide concentration and the intensity of odour perceived during a two-hour experiment in a climatic chamber. Conditions of the experiment: 4 persons in a sitting position in a room of 30 m^3 with an air exchange rate of 6 m^3 per person per hour. Judgement of intensity of odour by 4 test persons who are outside the room (100 units correspond to a reference of 365 p.p.b. pyridine).*

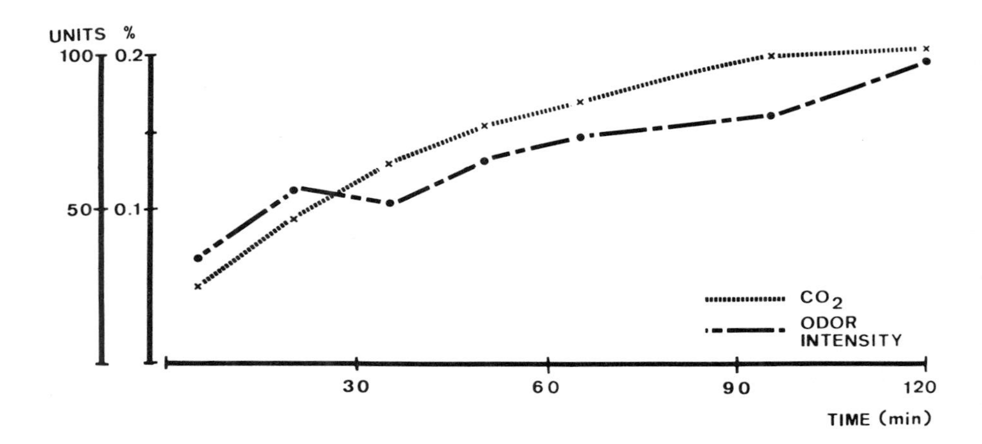

of ventilation of the room, pollutants can reach a concentration which is not only annoying but which can also cause health damage to particularly sensitive people such as those suffering from heart and circulatory diseases.

In order to judge the acceptable concentration for a non-smoker, besides the mentioned direct health effects, the annoyance caused has to be taken into consideration. Investigations under standardized conditions showed that the most sensible criteria were the degree of annoyance caused by a general reduction of air quality and the intensity of the eye irritations (Muramatsu *et al.*, 1983; Weber *et al.*, 1980).

Sources of biological contaminants which can cause allergies are located not only in ventilation systems but also within the rooms. Important in this category are cold mist vaporizers used as humidifiers. Among the well known allergens are the mites which can grow at a relative humidity of 50–55%. An additional source—dependent on human activities—is gas cooking and heating. It is important here to note that during combustion nitrogen dioxide is produced, which can cause an increased sensitivity to respiratory diseases.

In offices, contaminants can be produced which depend on current technological developments and have not yet been researched. Different types of solvent belong to this category—those used for cleaning as well as those used in certain types of photocopying techniques. Photocopying equipment, particularly types using dry ink, can be dangerous. Important here are also ozone and photochemical smog formations, which are a product of the interaction of ultraviolet rays (emitted from fluorescent lights, photocopying machines and video display terminals) and formaldehyde, nitrogen dioxide and hydrocarbons. The substances thus formed can cause irritations of the eyes as well as headaches (Sterling *et al.*, 1983).

2.5. Building Materials

In the case of contaminants which are continuously emitted by building construction materials, they are mostly residues of solvents. A very common contaminant present in indoor air is formaldehyde. The possible sources of this pollutant are manufactured wood products (particle boards, furniture, plywood, panelling), ceiling tiles, draperies, carpets and urea-formaldehyde foam insulations. The quantity of formaldehyde emitted depends on the composition, the processing technique and the age of the materials. Higher quantities of formaldehyde appear mostly in new constructions during the summer months.

The results of recent investigations show that in modern buildings formaldehyde concentrations are emitted which—seen from the point of view of health standards—are clearly higher than any acceptable limit (Kuhn *et al.*, 1983; Figure 3). Depending on its concentration and the duration of exposure, formaldehyde irritates at low concentration the skin, the eyes and mucous membranes, and causes at higher concentrations, nausea, headaches, dizziness, vomiting and coughing. The continuous presence of formaldehyde—especially when combined with other irritant gases and respirable particulates—can lead to respiratory diseases.

Other contaminants are aliphatic and aromatic hydrocarbons from paints, solvents and

FIGURE 3. *Formaldehyde content in the indoor air of new buildings. Measurements were carried out in all tests objects at intervals of four months; the first measurement was done before occupancy of the buildings in spring 1981 (= 0), the second measurement in summer, the third in winter and the fourth in spring 1982. Recorded were the values achieved after a few hours with closed windows. All the values attained were above the recommended limit of 0·1 p.p.m.*

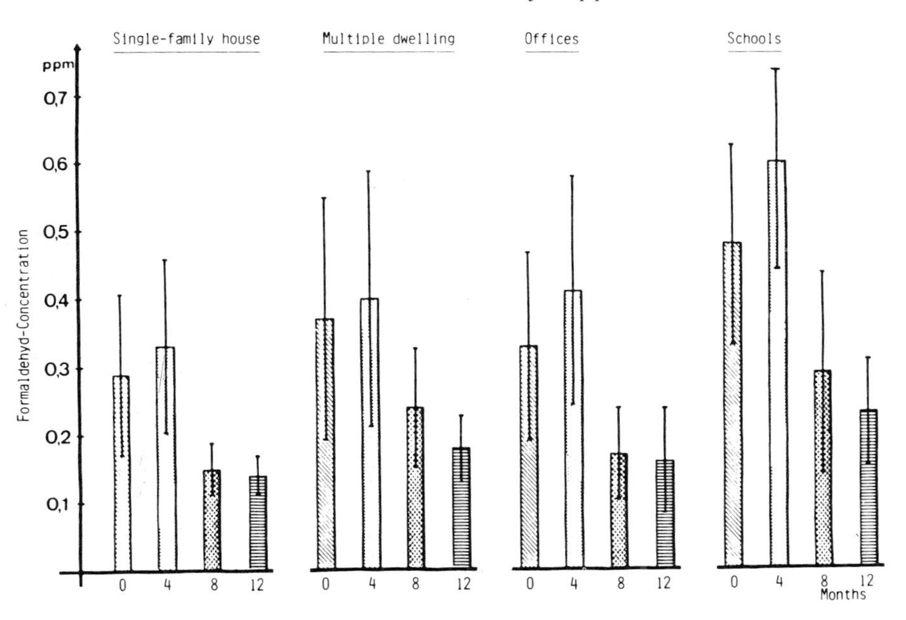

wood preservatives, and chlorinated hydrocarbons used as pesticides. There are only few investigations on the extent of the contamination caused by these substances; special attention must be paid in this case to the possible long-term effects.

Radon is a radioactive gas created by the decay of radium, a substance found in moist soil, rocks, mineral deposits, concrete, bricks and gypsum boards. Investigations in the United States and in Sweden show that in residential premises there is an increase in the radon concentration when ventilation is greatly reduced.

Pollution can also be caused by asbestos. Possible sources are pipe- and duct-insulations, firewalls, ceilings and wall insulations. Asbestos gets into the air of the room when these construction and insulation materials are used improperly.

3. *Evaluation of the Health Hazard*

How can the adverse effects to health and well-being of the numerous types of contaminant which pollute the indoor air be evaluated? We must make a difference between 'short-term' and 'long-term' emissions: most of the contaminants caused by human activities of short duration—as for example by cleaning activities—lead to annoyance or momentary irritation of the eyes and of the mucous membranes. When occurrences of such contaminations are not very frequent and when the products containing the contaminants are used properly, there is no danger to health. Moreover, it is, in this case, up to the individual whose actions pollute his indoor air to reduce such contamination. It is also up to him to get rid of the pollution by airing the room.

What is much more critical is the situation in which contaminants are emitted by constant sources such as building materials, insulations, paints, gas stoves or air-conditioning systems. Such 'long-term' emissions become then unavoidable for the occupants of the room. It is the exposure over a long period of time which becomes a real health hazard. Disquieting in this case are the constant emissions of formaldehyde and solvents, against which the occupant of a room can do absolutely nothing.

4. *Measures Against Indoor Pollution*

All pollutants emitted from the mentioned sources can be, if not eliminated, at least reduced. In the construction and operating of air-conditioning systems the following points must be carefully observed: the air inlet must be placed so as to prevent any contaminations by the street traffic, house heating equipment, industry as well as any sources emitting odours. Critical sources are humidifiers and cooling systems; these must be cleaned and disinfected regularly; filters are also to be controlled periodically.

Indoor pollutants originating from building and furniture materials must be avoided as much as possible. For this purpose there must be strict regulations regarding the licensing and control of the production of materials used in living and working rooms. Emissions should be limited so as to prevent any concentration of contaminants which could become hazardous to health by normal use of the materials. Furthermore, contaminants with high toxic effects, e.g. chlorinated pesticides and all carcinogenics, should be definitively banned from use in building materials.

These are meant to be preventive measures. Any necessary increase in the cost of production or control caused by these restrictive measures is certainly justified, since subsequent attempts to eliminate or reduce such emissions become extremely costly. But on the other hand, severe regulations and increase in control for building materials and for insulations should in no way restrict any necessary measures taken to save energy. For this purpose, only those materials and installations should be permitted which do not cause any health problems. This becomes especially important in private residence and working places, where people spend most of their time.

Furthermore, the following pollutions can be reduced at the source:

1. In dwellings smoking should be limited to the one room; children's playrooms should be kept smoke-free since children react much more sensitively to tobacco smoke. In offices there should be separate rooms for smokers and non-smokers; sensitive persons—e.g. patients suffering from asthma, heart and circulatory diseases—should not be exposed to tobacco smoke.
2. Solvents used for cleaning purposes should not be used in closed rooms, but with open windows. In this case the well-known basic rule of hygiene should be applied: any work causing pollution within a room should be done outside living quarters. Contamination by gas stoves or by photocopying equipment in offices can be avoided or at least reduced by means of local exhaust fans.
3. When air quality in a room has to be improved without frequent airing, air cleaners can be used to reduce the amount of respirable particulates, irritating gas and odours. But in order to be efficient these air cleaners must be equipped with filters and a sufficient air change capacity. A certain amount of the particulates can be reduced with negative ions—especially in combination with an air cleaner. The alleged positive effect of negative ions on the human productivity, efficiency, comfort and health is still very controversial. The reports here under consideration often contradict each other and in many cases they lack important data which could permit the reader to make an objective judgement. There have also been no reproducible effects with the use of an artificial 'electric climate' with combined electric fields to improve human well-being (Wanner, 1980).

5. *Minimal Supply of Fresh Air*

As mentioned before, contamination caused by individual activities depends on the behaviour of the persons occupying the room. Contaminations which cannot be avoided must be eliminated by proper ventilation. By following certain simple and basic rules, the occupants can, without great effort, contribute to keeping the amount of impurity at a relatively low level.

Periodic airing is important not only to issue a sufficient amount of oxygen, but also to eliminate the carbon dioxide and body odours, and any tobacco smoke. Sufficient airing also helps to prevent the humidity from getting too high; when humidity reaches 55–60%, condensation can appear which causes material damage as well as facilitating the growth of fungi and mites.

How often and how long should a room be ventilated? In rooms without occupants there should be a minimum of 0·3 to 0·5 air changes per hour. If a room is occupied, the frequency of ventilation depends on the number of occupants as well as the purpose for which the room is being used: on average 12–15 m^3 of fresh air are sufficient (ASHRAE 62-1981; Huber, 1983; and Sundell, 1982). If there are people smoking or physical activities being done, the amount of fresh air needed is two or three times higher—this means 30–40 m^3 per person per hour.

In order to achieve the minimum amount of fresh air of 12–15 m^3 per person per hour in rooms with natural ventilation the so-called intermittent airing should be used: opening the windows for 3–5 minutes; this permits a complete air change with a minimal loss of heat. In rooms with mechanical ventilation or air conditioning systems the needed supply of fresh air can be controlled. In order to save energy the required amount of fresh air should be adapted to the occupancy and activities being carried out as promptly as possible. In addition, energy can also be saved by means of heat exchanges or recirculating air.

6. *Conclusions*

All of these recommendations take into account the requirements of energy conservation as well as those of hygiene. Minimum ventilation rates are, in this case, an optimization of two conflicting requirements: on the one side a reduction of heat loss and on the other a sufficient amount of fresh air essential for health and well-being.

The problems which must be given first priority in order to keep the indoor pollution under control and to optimize the possibilities of energy conservation are the following:

1. The elaboration of regulations to limit emissions caused by materials and insulations. Among the logical and relatively inexpensive modes of intervention are public information programmes, developing of simple warning devices, and product testing and labelling.
2. The elaboration of recommendations based on practical experiences for minimum ventilation rates for rooms with natural ventilation as well as for those with air conditioning.
3. Regulations concerning supplementary ventilation essential in certain cases, as for example when working with solvents or photocopying equipment.
4. It is essential to inquire into the consequences of an increase in 'relative humidity' due to a lack of ventilation. And this because of the possible material damages as well as the growth of micro-organisms. More knowledge is also needed on the pollution caused by radon, which could be a limiting factor for minimum ventilation rates.

The findings just mentioned as well as the results of current researches should be put into practice without further delay. Subsequent corrections are always very costly. In the case of indoor air, special attention should be paid to the principle of prevention from the point of view of health as well as energy conservation. A decrease in heating energy would subsequently further the aim of reducing the contamination of the outdoor air.

References

ASHRAE 62–1981, Ventilation for acceptable indoor air quality, The American Society of Heating, Refrigerating and Air Conditioning Engineers, Inc.

Huber, G., and Wanner, H.U., 1983, Indoor air quality and minimum ventilation rate. *Environment International*, **9**, 153–156.

Indoor Air Pollution, 1981, International Symposium in Amherst. *Environment International*, **8**, nos 1–6.

Fanger, P.O. and Valbjørn, O. (eds.), 1979, *Indoor Climate, Effects on Human Comfort Performance and Health*, Danish Building Research Institute, Copenhagen.

Kuhn, M. and Wanner, H.U., 1982, Belastung der Raumluft durch Formaldehyd. *Sozial- und Präventivmedizin*, **27**, 260–261.

Muramatsu, T., Weber, A., Muramatsu, S. and Akerman, F., 1983, An experimental study on irritation and annoyance due to passive smoking. *International Archives of Occupational Environmental Health*, **51**, 305–317.

Sterling, T.D., Sterling, E. and Dimich-Ward, H., 1983, Building Illness in the white-collar workplace. *International Journal of Health Service*, **13**, No. 2, 277–287.

Sundell, J., 1982, Guidelines for nordic building regulations regarding indoor air quality. *Environment International*, **8** 17–20.

Wanner, H.U., 1980, Gesundheitliche Aspekte der Klimatisierung. *Blätter für Heizung und Lüftung*, **47**, 17–22.

Weber, A. and Fischer, T., 1980, Passive smoking at work. *Intenational Archives of Occupational and Environmental Health*, **47**, 209–221.

World Health Organisation, Regional Office for Europe, 1979, 1982, *Report of the Working Group on Health Aspects Related to Indoor Air Quality*. Symposis in Bilthoven and in Nördlingen.

Yocom, J.E., 1982, Indoor-outdoor air quality relationships. A critical review. *Journal of the Air Pollution Control Association*, **32**, 500–520 and 'Discussion Papers', **32**, 904–920.

Irritating and Annoying Effects of Passive Smoking

ANNETTA WEBER
Department of Hygiene and Ergonomics, ETH-Centre, CH–8092 Zurich, Switzerland

Abstract

The acute irritating and annoying effects have been investigated in a field and laboratory study in relation to the concentration of some smoke components in the air. Subjective eye, nose and throat irritations and eye-blink rate increase with increasing smoke concentration as well as with increasing exposure duration. At the workplace 30–70% of the indoor CO, NO and particle concentrations are due to tobacco smoke; 25–40% of the employees are disturbed and annoyed by smoke, and one-quarter suffer from eye irritations at work. It is concluded that an average healthy person can be exposed to an acceptable cigarette smoke level which produces a carbon monoxide concentration of 1·5 to 2·0 p.p.m. Above these limits, countermeasures to protect passive smokers are necessary. The required fresh air supply values are presented.

1. *Introduction*

The aim of our studies was to investigate the air pollution due to tobacco smoke and its acute effects, in order to draw some conclusions from a possible dose–response relationship on the environmental tobacco smoke concentration tolerable for men. Such a limit of tolerance could give some indications for measures to protect non-smokers, such as quantifying the fresh air supply to a room.

2. *Methods*

A field study in 44 workrooms and a laboratory study in a climatic chamber, in which cigarettes were smoked by a smoking machine, were carried out. The degree of air pollution due to tobacco smoke was evaluated by measuring the concentrations of carbon monoxide (CO), nitric oxide (NO), formaldehyde, acrolein, particles (PM) and nicotine in the air according to the methods described by Muramatsu *et al.* (1983) and Weber *et al.*

(1980, 1976). The part of the pollutants due to tobacco smoke was obtained by subtracting the background levels before smoking from the concentrations during smoking. These difference values are hereafter called ΔCO, ΔNO and so on.

The degree of acute irritating and annoying effects to exposed persons was simultaneously determined by means of questionnaires and measurements of eye-blink rate, considered as an objective measure for eye irritations (for more details, see Weber *et al.*, 1980 and Muramatsu *et al.*, 1983).

3. *Field Study: Results and Discussion*

Table 1 summarizes some results of the chemical measurements. The comparison of these Δ-values with the measured absolute indoor concentrations reveals that 30–70% of the measured indoor concentrations of CO, NO and particles are due to tobacco smoke.

FIGURE 1. *Evaluation of air quality and effects due to environmental tobacco smoke at the workplace. Results of 472 employees in 44 workrooms.*

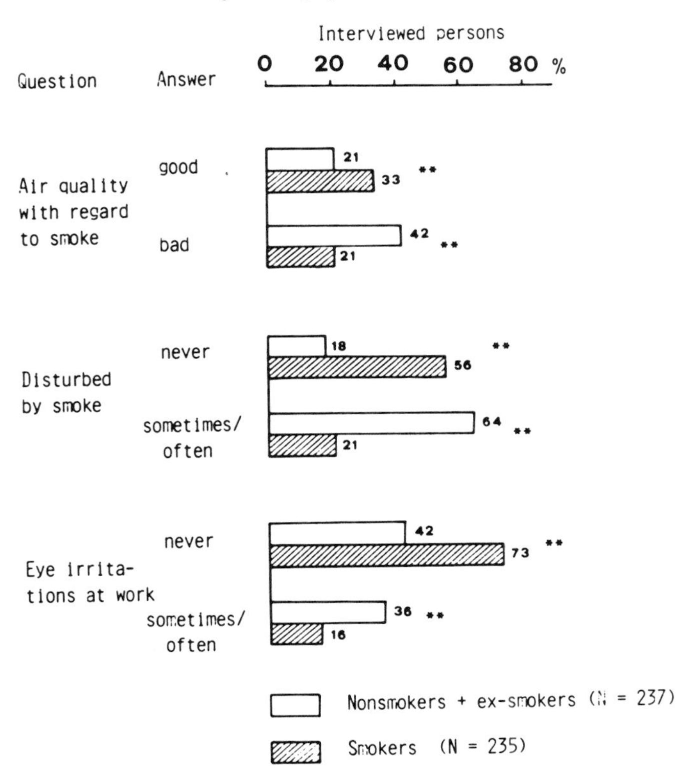

FIGURE 2. *Mean subjective eye irritations due to environmental tobacco smoke, related to smoke concentration and duration of exposure. ΔCO = CO level during smoke production minus background level before smoke production. 32 to 43 subjects. 0 min = measurement before smoke production. Period 0 to 5 min = increasing smoke concentration. Period 6 to 60 min = constant smoke concentration.*

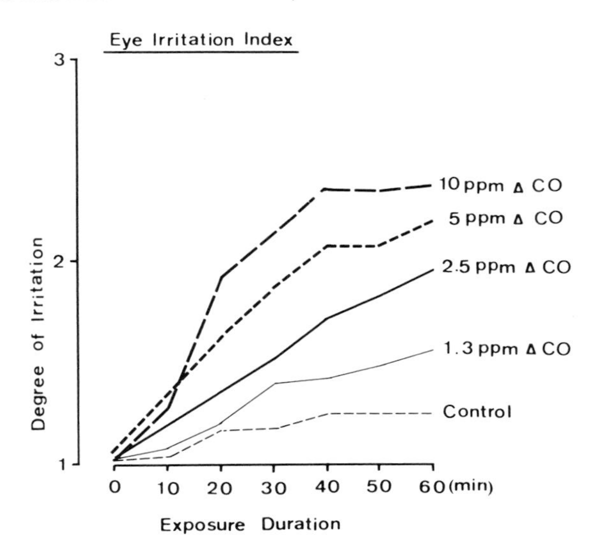

TABLE I. *Air pollution due to tobacco smoke in 44 workrooms.*
Δ-value = 'indoor concentration during work' minus 'indoor concentration before work'.

Component	Number of samples	Mean	Standard	Maximum
ΔCO (p.p.m.)	353	1·1	1·3	6·5
ΔNO (p.p.b.)	348	32	60	280
ΔPM ($\mu g/m^3$)	429	133	130	962
Δnicotine ($\mu g/m^3$)	140	0·9	1·9	13·8

The correlation between the gas phase components ΔCO and ΔNO is relatively high (Pearson correlation coefficient, r = 0·73). However, the correlations with Δnicotine and ΔPM are low. Therefore carbon monoxide can be considered as a useful indicator for nitric oxide, but neither for the particular matter nor for nicotine.

In Figure 1 some results of the interview are represented, separated into groups of smokers and non-smokers.

From Figure 1 it can be deduced that:

1. Approximately one-third of the employees qualified the air at work with regard to smoke as bad.

FIGURE 3. *Mean effects of environmental tobacco smoke on eye blink rate. ΔCO = CO level during smoke production minus background level before smoke production. 32 to 43 subjects. 0 min = measurement before smoke production. Period 0 to 5 min = increasing smoke concentration. Period 6 to 60 min = constant smoke concentration.*

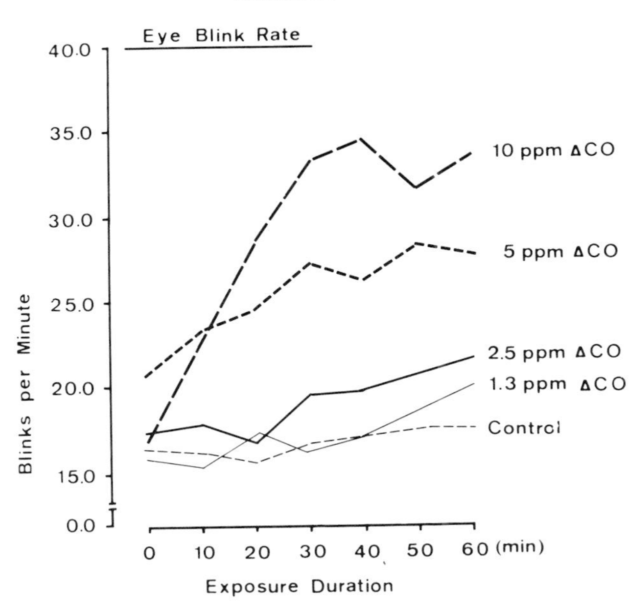

2. Forty per cent were disturbed by smoke.
3. One-quarter of the persons reported eye irritations at work.
4. Non-smokers reacted significantly more to environmental tobacco smoke than smokers.

Employees suffering from hay fever reported significantly more eye irritations at work than those without hay fever. Furthermore, 72% of the interviewed non-smokers and 67% of the smokers are in favour of a separation into smoker and non-smoker workrooms; 49% even support a partial or total prohibition of smoking at work.

4. *Laboratory Study: Results and Discussion*

The analysis of the questionnaire showed that the irritating effects are most pronounced on the eyes, followed by the nose and finally by the throat.

Figures 2 and 3 illustrate the results obtained for subjective eye irritations and eye-blink rate of subjects being exposed to different smoke concentrations which were kept constant for one hour.

FIGURE 4. *Percentage of persons with strong or very strong eye irritation reactions related to the degree and duration of exposure.*

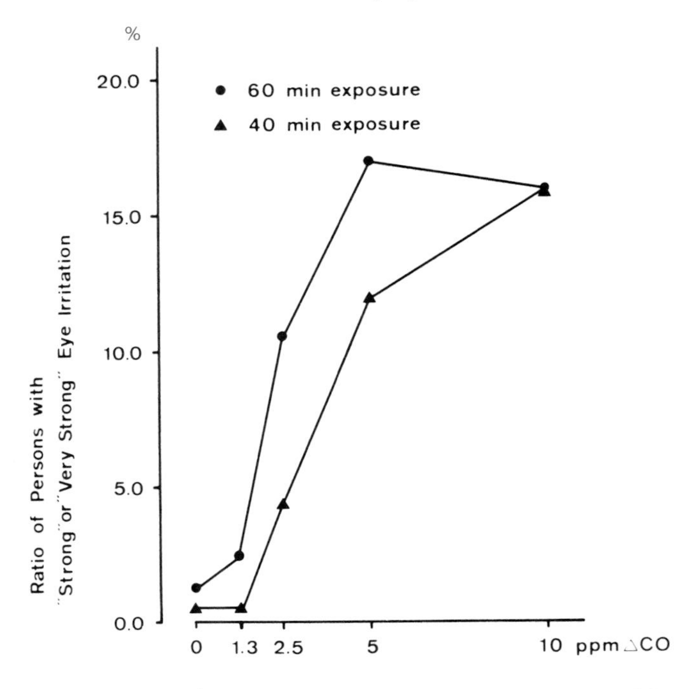

Two facts are obvious:

1. The mean eye irritating effect increases with increasing smoke concentration, determined by ΔCO levels.
2. The mean eye irritation increases with the duration of exposure in spite of constant smoke concentration.

The same, but less pronounced result, has been observed for nose and throat irritations. Annoyance—which has also been determined by means of a questionnaire—shows a different development than irritations: it increases rapidly as soon as smoke production begins for 10 to 15 minutes; it remains approximately constant during the rest of the exposure. Thus the duration of exposure has nearly no influence on the degree of annoyance.

The mean incidences of people with strong and very strong subjective eye irritations are reported in Figure 4.

Figure 4 clearly discloses that there is a marked increase of the incidence of strong eye irritations between the smoke levels corresponding to 1·3 and 2·5 p.p.m. ΔCO. Based on these results, a possible limit to protect healthy people in their everyday environment against impairment of well-being by environmental tobacco smoke should lie in this range, i.e. between 1·5 and 2·0 p.p.m. ΔCO. Indeed a marked increase from about 3 to over 10% of subjects with strong eye irritations appears therein. The upper concentration limit of 2·0 p.p.m. ΔCO is for instance already reached when 2 cigarettes are smoked per hour in a room of 80 m^3 with a single airchange. Hence countermeasures to protect passive smokers are desirable when the ΔCO level reaches 1·5 p.p.m. and are necessary when it hits 2·0 p.p.m. The lower limit should be applied to workplaces where passive smokers can hardly escape the exposure, and the upper limit to restaurants and other places, where people usually go voluntarily and for a shorter lapse of time.

Calculations show that a fresh air supply of 33 m^3 per hour and per smoked cigarette is necessary to keep the ΔCO concentration below the proposed upper limit of 2·0 p.p.m.; for the lower limit, 50 m^3 per hour and per smoked cigarette are required. Depending on the number of persons present in a room, a fresh air supply of 25 to 45 m^3 per hour and per person is necessary in order not to exceed the upper limit. In other words: the ventilation has to be 2 to 4 times higher than in a room where nobody smokes (in which only 12 to 15 m^3 per hour and person are required). For that reason the increased ventilation as a measure to protect passive smokers is not recommendable from the energetical and economical point of view. Therefore, whenever possible, organizational measures, such as separation into smoker and non-smoker rooms or a prohibition of smoking, rather than an increased fresh air supply, should be taken into consideration.

References

Muramatsu, T., Weber, A., Muramatsu S. and Akermann, F., 1983, Experimental study on irritations and annoyance due to passive smoking. *International Archives of Occupational and Environmental Health*, **51**, 305–317.

Weber, A., Fischer, T. and Grandjean, E., 1976, Objektive und subjektive physiologische Wirkungen des Passivrauchens. *International Archives of Occupational and Environmental Health*, **37**, 277–288.

Weber, A. and Fischer, T., 1980, Passive smoking at work. *International Archives of Occupational and Environmental Health*, **47**, 209–221.

Comparison of Non-Smokers' and Smokers' Perceptions of Environmental Conditions and Health and Comfort Symptoms in Office Environments With and Without Smoking

T.D. Sterling and E.M. Sterling*
Simon Fraser University, Burnaby, BC, Canada V5A 1S6

Abstract

One thousand and one hundred branch members of the New York branch of the Office and Professional Employees International Union working in nine office buildings filled out a detailed questionnaire on working conditions and health comfort complaints. Data were classified according to smoking habits of respondents and office rules regulating smoking. Neither smokers nor non-smokers differed in prevalence of complaints for a large variety of symptoms by smoking conditions in the office but more non-smokers complained about stressful conditions in offices where smoking was restricted or prohibited than where smoking was permitted.

Lack of differences in comfort complaints between smoking and non-smoking offices does not contradict findings of irritated responses due to passive exposure to smoke in controlled, especially chamber studies. Responses of the OPEIU members were taken under normal conditions of ventilation and lighting and no specific attention was drawn to the presence or absence of smokers. The findings are also in agreement with a study conducted by the US National Institute for Occupational Safety and Health where no association was found between density of smokers and levels of complaints.

1. *Introduction*

A great deal of attention has been concentrated on smoking as a source of indoor airborne substances and as a possible cause for building illness. However, modern buildings tend to generate a large variety of pollutants as well as entrapping large numbers of them which

* Director of Building Research, TDS Limited, 1507 West 12th Ave, Vancouver, BC, Canada, V6J 2E2.

penetrate from the outside (Sterling, 1977; Yocum, 1982). Also, review of the literature has shown that elevated levels of particulates and gases related to smoking have been measured indoors almost exclusively in experimental situations using special chambers, in the absence of ventilation or while excessive amounts of cigarettes were smoked. But for those studies conducted under normal conditions of occupation, smoking and ventilation, indoor levels of contaminants do not exceed substantially those found outdoors (Sterling, 1982).

Corroborating evidence for this also has come from a recent review of some 143 building studies. When pollutant levels are compared in offices where smoking is restricted and/or prohibited and offices where smoking is permitted, no differences are found under normal conditions of occupancy (Sterling, 1983; Sterling, 1983). At the same time, it has been demonstrated that various types of symptoms such as eye irritation or respiratory distress are reported when non-smokers (or smokers, for that matter) are exposed to high levels of tobacco smoke in unventilated offices or chambers (Weber, 1980, 1983). But does smoking affect health and comfort of office workers under normal conditions of office use?

2. *Method*

A computer-readable, self-administered work environment questionnaire was given to approximately 1 100 members in 9 buildings of the Office and Professional Employees International Union, Local 153, New York City. The questionnaire was constructed to document perceived environmental conditions, symptoms and complaints related to building illness among building occupants.

The Health and Work Environment Survey questionnaire contained detailed information about:

1. Environmental conditions.
2. Health-related symptoms.
3. Life-style factors.
4. Stress factors.
5. Questions about equipment use, employment history, types of appliances used at home, and others.

3. *Results*

The major findings of our study (reported elsewhere) were a highly substantial and statistically significant association between all indices of health and disease, on the one hand, and conditions of ventilation, lighting, VDT and CRT use and stress (possibly in that order). The analyses of environment, health and stress indices in relation to smoking are of special relevance to our inquiry.

The reactions of 469 non-smokers, 286 former smokers and 326 present-smokers were compared for a list of ergonomic indices: ventilation, temperature, humidity, lighting, and on specific responses to questions about air movement (too little and too much), lighting levels (too dim, too bright, too much glare), temperature (too hot, too cold), humidity (too dry, too moist), air (too smoky, too stuffy); for a list of health and comfort indices: building illness, neurophysiological, cardiorespiratory, musculoskeletal (related also to seating comfort), visual health; and finally a number of stress related indices: decision-making, job security, physical stress and relationship stress. With one single exception, all these distributions turned out to be similar and not statistically significant for non-smokers, former smokers and present-smokers. That exception, as may well be expected, was that of the specific judgement of 'smokiness of the air' (but not observed for 'stuffiness of air'). Sixty-three of the smokers compared with 52% of the non-smokers felt the air not to be too smoky in their environment. On the other hand, 16·7% of the non-smokers as compared to 6·6% of the smokers found the air very often to be too smoky (with $\chi^2 = 18{\cdot}74$, d.f. $= 4$, $p \leq 0{\cdot}001$). While the differences between groups are small, they do indicate that smokers have the greater tolerance for cigarette smoke than do non-smokers (with former smokers falling in between never- and present-smokers).

Some of the workers smoked and some of them did not. Some of them worked in places where smoking was permitted, some in places where smoking was prohibited, and some in places where smoking was restricted.

TABLE I. *Percentage distribution for 'ventilation index' responses for non-smokers and smokers working where smoking was permitted and restricted or prohibited.*

Ventilation index	1A: Non-smokers working where smoking is		1B: Smokers working where smoking is	
	Permitted	Prohibited or Restricted	Permitted	Prohibited or Restricted
Good	11·7	10·8	9·7	14·8
Average	56·3	63·1	62·0	69·7
Poor	32·0	26·2	28·3	15·5
	100%	100%	100%	100%
	$\chi^2 = 1{\cdot}17$	$p \leq 0{\cdot}88$	$\chi^2 = 10{\cdot}23$	$p \leq 0{\cdot}04$

Tables are organized as percentages in such a way that direct comparisons can be made. Each column gives the proportion of individuals who rate their environment, health or stress conditions as good, average or poor. Thus Table 1A shows that of the non-smokers working in places where smoking was permitted, 11·7% rated their ventilation conditions (as measured by the 'ventilation index') as good, and 32·0% as poor, while 10·8% of non-smokers working in places where smoking was either restricted or prohibited, rated their ventilation conditions as good, and 26·2% as poor. Among smokers working in environments where smoking was permitted, we find that 9·7% rated ventilation as good while 14·8% of smokers did so who worked where smoking was either restricted or prohibited (Table 1B). (The hypothesis of statistical independence between the frequencies was computed for 4 d.f., between groups in smoking permitted, restricted and prohibited offices using chi square statistics for non-smokers and smokers separately. Please note that in order to save space, tables here combine the restricted and prohibited groups.)

For non-smokers, the distribution of responses to questions assessing the quality of environmental conditions of ventilation (Table 1A), temperature (Table 2A), humidity (Table 3A, lighting and odour (not shown here) show no differences between environments with and without smoking. However, the same is not true for smokers who appear to be responsive to variations in ventilation and associated perceived temperature and humidity measures when smoking is restricted or prohibited (Tables 1B, 2B and 3B).

TABLE 2. *Percentage distribution for 'temperature index' responses for non-smokers and smokers working where smoking was permitted and restricted or prohibited.*

	2A: Non-smokers working where smoking is		2B: Smokers working where smoking is	
Temperature index	Permitted	Prohibited or Restricted	Permitted	Prohibited or Restricted
Good	13·1	4·8	10·1	15·0
Average	69·2	85·7	71·5	76·2
Poor	17·7	9·5	18·4	8·8
	100%	100%	100%	100%
	$\chi^2 = 8{\cdot}06$	$p \leq 0{\cdot}09$	$\chi^2 = 8{\cdot}84$	$p \leq 0{\cdot}07$

TABLE 3. *Percentage distribution for 'humidity index' responses for non-smokers and smokers working where smoking was permitted and restricted or prohibited.*

	3A: Non-smokers working where smoking is		3B: Smokers working where smoking is	
Humidity index	Permitted	Prohibited or Restricted	Permitted	Prohibited or Restricted
Good	30·2	39·1	28·0	40·3
Average	63·6	59·4	69·2	54·9
Poor	6·2	1·6	2·8	4·9
	100%	100%	100%	100%
	$\chi^2 = 3{\cdot}9$	$p \leq 0{\cdot}42$	$\chi^2 = 10{\cdot}28$	$p \leq 0{\cdot}04$

The distribution of responses to questions assessing the presence of symptoms related to building illness, visual health, neurological, cardiorespiratory, musculoskeletal and somatic health shows no difference between environments with and without smoking, but this time for neither smokers nor non-smokers (not shown here, except building illness). The 'building illness index' is perhaps the most relevant of all these health indices used (Table 4). It summarizes symptoms most often listed in connection with epidemics of health-related complaints in modern sealed structures (e.g. headache, fatigue, nose, throat and eye irritation, sore throat and cold symptoms). Table 5 shows that the percentage distribution of building illness related responses is not related to either smoking practices of the employee or of the workplace. In fact, health-related complaints as a group show no significant relationship or even a consistent trend between the percentage of employees who report health-related symptoms and either their or their workplace's smoking status. These observations are further supported by the lack of differences in the proportion of absenteeism in workplaces with and without smoking (Table 5).

TABLE 4. *Percentage distribution for 'building illness index' responses for non-smokers and smokers working where smoking was permitted and restricted or prohibited.*

	4A: Non-smokers working where smoking is		4B: Smokers working where smoking is	
Building illness index	Permitted	Prohibited or Restricted	Permitted	Prohibited or Restricted
Good	43·1	37·9	38·9	43·0
Average	38·7	51·5	46·5	44·4
Poor	18·2	10·6	14·6	12·6
	100%	100%	100%	100%
	$\chi^2 = 6{\cdot}49$	$p \leq 0{\cdot}16$	$\chi^2 = 1{\cdot}087$	$p \leq 0{\cdot}90$

TABLE 5. *Percentage distribution for 'absenteeism index' responses for non-smokers and smokers working where smoking was permitted and restricted or prohibited.*

	5A: Non-smokers working where smoking is		5B: Smokers working where smoking is	
Absenteeism index	Permitted	Prohibited or Restricted	Permitted	Prohibited or Restricted
Good	88·8	82·8	87·6	83·0
Average	10·4	17·2	11·4	14·0
Poor	0·7	0	1·0	2·7
	100%	100%	100%	100%
	$\chi^2 = 2{\cdot}47$	$p \leq 0{\cdot}65$	$\chi^2 = 7{\cdot}26$	$p \leq 0{\cdot}12$

Table 6 shows the relationship of smoking to perceived stress. Employees working in places where smoking is permitted report substantially less stress than employees in smoking restricted or prohibited workplaces. It is likely that this relationship simply reflects a more permissive and tolerant attitude by the employers.

TABLE 6. *Percentage distribution for 'job security index' responses for non-smokers and smokers working where smoking was permitted and restricted or prohibited.*

Job security index	6A: Non-smokers working where smoking is		6B: Smokers working where smoking is	
	Permitted	Prohibited or Restricted	Permitted	Prohibited or Restricted
Good	66·4	35·9	68·5	46·9
Average	32·8	53·1	29·6	44·8
Poor	0·7	10·9	2·0	8·4
	100%	100%	100%	100%
	$\chi^2 = 25{\cdot}11$	$p \leq 0{\cdot}001$	$\chi^2 = 27{\cdot}29$	$p \leq 0{\cdot}001$

4. *Discussion*

The review of available studies does not provide any objective evidence that either pollution levels or patterns of health-related complaints differ in some remarkable way between locations with or without smoking restrictions. But it must be stressed that these observations were obtained during normal working conditions. For this reason, these findings are not comparable to reports of responses to smoking among non-smoking office workers obtained under special conditions (Barad, 1979; Working Women, 1981; Weber, 1980, 1983). Where smoking is felt to be irritating, such complaints ought to be attended to. Yet a question might be asked whether it is the presence of smokers or the act of smoking that causes the complaints about smoking?

In summary, a careful review of available evidence contradicts the belief that smoking is the or even a pivotal source of indoor pollution or of health-related building complaints. Modern buildings tend to generate and entrap pollutants from numerous sources. Under inadequate ventilation, conditions may be created where discomfort and illness result irrespective of whether or not smoking is permitted.

References

Barad, C.B., 1979, Smoking on the job: the controversy heats up. *Occupational Health and Safety*, **48**, 21–24.

Salisbury, S.K., Kelter, A., Miller, B. and Roper, P., 1982, TA 80–122–1117. 101 Marietta Tower Building, Atlanta, Georgia, US. National Institute for Occupational Safety and Health, Health Hazard Evaluation, Cincinnati, Ohio.

Sterling, E., McIntyre, D. and Sterling, T., 1984, The effects of sealed office buildings on the ambient environment of office workers. This vol., pp.70–76.

Sterling, E. and Sterling, T., 1981, The impact of different ventilation and lighting levels on building illness: an experimental study. Proceedings of the International Symposium on Indoor Air Pollution, Health and Energy Conservation, 13–16 October, 1981, University of Massachusetts, Amherst, Massachusetts. *Canadian Journal of Public Health* (in press).

Sterling, T.D., Dimich, H. and Kobayashi, D., 1982, Indoor byproduct levels of tobacco smoke: a critical review of the literature. *Journal of the Air Pollution Control Association*, **32**, 250–259.

Sterling, T.D. and Kobayashi, D., 1977, Exposure to pollutants in enclosed living spaces. *Environmental Research*, **13**, 1–35.

Sterling, T., Sterling, E. and Dimich-Ward, H., 1983, Air quality in public buildings with health complaints. *ASHRAE Transactions*, 89, 2A and B.

Weber, A., and Fischer, T., 1980, Passive smoking at work. *International Archives of Occupational and Environmental Health*, **47**, 209.

Weber, A., 1983, Irritating and annoying effects of passive smoking. This vol., pp.28–33.

Working Women, April 1981, *Health Hazard for Office Workers*, A Report by Working Women Education Fund, 1224 Huron Road, Cleveland, Ohio 44115.

Yocum, J.E., 1982, Indoor-outdoor air quality relationships. *Journal of the Air Pollution Control Association*, 32, 500–520.

Ocular Annoyance Due to Improper Air-Conditioning in a New VDT Office Environment

F. MAULI
INPS Medical Service and USL 25 WHO, Verona, Italy
AND R. BELLUCCI
Department of Ophthalmology, University of Verona, Italy

Abstract

The aim of this study was to formulate hypotheses regarding the cause of the continuing presence of visual disturbances in VDT operators in an environment adequately designed in terms of visual ergonomics.

We therefore studied the air-conditioning system of an office environment, analysing the following parameters: temperature, relative humidity, ventilation, air speed, chemical and bacterial composition of the air.

Lastly, we postulated possible causal relationships between onset of ocular disturbances and improper characteristics of the air-conditioning system, stressing the need for a more thorough approach to the biological aspects.

1. *Introduction*

A VDT section was organized for the first time in 1979 in a local department of a major public authority. The operators began very soon to manifest ocular disturbances, which they had not experienced before, and complained particularly about the visual ergonomics of the new equipment.

The VDTs were located in a small room originally given over to normal office work, with south-facing windows occupying one entire wall. The ergonomic aspects of the new activity had not been considered.

As a result of the complaints, and the organizational requirements of the authority, the section was completely reorganized. The new rooms are spacious, with windows facing north, in anticipation of an increase in activity.

The suggestions made by our team have led to the creation of an efficient lay-out as regards visual ergonomics. Work organization, task and operators are unchanged. The microclimatic conditions on the other hand are appreciably different owing to the installation of a new air-conditioning system in these new rooms. At the present time the operators are not allowed to switch off the system and/or open the windows.

In these new work conditions, eye symptoms on the whole diminish, though there is a relative increase in conjunctival symptoms. Given the same subjects, visual performance and visual task, comparison between the two sets of symptoms led us to study thoroughly the new microclimatic conditions, which we had had no opportunity to influence previously. We were encouraged in our research by the criticisms made by the operators with regard to the air conditioning system.

2. *Methods*

We therefore carried out an environmental survey, comparing the microclimatic conditions existing simultaneously outdoors and in the other naturally ventilated rooms in the same building. In addition to the normal parameters used in this type of research, we also studied the quality of the air from both the chemical and the biological point of view (e.g. temperature, relative humidity, air speed, ventilation, air quality).

It should be pointed out that the printing machine (PM) room communicates with the VDT room via a door, and the operators have to pass to and fro regularly in the course of their work. The two rooms have their own air-conditioning systems: the PM room system has a capacity of approximately 50 000 m^3 per hour, and the room measures approximately 500 m^3, while the VDT room system has a capacity of 4500 m^3 per hour, and the room measures approximately 450 m^3.

It can easily be imagined that such a powerful air-conditioning system requires periodic checks on the microclimatic variables.

3. *Results*

We shall now examine the salient data relating to our measurements, according to standard methods. As may be seen from the values given in Table 1, the values on the whole are within the accepted ranges for areas of thermal comfort. A number of problems derive from the fact that the operators have to go backwards and forwards fairly often from the VDT room to the PM room and vice versa.

The relative humidity values encountered in the various different rooms reveal that the levels in the air-conditioned rooms are below those recommended in published studies from work of this type. We should also consider that the mean natural humidity is fairly high. In addition to the subjective sensations of the operators, there was another finding of an empiric type suggesting an excessively low relative humidity percentage in the new rooms: the office plants which thrived in the previous work environment, showed evident signs of distress shortly after being transferred to the new rooms.

TABLE 1. *Temperature and relative humidity inside and outside VDT working environment.*

	Conditioned rooms			
	Printing machine	VDT workrooms	Not conditioned	Outdoors
Air temperature (°C)	21·8	24·0	21·0	20·5
Effective temperature (°C)	19·2	21·1	18·8	–
Relative humidity (%)	48	43	52	71

The number of air changes in the conditioned rooms appears to be excessive in relation to the requirements of the operators, particularly in the PM room (Table 2). We find, however, no evidence of international standards laying down upper limits in this respect.

The air speed values recorded, as given in Table 2, are directly related to the capacity of the two air-conditioning systems, and proved to be higher than those normally recommended (approximately 0·1 m/s) for this type of work. In the 'under' conditioning system used in this case, the air is delivered via grids in a built-up floor (Figure 1A). In addition, the layout of the grids can be rearranged according to needs or preferences. This system makes it possible to reduce draughts as there are no special air ducts and the air delivered by the conditioner spreads in the space between the original floor and the built-up floor, and penetrates into the room via the grids. In the initial project (Figure 1B) the grids were to be distributed geometrically over the floor area. Later they were installed along the walls as far as possible from the work places, so as to reduce the annoyance created by excessive air spread.

TABLE 2. *Ventilation and air speed in the rooms.*

	Conditioned rooms		
	Printing machine	VDT workrooms	Not conditioned
Ventilation (m^3/h)	30	9	–
Air speed (m/s)	0·4–0·5	0·2–0·4	0·2–0·4

For reasons of completeness and in order not to neglect those possible causes of ocular disturbances still obtained in a context of good visual ergonomics and work organization, we carried out biological and qualitative chemical analyses of the air, according to the same procedures previously adopted.

We studied the air from a chemical point of view, by means of IR spectrophotometry. The investigation was carried out at different times of day, so as to take account of the possible effects of traffic on the roads outside the office. We decided not to carry out quantitative analysis in view of the large numbers of air changes and the high degree of efficiency

FIGURE 1. *System of ventilation of the VDT room.*

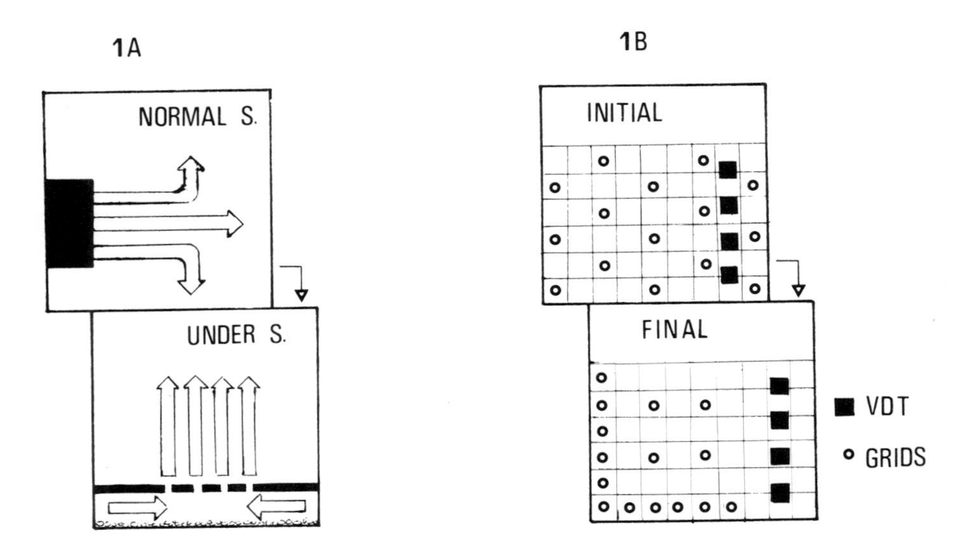

of the filters used in the system (94%). A spectrum highly similar to the normal reference spectrum was obtained.

The results of the air analysis from the microbiological point of view (Table 3) also appear substantially affected by the capacity of the system and the efficiency of the filters. It is immediately clear that the bacterial and mycotic counts are extremely low. For purposes of comparison, 35/70 micro-organisms per cubic metre are commonly accepted as the limits for operating theatres and 150/350 micro-organisms per cubic metre in hospital wards. In the VDT room the relatively higher values as compared to the PM room may probably be related to the inferior capacity of the air-conditioning system and the larger number of operators in this room. No proper air-conditioning system should neglect the aim of keeping the natural defence mechanisms of the body active. In our case, however, the air seems to be excessively sterile and thus inadequate from this point of view. As regards the eye, for instance, in the course of time there could be a quantitative reduction in antibodies in the lacrymal film. We therefore suggest that the biological aspect of air-conditioning warrants more thorough consideration.

TABLE 3. *Microbiological analysis of the environments studied—each value is the mean of 30 measurements.*

	Conditioned rooms			
	Printing machine	VDT workrooms	Not conditioned	Outdoors
Bacteria (C.F.U./m^3)	16	50	127	583
Fungi (C.F.U./m^3)	33	27	83	705

4. *Discussion*

The results of our study confirm that on examining the visual symptoms of a sample of VDT operators it is not enough to consider the general and visual ergonomic conditions or the visual performance, but ambient air conditions should also be taken into account. In this connection it should be stressed that the use of conditioned air is becoming increasingly widespread, in response to various thermal, hygrometric and other control requirements.

The eye may be affected in various different ways. Temperature affects the state of conjunctival perfusion and reduction in humidity of the air and/or the presence of draughts may increase the evaporation of the lacrymal film, at times making it impossible to use contact lenses. The problems appear to give rise to symptoms particularly when the reduction in the relative humidity is high as compared to the outside air. The presence of draughts may make for a worsening of the situation; they should therefore be kept to a minimum, and their direction controlled.

A proper air-conditioning system should clearly take into greater account the adaptation of those working on the premises to outside conditions. With this in mind, we hold that the biological aspects of air-conditioning are of fundamental importance and warrant further more thorough study, both as regards the method to be adopted and as regards possible effects on the body.

References

ACGIH, 1980, Threshold limit values for chemical substances and physical agents in the work environment with intended changes for 1980.

ASHRAE, 1977, *Handbook of Fundamentals* (New York: American Society of Heating, Refrigerating and Air Conditioning Engineers).

Cakir, A., Hart, D.J. and Stewart, T.F.M., 1980, *Visual Display Terminals* (Chichester: J. Wiley).

Grandjean, E. and Vigliani, E., (eds), 1980, *Ergonomic Aspects of Visual Display Terminals* (London: Taylor & Francis).

Kraiss, K.F. and Moraal, J., 1976, *Introduction to Human Engineering* (Cologne: TüV Rheinland).

Pizzetti, C., 1980, *Condizionamento dell'Aria e Refrigerazione* (Milan: Masson Italia).

Ruggeri, R., 1981, *Microclima degli Ambienti Civili ed Industriali* (Milan: CLUP).

Tintori Pisano, E., 1982, *Il Posto di Lavoro Ergonomico per Operatori di Terminali Video* (Milan: Clinica del Lavoro 'L. Devoto').

Ill Health Among Office Workers: an Examination of the Relationship Between Office Design and Employee Well-Being

ALAN HEDGE
Department of Applied Psychology, University of Aston in Birmingham, Birmingham B4 7ET, UK

Abstract

Data on employees' self-reported health problems are described from six user surveys conducted in three office buildings (conventional, enclosed offices; non-air-conditioned and air-conditioned open-plan offices). Overall a significantly higher incidence of reported headaches was found amongst staff working in open-plan offices compared with those in conventional offices and this seems to be related to the incidence of problems of reflected glare. Problems of eye irritation and respiratory complaints (coughs, sore throats, etc.) are similarly found to be most common among staff in open-plan offices, but importantly, only when these are air-conditioned and have poor daylight penetration, otherwise the incidence of health problems is comparable with that for conventional accommodation. In the two air-conditioned open-plan offices more women than men reported frequent headaches, eye problems, and u.r.t. infections. These health problems do not appear to be age dependent. The implications of these findings and those of other recent studies of health issues in offices are discussed.

1. *Introduction*

There is a growing body of evidence which suggests that certain aspects of modern office environments may adversely affect the health and productivity of employees (Craig, 1981). Frequent headaches, complaints of eye irritation, and a high incidence of upper respiratory tract complaints (sore throats, coughs, colds, etc.) are typical of the problems reported. Moreover, such problems are frequently thought to be linked with the adoption of open-plan offices characteristic of the institutional and commercial offices of many western countries. In a recent study Turiel *et al.* (1983) report a detailed comparative study of two office buildings in San Francisco, one an air-conditioned building where a high incidence of health-related complaints had been registered by staff and a nearby naturally ventilated 'control' office building, the former being 'open-plan', the latter more of a 'conventional' layout (personal communication). Indoor air quality was monitored in the open-plan building under two different ventilation modes (100% outside air, and recirculation mode—approximately 15% outside air) and a wide variety of measures of this were taken (air flow rate, temperature, relative humidity, odour perception, microbial burden, particulate mass, carbon monoxide and dioxide levels, nitrogen dioxide, formaldehyde and 28 other organic compounds). No single contaminant exceeded current health standards and there were no differences between the employees in each building in terms of their racial composition, personal profiles (age, sex, smoker/non-smoker, etc.) or medical histories, however significantly more complaints of health problems (eye irritation/itching, nose/throat irritation, shortness of breath, chest tightness, and eye inflammation/infection) were found among those working in the air conditioned open-plan offices.

While these results present cause for concern what is not at all clear is whether or not such health problems necessarily must follow from working in open-plan, because the influence of this design is usually confounded by the presence of other design features, such as the use of air-conditioning. To further explore these problems six surveys of employees' reactions to a wide variety of features of their work and working conditions, including the incidence of certain health problems, in conventional, non-air-conditioned open-plan, and air-conditioned open-plan offices are reported, and the interrelationships between these health problems and the characteristics of those employees suffering these are also investigated.

2. *Method*

All of the surveys to be described relied on self-completion questionnaires for gathering information on the incidence of health and other issues in the offices. Specific details of each of these surveys are as follows:

2.1. **Air-Conditioned Open-Plan Offices**

O_1 (n = 649): These are the offices of a Metropolitan County Council completed for occupation in 1972. Some results of this survey have previously been reported (Hedge, 1980, 1982), however, a detailed re-analysis of the health items is given here. Of the respondents 452 were men and 192 were women (5 did not state their sex).

O_2 (n = 359): These are the offices of a Metropolitan District Council completed for occupation in 1976. Again a report of these results has been published (Hedge, 1983) but detailed issues of health are explored more closely here. Of the respondents 189 were men and 170 were women.

2.2. Non-Air-Conditioned Open-Plan Offices

O_{3a} (n = 118): These are the offices of a District Council completed for occupation in 1982. The building comprises predominantly open-plan offices. It is not air-conditioned, has opening windows, and is designed for maximum daylight penetration. Of these respondents 105 had worked previously in the conventional offices surveyed in C_2. Of the sample 63 were men and 55 were women.
O_{3b} (n = 127): This was a repeat survey 15 months later, of the above offices. Here a more extensive questionnaire was used to gather further detailed information on the incidence of health and other issues. The sample comprised 67 men and 60 women.

2.3. Conventional Offices

C_2 (n = 59): These are conventional, cellular offices linked with the offices surveyed in O_2. Staff in these offices completed the same questionnaire at the same time as those in the O_2 survey. In total 33 men and 26 women were surveyed.
C_3 (n = 147): These were a variety of conventional style offices occupied by those staff who moved to the new open-plan offices surveyed in O_3. There were 74 men and 73 women respondents.

3. *Results*

Data on the health problems reported by employees at work are presented below. The questionnaire items on health to be discussed were tension, frequency of headaches, problems of eye irritation/soreness, and incidence of upper respiratory tract (u.r.t.) complaints, such as sore throats, coughs, colds, etc. The results for each of these problems are described separately before looking at their interrelationships.

3.1. Tension

Self-reports of tension at work showed that for all of the offices surveyed around one-third of employees complained that such a feeling was a frequent occurrence at work (Table 1). However, there was no apparent variation of the incidence of this with office type and it may be that 'tension at work' is more closely related to job factors than to environmental conditions in the office.

3.2. Headaches

Complaints of frequent headaches at work clearly show a relationship with office type (Table 1). Around twice the incidence of headaches were found amongst those in open-plan

TABLE I. *The percentages of employees complaining of health problems in each of the office surveys.*

	Cellular offices		Open-plan (air-conditioned)		Open plan (naturally ventilated)	
	C_2 (n = 59)	C_3 (n = 147)	O_1 (n = 649)	O_2 (n = 359)	O_{3a} (n = 118)	O_{3b} (n = 127)
Tension	32	34	36	38	37	41
Headaches	16	21	39	39	36	29
Eye irritation	17	17	50	45	20	34
U.r.t. infections	24	23	40	46	21	47

compared with employees working in conventional offices, and whether or not the open-plan offices are air-conditioned does not seem to affect this result. For all three open-plan offices these responses correlate significantly (Pearson's 'r') with complaints about glare from the fluorescent lighting (O_1—r = 0·41, n = 634, $p < 0{\cdot}001$; O_2—r = 0·38, n = 356, $p < 0{\cdot}001$; O_{3a}—r = 0·44, n = 118, $p < 0{\cdot}001$; O_{3b}—r = 0·39, n = 123, $p < 0{\cdot}001$). In O_1 complaints about frequent headaches also significantly correlated with frequent feelings of nausea (r = 0·49, n = 634, $p < 0{\cdot}001$) and with a deterioration in eyesight since starting to work in the office (r = 0·37, n = 646, $p < 0{\cdot}001$), but unfortunately these two complaints were not investigated in the other surveys. The highest incidence of self-reported headaches was in the O_1 and O_2 offices where around twice the percentage of women compared with men reported these (O_1—men = 31%, women = 59%; O_2—men = 28%, women = 54%). Overall these sex differences were highly significant (O_1—χ^2 = 46·29, d.f. = 1, $p < 0{\cdot}001$; O_2—χ^2 = 28·64, d.f. = 1, $p < 0{\cdot}001$). However, complaints of headaches were not age related. Data on both the severity of headaches and incidence of these outside work were collected in the O_{3b} survey. This showed that while the headaches are mostly rated as mild, only 2% of the sample report these outside of work. Interestingly, 32 of the 37 people reporting frequent headaches in this survey also said that they frequently used photocopying machinery. The possibility of a link here is worthy of further investigation.

3.3. Eye Irritation

A high incidence of problems of eye irritation was found in both of the air conditioned open-plan offices but, surprisingly, the proportion of such problems in the non-air-conditioned open-plan office was similar to that for conventional, cellular accommodation (Table 1). In both air-conditioned offices complaints of eye irritation/soreness were significantly correlated with complaints of glare from the lighting (O_1—r = 0·46, n = 634, $p < 0{\cdot}001$; O_2—r = 0·46, n = 355, $p < 0{\cdot}001$) and, for O_1, with eyesight deterioration (r = 0·54, n = 645, $p < 0{\cdot}001$), while for O_2 this correlated negatively with views on the adequacy of the lighting system (r = 0·45, n = 356, $p < 0{\cdot}001$). For the O_1 offices there were also significant sex differences (O_1—$\chi^2 = 40{\cdot}25$, d.f. = 1, $p < 0{\cdot}001$) with women again suffering more frequent problems regardless of age (men = 42%; women = 70%).

3.4. U.r.t. Complaints

Frequent problems of coughs, colds, sore throats, etc. at work were reported by more employees in the air-conditioned open-plan offices than in either the non-air-conditioned open-plan or conventional offices (Table 1), and once more there were significant sex differences in the frequency of complaints (O_1—$\chi^2 = 23{\cdot}57$, d.f. = 1, $p < 0{\cdot}001$; O_2—$\chi^2 = 11{\cdot}40$, d.f. = 1, $p < 0{\cdot}01$) which again were not age dependent.

3.5. Multiple Health Problems

To test for the possibility of individuals suffering multiple health problems, responses to the three health items were intercorrelated for each of the air-conditioned open-plan offices, and all of these correlations were highly significant (Table 2). Moreover, when these data were crosstabulated a clear relationship between employee's sex and multiple health problems emerged with women being at least twice as likely to suffer these as men (O_1—men = 14% (63); women = 32% (60); O_2—men = 14% (24); women = 26% (43)).

Finally, no evidence was found that any of the health complaints were related to the incidence of smoking in the offices.

TABLE 2. *Pearson product–moment correlations of health items for the O_1 and O_2 survey.*

Offices		Eye irritation	u.r.t. complaints
O_1	Headaches	$0{\cdot}54^a$	$0{\cdot}43^a$
	Eye Irritation		$0{\cdot}44^a$
O_2	Headaches	$0{\cdot}46^a$	$0{\cdot}38^a$
	Eye Irritation		$0{\cdot}40^a$

$^a p < 0{\cdot}001$

4. *Discussion*

The results of the surveys agree with those of Turiel *et al.* (1983) that more employees working in air-conditioned open-plan offices report health problems than those working in naturally ventilated offices. The frequency of headaches was consistently higher in the open-plan offices, regardless of mode of ventilation, than in the conventional offices. This seems, in part, a consequence of the frequent glare problems reported in these. Even when the open-plan offices have better daylight penetration than is common in many office buildings (as in O_3) the incidence of headaches is only marginally reduced. However, the results also clearly show that the occurrence of other health problems (eye irritation, u.r.t. complaints) is not an inevitable consequence of open-plan but depends on whether or not the offices are also air-conditioned. Proportionally more women than men suffer one or more of these problems irrespective of age, and this is also true of the sizeable percentage of employees in the two air-conditioned open-plan offices suffering multiple health problems.

While this evidence is suggestive of widespread health problems amongst office staff, caution needs to be exercised because these data were based on self-reports of ill-health and not on actual illness rates. Information on the actual frequency and severity of health problems needs to be systematically gathered over quite long timescales before concluding that air-conditioned open-plan offices really are less healthy workplaces than are other types of offices. The apparently higher incidence of these problems in women is also worthy of closer study as are other questions such as what factors might precipitate these health problems, and what might differentiate sufferers from non-sufferers? Although the health problems which have been investigated may not result in acute illness, they do represent chronic and debilitating problems for sufferers, and they highlight the need for future studies on improvements to the present quality of many office environments.

References

Craig, M., 1981, *Office Workers' Survival Handbook: a Guide to Fighting Health Hazards in the Office* (London: BSSRS publications).

Hedge, A., 1980, Office design: user reactions to open plan, in *People and the Man-Made Environment: Building, Urban and Landscape Design related to Human Behaviour*, R. Thorne and S. Arden (eds) (Sydney: University of Sydney), pp.57–68.

Hedge, A., 1982, The open plan office: a systematic investigation of employee reactions to their work environment. *Environment and Behavior*, **14**, 519–542.

Hedge, A., 1984, 'Open' vs 'enclosed' workspaces: the impact of design on employee reactions to their offices, in *Behaviour Issues in Office Design*, J. Wineman (ed.) (New York: Van Nostrand–Rhinehold).

Turiel, I., Hollowell, C.D., Miksch, R.R., Rudy, J.V. and Young, R.A., 1983, 'The effects of reduced ventilation on indoor air quality in an office building. *Atmospheric Environment*, **17**, 51–64.

Collective Dermatitis in a Modern Office

M. Lob, M. Guillemin, P. Madelaine and M.-A. Boillat
Institut universitaire de médecine du travail et d'hygiène industrielle, Av. César-Roux 18, 1005 Lausanne, Switzerland
and F. Baudraz
Service universitaire de dermatologie, CHUV, 1011 Lausanne, Switzerland

Abstract

Five cases of dermatitis of the face and neck are described. They were found in civil servants shortly after they had moved into a new office. We found that the isolation material had probably not been correctly installed and that there was a small amount of glass fibre in the atmosphere. Glass fibres were isolated in the conjunctival liquid of one of the patients suffering from conjunctivitis. Histological examinations of skin biopsy are compatible with irritative dermatitis. Action was taken to prevent liberation of glass fibres into the atmosphere.

1. *Introduction*

Many studies have been published about the effect of fibreglass (FG) on man. Several papers summarize well the present knowledge in this field (Current Status of Health Aspects of Fibrous Glass, 1982; Hill, 1978, Occupational Exposure to Fibrous Glass, 1977). Others give statistics of mortality correlated to FG (Hill, 1978, Morgan *et al.* 1981). The irritative properties of FG are well known. Two important points are actually being discussed and opinions differ: firstly with regard to the possible sensitizing properties, probably due to the resins added in the fibres (Bjornberg *et al.*, 1979; Cuypers *et al.*, 1975, Fisher and Warkentin, 1969; Grzegorczyk, 1982; Heisel and Hant, 1968; Maggioni *et al.*, 1980; McKenna *et al.*, 1958; Sulzberger and Baer, 1942); secondly, with regard to the possible carcinogenic properties (Bernstein *et al.*, 1980; Gross, 1982; Montague, 1974).

Most papers dealing with FG discuss either experimental studies or the action of FG on workers who manufacture, handle or deal with this material. On the other hand, there are very few papers concerning the effect of FG on other people who may happen to come into contact with FG particles. We will briefly summarize the few publications on this matter.

Newball and Brahim (1976) described severe respiratory troubles that appeared in four members of a family that were attributed to the inhalation of FG.

Eby *et al.* (1972) described the case of an eight-year-old negro girl who developed a pruritic symmetrical dermatitis over the mid-posterior section of her thighs two days after her enrolment in an elementary school with FG reinforced plastic school desks. Lechner and Hartmann (1979) described multiple foreign-body granulomas induces by FG in a six-year-old girl, localized on the gluteal region, on the flexor sides of the legs and on the feet. The girl often played in a barn where FG was stored.

Many papers deal with itching or rashes caused by clothes washed in the same machine as FG curtains (Madoff, 1962; Abel, 1966; Peachey, 1967; Fisher *et al.*, 1969).

2. *Observations*

Owing to the scarcity of these observations we thought it would be interesting to describe a type of dermatitis epidemic located on the neck and face of uniformed civil servants working in a newly erected building. The alarm was given to us by one of the heads of this centre. These civil servants worked for 28-day periods, spending two-thirds of this time in the offices, the remainder outdoors. About 10 of them suffered itchy eruptions (spots) a

FIGURE 1. *Birefringent particle in the conjunctival secretions (case no. 2).*

TABLE 1. *Summary of the results.*

No.	Age (years)	Past history	Latency (days)	Symptoms	Diagnosis
1	32	In good health	5	Itching, irritation of the skin. Trouble after shaving.	Dry dermatitis of the face, neck and chin, with macular erythematous papulae. Light conjunctivitis. Biopsy: dermo-epidermitis. No FG seen under microscope.
2	40	In good health	21	Itching and erythematous spots on the face. Conjunctivitis.	Purulent conjunctivitis[a] and dermatitis surinfected by scratching. Changing his work place brought no improvement. Two months later: *status quo*. Skin biopsy: irritative (probably) dermatitis as in case no. 1. Slit lamp examination: a few sparking particles, moving in the lacrymal film. In the secretions of the conjunctiva few birefringent particles were apparent. These were elongated and slightly curved (Figure 1) corresponding very probably to FG and quite similar to the FG found in the ambient air (see later).
3	52	Nervousness	15	Itching and red spots on neck and face. Light prickling of conjunctiva. Improvement when on leave.	Irritative dermatitis.
4	24	In good health	2–3	Itching of the neck. Gave up shaving. Troubles disappeared after two days leave.	Light dermatitis. No FG seen on the skin biopsy.
5	45	In good health	10	Red and bloody spots in the neck. Self-treatment with corticoid ointment.	Only a few red spots. Irritative dermatitis.

[a] We thank the ophthalmologic clinic for supplying this information.

few days after the beginning of a working period in the centre; these disappeared when they were on leave or when they had finished their working period. The itching and the spots were more marked when the people remained a long time in the offices or in the dormitories. An alimentary origin seemed very improbable, as the centre has no canteen and each worker brings his own meal.

Considering the history, an occupational origin or at least an indirect correlation with the working place seemed very probable. We therefore visited the various offices in the centre. Our attention was immediately drawn by the ceilings composed of metallic parallel slats 8·5 cm wide, separated from one another by 1·5 cm intervals. Under the slats a black insulating material forming a fibrous 'mattress' could be seen. From the manufacturers it was found that this material was composed of FG plus a binding product (formolphenol). During two months we examined five civil servants. The results are summarized in Table 1.

Considering the course of the illness, its irritative character and the arrangement of the insulating material under the ceiling slats, it seemed necessary to search for FG in the office atmosphere. Three types of sampling were performed:

1. Sampling of total dust in air by weighing the filters. Flow: 2–3 l/min. Membrane–filter Millipore HAWP, diameter 37 mm.
2. Sampling for counting the fibres in air with optical microscope at phase contrast (NIOSH method). Flow 1·7–1·91 l/min. Aspiration on membrane-filter Millipore MAWG, diameter 37 mm.
3. Automatic continuous counting of fibres in air with 'fibrous aerosol monitor' (FAM) and counting of the incorporated filter, same type as in 2.

Table 2 shows the results obtained by counting FG and weighing total dust from an air sampling filter.

TABLE 2. *Results of counting and weighing (summary).*

Sampled volume (l)	Fibres/ml[a]	C.I.[b] (%)	Total dust (mg/m^3)
3715	0·03	20	–
1289	–	–	0·31
1276	0·04	30	–
1079	0·03	40	–
1255	0·05	30	–
1661	–	–	0·30

[a] Fibres 3–5 μm diameter.
[b] C.I.: Confidence interval at 95%.

Figure 2 shows the fluctuation of the FG concentration (in fibres/ml) (sampling 11–12 March 1983).

3. *Discussion*

At first sight these results seem to be satisfactory as they are within the limits set by Swiss law (10 mg/m^3) concerning mineral fibres (with the exception of asbestos). It was difficult to believe that such a small amount in the air could be the cause of these troubles. However, after discussion with the occupational physician and the technical director of the factories manufacturing the insulating material used in this centre our opinion changed. The concentrations found are identical to those found in the factory workrooms, where it had been commonly observed that new workers presented exactly the same troubles as those described above in civil servants.

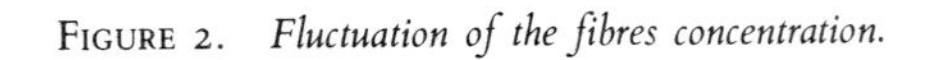

FIGURE 2. *Fluctuation of the fibres concentration.*

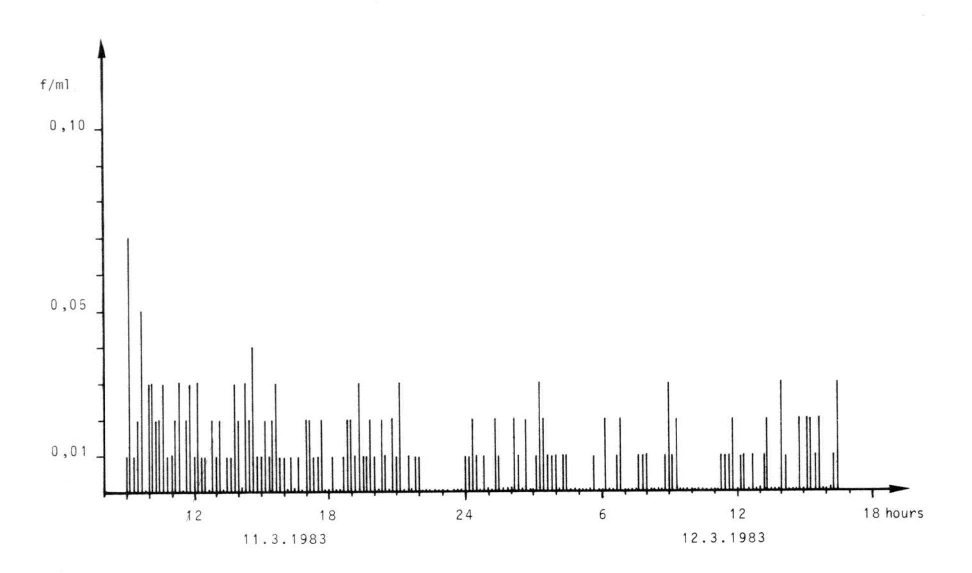

After a few weeks the dermatitis healed without relapsing, due to tolerance or resistance, or to occupational hygiene measures. The physician of the factory considered that the concentrations of FG in the offices of the centre were quite unacceptable and that the insulating material had most probably been incorrectly positioned.

We examined eight other people working in the same building but only concerned with material maintenance. None complained of the troubles described above. These people are permanent workers, spending most of their working time in the workrooms and not in the offices. They have an open collar and wear no tie. The civil servants on the other hand remained only 28 days in the offices, well dressed in their uniforms, wearing a collar and tie.

In the literature concerning skin diseases related to FG we have only found three references reporting events the same as those we have investigated. Makower (1981) reported that 60 employees of NBC in New York complained of headaches, hives and scratchy throat. The investigators concluded that the newly renovated offices contained low levels of FG particles shaken loose from the ceiling during remodelling. As in our observations the level of the glass fibres do not exceed the TLV.

Soon after the extension of a building came into use, Verbeck *et al.* (1981) observed that nearly all of the 13 office workers in the rooms of this extension presented an itchy rash, mainly on the trunk, and had burning eyes. Air samples revealed long glass fibres derived from the ceiling insulation material. The cause was a fault in the construction plan facilitating entry of FG into the indoor air.

Farkas (1983) observed FG dermatitis in office workers employed in a new building where the ceilings were covered with aluminium boards with circular openings, above which were laid packets of glass wool in plastic bags which had been partially torn when being laid. In windy weather and in a draught fragments of FG fell on the furniture and the employees.

With regard to the effect of FG on the eyes Longley and Jones (1966) reported the history of a 48-year-old white woman who complained of a 'dreadful pain' in the eyes. She informed the ophthalmologist that she processed FG insulated cable one day per week for a period of eight or nine months. The diagnosis of keratitis was made after finding minute fragments of refractile material adhering lightly to the conjunctiva. Stockholm *et al.* (1982) investigated the effect of man-made mineral fibres on the human eye in a cross-sectional study of 15 workers exposed to Rockwool and a matched reference group of 15 people. A significantly higher frequency of eye symptoms related to work conditions was found among exposed workers.

We have not thoroughly investigated whether the dermatitis described was of the irritative or allergic types or both, as no patch-tests were done.

Two arguments favour a purely irritative action. First, resin, clearly visible at the intersection of the fibres (electron microscope), is polymerized at a temperature of 280°C when insulating material is manufactured. Its state is then inert. Second, if it was of the allergic type it would be difficult to explain how so many people suffered from the disease and then became 'resistant'.

With regard to the offices, the present situation needs to be corrected. Specialists are considering this operation.

References

Abel, R.R., 1966, Washing machine and fiberglass. *Archives of Dermatology*, **93**, 78.

Bernstein, D.M., Drew, R.T. and Kuschner, M., 1980, Experimental approaches for exposure to sized glass fiber. *Environmental Health Perspectives*, **34** 47–57.

Bjornberg, A., Lowhagen, G.B., and Tengberg, J.E., 1979, Skin reactivity in workers with and without itching from occupational exposure to glass/fibres. *Acta Dermatovener* (Stockholm), **59**, 49–53.

Current status of health aspects of fibrous glass and other man-made vitreous fiber, 1982, (An Annotated Bibliography), *Medical Series Bulletin*, 21–82. Industrial Health Foundation Inc.

Cuypers, J.M.C., Bleumink, E. and Nater, J.P., 1975, Dermatologische Aspekte des Glasfaserfabrikation. *Berufsdermatosen*, **23**, 143–154.

Eby, C.S. and Jetton, R.L., 1972, School desk dermatitis. *Archives of Dermatology*, **105**, 890–891.

Enterline, P.E. and Henderson, V., 1975, The health of retired fibrous glass workers. *Archives of Environmental Health*, **30**, 113–116.

Farkas, J., 1983, Fibreglass dermatitis in employees of a project-office in a new building. *Contact Dermatitis*, **9**, 79.

Fisher, A.A., 1968, *Contact Dermatitis* (Philadelphia: Lea and Febiger).

Fisher, B.K. and Warkentin, J.D., 1969, Fiber glass dermatitis. *Archives of Dermatology*, **99**, 717–719.

Gross, P., 1982, Man-made vitreous fibers: present status of research on health effects. *International Archives of Occupational and Environmental Health*, **50**, 103–112.

Grzegorczyk, L., 1982, Zur Bedeutung physikalischer Faktoren für die Entstehung von Berufsdermatosen. *Dermatosen*, **30**, 179–181.

Heisel, E.B. and Hunt, F.E., 1968, Further studies in cutaneous reactions to glass fibers. *Archives of Environmental Health*, **17**, 705–711.

Hill, J.W., 1978, Man-made mineral-fibres. *Journal of the Society of Occupational Medicine*, **28**, 134–141.

Lechner, W. and Hartmann, A.A., 1979, Glasfaser induzierte Fremdkörpergranulome. *Der Hautarzt*, **30**, 100–101.

Longley, E.O. and Jones, R.C., 1966, Fiberglass conjunctivitis and keratitis. *Archives of Environmental Health*, **13**, 790–793.

Madoff, M.A., 1962, Dermatitis associated with fibrous glass material. *Tufts Folia Medica*, **8**, 100–101.

Maggioni, A., Meregalli, G., Sala, C., and Riva, M., 1980, Patologia respiratoria e cutanea negli addetti alla produzione di fibre di vetro (filiato). *Medicina del Lavoro*, **71**, 216–227.

Makower, J., 1981, *Office Hazards: How Your Job Can Make You Sick* (Washington, D.C.: Tilden Press).

McKenna, W.B., Smith, J.F.F. and Maclean, D.A., 1958, Dermatoses in the manufacture of glass fibre. *British Journal of Industrial Medicine*, **15**, 47–51.

McRae, M.E., 1970, Fibre glass dermatitis. *Cutis*, **6**, 1234–1236.

Montague, K. and Montague, P., 1974, Fiber glass. *Environment*, **16**, No. 7, 6–9.

Morgan, R.W., Kaplan, S.D. and Bratsberg, J.A., 1981, Mortality study of fibrous glass production workers. *Archives of Environmental Health*, **36**, 179–183.

Newball, H.H. and Brahim, S.A., 1976, Respiratory response to domestic fibrous glass exposure. *Environmental Research*, **12**, 201–207.

NIOSH, 1977, *Occupational exposure to fibrous glass. Criteria for a recommended standard* (US Department of Health, Education and Welfare, NIOSH), 189 p.

Peachey, R.D.G., 1967, Glass-fibre itch: a modern washday hazard. *British Medical Journal*, **2**, 221–222.

Stockholm, J., Norn, M. and Schneider, T., 1982, Ophthalmologic effects of man-made mineral fibers. *Scandinavian Journal of Work, Environment and Health*, **8**, 185–190.

Sulzberger, M.B. and Baer, R.L., 1942, The effects of fiberglass on animal and human skin. *Industrial Medicine*, **11**, 482–484.

Wall Street Journal, 14 Jan 1981, Clearing the air on illness at NBC offices.

Verbeck, S.J.A., Buise-Van Unnik, and Malten, K.E., 1981, Itching in office workers from glass fibres. *Contact Dermatitis*, **8**, 354.

Relationship Between Environmental Factors, Job Satisfaction and Mental Strain in an Open-Plan Drafting Office

K. LINDSTRÖM AND J. VUORI
Department of Psychology, Institute of Occupational Health, Haartmaninkatu 1, 00290 Helsinki 29, Finland

Abstract

The relationships between environmental and individual factors and job satisfaction and psychological strain symptoms were studied among a group of workers in an open-plan office. Particular attention was paid to noise-induced problems. The subjects were mainly ship designers and draftsmen (n = 38) whose mean age was 32 years. The methods applied were a structured questionnaire and two personality inventories. Job satisfaction was related to higher cognitive task demands and low neuroticism. Disturbances in work performance due to noise were greater especially when the estimated noise level was higher, when the worksite was impractical, and when sociability was high. Symptoms of subjective strain were associated with poor ergonomics of the worksite and sudden loud noises in the work environment. Higher impulsiveness and social withdrawal were also related to a higher frequency of symptoms.

1. *Introduction*

Annoyance due to background noise and contaminated ambient air were the problems most often spontaneously mentioned by men working in the open-plan drafting office of a shipyard. It is well known that disturbances due to noise are common in work environments such as open-plan offices. It can be especially distracting with respect to work performance that demands concentration (e.g. Nemecek and Grandjean, 1973; Nemecek *et al.*, 1976).

In connection with an extensive study on the psychological effects of industrial noise (Vuori *et al.*, 1982), we investigated how job satisfaction, disturbances due to noise, and subjective symptoms were related to various environmental factors, to the work itself, and to certain personality characteristics of workers.

2. *Methods*

Our subjects were 38 men working in an open plan drafting office of a shipyard. Their mean age was 31·6 years (S.D. = 6·3). They worked at their present workplace an average of 6·9 years (S.D. = 4·9). The group comprised 15 ship designers, 10 draftsmen, 3 drafting assistants, and 10 men with various other occupations. Thirteen had the obligatory basic education, 14 middle-level education, and 11 had passed the matriculation examination.

Both the building and the work facilities were built in 1974. A total of 120 persons (100 men and 20 women) were working in this open-plan office. The level of noise in the work environment was 50–60 dB(A).

Men less than 45 years old and who had work related to ship planning and drafting participated in our study. They filled out a questionnaire dealing with the characteristics of the work environment, the work itself, job satisfaction, subjective psychological and psychosomatic symptoms, and disturbances due to noise. Eysenck's Personality Inventory (EPI) and Weinstein's Noise Sensitivity Scale were administered for personality assessment.

Ten of the 38 men also participated in measurements of daily strain indicated by fluctuations in their levels of arousal and activation during the workday. Reaction times to a series of 200 visual stimuli were measured in the morning before the shift, at the middle of it, and at the end of the workday.

3. *Results*

3.1. **Descriptive Results**

The subjects perceived their work as physically light but mentally moderately heavy. Twenty-two of them rated the work environment as comfortable or satisfactory, and nine as uncomfortable. The ratings concerning harmful factors in the work environment are seen in Table 1. Noise in the environment and contaminated ambient air were the harmful factors mentioned most often. The work itself was perceived to a great extent as cognitively

TABLE 1. *Perceived work environment.*

Factor	Moderate or heavy stressor (n = 38)
Noisy environment	29
Contaminated ambient air	19
Noise level	18
Lighting	15
Sudden loud noises	14
Impractical worksite	14
Crowded environment	9

demanding in 20 cases and as cognitively moderately demanding in 17 cases. The time pressure was seen as a slight stressor by 27 and as a great stressor by 1.

Job satisfaction was evaluated as good by 10, as moderate by 26, and as poor by 2 subjects. Noise was experienced by 9 subjects as disturbing their work performance somewhat, and 3 experienced it as a factor disturbing work performance to a great extent.

The most common of the 29 subjective symptoms are given in Table 2.

TABLE 2. *Commonest subjective symptoms of strain.*

Symptom	Number of subjects (n = 38)
Getting lost in one's own thoughts	18
Concentration difficulties	16
Irritability	16
Unusual fatigue	16
Sweating	12
Memory difficulties	12

3.2. **Analytical Results; Interrelations**

For the analysis of the relationships between the perceived charcteristics of work environment, the work itself, and individual factors and job satisfaction, work disturbance due to noise, and subjective symptoms, both correlation analysis (Pearson's 'r') and regression analysis were used.

Job Satisfaction

This correlated significantly with high cognitive demands of work, a low level of neuroticism (EPI), and low impulsiveness (EPI). In regression analysis high cognitive task demands and high educational level explained 15% of the variance in job satisfaction (r = 0·38, $p < 0{\cdot}05$).

In regression analysis a combination of cognitive task demands, the estimation of the noise level, impractical worksite, high neuroticism, and high sociability were related to perceived work disturbances due to noise (r = 0·63, $p < 0{\cdot}01$).

Subjective Symptoms

These were classified by means of factor analysis to four categories: somatic symptoms; sleep disturbances and depressive mood; fatigue and concentration difficulties. The relationships between these sum variables formed according to the aforementioned factors and perceived environmental and individual characteristics were analysed by means of regression analysis.

Though somatic symptoms were related to high perceived mental work load and to the sudden, loud noises, the correlation was not statistically significant.

Sleep disturbances and depressive mood were related to younger age, lower educational level, lower cognitive task demands, improper ergonomics of the worksite, social withdrawal, and higher impulsiveness ($r = 0{\cdot}71$, $p < 0{\cdot}01$).

Symptoms of fatigue were related to lower education, sudden loud noise in the work environment, and higher impulsiveness ($r = 0{\cdot}54$, $p < 0{\cdot}01$).

Concentration difficulties were associated with low perceived physical work load, high cognitive task demands, impractical work site, social withdrawal, and high impulsiveness ($r = 0{\cdot}66$, $p < 0{\cdot}01$).

Acute Strain

Reaction times were slowest in the morning before the work shift and speeded up during the workday. At the end of the workday they were the shortest. At the individual level longer reaction times were related to symptoms of fatigue.

4. *Discussion*

Draftsmen and ship designers working in an open-plan office perceived their work environment as rather comfortable, although they complained rather often about the noise and the quality of the ambient air. The work itself was considered mentally moderately heavy and cognitively more demanding. Job satisfaction was quite good and was related to higher cognitive demands of the work and lower neuroticism.

The disturbing effects of noise on work performance were related mainly to the estimated noise level, higher cognitive demands and higher sociability (EPI).

The subjective symptoms seemed to be related to both occupational and individual characteristics. Sudden, loud noises in the work environment and poor ergonomics of the worksite were important. Those who were more impulsive and socially withdrawn had more symptoms of psychological strain.

Our results were obtained with a small but quite homogeneous group of workers. They indicate that work that is more cognitively demanding is disturbed in open plan offices mainly because of disturbing noise. Our findings agree with previous results (e.g. Nemecek and Grandjean 1973). Fatigue and somatic symptoms also depended partly on sudden, loud noises in the environment.

References

Nemecek, J. and Grandjean, E., 1973, Noise in landscaped office. *Applied Ergonomics*, **4**, No. 1, 19–22.

Nemecek, J., Turrian, V. and Sancin, E., 1976, Lärmstörungen in Büros. *Sozial- und Präventiv Medicin*, **21**, 133–134.
Vuori, J., Lindström, K. and Mäntysalo, S., 1982, Psychological effects of occupational exposure to noise. *Työterveyslaitoksen Tutkimuksia*, no. 193 (Helsinki: Institute of Occupational Health). In Finnish with English summary.

Music During Office Work

J. Nemecek
Swiss Federal Institute of Technology, Department of Hygiene and Applied Physiology,
CH–8092 Zurich, Switzerland

Abstract

To judge the positive and negative effects of music while doing mental work a study by questionnaire was carried out, covering 217 office employees in 16 bank branches. The results show that the type of work, the position of the employee, his education, his age and also the type of music determine the degree of annoyance possibly caused by music. It also appears that music is not among the most important factors which are responsible for 'well-being' at the office.

1. *Introduction*

In the past years in many offices, especially in offices of the services sector, background music has been introduced. The following considerations support the concept of presenting background music: the introduction of new technologies in office work has led to stereotype working procedures for the individual, the freedom of work, i.e. the scope of work, the personal decision-making etc. is restricted and thus work becomes more and more monotonous. Music is thought to improve the motivation of the employees, but it also has the advantage of masking the understandability of conversations between co-workers and thus reduces disturbances during work. Nevertheless, music in offices is a rather controversial subject, distractions and negative effects on concentration being feared.

2. *Methods*

To gain a better understanding of the positive and negative aspects of music while doing mental work a study was carried out in 16 branch offices of a major Swiss bank, located both in rural and city areas (Bee *et al.*, 1982). The study consisted in an interrogation of the persons working in these offices. To this purpose a questionnaire was to be filled out by the employees. The questions covered personal data such as age, position, etc. as well as the person's opinion about the effect of the music.

Altogether 217 office employees were interrogated, 55% men and 45% women. The mean age of the subjects was 32 years. Three-quarters of the employees questioned belonged to the medium and lower staff, secretaries, etc., one-quarter represented department managers and team leaders. The ratio of employees helping customers at the bank counter to other employees engaged in administration was about 50–50. In all 16 branch offices equipment for music transmission had been installed according to the wish of the staff. The frequency and duration of the music transmissions was different for each office but in none of them did it last uninterruptedly during the whole work hours. Also the type of music was individual for each office: it varied between light, classical or pop music, radio music or so-called functional music (i.e. background music of a non-dynamic nature). About 85% of the employees interrogated could turn on or stop the music themselves after agreement with their colleagues sitting close by.

3. *Results*

Only the most interesting results of the interrogations will be mentioned hereafter. A general question was asked about the advantage of music at work, to which a choice of answers was proposed in the questionnaire. Most employees selected the item 'working climate' (64%). Next ranked the point 'masking office noise' (52%), then 'better relaxation during short breaks' (42%), 'masking conversations of the others' (11%) and finally 'positive influences upon bank customers' (6%).

The next question dealt with the nature of annoying noises in offices. For each type of noise five intensity degrees of disturbance were offered. It appeared that a feeling of 'strong' and 'moderately strong' disturbance is mostly caused by 'office machines' (42%), followed by 'telephone' (32%), 'conversations' (17%), 'traffic noise' (9%), etc. It is interesting to note that the statement 'strong' or 'moderately strong disturbance by conversations' was indicated by 17% only, as in previous investigations about office noise this item 'disturbing conversations' had ranged as the top annoying factor with a frequency of about 50% (Nemecek, 1983). But in these offices no music was available and so it seems that disturbing conversations are partly masked by the music.

Answers were also proposed to the question about disadvantages of music at work. In first place ranks 'I dislike the type of music transmitted' (31%), followed by 'I must concentrate more to avoid mistakes' (21%), then 'distraction of my attention' (19%), 'mental drifting away due to associations (films, trips)' (9%), etc.

In the next step the problem of annoyance by music was treated. The answers given to the question about the degree of disturbance caused by music show that nearly 20% of all employees questioned are 'highly' or 'rather' disturbed, while 50% only are 'not at all' disturbed. When these answers were related to the person's functional position in the office it appeared that the number of managers and team leaders disturbed by music is twice as high as that of medium and lower staff. On the other hand, the medium and lower staff is twice as much 'not at all disturbed' as the managers. So it may be assumed that the determining factor for annoyance through music is the demand of the office task.

This consideration led to a next step, in which the same answers as above were considered in relation to the schooling of employees. We chose the following groups: no qualifications, medium qualifications, special training in banking and college certificate or university degree (Figure 1). Here the degree of annoyance by music rises with increased schooling while the percentage of persons 'not being disturbed' by music rises with decreasing schooling.

FIGURE 1. *Disturbances by music during office work in relation to the schooling of the employees.*

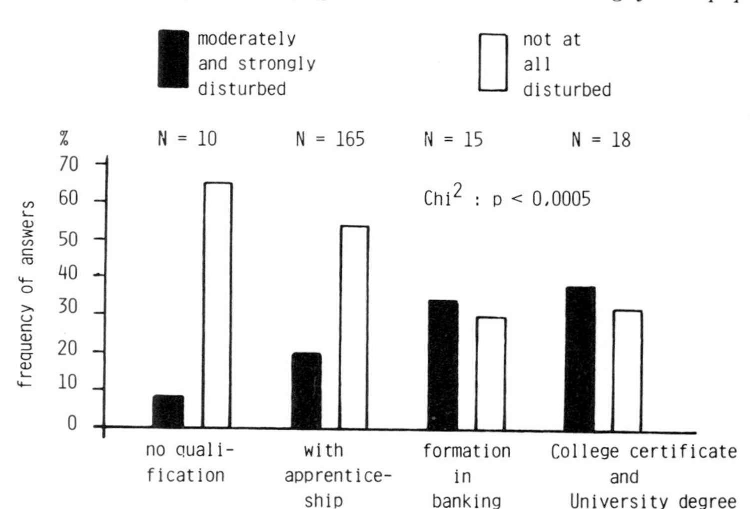

Further a possible relationship between annoyance by music and age of the persons questioned was investigated. It shows that disapproval of music rises with increasing age but only until the age of fifty and then decreases again. But here a participation of the professional strain cannot be excluded: older employees are apt to hold more responsible positions and work on more responsible tasks than the younger. A similar relationship with a break after fifty years has been observed in previous studies of annoyances caused by office noise (Nemecek and Turrian, 1978).

To investigate the relationship between annoyance caused by music and the character of the work in the office a bipolar set-up was introduced. The employees were asked to mark their appreciation of the demand of their work as described by different items and also to indicate their appreciation of the structure of their work. These assessments of work demand were related to an annoyance index which was calculated from the annoyance caused by music, as indicated by each employee. There seems to be a tendency showing that:

1. Music is disturbing when doing mentally demanding work.

2. Music as an annoying factor is of little importance when the task is interesting and straightforward and the work as a whole is satisfying—it seems that music then is not heard.

These results demonstrate that tasks demanding high concentration and strenuous mental work need a quiet environment in the office without music or noise, while tasks demanding less concentration and less mental load allow more tolerance for music. On the other hand, an interesting and independent task, giving the employee a feeling of satisfaction and demanding his full attention appears to protect him from music annoyance. These considerations generally correspond very much with our previous findings about noise annoyance during office work (Nemecek and Turrian, 1978).

With another question it was investigated if different types of music have any influence on the annoyance. Figure 2 shows the frequency of complaints about annoyance, and the indifference towards music depending on five different types of music. It reveals that the most accepted type of music during office work is light music. This corresponds very much with knowledge and experiences gained a long time ago about music transmission during monotonous work in factories (Grandjean, 1980). The same figure shows that the four other types of music appear to be equal in the judgement of the office employees.

FIGURE 2. *Degree of disturbance by different types of music.*

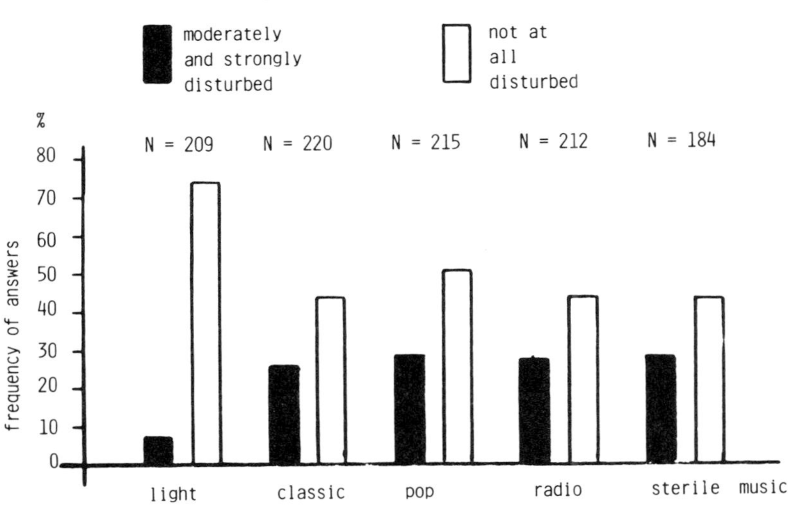

Figure 3 points to the great differences in music appreciation by managers on one side and by medium and lower staff on the other. While the managers are mostly annoyed by pop, radio and sterile music, the highest annoyance expressed by medium staff occurs when listening to sterile or classical music. Anyway light music is the least disturbing type both for managers and the staff.

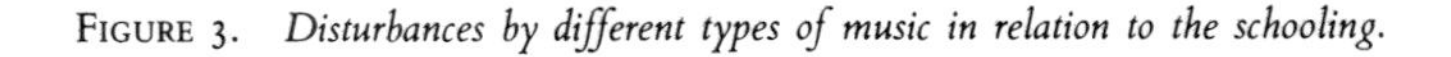

FIGURE 3. *Disturbances by different types of music in relation to the schooling.*

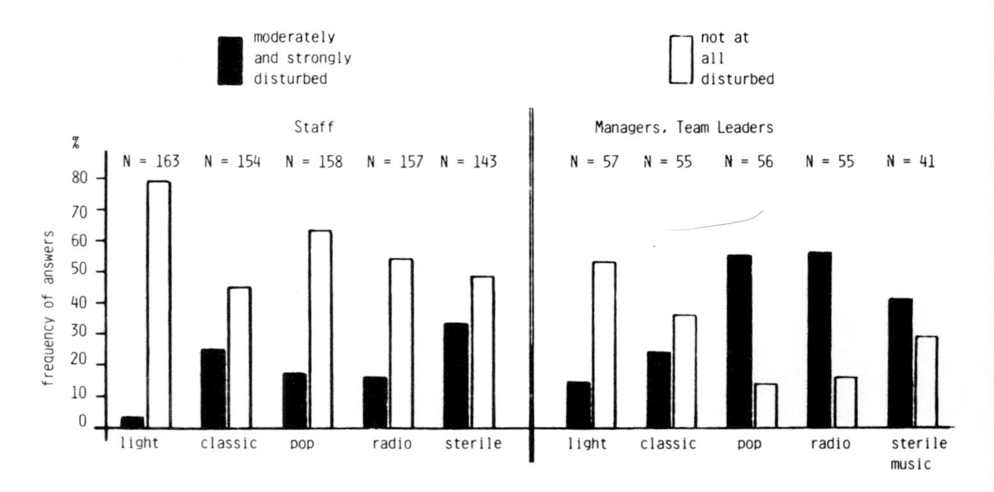

As mentioned at the beginning of this paper, in most of the bank branches investigated equipment for music transmission had been introduced following the wishes of the employees. So it was possible to ask them of which importance they considered music at work as compared to other conditioning factors of their office.

TABLE 1. *Determining factors for the well-being in the office.*

Factors	More important than music (no. of times)
General office climate	14·5
Salary	4·3
Way to the office	3·3
Condition of the work place	2·5
Dimension of room	1·3
Music at work	
Plants	0·8
Pictures	0·4
Breaks	0·1

In Table 1 the rating of different factors describing 'well-being' in the office is compared to the factor 'music at work'. It shows that the predominant character is the general 'work climate', which is nearly 15 times more important than the item 'music at work'. 'Music at work' ranged in the sixth place only in a total of nine.

4. *Summary*

1. Only 50% of all bank employees questioned were not at all annoyed by music during work.
2. The music may cause annoyance:
   a. when doing office work demanding more concentration and more mental efficiency;
   b. to the managers and team leaders;
   c. to the employees with better schooling;
   d. to the employees between 40 and 50 years of age.
3. Office staff in medium and lower positions as well as employees younger than 30 years complain less about disturbance by music.
4. The most suitable type of music for office work is 'light music'.
5. Pop music, classical music, sterile and radio music appear to be unsuitable.
6. Among the factors determining 'well-being' at work, music ranges in the sixth place only, with a weighting about 15 times less than the factor 'work climate', which appears to be the most important.

References

Bee, M., Kuendig, J. and Zeier, M., 1982, *Musik am Arbeitsplatz in einer Bank* (Zürich: Gruppendiplomarbeit, HWV).
Grandjean, E., 1980, *Fitting the Task to the Man—an Ergonomic Approach* (London: Taylor & Francis).
Nemecek, J., 1983, *Lärm am Arbeitsplatz* (Ludwigshafen/Rhein: F. Kiehl Verlag).
Nemecek, J. and Turrian, V., 1978, Untersuchungen von Störwirkungen durch Lärm in Büros. *Zeitschrift für Arbeitswissenschaften*, **32**, 21–24.

The Effects of Sealed Office Buildings on the Ambient Environment of Office Workers

E.M. Sterling, E.D. McIntyre and T.D. Sterling
Building Research, TDS Limited, 70–1057 West 12th Ave, Vancouver BC, Canada V6J 2E2

Abstract

We have recently completed a review of 116 investigated incidents of building illness among office workers in North America and 27 buildings selected for investigations for reasons other than building illness. Data extracted from these 143 studies form a valuable archive of information about the pollutant levels observed in buildings with and without persistent health and environment-related complaints.

One hundred and thirty-six different pollutants have been measured in more than one building. Pollutant levels in these buildings turned out to be no higher than levels measured in similar buildings with no record of health-related complaints. Also pollutant levels do not seem to be affected by office smoking policies. There is some evidence that relatively high levels of indoor organic fractions provide a source for the formation of irritating photochemical oxidants. This process may be enhanced by ultraviolet emissions of fluorescent lighting.

1. *Introduction*

Paralleling reductions in ventilation undertaken to conserve energy has been an increase of requests for health hazard evaluations initiated by occupants of sealed, air-conditioned buildings who believe their office or work environment to be hazardous and their symptoms to be building related. In fact, the term 'building illness' has been suggested for these incidents. A large number of reports from such investigations may now be available in the United States and Canada in addition to similarly motivated European and Japanese studies and present a potentially invaluable source of information on building ventilation, industrial hygiene measures, indoor air quality, health and occupant comfort.

This report is based on the information obtained from 143 building studies made available through the US National Institute for Occupational Safety and Health, the Centres for Disease Control and other investigators.

2. *Air Quality in Working Buildings*

There are a small number of incidents of building illness for which a clear-cut cause can be established. Elimination of that cause also eliminates the health related complaints. Respiratory symptoms were related to toxic dusts left as detergent residues from industrial carpet shampoos (Kreiss, 1981). Burning eyes, coughing, breathing difficulties, nausea and dizziness were traced to formaldehyde off-gassing from interior materials (Makower, 1981). Possibly, the most notorious and dangerous examples have been outbreaks of Legionnaire's disease and hypersensitivity pneumonitis linked to viruses, bacteria and fungi from air ventilation and hot water systems (Broome, 1979; Salvaggio, 1979; Banaszak, 1970; Fink, 1971). However, most studies of incidents of building illness failed to locate a direct cause for the experienced symptoms of discomfort and illness.

Most studies of buildings with illness complaints seem to have explored the possibility that heightened levels of indoor pollutants were the cause of the problems and obtained a series of measurements of at least indicators of pollution levels such as carbon dioxide, carbon monoxide, formaldehyde, ozone and particulates. Similar measures were obtained also from buildings studied for reasons other than comfort or illness problems. Information contained in 143 such investigations has been extracted so far into a computer-based data archive. One hundred and sixteen investigations were undertaken of buildings troubled by health and comfort complaints and 27 investigations were conducted for research purposes. 132 different chemicals and 12 other observations such as noise or bacteria are cited at least once. The archive contains an adequate number of measurements for many pollutants to provide information on the pattern of pollutant levels found in modern sealed buildings.

FIGURE 1. *Carbon monoxide and buildings investigated for health complaints.*

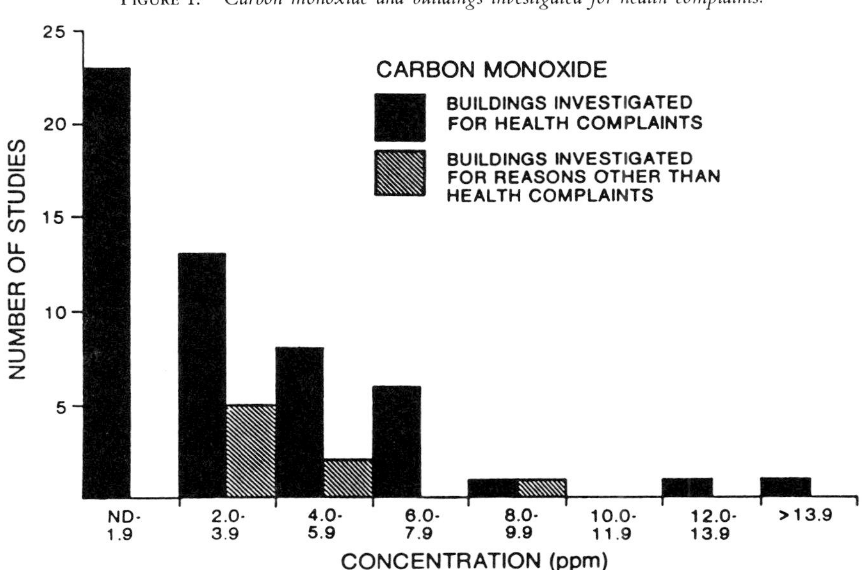

The distributions of pollution levels of all buildings are strongly skewed toward low values. Distributions of observed concentrations of pollutants overlap between the 116 buildings investigated for health and the 27 buildings investigated for reasons other than health. Figures 1 and 2 show the typical distribution of concentrations and overlap for both carbon monoxide and particulates. Similar distributions of concentrations are observed for all other pollutants (not shown here). It may be concluded, therefore, that pollution levels in buildings investigated for health complaints do not differ from those found in buildings investigated for other reasons. For purposes here, all buildings are combined.

FIGURE 2. *Particulates and buildings investigated for health complaints.*

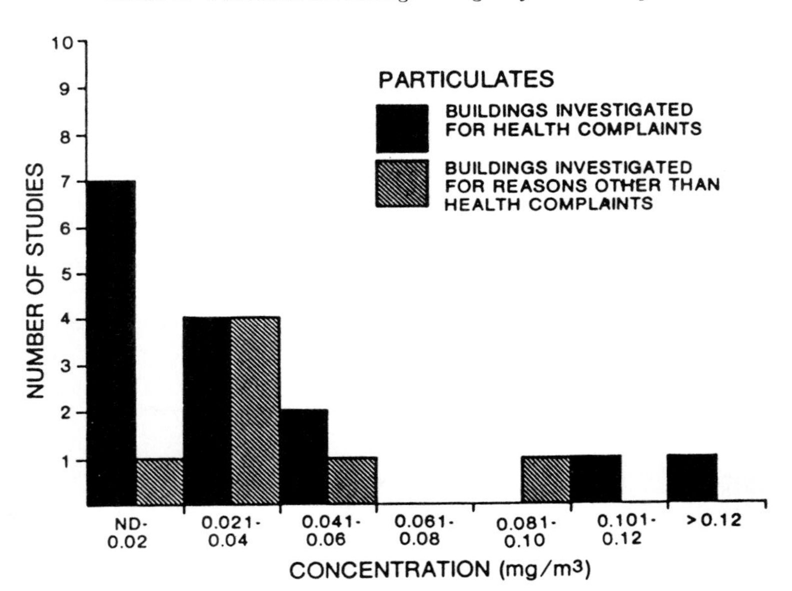

Table 1 presents information for those 16 most frequently measured pollutants from 143 buildings; it shows the number of buildings from which data points were obtained and the median levels measured. (Medians are given to adjust for the many reports of 'not detectable' (ND) levels, which in most cases may represent levels lower than the sensitivity of measurement procedures, and of 'trace', for which no value can be assigned.)

The average value of pollutants reported in modern buildings does not exceed levels deemed to be hazardous by occupational or industrial standards. Many occurred in such low levels that no detectable (ND) or barely detectable (trace) amounts could be found. The carbon monoxide median level was 2·54 p.p.m. based on 61 buildings. The carbon dioxide median level of 400 p.p.m. was based on 26 buildings. The particulate median level of 0·029 mg/m^3 was based on 22 buildings. Formaldehyde was measured in 44 buildings with a median value of 0·02 p.p.m.. In general, measured levels of indoor pollutants were no larger than those already reported in the literature. The many reports that have measured but not detected various pollutants indicate that these pollutants occur, if they do, in values not now considered hazardous. The same is true of the many other pollutants measured in only one or two buildings and not listed in Table 1.

TABLE I. *Average levels of 17 pollutants measured most frequently in buildings investigated for health complaints.*

Pollutant	All buildings	Number of reports
Acids	ND[a]	14
Aldehydes[b]	ND	8
Amines	ND	10
Ammonia	ND	9
Aromatic hydrocarbons[c]	Trace	55
Carbon dioxide	400 p.p.m.	26
Carbon monoxide	2·54	61
Formaldehyde	0·02 p.p.m.	44
Hydrazine	ND	6
Hydrogen sulphide	ND	9
Hydrocarbons	Trace	77
Metals	Trace	8
Nitrous and nitric oxides	ND	31
Nitrogen dioxide	ND	13
Ozone	ND	27
Particulates	0·029 mg/m^3	22
Sulphur dioxide	ND	20

[a] Where the median equals ND or trace, over 50% of investigators tested for that pollutant and reported ND or trace amounts.
[b] Not including formaldehyde.
[c] Hospitals not included here.

Contaminant levels in buildings with and without smoking restrictions were also compared to determine the added burden tobacco smoke might contribute to indoor air quality. Both comparison of median levels and of detailed graphs of range distributions (not shown here) showed no significant difference between buildings where smoking was allowed and where it was restricted or prohibited (Table 2).

3. *Discussion*

All buildings with health related complaints were sealed structures depending on mechanical ventilation and air-conditioning for thermal comfort and air quality and all appear to be lit by fluorescent lights. One possible hypothesis explaining the incidents of building related symptoms occurring in some sealed buildings is that, as in a sealed test tube, the many pollutants present, especially hydrocarbon vapours, interact and combine to create irritating byproducts similar to photochemical smog. Smog measured outdoors has been shown to be associated with a symptom complex similar to that reported in building illness studies, especially the ever present eye, nose and throat irritation. Photochemical smog also has been shown to be related to many of the same vapours, nitrates and enzymes found inside buildings (Altshuller, 1978). It is also known that the formation of photochemical oxidants is accelerated by ultraviolet light. Many fluorescent lamps in buildings have detectable ultraviolet emissions (Duro Test Corp., 1978).

TABLE 2. *Average levels of 17 pollutants measured most frequently in buildings investigated for health complaints categorized by smoking restriction.*

Pollutant	Buildings with no smoking restrictions	Number of reports	Buildings with smoking restrictions	Number of reports
Acids	ND	13	ND	1
Aldehydes[a]	ND	7	0·052 mg/m^3	1
Amines	ND	10	–	0
Ammonia	ND	8	ND	1
Aromatic hydrocarbons[b]	Trace	54	0·82 mg/m^3	1
Carbon dioxide	440 p.p.m.	23	613 p.p.m.	3
Carbon monoxide	2·31 p.p.m.	52	4·0 p.p.m.	9
Formaldehyde	0·021	39	ND	5
Hydrazine	ND	4	ND	2
Hydrocarbons	Trace	75	ND	2
Hydrogen sulphide	ND	8	ND	1
Metals	Trace	7	ND	1
Nitrous and nitric oxides	ND	29	13 p.p.b.	2
Nitrogen dioxide	ND	13	–	0
Ozone	ND	23	0·015 p.p.m.	4
Particulates	0·036 mg/m^3	20	0·021 mg/m^3	2
Sulphur dioxide	ND	17	0·011 p.p.m.	3

[a] Not including formaldehyde.
[b] Hospitals not included here.

Recent measurement studies by Turiel *et al.* (1982), Hicks (1980), and Hollowell and Miksch (1981) provide additional support by demonstrations that the number and concentration of organic contaminants in tight buildings with complaints exceeded that of outdoor air.

All the necessary conditions exist in offices to produce photochemical smog. Experimental evidence for formation of photochemical smog also is offered by Sterling and Sterling (1983). An office floor was experimentally manipulated. While eye irritation decreased in areas when either fresh air was increased or fluorescent lights were replaced, there was a dramatic improvement when fresh air was increased and ultraviolet levels were reduced simultaneously. Eye irritation returned to previous prevalence when the original conditions of lighting and ventilation were restored.

4. *Conclusion*

There is a significant amount of information now available on the many architectural and engineering factors which affect the health and comfort of office workers. It is becoming evident that totally sealed buildings dependent on sophisticated mechanical systems have been unable to provide acceptable conditions for human occupation without unacceptable energy costs.

A new ergonomic architecture that is responsive to human health and comfort in addition to energy conservation, new technology and new materials is now required for design of office buildings fit for human occupation.

References

American Society of Heating, Refrigerating and Air Conditioning Engineers, 1974, *Energy Conservation in New Building Design*, ASHRAE Standard 90–75, New York.

Altshuller, A.P., 1978, Assessment of the contribution of chemical species to the eye irritation potential of photochemical smog. *Journal of the Air Pollution Control Association*, **28**, 594–598.

Banaszak, E.F., Thiede, W.H. and Fink, J.N., 1970, Hypersensitivity pneumonitis due to contamination of an air conditioner. *New England Journal of Medicine*, **283**, 271–276.

Broome, C.V. and Fraser, D.W., 1979, Epidemiologic aspects of Legionellosis. *Epidemiologic Reviews*, **1**, 1–16.

Duro-Test Corp., 1978, Vita-Lite, sunlight indoors from DuroTest. Corporation form 724–7811u5M.

Fink, J.N. *et al.*, 1971, Interstitial pneumonitis due to hypersensitivity to an organism contaminating a heating system. *Annals of Internal Medicine*, **74**, 80–83.

Hicks, J.B., 1980, *Tight Building Syndrome: Summary*. San Rafael California, Firemen's Fund Insurance Companies, Sacramento, California.

Hollowell, C.D. and Miksch, R.R., 1981, Sources and concentrations of organic compounds in indoor environments. *Bulletin of the New York Academy of Medicine*, **57**, 962.

Karapinski, V.J., 1980, *Formaldehyde in Office and Commercial Environments*, Indiana State Board of Health, Indianapolis, Indiana.

Kreiss, K., Gonzalez, M.G., Conwright, K.I. and Scheere, A.R., 1981, Respiratory irritation due to carpet shampoos. *Proceedings International Symposium on Indoor Pollution, Health and Energy Conservation*, 13–16 October, University of Massachusetts, Amherst, Massachusetts.

Makower, J., 1981, *Office Hazards: How Your Job Can Make You Sick* (Washington, DC: Tilden Press).

Salvaggio, J.E. and Karr, R.M., 1979, Hypersensitivity pneumonitis: state of the art. *Chest*, **75**, 270–274.

Schmidt, H.E., Hollowell, C.D., Miksch, R.R. and Newton, A.S., 1980, Trace organics in offices. Building Ventilation and Indoor Air Quality Program, Energy and Environment Division, Lawrence Berkeley Laboratory, University of California, Berkeley.

Sterling, E. and Sterling, T., 1984. The impact of different ventilation levels and fluorescent lighting types on building illness: an experimental study. *Canadian Journal of Public Health* (in press).

Sterling, T.D. and Sterling E.M., 1984, Comparison of non-smokers' and smokers' perceptions of environmental conditions and health and comfort symptoms in office environments with and without smoking. This vol., pp.34–40.

Turiel, I., Hollowell, C.D., Miksch, R.R., Rudy, J.V., Young, R.A., and Coye, M.J., 1982, The effects of reduced ventilation on indoor quality in an office building, *Atmospheric Environment*, **17**, 51–64.

Noise, Lighting and Climate Inside Different Office Work Places

G. Costa, P. Apostoli
Istituto di Medicina del Lavoro, Università di Verona, Italy
and A. Peretti
Fondazione Clinica del Lavoro, Padova, Italy

Abstract

The study examined seven bank offices, four mechanographic centres, one telephone exchange network and one postal sorting office. The recorded noise levels were rather different among the offices in relation to the kind and the number of machines utilized. More specifically, they fluctuated between 65 and 95 Leq dB (A), exceeding, in most cases, the limits proposed for the various office activities, causing interference with intellectual performance, understanding of speech and also exposure to risk of hearing impairment. The kind and the intensity of illumination were quite similar inside the different offices, almost all characterized by a poor contribution of solar light, incorrect use of artificial lighting and lack of care in room design and in the arrangement of work places. The indoor climate was comfortable for all working conditions as a result of the use of air-conditioning systems.

1. *Introduction*

In many modern offices environmental problems related to some 'traditional' factors, such as noise, lighting and climate, are still present and, in some cases, increasing side by side with those connected with the introduction of new technologies and new forms of work organization.

This study attempts to show the importance these factors still have in determining uncomfortable conditions and, to some extent, as a health risk.

2. *Methods*

Measurements were made in four types of offices:

1. Seven suburban bank branches with large, separate rooms varying between 100 and 400 m^2, with a total number of clerks varying from 6 to 30 in all, where normal operations and tax collection were carried out. All offices were protected by unbreakable glass systems. The following places were studied: reception area, telex computers and desks.
2. Four mechanographic centres of a bank, a hospital and two private survey offices, of separate rooms varying from 50 to 200 m^2, where we investigated the following positions: card punching, card reading, VDU keyboards, printing, enclosing and labelling.
3. An international telephone exchange network, in which 50 people, in one single unit of 550 m^2, worked sitting next one to another for manual switchboard management and, in another room of the same area, 4 people were in charge of checking the automatic exchange unit.
4. One post office of about 800 m^2 in which 'sorting' was carried out by means of a new automatic machine line.

3. *Results*

In Table 1 the equivalent noise levels, recorded in the various work places studied, are expressed. The average noise levels in the 'central points' of the four sectors under study

TABLE I. *Noise levels in different offices and work places.*

Office	Work places	Leq dB(A)	
		$\bar{x}$	S.D.
Bank branch	Central points	67·5	3·6
	Computers	74·1	2·9
	VDU keyboards	75·0	0·6
	Mechanical typewriters	72·3	1·1
Mechanographic centre	Central points	73·2	4·8
	Programming	73·2	1·8
	Reading without cards	81·7	1·3
	,, with cards	91·0	1·4
	Printing (closed)	79·6	3·5
	,, (open)	95·0	0·8
	Punching	73·6	4·3
	VDU keyboards	72·8	1·8
	Enclosing—Labelling	80·3	2·9
Telephone exchange	Manual (central point)	63·7	
	,, (desks)	68·3	4·1
	Automatic (central point)	57·4	
	,, (desks)	64·4	2·9
Postal sorting office	Central point	74·0	2·8
	Controls	76·6	1·4
	Sorting	82·1	2·0

vary between 57 and 74 dB(A). The findings in the single work places prove how noise level is directly influenced by the type and the number of machines operated and by the method of use. For example, in the mechanographic centres, the noise range is remarkably high (between 73 and 91 dB(A)) and in the case of a particular operation (printing by an 'open' machine) the noise level reaches 95 dB(A).

In Table 2 the values of incident lighting on work surfaces are shown. It shows very low illumination values, particularly in some working conditions such as telex operating and computing in bank offices and VDU keyboards in mechanographic centres. The sources of lighting were almost completely artificial, made of fluorescent tubes fixed to the ceiling and partially screened by a visible plastic grilling. In most cases the allocation of work places was not calculated according to the position of lighting and only in few cases additional sources of lighting by electric filament lamps were provided.

TABLE 2. *Illumination levels in different offices and work places.*

Office	Work places	Lux	
		$\bar{x}$	S.D.
Bank branch	Reception	268	86
	Desks	307	75
	Computers	157	55
Mechanographic centre	VDU keyboards	214	88
	Programming	283	150
	Desks	350	207
Telephone exchange	Desks	306	59
Postal sorting office	Controls	230	93
	Sorting	260	110

As shown in Table 3, the climatic conditions fall within the limits set by ASHRAE (1972) in relation to the thermal comfort during different periods of the year. This is due to the general utilization of air-conditioning plants with automatic programming by choice.

TABLE 3. *Climatic conditions in different offices.*

Office	Air temperature (°C)	Relative humidity (%)	Ventilation (m/s)
Bank branch	21·2–22·4	39–59	< 0·1
Mechanographic centre	22·3–25·2	53–70	0·1–1·4
Telephone exchange	21·0–22·0	50–58	< 0·1
Postal sorting office	21·4–24·1	45–64	0·1–0·5

4. *Discussion*

Comparing the noise levels recorded with the various 'acceptable' limits proposed by several international norms and guidelines (reported by Cosa, 1980; and Merluzzi, 1981) one can confirm that our data exceed the limit of 55 dB(A) fixed for offices with prevalent intellectual activities and, in all except three cases, are also above the limit of 65 dB(A) suggested for typewriting and telex-operating activities. Most of our observations are between the 70 dB(A) limit set for highly mechanized offices and the 85 dB(A) limit set by ACGIH (1983) for general work environments. Besides taking into consideration that such noises are prevalently discontinuous, one must say that in many cases the noise level is so high as to cause not only intellectual disturbance and general discomfort but also impede understanding of speech and, in some work places, such as card-printing, also expose people to serious risk of hearing impairment.

We have also found that the spectral composition of recorded noise varies notably from one office to another: in fact it goes from spectra characterized by higher peaks for human voice frequencies (the room of the manual telephone exchange) to spectra characterized by high levels at higher frequencies (coding machines and typewriters). Therefore, we can show that the introduction of new technologies or machines is not always accompanied by a correct estimation of noise pollution. This is typified by the various large machines, especially the printers introduced in the mechanographic centres and post offices; on a lower scale it is also true for electric typewriters, whose levels of noise are still much higher than the suggested levels (64–67 dB(A), Pinferi, 1980) for single sources (Table 4). The possibility of reducing significantly the noise level of these machines is shown on the one hand by data furnished by the same producers and, on the other hand, by the levels given in Table 4: the utilization of 'silent' typewriters reduced the noise level by as much as 6–8 dB(A) with respect to other electric/electronic typewriters.

TABLE 4. *Noise levels of different typewriters.*

Typewriter	Leq dB(A)
Mechanical	71
Electric	74
Electronic, with revolving head	76
Electronic, 'silent type'	68

For lighting, the comparison with the suggested levels of lighting in office (see Grandjean, 1980, and Rey, 1981) shows the existence of poor lighting. In our opinion this is due to a lack of care in room design and in the arrangement of work places as well as inappropriate installation of general and local systems of artificial lighting in relation to the work places and sources of natural light. Most bank offices had windows on one side of the room and most work places on the opposite side, and the artificial lighting was strongly fixed to the ceiling without consideration of the position of the work places. More often than not the installation of the artificial lighting and of the work positions satisfied more the external aspect than the ergo-ophthalmic needs.

Regarding the climatic conditions, they are comfortable because of the air-conditioning systems available, but we have observed some situations which can cause discomfort to office workers. In particular one regards the inadequate ventilation, accompanied by a sense of isolation, inside the bank branches protected by unbreakable glass fixtures, and also the presence of 'draughts' due to incorrect position of air-conditioning inlets in relation to work places.

In conclusion, it seems that too often in offices not enough consideration is given to a 'traditional' risk or discomfort factor, in particular noise and lighting. Such problems in many cases can be solved quite simply by taking into greater consideration the arrangement of the work rooms according to ergonomic solutions and the manufacturing of machines which not only respond to productive needs but also to the comfort and well-being of the users.

References

ACGIH, 1983, *TLVs for Chemical Substances and Physical Agents in the Work Environment for 1983–1984*, Cincinnati, Ohio: ACGIH.

ASHRAE, 1972, *Handbook of Fundamentals*, New York: ASHRAE.

Cosa, M., 1980, *Il rumore urbano e industriale*. Rome: Istituto Italiano di Medicina Sociale.

Grandjean, E., 1980, *Fitting the Task to the Man: an Ergonomic Approach* (London: Taylor & Francis).

Merluzzi, F., 1981, Patologia da rumore. In *Trattato di Medicina del Lavoro*, edited by E. Sartorelli (Padova: Piccin), pp.1119–1150.

Pinferi, U., 1980, Il rumore nelle macchine per ufficio. *Inquinamento*, **2**, 53–55.

Rey, P., and Meyer, J.J., 1981, Vision et éclairage. In *Précis de Physiologie du Travail; Notions d'Ergonomie*, edited by J. Scherrer (Paris: Masson) pp.429–472.

Possible Hazards from Laser Printers

A. SONNINO AND I. PAVAN
Insitute of Occupational Health, University of Turin, Italy

Abstract

Five laser printers were tested. All of them were proved to be safe and do not constitute any health risk resulting from the presence of lasers. Therefore the whole apparatus may be put into Class 1 of the laser classification ('exempt') of the Bureau of Radiological Health (BRH).

As far as chemical risk is concerned our test did not show any excess of accepted limits reported by ACGIH. The only machine that used TFN as a photosensitive material was modified and no longer constitutes a risk.

1. *Introduction*

Today the importance of computers in running companies and public services has led to the formation of computer centres of ever-increasing dimensions. At the same time it has become necessary to install large centres to meet printing needs.

It was, therefore, necessary to improve and develop the technology to solve this problem, resulting in the setting up of new, usually mechanical, line printers which became noisier the faster they became. This increase in noise was particularly evident when a large number of machines were placed together. Apart from noise there are several other difficulties, one being economic, and that is that a fast mechanical printing machine may cost several thousand dollars and more often than not requires costly control units.

We also have the problem of stock. In order to run a computer centre the manager is often obliged to buy in large quantities of various types of forms which sometimes lie idle or even become obsolete before they can be used.

Because of these problems, several new printing systems were developed, one being a system based on photocopying machines, using powdered ink.

Since printing machines based on this principle have now been introduced onto the Italian market, we have investigated the possible effects on operators due to the chemical and physical agents used. A point worth noting is that in traditionally aseptic no-risk environments, for the first time chemical substances were introduced because of this technology.

2. *Description of a Laser Printer*

The printer employs electro-optical and laser technology.

A high voltage generator (called corotron) charges the surface of a continuously rotating photoconductor drum. The surface is then exposed to the scanning laser beam which discharges the photoconductor surface selectively in accordance with the character matrix pattern to be printed. A black thermoplastic powder, the toner, is charged with an electrostatic potential and is attracted to the drum surface in the developer station. A negative image of the information to be printed now exists on the drum surface. The paper is brought into contact with the drum surface and the toner is transferred onto it by means of a second corotron.

The toner image on the paper is then made fast in a fusing station by means of heat or chemical agents. Some printing machines have a facility which prints, simultaneously with the text, forms layout or other graphic data.

3. *Experiment and Discussion*

Our aim was to investigate the hypothetical risks brought about by laser printers.

When dealing with physical agents we noted that a laser He–Ne beam was always used, giving off a red visible radiation. The power source was within the range of 5–15 mW.

According to international studies such beams have a destructive capacity to the human eye when exposed.

Therefore we investigated the risk to the operators, especially when they carried out non-routine jobs. In the five laser printers tested, the beams were covered by shields which protected the operator both from direct and reflected exposure.

Usually, on removing the shield or breaking the beam with any tool the laser source was interrupted either by means of an electrical or a mechanical system.

As a result it was decided that the presence of the laser beam in the printing machines tested did not represent an element of risk in as much as it is inaccessible.

As in the photocopying systems the printing is fixed either by fusion or by fusion plus pressure of the toner into the paper fibres. Sometimes hot Teflon rollers are used and sometimes quartz incandescent bulbs, the latter may also emit UV rays, which are potentially dangerous to eyes.

One out of five of these printing machines used a new type of toner fixing by means of chemicals, the other four used a traditional toner fusing system. Three of these used pressure and heat fixing, produced by electrical elements, while the remaining one used a quartz incandescent lamp, which was interrupted by a series of interlocks when the machine was opened. Therefore, in our opinion, there is no significant risk of exposure to UV radiation.

When dealing with chemical agents the laser printers are similar to the photocopying machine as far as photosensitive materials and toners are concerned.

Chemical analysis broke down the toner used by a laser printing machine into the following base chemical formula:

1. Black carbon: about 10%.
2. A charge control agent, usually an organic chemical compound deriving from diphenyl hydrazones: about 5%.
3. The printing binder, a styrene–acrylate copolymer: about 90%.

Photosensitive materials used on the drum were arsenic selenide, cadmium sulphide, trinitrofluorenone (TNF) and chlorodiane blue (a diazo compound).

In some cases the drum surface was covered by a sheet of plastic (usually mylar). TNF was substituted by chlorodiane blue because it is a suspected human carcinogenic substance.

A source of air pollution could be a result of the pyrolysis of toners when heated in the fusing station. A recent work (Sonnino *et al.*, 1983) excluded any possibility of risk for operators.

However, the new fixing technique mentioned earlier was developed in order to meet the needs of people that use papers which are heat sensitive, such as bank cheques. This technology is based on the evaporation of a freon TA mixture composed of 90% (by weight) freon 113 (1,1,2-trichloro-1,2,2-trifluoroethane) and 10% (by weight) acetone in a closed system.

As the fixing chamber is not sealed, due to paper circulation, we assumed that there was a possible leakage of the solvent mixture into the environment. Freon and acetone, although commonly used industrially and domestically, are considered toxics by the American Conference of Governmental Industrial Hygienists (ACGIH, 1973) which has suggested the following threshold limits: acetone: 750 p.p.m., and freon 113: 1000 p.p.m..

We analysed the acetone and freon quantities present in the working environment.

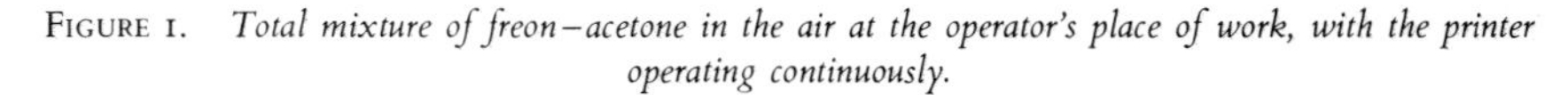
FIGURE I. *Total mixture of freon–acetone in the air at the operator's place of work, with the printer operating continuously.*

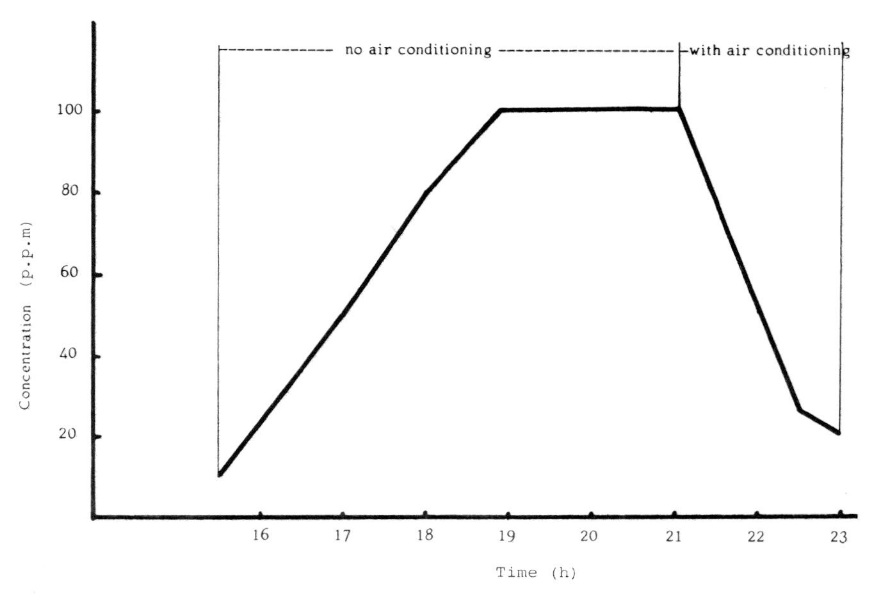

These analyses were carried out both in ventilated and non-ventilated rooms, even though laser printer manufacturers clearly state that these machines must be used in a ventilated room at 550 m^3/h (which is the standard value in conditioning computer centres).

TABLE 1. *Mean freon and acetone concentration in the air.*

Sampling hours	Note	Acetone (p.p.m.)	Freon TF (p.p.m.)
17–20	No air conditioning	10	53
20	Start air condition		
21–23	Air condition	2	18
23–24	Air conditioning	1	10

The results can be seen in Figure 1 and Table 1.

References

ACGIH, 1973, *A Guide for Control of Laser Hazards* (Cincinnati, Ohio: ACGIH).

IBM, Scientific report. *Direction of Product Safety Operations* (Armonk, NY: IBM).

Sliney, D. and Wolbarsht, M., 1980, *Safety with Laser and Other Optical Sources* (New York: Plenum Press).

Sonnino, A., Pavan, I., Scansetti, G. and Rubino, G.F., 1983, Rischi connessi all'uso di stampanti laser, *La Medicina del Lavoro*, **74**, 211–216.

Von Burg, R., 1981, Toxicology updates: trinitrofluorenone, *Journal of Applied Toxicology*, **1**, 50.

Environmental Design Trends for Modern Office Work

P.G. Cane, P.F. Castre, E. Tamagno and E. Tintori Pisano
Department of Architectural Design, Polytechnic of Turin, Italy

Abstract

An investigation of the Telephone Information Service using VDTs was carried out. Environmental conditions were found insufficient, but no specific harmful factors could be detected. A large survey is planned for 1984.

The main purpose of this investigation is to compare subjective assessment of the environment in traditional situations and when new technology is introduced and to research into workers' needs in both situations in order to see whether and how human demands on the environment are changing.

Subjective assessment is achieved by means of interviews; and results about surface requirements, lay-out aspects, aesthetic quality in the environment, as well as illumination, acoustic and thermal variables, are meant to give a contribution to outline present trends in the ergonomic design of work environment, either due to computerization or independent of it.

1. *Introduction*

The present investigation is part of a project by two students who are workers in the company where they have carried on the inquiry among their workmates. The investigation has in no way been formalized with the management and its peculiarity lies in the fact that field work is coupled with university work. It has been originated by the need to outline guiding principles in restoring a small building, about 80 years old, which through time has already undergone some transformations and where an automated data bank is supposed to be established.

Since the early 1970s the building has been the place of the Design Division of a big company: tasks performed there are mainly concerned with drafting, computation, experimentation, testing, design and research about engines. A hundred and seventy designers and technicians are employed there, 29 of which are graduates and 62 having a high school diploma.

Computerization began in 1977 and started with tasks concerning computation, simulated engine operations, components checkout; VDTs used on this purpose were two in 1981, are nine in 1983 and will be increased up to twenty. These VDT work-stations are spread all over the drafting units. Other VDT stations concerned with computer-aided design and computer graphics are more recent and are connected to their own system. Only the CAD computer is located in the building considered.

2. *Aim*

The purpose of the investigation is to examine workers' attitude towards their work environment, in order to see whether and how their needs change as long as tasks are computerized.

The investigation concentrates on physical variables in the environment even if workers have reported about variables of more dramatic concern, such as those having to do with work-stations, equipment, software, job content and work organization. The only non-physical variable on which subjective assessment is accounted here is the one concerning interpersonal relations as in some respects it can be induced by physical variables.

3. *Methods*

Two types of interview have been used by the investigators. They have started with an unstructured interview to all the workers by means of which they have been encouraged to express their feelings in relation to their environment situation and to point out the factors to which they ascribe a given level of acceptability and comfort. At this stage specific questions on single variables have been avoided in order not to influence either analysis and definition of them or relations among them.

Environmental variables have thus been defined by virtue of subjects' answers, according to the frequency critical factors or pertinent matters where occurring.

These variables have become the subject of a more standardized interview to 36 subjects who in the former stage had been shown to be highly motivated in the investigation. The subjects have been split up into three groups in reference to the techniques and instruments used in their jobs, classified as traditional (T), computerized (C) and mixed (M) when being in course of change, so that the assessment of the environment can be compared in the three situations. They have been asked to evaluate each variable by means of a three point scale: satisfied, uncertain, unsatisfied.

Objective measurements were planned at the beginning but then proved to be not realistic, because of bureaucratic delays in authorizing the introduction of measuring devices. It must be said, however, that lack of objective data does not appear so relevant in this case, as quantitative characteristics seem quite homogeneous in different parts of the whole building, so that comparison of the three groups can be said to take place in the same environmental condition; the only exception is climatization in rooms where C work is prevailing.

4. *Results*

At the free interview stage subjects have put their stress on eight variables, three of which are related to the building itself, and the way it is organized and arranged, four at its technological equipment; the last concerns some social aspects in the environment inasmuch as they are induced by human perception of physical variables as a whole. The first three deal with space organization, personal work area (surface) and furnishing; the other four with illumination, thermal conditions, noise and air quality.

FIGURE 1. *Traditional work (T): Unsatisfied subjects.*

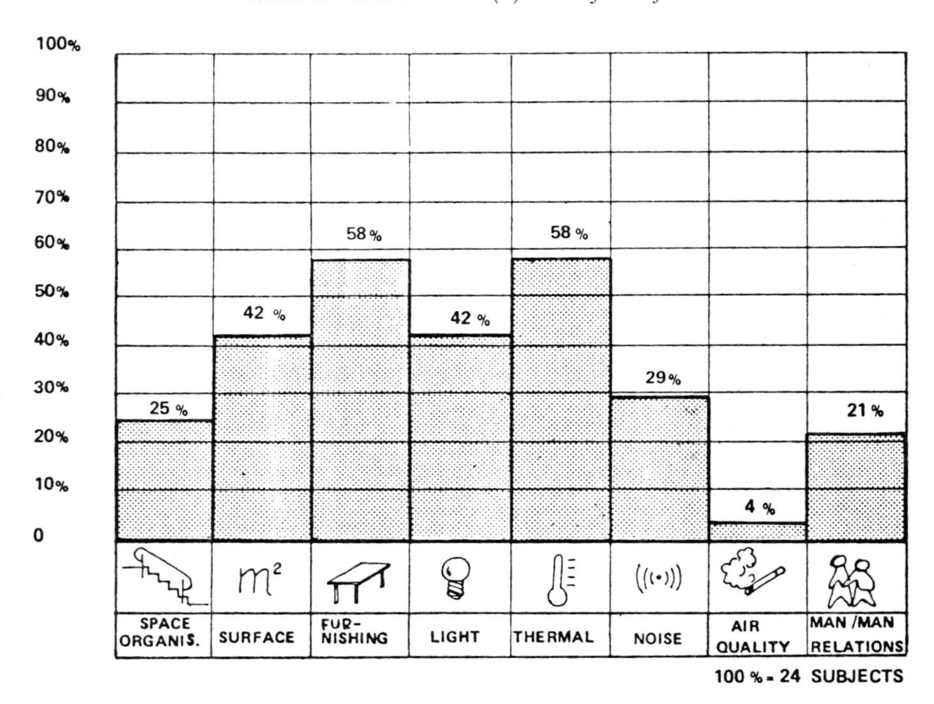

In the second phase of the enquiry subjects' complaints about each variable are distributed as shown in Figures 1, 2 and 3 and the conclusive remark is that there is no significant difference among unsatisfied workers, group T on one side, groups C and M, as well as C + M, on the other side (Fisher φ = 0·0). More detailed comments are related to:

1. Space organization: is more a problem for T subjects than the others.
2. Work area problems: are not mentioned by M subjects; 67% VDT operators, on the contrary, underline unacceptable density, which is complained also by 42% of T subjects.
3. Furniture discontent: is deeper among M subjects 67%, even if it has a strong presence among T and C subjects, 58% and 50% respectively.

FIGURE 2. *Computerized work (C): Unsatisfied subjects.*

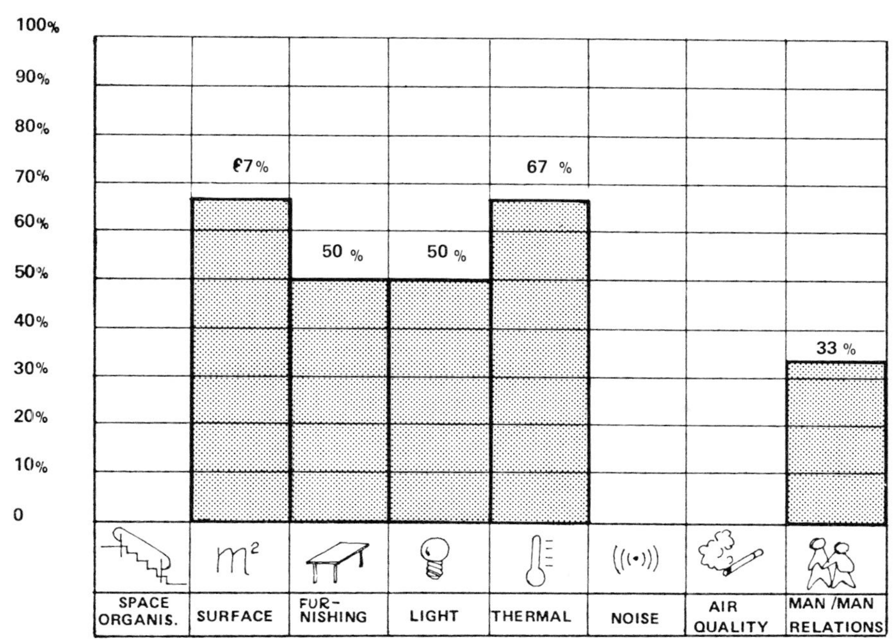

4. Light: is more satisfactory to M subjects (only 33% unsatisfied) than the other two groups, which differ very little, 50% unsatisfied among C subjects versus 42% among T subjects.
5. Thermal variables: are not satisfactory to all the subjects, who have mainly complained of draughts originated from leakage in doors and windows; the highest percentage of complaints, 67%, comes from C subjects whose majority enjoy air conditioning but do not judge it specifically responsible for their discomfort.
6. Acoustic variables: appear not to be a problem once computerization has been introduced or when it is in progress; among T subjects noise is mainly of human origin, due to verbal communication, movement, manual operations, which to some extent is in agreement with poorness of interpersonal relations as complained by 33% among C subjects versus 21% among T subjects, while M subjects are 50%.

5. *Discussion*

C subjects' complaints concentrate on fewer variables than T subjects', more exactly on space dimensions, furnishing, light and thermal conditions; M subjects confirm the same

FIGURE 3. *Jobs being computerized (M): Unsatisfied subjects.*

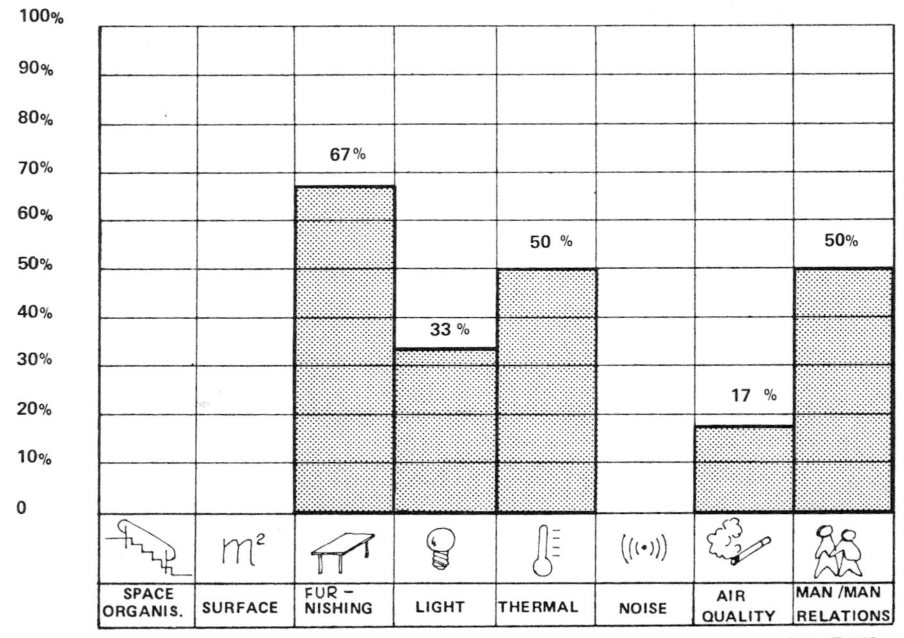

trend. Moreover, the percentage of complaints about these variables is higher among C subjects than T subjects.

The only exception is furnishing where more functional characteristics seem to reduce discontent, which is in any case reported: T subjects regret their furniture has no or little concern with human operations to be performed; C subjects are due for a rationalization of furniture design because of VDTs, but realize that pieces of furniture are nothing but a work device, irrespective of the other functions they can have in characterizing the environment, allowing people to recognize its peculiarities and its different parts, increasing or decreasing environmental stimulation according to levels of attention required by the job. Data about M subjects make them appear as impressed by lack of compensation related to T environment, because of dullness which is perceived as a heavy failure in the more functional C environment.

Complaints about light do not vary too much between T and C workers, showing that low level of acceptability deals much more with environmental than task variables, given the great concern of the latter with vision; on the other hand M subjects with their much lower percentage of complaints confirm that there is no evidence of light interfering with task performance, which they would have realized at the initial stage of their C work.

Among M subjects lack of environmental analysis seems quite probable: the high percentage of M subjects complaining about furniture could be a diversion from environmental dullness depending on levels of luminance, and not only on furnishing. The transitional phase these subjects are undergoing makes some of their problems heavier than the other subjects', while other problems appear underestimated maybe because of some Hawthorne effect: in the first case we can see interpersonal relations which half of them report as intolerable, versus 21% among T subjects and 33% among C subjects; in the second case we can mention space organization of which they do not complain at all.

About this variable C work is less demanding in terms of the human operator's movements and consequently in terms of functional characteristics of space related to communication, coordination or similar.

The last remark concerns interpersonal relations which are poorer among C subjects in spite of the fact that operators are physically much closer; it may be their concentration is captured by job content or procedure and too little is left to human communication.

Finally it could be underlined that complaints about thermal variables confirm once more how treatment of single parts in a system may be ineffective if system relations are not taken into due account.

It can be concluded that thermal variables are the only ones considered by workers on a functional point of view and T and C groups do not differ very much.

Other factors of dissatisfaction are much more related to perceptual qualities of the environment: if on one side these factors are not peculiar to technological changes, the sample considered shows that in front of these changes workers' demands tend to be stressed in the direction of variables which contribute to improve the symbolic and aesthetic quality of the environment.

How to Measure and Evaluate the Thermal Environment

N.K. Christensen and B.W. Olesen
Brüel & Kjaer, Naerum, Denmark

Abstract

Until now there have only been a few standards or recommendations on the indoor thermal environment. Recently, a couple of standards have been proposed by the International Organization for Standardization (ISO) and the American Society of Heating, Refrigerating and Air-Conditioning Engineers (ASHRAE). To verify the requirements, different thermal measurements have to be performed. Guidelines as to where and how to measure are given in the proposals. The parameters, definitions and methods are the same in the different standards, and the requirements are also comparable.

The present paper deals with the requirements and measurements that are relevant for moderate thermal environments in places of residence, offices, hospitals and light industry. For evaluation of very hot or very cold surroundings, other methods are required. Only measurement of parameters that influence the perception of the thermal surroundings are included.

1. *The General Thermal Condition*

Man's heat balance, and therefore his thermal sensation, is influenced by four environmental parameters: air temperature, mean radiant temperature, relative air velocity and water vapour pressure. In addition to these four environmental parameters, thermal comfort is also influenced by the amount of activity and clothing worn.

Fanger (1973) has developed a relationship between these six parameters and the 'Predicted Mean Vote' or PMV. The PMV is the predicted mean vote of a large group of people voting on the basis of the thermal sensation scale shown in Figure 1.

It is important to be able to predict the number of people who are feeling either uncomfortably warm or cold, because these people are likely to complain about the thermal environment. This prediction is achieved using the PPD (Predicted Percentage of Dissatisfied) index, which gives as a function of the PMV value, the percentage of people likely to feel either uncomfortably cool (thermal rating less than -1) or uncomfortably warm (thermal vote above $+1$). The relationship between PMV and PPD is shown in Figure 2.

FIGURE 1. *The PMV scale.*

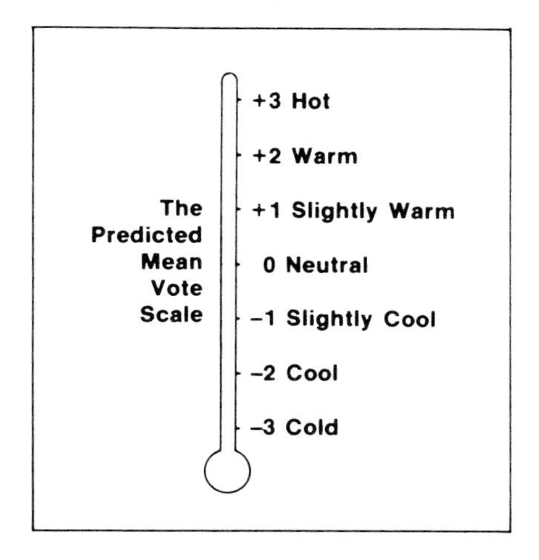

FIGURE 2. *The relationship between PPD (Predicted Percentage of Dissatisfied) and PMV (Predicted Mean Vote).*

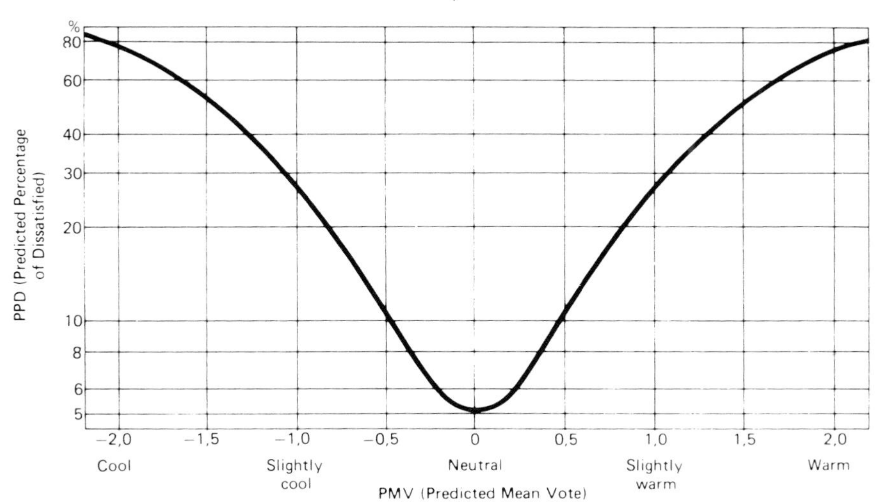

2. *Requirements for the Thermal Environment*

Because the PMV index takes into account all six parameters which influence the overall thermal sensation, it is convenient to express the requirements of the thermal environment by stating limits for the PMV value.

The ISO has proposed the use of the PMV–PPD index to evaluate the thermal comfort of environments. The new draft International Standard ISO/DIS 7730 (1982) recommends that the PPD be lower than 10% which corresponds to $-0\cdot5 < \text{PMV} < +0\cdot5$.

NKB (The Nordic Committee on Building Regulations, 1981) and ASHRAE (American Society of Heating, Refrigeration and Air Conditioning Engineers, 1981) have also specified recommended temperature intervals for winter and summer conditions corresponding to $-0\cdot5 < \text{PMV} < +0\cdot5$.

Limits for local thermal discomfort are also specified in ISO/DIS 7730. These limits are however, only valid for people engaged on light, mainly sedentary activity. During winter conditions (heating period) the limits are:

1. The operative temperature must be between 20°C and 24°C.
2. The vertical air temperature difference between 1·1 m and 0·1 m above floor (head and ankle level) must be less than 3°C.
3. The surface temperature of the floor must be between 19°C and 26°C, but floor heating systems may be designed for 29°C.
4. The mean air velocity must be less than 0·15 m/s.
5. The radiant temperature asymmetry from windows or other cold vertical surfaces must be less than 10°C (in relation to a small vertical plane 0·6 m above the floor).
6. The radiant temperature asymmetry from a warm (heated) ceiling must be less than 5°C (in relation to a small horizontal plane 0·6 m above the floor).

Similar requirements are given for light, mainly sedentary activity during summer conditions (cooling period).

Keeping the vertical air temperature difference between 1·1 m and 0·1 m above floor level to less than 3°C is sufficient to prevent more than 5% expressing dissatisfaction due to the vertical air temperature difference (see Figure 3, Olesen *et al.*, 1979).

FIGURE 3. *The percentage of dissatisfied as a function of air temperature difference between head (1·1 m) and ankles (0·1 m).*

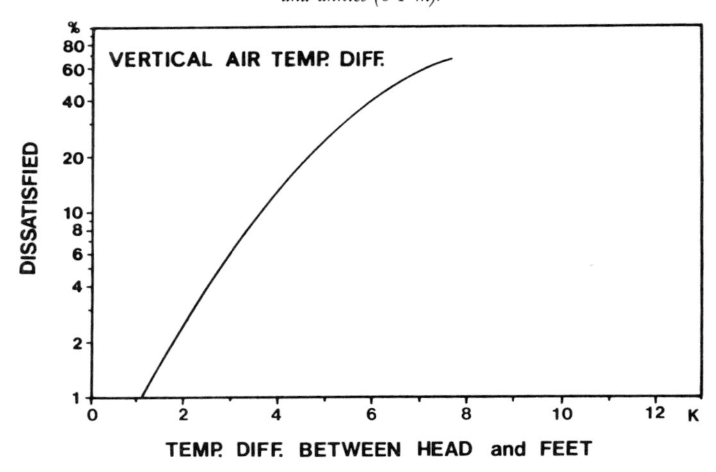

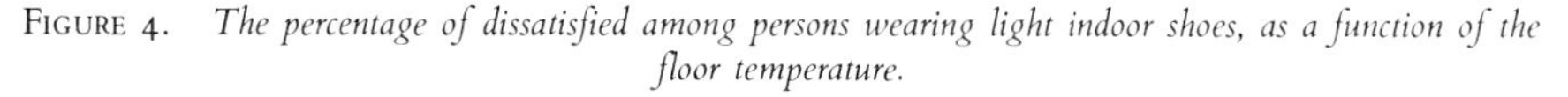

FIGURE 4. *The percentage of dissatisfied among persons wearing light indoor shoes, as a function of the floor temperature.*

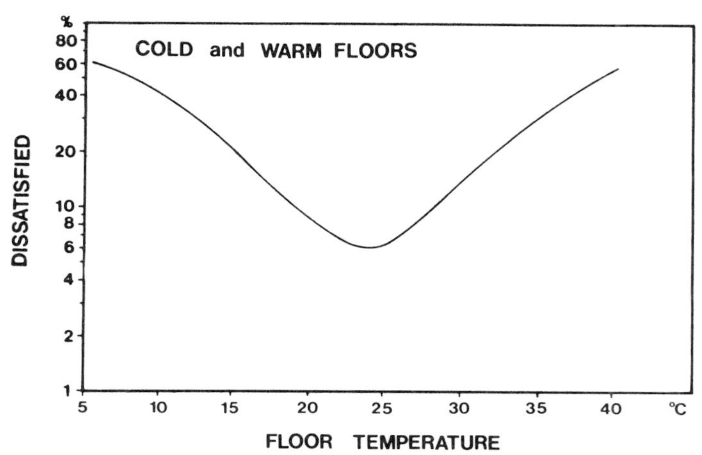

Maintaining the floor surface temperature within the recommended interval will ensure that less than 10% are dissatisfied because of either cold or warm feet (see Figure 4, Olesen *et al.*, 1977).

Based on new results from the Technical University of Denmark (Fanger *et al.*, 1980, Fanger, 1983), fulfilling the requirements for radiant temperature asymmetry will result in less than 5% being dissatisfied due to cold vertical surfaces and also less than 5% being dissatisfied due to warm ceilings (Figures 5 and 6).

FIGURE 5. *The percentage of dissatisfied as a function of the radiant temperature asymmetry caused by a cold vertical area, e.g. a window.*

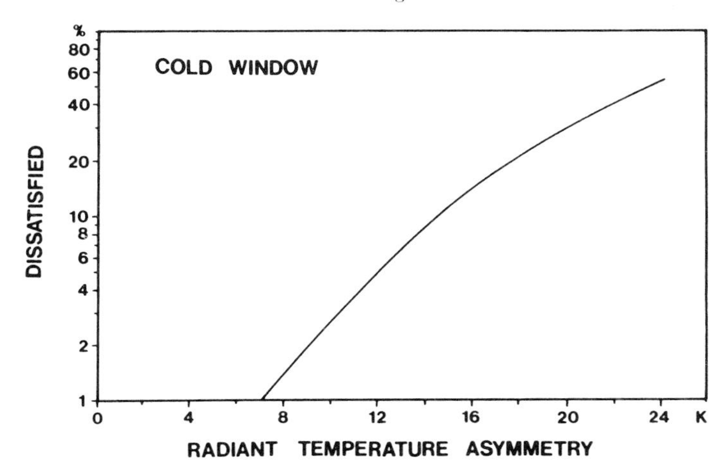

Additional recent research (Fanger, 1983) has shown that a mean air velocity of 0·15 m/s can produce up to 20% dissatisfied because of draughts (Figure 7).

These results are based on experiments with 100 people (tested individually) exposed to fluctuating air velocities at different air temperatures. Figure 8 shows a typical example of the fluctuations in air velocity in the test chamber. The fluctuations were similar to those normally occurring in ventilated spaces. This is an important factor because earlier results have shown that the fluctuations influence the sensation of draught (Fanger *et al.*, 1977).

FIGURE 6. *The percentage of dissatisfied as a function of the radiant temperature asymmetry caused by a heated ceiling.*

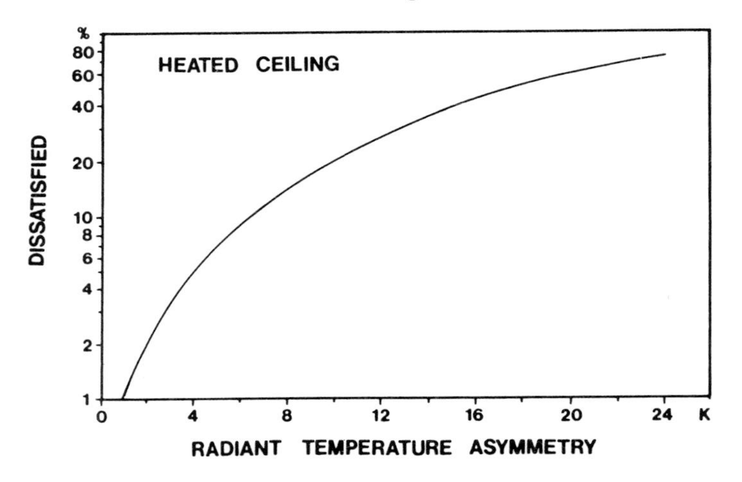

FIGURE 7. *The percentage of dissatisfied (i.e. those feeling a draught) as a function of the mean air velocity at three different air temperatures.*

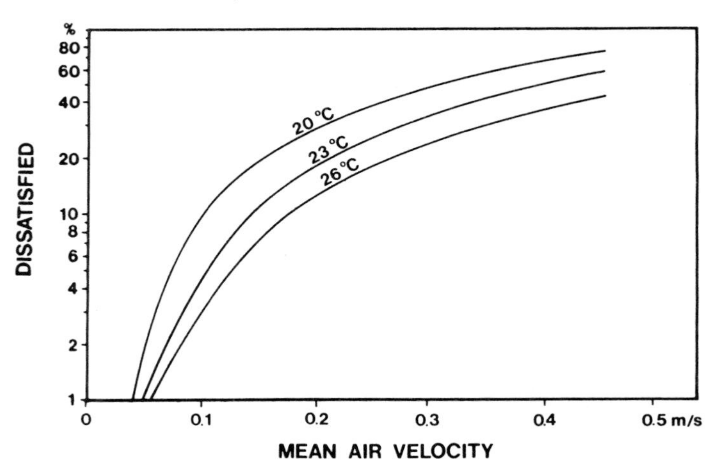

FIGURE 8. *An example of the fluctuations in air velocity.*

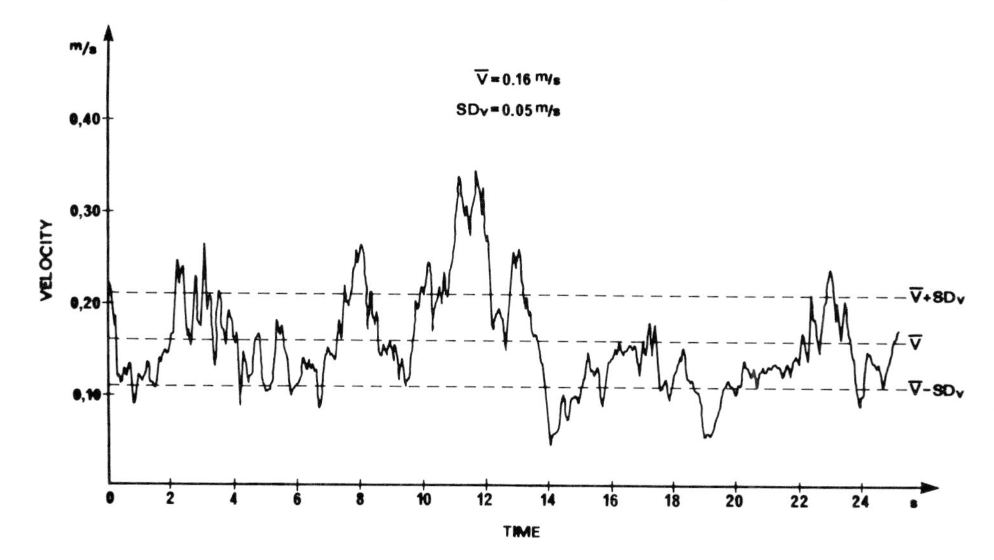

The new results have shown that people are more sensitive to draught than expected. To ensure that less than 5% of the occupants will complain about draught, the limits for mean air velocity will therefore have to be lowered.

It is important to emphasize that, if more than one factor causing local discomfort is present the percentages of dissatisfied due to the different factors cannot just be added together. At this moment no data are available to predict the percentage of dissatisfied, e.g. when people are exposed to radiant temperature asymmetry from a cold surface and at the same time are exposed to draught.

3. *Measurements*

To verify the limits for an unacceptable thermal environment, it is necessary to estimate the PMV–PPD index.

This can be done by measuring the four environmental parameters (air temperature, mean radiant temperature, air velocity and air humidity) and using tables to estimate the activity level and the amount of clothing worn. After having determined these six parameters, the PMV–PPD index can be calculated or found in a table.

An easier and more direct method is to use a measuring principle that integrates the influence of the air temperature, mean radiant temperature and air velocity. This principle, based on Fanger's comfort equation has been developed at the Technical University of Denmark (Madsen, 1976 and 1979) and incorporated into a single instrument (Figure 9). The clothing value, activity level and air humidity are selected and then set on the instrument, so that

FIGURE 9. *The instrument for direct measurement of the thermal comfort condition (PMV–PPD index).*

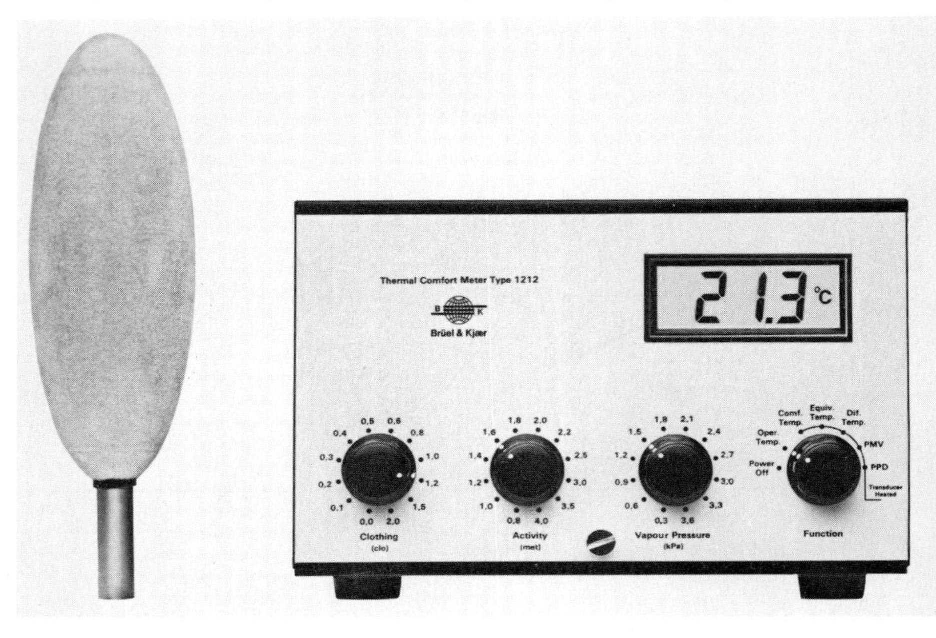

a PMV value can be obtained directly.

The draft International Standard ISO/DIS 7726 recommends that the PMV value be found at a height of 0·6 m for a sitting person and 1·1 m for standing person.

To verify the requirements with respect to local thermal discomfort, it is necessary to measure air temperature, radiant temperature asymmetry, floor surface temperature and air velocity.

The vertical temperature difference is the difference between air temperature 1·1 m above the floor and air temperature 0·1 m above the floor. The air temperature must therefore be measured at these two heights. Air velocity must be measured 0·1 m and 1·1 m above the floor for a seated person and 0·1 m and 1·7 m above the floor for a standing person. These heights correspond to the height of ankle and neck respectively, since these are the parts of a normally clothed body which are most sensitive to draught. Finally the radiant temperature asymmetry must be measured at the same heights as the PMV value.

4. *Conclusion*

The described standards and the new measuring procedure are valid in moderate thermal environments.

These new standards provide engineers, architects and occupational hygienists with useful guidelines for assessing the thermal environment.

References

ASHRAE, Standard 55–81, 1981, *Thermal Environmental Conditions for Human Occupancy*.
Draft International Standard ISO/DIS 7730, 1983.
Fanger, P.O., 1973, *Thermal Comfort* (Malabar, Florida: Robert E. Krieger Publishing Co., 1982).
Fanger, P.O., *Thermal Comfort Requirements*. Proceedings of the Second International Congress on Building Energy Management. Ames, Iowa, USA, 1983.
Fanger, P.O., Banhidi, L., Olesen, B.W. and Langkilde, G., 1980, *Comfort Limits for Heated Ceilings*, *ASHRAE Transactions*, 86.
Fanger, P.O. and Pedersen, C.J.K., 1977, *Discomfort Due to Air Velocities in Spaces*, IIR Commission E1 Belgrade. Nov.
Madsen, T.L., 1976, *Thermal Comfort Measurements*. *ASHRAE Transactions*, **82** (1).
Madsen, T.L., 1979, Measurement of thermal comfort and discomfort, in *Indoor Climate*, edited by P.O. Fanger and O. Valbjørn (Copenhagen: Danish Building Research Institute).
NKB (Nordisk Komite for Bygningsbestemmelser), 1981, *Inomhusklimat*. NKD-rapport nr. 40, May.
Olesen, B.W., 1977, *Thermal Comfort Requirements for Floors*. Proc. of the meeting of Commission B1, B2, E1, of the IIR, Belgrade, pp.307–313.
Olesen, B.W., Schøler, M. and Fanger, P.O., 1979, Discomfort caused by vertical air temperature differences, in *Indoor Climate*, edited by P.O. Fanger and O. Valbjørn (Copenhagen: Danish Building Research Institute).

From Evaluation to User Functional Requirements

G. Davis and F. Szigeti
TEAG—The Environmental Analysis Group, PO Box 1088, Station B, Ottawa, Canada K1P 5R1

Abstract

In the traditional approach for establishing user requirements for the office work environment, there was little or no explicit evaluation of facilities. In the directed approach, some limited evaluation is common. In joint planning the knowledge and experience of the occupants/users is also utilized. Formal technical studies may be conducted depending on the environmental awareness of the occupants/users or decision-makers. When issues arise out of the knowledge gained in evaluation, options and recommendations are presented to management, and decisions are obtained.

1. *Approaches to Establishing User Functional Requirements for Office Work Environments*

Planning for the office work environment has become increasingly complex in recent decades (Davis and Szigeti, 1982; Glover, 1976).

1.1. **Traditional**

The traditional approach for establishing user requirements has been for professionals, such as programmers, architects, space planners and/or interior designers to visit an organization's existing facilities, interview managers and the facility operations group, and then specify what should be provided. Sometimes a manager would sketch what she or he wanted, and then designers or draftsmen would prepare formal layouts based on the manager's instructions. The organization's facilities service group would direct and support the process.

If assessments of existing or surrogate facilities are conducted at all, they are likely done informally, without explicit record-keeping. The programmer or architect would walk through the existing offices with the manager or supervisor, asking questions. Sometimes he might have an assistant along to take notes and sketch the existing layouts. The manager is usually the prime source of information. If he does not have the necessary information, the programmers or architects would have to base their recommendations more on their own judgement and prior experience than on the projected requirements for that particular organization. Independent facility programmers are seldom involved.

1.2. Directed

In recent years a directed approach has gained acceptance for determining user functional requirements for the office work environment. In the directed process, corporate planners and operating managers provide direction to the facilities service group, regarding planning and functional programming. Actual space programming and planning is usually done by professional specialists, either from within the organization or consultants to it.

Involvement of subordinate staff is typically limited to responding to a questionnaire. A small number of staff and/or visitors may be interviewed by the programming and/or design professionals. Managers and supervisors are usually interviewed, to learn what they consider should be provided for their staff. The data-gathering is often informal. Although subordinate staff may be questioned about specific equipment needs at their work station, their relationship to others, their paper and work flows, it is unusual for programmers to interview even a sample of subordinate staff about their ergonomic and health requirements.

1.3. Joint

A more recent approach for determining user functional requirements for the office work environment is called joint planning. This approach is compatible with the emerging 'quality of working life' movement, and utilizes the knowledge that staff have of existing facilities and work. Planning is done by a coalition that includes actual users, often acting as delegates of their work groups, in addition to the typical technical experts such as architects, engineers and space planners. Delegates can often efficiently identify ergonomic and health issues for which evaluation of facilities is required, or for which existing standards have proven inappropriate. For instance, although there has been considerable study of the ergonomics of individual pieces of furniture, especially the chair, there is a need for better understanding of the synergism among several units of furniture and equipment, assembled together at a single work-station. As workers gain greater environmental competence, union officials will likely become more demanding in their requirements for explicit data on the ergonomic and health aspects of the office settings in which they and their members must work.

1.4. Technical and Laboratory Testing

Whichever of the prior approaches is used, more and more technical and laboratory tests are being used during evaluations. When careful tests are conducted, the results can surprise the investigators. In one open-plan office, it had been assumed that low speech privacy was due mainly to sound travelling over and under the office landscape screens. However, the tests in this study demonstrated that more sound passed directly through the furniture screens than had been expected. The standards for sound transmission had been set too low for this situation, and needed to be changed.

It is now practicable to evaluate health aspects of the office environment, conveniently and at low cost. Portable gas chromatographs can be used to identify the 'signature' of specific pollutants for which they have previously been calibrated, although those now available are not capable of sophisticated analyses. With such a unit it is practicable to go to the air-conditioning equipment in the roof of a building and in real time, test the air being returned from the office space. With this equipment, it has been possible to identify the brands of photocopiers being used in an office building, without ever having to go onto the office floors to actually see the equipment. Formal laboratory studies may also be conducted, such as analysis of composition of particulate and other pollutants in air samples.

Professionals can also draw on a continuing, growing, published body of theoretical insights and empirical knowledge. For instance, the non-visual effects of light and color are becoming understood (Küller, 1981), and environmental psychology is now an established sub-discipline (Canter and Craik, 1981; Holahan, 1982).

2. *Specifying the Requirements for General-Purpose Office Facilities to Accommodate Varying Levels of Information Technology*

2.1. **Assignment**

After the authors had recently led and facilitated joint planning experiences, they were asked during 1982 to use the directed process to develop the user functional requirements for general purpose office accommodation that would accept varying levels of information technology. The requirements were to take into account: the findings from the authors' own evaluations; lessons learned in the joint planning experiences; findings from current research by others; the impact of information technology on the building structure and services; and the need to consider the overall performance of the building. They dealt explicitly with several issues affecting the health, well-being and effectiveness of office workers, including:

2.2. **Issues**

1. What basic functional capability and level of service is appropriate for the general purpose office spaces?

   In base building, the facility should accept typical on-floor electrical and communications wiring, floor loads, heating, ventilating, compartmentalization, acoustic treatment, and so on, at no special cost for fit-up.

2. What is the threshold of 'acceptable' specialness of tenant requirements?
   In the general-purpose office areas, 'equipment zones' are required, in which special-purpose office requirements can be accommodated. Examples of the range of acceptable special-purpose office requirements are:
   (a) Mini-computers and their peripheral equipment, such as terminals and printers, and related storage of disks, tapes and other media.
   (b) Convenience photocopiers, subject to not exceeding a specified air exhaust capacity per 1000 m^2.
   (c) Non-hazardous dry-lab functions such as instrument checking and minor maintenance.
   (d) Other equipment which requires dedicated or conditioned electrical power circuits, with steady voltage and without spikes.

3. What specialness is not permitted?
   The following categories of special requirements are not permitted within the building:
   (a) Dangerous or hazardous materials, which would have to be exhausted through fume-hoods or other special air ducts, and which would present a health hazard if any recirculation occurred in the building; liquid waste which requires special piping because of corrosiveness or health hazard; and so on.
   (b) Requirements for special plumbing, such as domestic-type water supply and waste, except at certain designated locations on each floor.
   (c) Wet labs for chemical tests and analyses, requiring special air supply or exhaust.

4. Should the planning and design of the facility respond to what is being learned about 'overall building performance', without waiting for this knowledge to be codified and issued as standards?
   The functional performance requirements specified for this building reflect data on overall building performance obtained during investigative and analytical evaluations which the authors led or participated in, of other office buildings owned by the same organization. These evaluations were consistent with the draft standard guide for evaluation of overall building performance now being developed within ASTM Subcommittee E6.25 on Overall Performance of Buildings.

5. Several additional issues were addressed, and other user functional performance requirements were specified, such as:
   (a) What is required for the character and image of a building as a whole and for the office workspaces?
   (b) For what type of space-planning and for what degree of openness/compartmentalization should the office space be planned, and what effect does this decision have on building form and function?
   (c) How should tradeoffs be made on the design features of buildings which constrain space planning and use?
   (d) And so on.

2.3. Decisions

A total of twelve issues were formally analysed and presented to management. Decisions were obtained on all issues presented. For many of the issues the presentation included:

1. Statement of the issue. A sentence or a short paragraph identifying the subject about which a decision was needed.
2. Discussion. As appropriate, context, background and discussion of the issue and its implications.
3. Options. The principal options that were considered.
4. Recommendations. Management was presented with specific guidance. In most instances management accepted the authors' recommendation, but for two issues, the authors were directed to adapt or merge certain options in accordance with evolving priorities of the organization.

References

Canter, D.V., and Craik, K.H., 1981, Environmental psychology. *Journal of Environmental Psychology*, **1**, 1–11.

Davis, G., and Szigeti, F., 1982, Programming, space planning and office design. *Environment and Behavior*, **14** (3), 299–317.

Glover, M., 1976, Alternative processes: building procurement, design and construction. *IF Occasional Paper 2* (Montreal: Industrialization Forum, University of Montreal and Champaign/Urbana: University of Illinois).

Holahan, C.J., 1982, *Environmental Psychology* (New York: Random House) 422 pp.

Küller, R., 1981, *Non-visual Effects of Light and Colour* (Stockholm, Sweden: Swedish National Council for Building Research) 239 pp.

The Magic of Control Groups in VDT Field Studies (Introductory paper)

T. LÄUBLI AND E. GRANDJEAN
Department of Hygiene and Ergonomics, Swiss Federal Institute of Technology, ETH-Centre, 8092 Zurich, Switzerland

1. *VDT and Non-VDT Work*

The introduction of VDTs into offices is normally accompanied by changes in task design, organization, work environment and productivity. 'Use'/'non-use' of VDTs is not an independent criterion. All the above mentioned factors depend directly on the introduction of VDTs, and in real work situations they cannot be separated. This interdependence is of crucial importance if we are looking for control groups.

1.1. **Payment Transactions**

In the field studies by Läubli (1981) two groups dealing with payment transactions were compared. They worked in the same bank, using the same computer systems. One group was using VDTs that had been introduced some years ago. For the other group, installation was planned within the following years.

TABLE I. *Payment transactions with or without VDTs.*

	Without VDT (n = 54)	With VDT (n = 55)
estimated productivity (daily transactions/worker)	30	300
feelings of 'burning eyes' (% of subjects)	41	60
1000 transactions cause 'burning eyes' in . . . people	14	2

The introduction of the VDTs led to an increase in productivity and output. If output were held constant, the weekly working time would be reduced from 44 hours to 5 hours. It is obvious that in such a case health problems would not arise. However, in reality the introduction of VDTs was combined with an increase in productivity and output, and it was found that people working with VDTs suffered more often from 'burning eyes'.

1.2. Telephone Operators

Starr (1982) studied telephone operators who spent their entire working time answering customer requests for telephone numbers. Using 145 operators, listings were retrieved on a VDT; in the control group (n = 105) listings were printed in paper books. Productivity was equal.

TABLE 2. *Discomfort in telephone operators (% operators reporting discomfort).*

	Using VDTs (n = 145)	Using paper (n = 105)	Significance
'burning, tearing or itching eyes'	61	55	n.s.
'discomfort in shoulders'	48	37	n.s.

The demonstrated figures about complaints are typical for the total result, a lot of other variables show the same minimal differences between the two groups. The author's conclusion was that the introduction of VDTs to other office jobs need not be accompanied by increased health impairments. We would interpret these findings differently. Telephone operators have generally a strenuous job (Grandjean, 1959). The incidence of complaints was high in both groups. The introduction of VDTs was followed by an insignificant increase in impairments. The questions arise whether impairments are caused in both groups by the equally high intensity of work or by unsuitable equipment. In the latter case it may be a badly printed telephone book or in the VDT-group an insufficient visual screen. Here again the use of VDTs is combined with a high incidence of complaints.

2. *Eye-strain*

2.1. 'Eye-strain' in Field Studies Using Control Groups

Comparing the various field studies using control groups, we may get some evidence of the severity of findings. We hope to cover the relevant literature and refer partly to Dainoff (1982).

TABLE 3. *Survey over field studies using control groups.*

Occupation	Using VDTs	'Eye-strain' (%)	'Burning eyes' (%)	Sample	Author
sample of office work	yes	51		257	Coe, 1980
sample of office work	no	45		129	
shop girls	no	42		119	Maeda, 1980
accounting-machine	no	75		57	
data entry	yes	72	59	53	Läubli, 1981
payments transactions 'A'	yes	72	45	55	
payment transactions 'B'	yes	70	60	54	
payment transactions	no	50	41	55	
fulltime typists	no	60	56	78	
sample of clerical work	yes	72		} 312	Rey, 1980
sample of clerical work	no	50			
office worker	yes		71	} 333	Sauter, 1983
office worker	no		56		
sample of clerical work	yes	91	80	254	Smith, 1981
sample of clerical work	no	60	44	158	
telephone operators	yes		61	250	Starr, 1982
telephone operators	no		55	105	

Complaints about 'eye-strain' were generally frequent. Even in the control group showing the lowest incidence, 42% complained about eye-strain. The authors reported more eye-strain in the VDT-groups than in their counterparts. Yet the strenuous work of accounting machine operation seemed to produce even more eye-fatigue than work at VDTs.

2.2. 'Eye-strain' and Daily Working Time on VDTs

Several field studies reported the incidence of complaints called 'eye-strain' and the range or mean of the daily time spent on VDTs. The available data are plotted in Figure 1.

The definitions of 'eye-strain' were not clear and could have significantly differed among the cited studies; the calculation of the daily working time at VDTs might not have been consistent either. Nevertheless, the plotted picture is illustrative. It shows a linear correlation between the incidence of 'eye-strain' and the daily working time at VDTs. This relation could just as well be caused by a relation between length of VDT-use and the uniformity of work. Gunnarson and Söderberg (1983) found that an enlargement of the time that was spent on VDTs during the unchanged total working time caused an increase of eye-fatigue.

2.3. Physiological Reactions Using VDTs

Physiological changes that might eventually be caused by VDT-work have seldom been measured in field studies. Marvin (1981) did not find any before or after effects using visual

Figure 1. *Eye-strain and daily working time at VDTs—a representation of various field studies.*

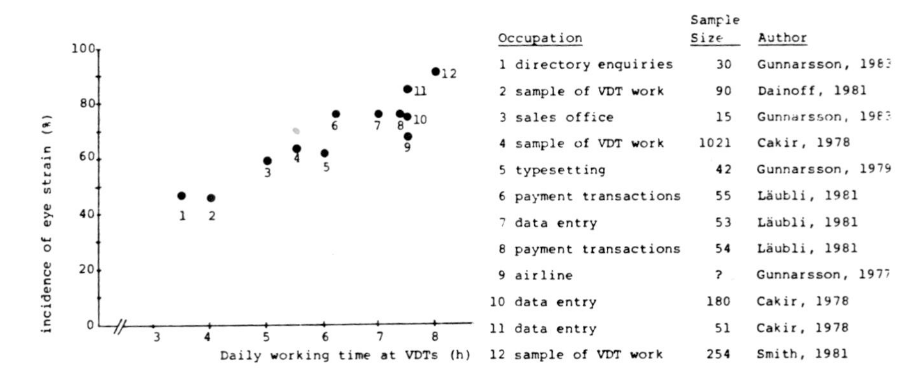

	Occupation	Sample Size	Author
1	directory enquiries	30	Gunnarsson, 1983
2	sample of VDT work	90	Dainoff, 1981
3	sales office	15	Gunnarsson, 1983
4	sample of VDT work	1021	Cakir, 1978
5	typesetting	42	Gunnarsson, 1979
6	payment transactions	55	Läubli, 1981
7	data entry	53	Läubli, 1981
8	payment transactions	54	Läubli, 1981
9	airline	?	Gunnarsson, 1977
10	data entry	180	Cakir, 1978
11	data entry	51	Cakir, 1978
12	sample of VDT work	254	Smith, 1981

screening tests (acuity, foria). Some rough measurements of visual acuity, foria and refraction, did not give indications for decreased visual functions compared to control groups not working at VDTs (Läubli, 1981). Gunnarson and Söderberg (1983) found a more pronounced decrease of the accommodation power when the daily portion of VDT-work was increased.

Laboratory experiments by Haider (1980) and by Östberg (1980) showed a greater influence on accommodation by VDT than by non-VDT-work. Muter *et al.* (1982) and Gould and Grischkowsky (1982) found a decreased reading speed from VDT compared to printed text.

In a recent laboratory experiment (Läubli and Hünting, 1984), 29 subjects had to read from a simulated visual display apparatus during 3 hours. The reading task itself led to an important increase of eye-fatigue which was accompanied by a decrease of several visual functions.

2.4. Medical Aspects

In ophthalmology only little is known about 'chronic astenopia'. In spite of the world-wide discussions about health hazards caused by VDTs, there exists no study examining these suspicions. Grootendorst (1983) measured visual acuity of 33 cm in 259 VDT operators. Follow-up studies 1½ and 3 years later showed a decrease of visual acuity. The author explained this finding by methodologically-caused falsifications and concluded that VDT-work did not cause a deterioration of vision. We consider these findings to be inconclusive and would like to suggest that a real decrease of near vision was detected.

2.5. Causative Factors

Considering eye-strain as an indicator of overload of vision, relations to specific working conditions may be elaborated. As an example Läubli (1981) found that in a group having high contrasts between the source document and the screen, long-lasting eye-fatigue appeared more often than in a group with moderate contrasts. In the laboratory, Bauer (1980) showed that black characters on light background were favourable to the reverse contrast. It was evaluated by error-rate and reading speed. But note that it was necessary to increase the phosphor-refresh rate to 90 Hz, to prevent visible flicker.

3. *Constrained Postures*

3.1. Swiss Studies on 'Daily Pains' in Various Parts of the Body

We have adapted the Japanese Questionnaire on Occupational Cervicobrachial Disorders to the Swiss situation. This questionnaire has now been used in eight different occupational groups. Subjects are asked about various feelings of discomfort in specific parts of the body. A survey of the results is presented in Figure 2.

FIGURE 2. *Daily pain in the neck or arm: results using the Swiss form of the Japanese questionnaire on OCD.*

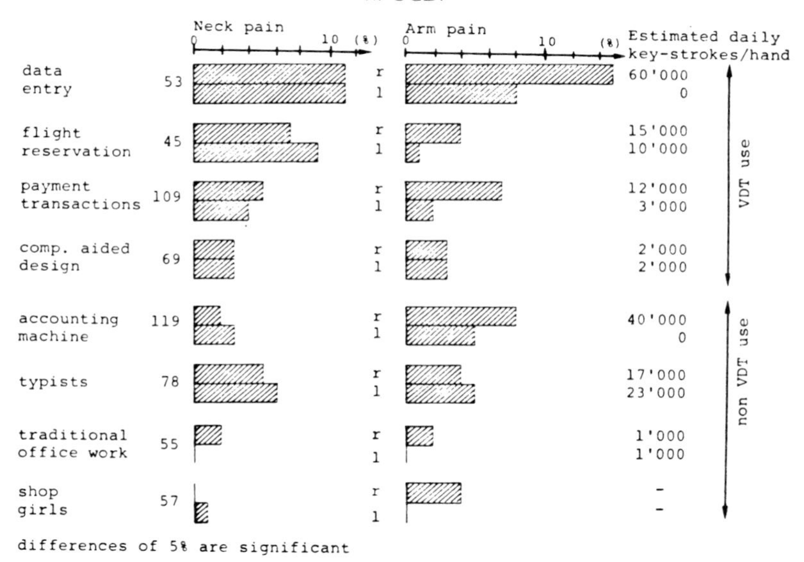

Comparing the incidence of arm pains and the daily key-strokes, we may note a certain relationship. In the group 'Payment transactions' the introduction of VDTs has led to a higher productivity and therefore to a higher keying rate although the number of key-strokes per transaction has been reduced.

3.2. Medical Findings

The medical impact of the presented musculoskeletal complaints is clearer than in the case of eye-strain. Some partly unpublished data by Läubli (1982) may illustrate the medical importance of the subjective complaints (see Table 4).

TABLE 4. *Anamnestic and clinical findings of rheumatic disorders in office work* (% of subjects).

Discomfort in the neck–shoulder–arm region	VDT-use		Non-VDT-use	
	Data entry (*n* = 53)	Payment transact. (*n* = 109)	Typists (*n* = 78)	Payment transact. (*n* = 55)
work-dependent pains or stiffness	73	31	30	25
clinical finding of tendo-myotic pressure pains	38	28	35	11
cause to visit medical doctor	19	21	27	13

As in the case for eye-strain, feelings of discomfort in the musculoskeletal system are generally frequent. In the greater part (95–60%) subjects judged the disorders to be work-dependent. A medical examination revealed the corresponding clinical symptoms of chronic high muscle tension and of irritations of tendon insertions. About every third person who had some problems, was consulting a medical doctor to cure his pains.

Incidences of neck or shoulder problems were reported by Starr *et al.* (1982), Sauter (1983) and Smith (1981). They found an incidence of 65% to 81% in the VDT-groups and of 48% to 55% in their controls.

3.3. Causative Factors

Sometimes well-known ergonomic rules are not followed in practice. In a field study (Läubli, 1981) the result of restricted leg room is illustrated (Figure 3).

In another study (Läubli and Hünting, 1984) back pains were less frequent in a subgroup working on a new and better-designed VDT workstation compared to a control using an older model with fixed keyboards.

4. *Conclusions*

VDTs themselves do not cause any health injuries, but as a result of the way in which they are used in practice, operators often suffer from eye-strain and constrained postures. In extreme cases a health risk must be prevented.

FIGURE 3. *Leg room and back pain in payment transactions ($n = 53$).*

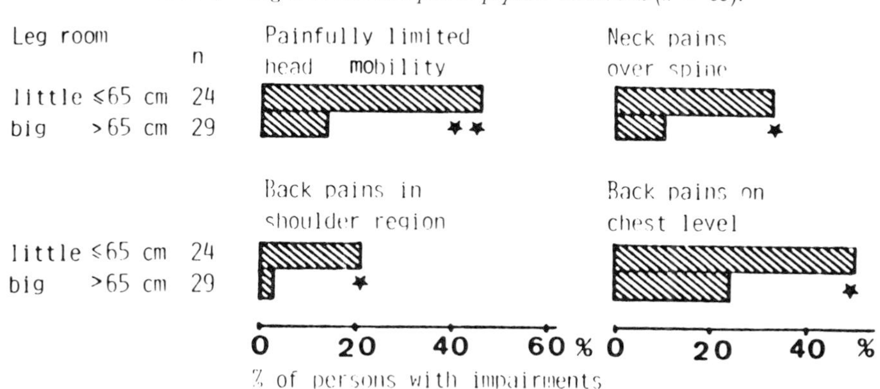

$^{*}p < 0{\cdot}05$; $^{**}p < 0{\cdot}01$; Mann–Whitney U-Test.

Because the introduction of VDTs into offices is always accompanied by various changes in the whole work situation, a comparison with control groups can never prove that the VDT, regarded as an isolated factor, is the cause of different findings. Control groups are necessary to validate the measuring procedures. Without a control group, an evaluation of subjective complaints about fatigue, eye-strain and pains is dubious.

The synopsis of the presented field studies shows that VDT-work is often accompanied by eye impairments and constrained postures. However, there may be other occupations which do not involve the use of a VDT, in which there are working conditions that are unavoidable and can cause even more problems. We still need a well-designed study that could throw light on the health injuries that it is suspected are caused by VDTs (or similar occupations). The consequences of 'chronic fatigue' are not clear.

When discussing possible health hazards, we should not forget that ergonomics offers several clues to improve working conditions. The visual displays are being rapidly improved. Studying people's behaviour working on VDTs, we may develop proper workstation designs (Grandjean, 1982). Sometimes a reduction of the working time may be adequate to compensate for an increased productivity. A good work organization should enable the personnel to take part in a variety of activities.

References

References that are listed in the survey by Dainoff (1982) are not printed here.

Dainoff, M.J., 1982, Occupational stress factors in visual display terminals (VDT) operation: a review of empirical research. *Behaviour and Information Technology*, **1** (2), 141–176.

Gould, J.D. and Grischkowsky, N., 1982, Doing the same work with hardcopy and with cathode ray tube computer terminals. *Proceedings Human Factors Annual Meeting*, 165–166.

Grandjean, E., 1959, Physiologische Untersuchungen über die nervöse Ermüdung bei Telephonistinnen und Büroangestellten. *Internationale Zeitschrift für angewandte Physiologie Einschliesslich Arbeitsphysiologie*, **17**, 400–418.

Grandjean, E., Hünting, W. and Pidermann, M., 1983, VDT workstation design: preferred settings and their effects. *Human Factors,* **25** (2), 161–175.

Grootendorst, G., 1983, Oog en Beeldscherm. *Tijdschrift voor Ergonomie*, **8** (2), 7–11.

Gunnarsson, E. and Söderberg, I., 1983, Eye strain resulting from VDT work at the Swedish Telecommunications Administration. *Applied Ergonomics*, **14**, 61–69.

Läubli, Th. and Hünting, W., 1984, Gesundheitsprobleme bei ganztägiger Bildschirmarbeit am Beispiel der Flugreservationskontrolle. In *Zeitschrift für Arbeitswissenschaft*, in the press.

Läubli, Th., Hünting, W. and Grandjean, E., 1982, Cervicobrachial syndrome in VDT operators. International Workshop on occupational neck and upper-limb disorders due to constrained work. Tokyo, Japan.

Maeda, K., Hünting, W. and Grandjean, E., 1980, Localized fatigue in accounting machine operators. *Journal of Occupational Medicine*, **22**, 810–816.

Muter, P., Latremouille, S.A., Theurniet, W.L. and Beam, P., 1982, Extended reading of continuous text on television screens. *Human Factors*, **24**, 501–508.

Sauter, St. L., Gottlieb, M.S., Jones, K.J., Dodson, V.N. and Rohrer, K.M., 1982, Job and health implications of VDT use: Initial results of the Wisconsin NIOSH Study. *Human Factors in Computer Systems Conference*, Gaithersburg.

Starr, St. J., Thompson, C.R. and Shute, St. J., 1982, Effects of video display terminals on telephone operators. *Human Factors*, **24**, 699–711.

Van der Heiden, G.H., Bräuniger, U., and Grandjean, E., 1984, Ergonomic studies on computer aided design. This vol., pp.119–128.

Health Aspects of VDT Operators in the Newspaper Industry

K. NISHIYAMA
Department of Preventive Medicine, Shiga University of Medical Science,
Tsukinowacho, Seta, Ohtsushi, Japan
M. NAKASEKO
Department of Hygiene, Kansai Medical University, Fumizonocho, Moriguchishi,
Osaka, Japan
AND T. UEHATA
Department of Hygiene, Kyorin Medical School, Shinkawa, Mitakashi, Japan

Abstract

A study on 437 VDT-operators in the newspaper industry disclosed that they have much more impairments of the eyes and of the body, more troubles of health and more stress factors in their work than keyboard operators. The majority of VDT-operators reported incorrect lighting characteristics of the VDT screens. Furthermore the wrong software for the operation of VDTs was pointed out. Eye impairments seem to be caused mainly by incorrect lighting characteristics of the VDTs. Body impairments also seem to be caused by them as well as by the non-adjustability of their chair heights.

1. *Introduction*

The present study on health aspects of VDT operators in the newspaper industry was carried out in spring 1983. This was the first large-scale investigation to analyse eye and physical impairments related to the display work compared to keyboard-operating work. Japanese VDT-operators are especially likely to have fatigue or impairments in the eyes and each part of the body because of the 'Kanji' characters that require more strokes and complicated patterns than alphanumeric characters used in most foreign countries.

2. *Methods*

Questionnaires were delivered to the editors and other workers of seven newspaper companies throughout the Japan Federation of Newpaper Workers' Union.

The 981 respondents were classified into three groups. The first group consisted of 437 VDT-operators, the second group of 122 keyboard-operators not using VDTs, and the third group consisted of 322 other workers with various jobs, some of which were quite different from the former groups.

3. *Results*

3.1. **Characteristics of Respondents**

Table 1 shows the characteristics of respondents in VDT-operators and keyboard-operators. Compared to keyboard-operators, VDT-operators were the young workers with short periods of employment years as well as the short-operating spell of the devices. With regard to other characteristics, there were no significant differences in body height, visual sight with and without glasses and the number of duty days per month.

TABLE I. *The characteristics and the working hours of the respondents.* **: $p < 0.01$ by Kolmogorov-Smirnov Two-sample tests.

	VDT-operators $n = 437$	Keyboard-operators $n = 122$
Age (years)	37·0	39·2**
Employment years	17·6	19·7**
Service length for present job (years)	7·9	11·1**
Operating minutes/day (usual)	278	262**
Operating minutes/spell	64	77***

3.2. **Impairments**

Nine items of eye impairments were significantly more complained of by the VDT-operators than keyboard-operators as shown in Figure 1.

Thirty-seven instances of eye impairment were classified into three factors by factor analysis. The first factor was associated with symptoms of eye-irritation and eye-pain, the second factor was associated with symptoms of visual distortion and colour vision, and the third factor was linked with symptoms of impaired accommodation. The VDT-operators

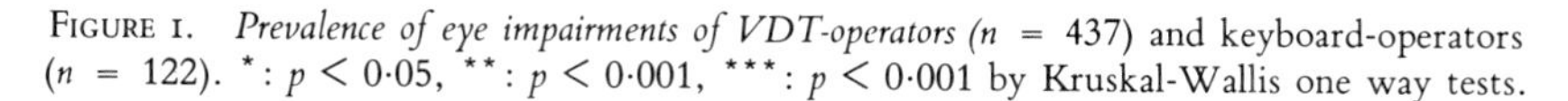

FIGURE 1. *Prevalence of eye impairments of VDT-operators* ($n = 437$) and keyboard-operators ($n = 122$). *: $p < 0{\cdot}05$, **: $p < 0{\cdot}001$, ***: $p < 0{\cdot}001$ by Kruskal-Wallis one way tests.

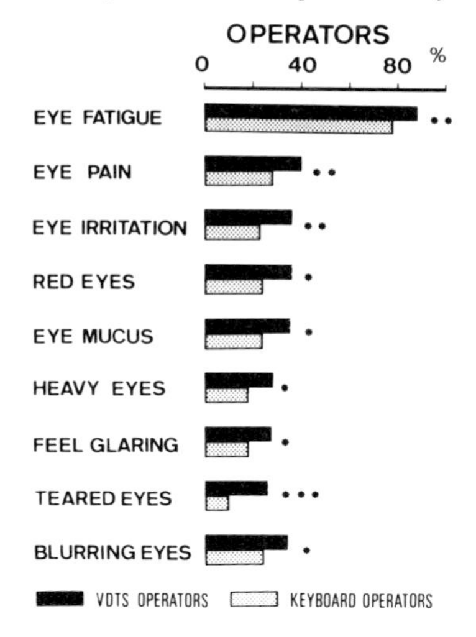

had, in general, the eye impairments relating to the factor of eye-irritation and eye-pain more than other factors.

Figures 2 and 3 show the percentages of VDT-operators and keyboard-operators with physical impairment (stiffness or dullness) of each part of the body.

As shown in Figure 2, there were significant statistical differences in the shoulders and upper back between VDT-operators and keyboard-operators. As far as stiffness or dullness in VDT-operators were concerned, the neck, shoulders, upper back and lower back were included

FIGURE 2. *Prevalence of stiffness or dullness of the area of neck, shoulders, upper back and lower back.* *: $p < 0{\cdot}10$, **: $p < 0{\cdot}05$ by Kruskal-Wallis one way tests.

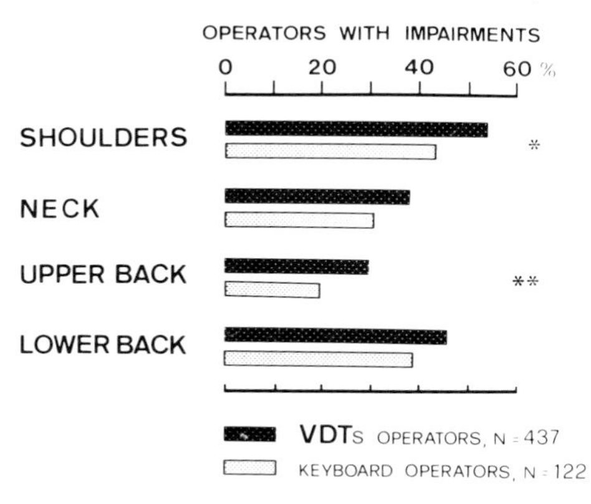

FIGURE 3. *Prevalence of dullness in the area of arms and fingers *: $p < 0{\cdot}05$* by Kruskal-Wallis one way tests.

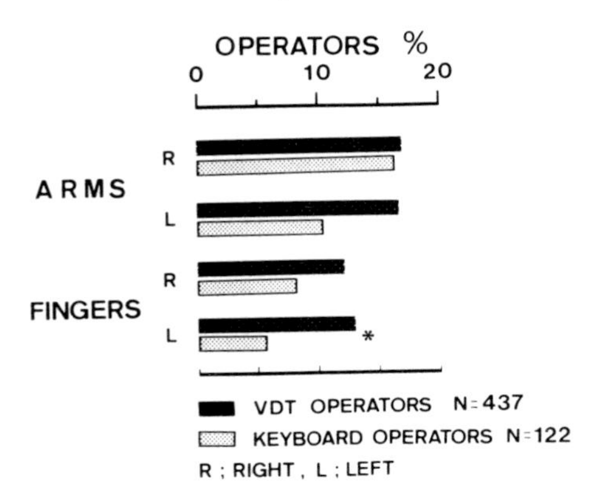

in one factor, and arms and fingers were included in another factor as shown in Figure 3. The first factor is related to the constrained posture from sitting in front of the display terminals. Another factor is linked with the keyboard-operating posture. The left hand was used for typing the keys while the right hand was used for operating the cursor keys or the joy stick.

3.3. Ergonomic Conditions

Of the operators 69–78% used pen touch instruments as the input devices and 45% of VDT-operators used the joy stick with their right hand. There are no significant differences in the ergonomic dimensions of the workstations reported between VDT-operators and keyboard-operators.

Of the VDT-operators 98% used the green characters on black backgrounds. Typical discomforts of the operators were linked with the lighting characteristics: flicker of the screen (54%), unclear characters displayed on the screen (57%), and distracting reflections on the face of the screen (57%). For software problems VDT-operators complained of long waiting times to get the output (34%), the frequent need to ask colleagues or refer to the operating manuals (34%), occurrences of incomprehensible errors (26%), and repetitive operations after computer troubles (19%).

3.4. Relation of the Impairments to the Ergonomic Conditions

Figures 4, 5 and Table 2 show that VDT-operators reported less impairments of the eyes than other VDT-operators, who were annoyed by the lighting characteristics of the display screens as well as ambient lighting conditions. The most interesting result is given in Figure 4.

FIGURE 4. *Prevalence of eye impairments of VDT-operators in relation to illumination of characters and their background on the screen.*

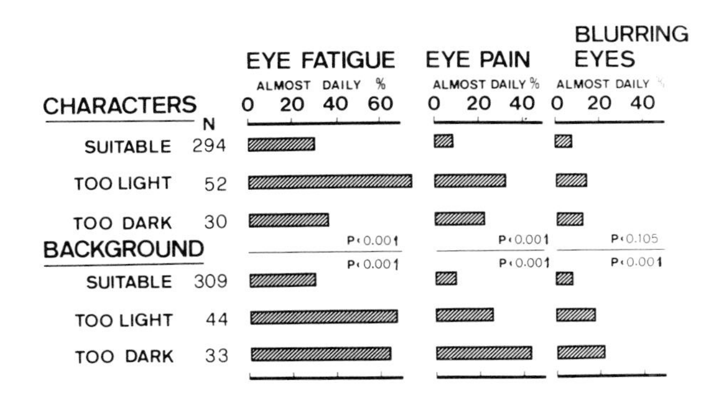

Too bright characters and too bright or too dark background resulted in the high prevalence of the impaired eyes of VDT-operators.

Table 2 shows that the illuminating characteristics of the characters displayed on the screen gave higher prevalence of eye-fatigue than of eye-pain and blurring eyes.

In addition to the display screen, Figure 5 shows that VDT-operators with annoying ambient lighting conditions had significantly higher prevalence of eye-fatigue and eye-pain.

Physical impairments in the neck, shoulders and back of VDT-operators were associated with whether or not VDT-operators were annoyed by the lighting characteristics of the screen.

Around 20% of VDT-operators were annoyed by the lighting characteristics of the screen and more than 20% of them complained of stiffness or dullness in the neck, shoulders, upper back and lower back. They complained more of constrained posture than the

TABLE 2. *Prevalence of eye impairments in relation to illuminating characteristics of VDT screen.* *: $p < 0{\cdot}05$, **: $p < 0{\cdot}01$, ***: $p < 0{\cdot}001$ by Kruskal-Wallis one way analyses.

		N	Eye fatigue	Eye pain	Blurring eyes
Flicker	Yes	208	49***	20***	13**
	No	176	26	8	6
Sharpness	Yes	206	50	21	13***
	No	179	26	8	7
Reflection	Yes	221	62***	22***	14**
	No	164	23	5	5
Character size	Yes	276	54***	24**	18
	No	108	33	11	7
Line space	Yes	297	50	26	15
	No	82	33	8	7

Figure 5. *Prevalence of eye impairments of VDT-operators in relation to ambient lighting condition.*

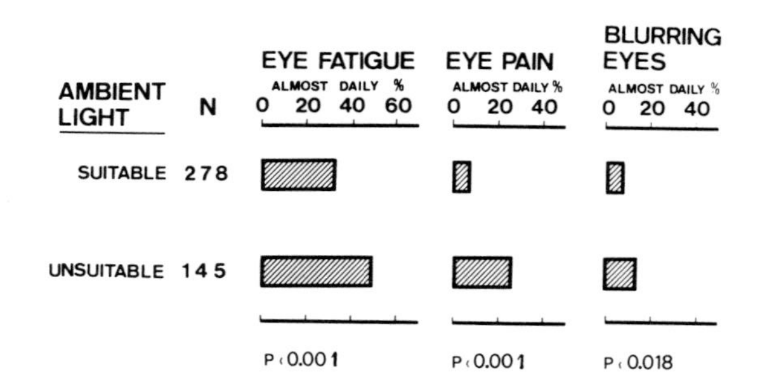

group with the suitable lighting characteristics of VDTs, although they were younger than others and there were no significant differences of dimensions about the workstations between the two groups. This suggests that unsuitable lighting characteristics of the VDT screen causes physical impairments of body parts maintaining the posture.

Ergonomic Studies on Computer Aided Design

G.H. van der Heiden, U. Bräuninger and E. Grandjean
Department of Ergonomics and Hygiene, Swiss Federal Institute of Technology,
8092 Zürich, Switzerland

Abstract

In a field study the hard- and software components of Computer Aided Design workstations were studied. A work-sampling study was carried out to characterize the use of keyboard, digitizer-tablet and screen. Subjective impressions of CAD software and CAD hardware, and of health aspects were collected by means of a questionnaire. Working-methods and working-postures were recorded on videotape. The two most important differences in comparing data- or word-processing terminals are as follows: (1) dynamic working methods result in the absence of constrained postures in CAD operators and allow frequent full-body exercise; (2) CAD-operators spend more time (46–68% of working hours) viewing the video display than the average data- or work-processing-operators. Some ergonomic recommendations have been deduced for the construction of CAD terminals, as well as for the ergonomic improvement of existing workplaces.

1. *The CAD-Terminal*

To obtain data on the ergonomic problem caused by the introduction of CAD terminals, a field study was carried out on engineering tasks using CAD in 1983. The study began with an investigation of how the workstations are presently used in various applications.

The 38 CAD terminals included in the study consisted of the basic workplace elements shown in Figure 1. These are:

1. A monochromatic graphic video display *screen* with a green phosphor. 27 were raster-scan displays, 11 vector-storage displays.
2. A digitizer command *tablet* with a pen as a cursor control to manipulate the screen graphics. The tablet could also be utilized as a user-definable menu to provide means of rapid command input.
3. An alphanumeric *keyboard*, used as a data or command entry device.

Of the 38 workstations 30 also included a 'hardcopy-printer unit'. A CAD-terminal table was a part of the standard-equipment, and it contained the CAD-system electronic

FIGURE 1. *The CAD-workstation (1 = graphic video display; 2 = digitizer command tablet and pen; 3 = alphanumeric keyboard; 4 = hardcopy unit; 5 = CAD-terminal table).*

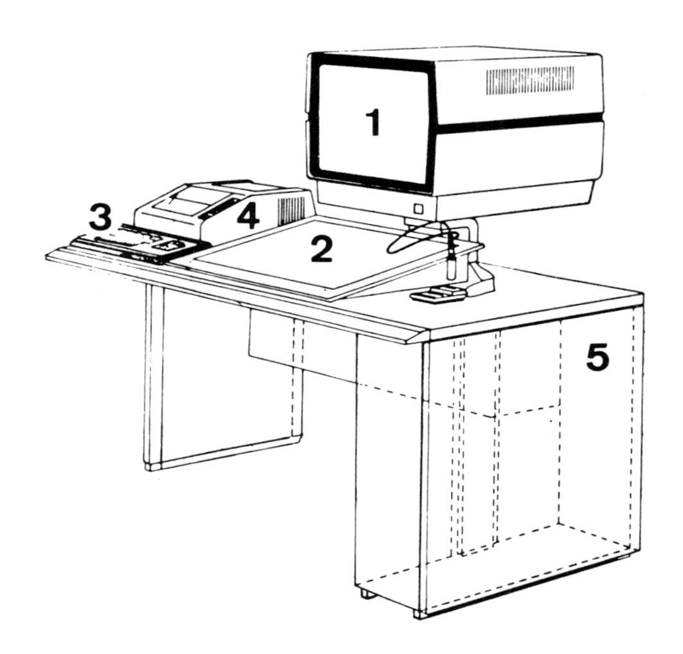

hardware. An extra reference table, as well as a special chair are being offered to CAD-users as an option. Of 38 workstations 35 had the reference table, 9 of the 38 had the CAD chair.

A maximum of four workstations were connected to one CAD central computer. It must be mentioned here that 65% of CAD-operators involved in the study also had their own drawing table. These were used primarily to prepare sketches for CAD drawing and to check CAD plots.

2. *Methods*

The field study, carried out in four Swiss industrial companies, consisted of three parts:

1. A work-sampling study to analyse the use of the basic workstation elements.
2. A survey of opinions of CAD-operators with questionnaires.
3. Measurements of lighting conditions as well as dimensions of CAD-workstations.

The aim of the time study (according to Haller-Wedel, 1962) was to analyse the direction of vision, the handling of documents, and the operating of tablet, pen and keyboard. Work sampling studies were carried out in the departments A, B and C, in order to reveal possible differences between CAD-applications.

Subjective experiences concerning health aspects, CAD hardware and CAD software, and workstation-layout were recorded by means of a questionnaire. Sixty-nine CAD-operators completed this questionnaire.

Measurements at the workplaces covered such dimensions as seat and table height above floor, seat-level-to-desk distance, and workspace and legroom provided. It also contained a determination of the ambient lighting conditions (illumination levels), and of the luminances and luminance contrasts in the visual field. The aim of these measurements was to compare objective measurement results with subjective impressions of the operators concerned, and with existing ergonomic guidelines.

3. *Results*

3.1. **Work-Sampling Studies**

Table 1 shows the results of the work-sampling studies for the departments A, B and C, each with a different CAD application. The work time of the operators is divided into 10 activities. Pauses and other absence times are excluded.

Table 2 comprises the total number of the activities: watching screen, operating keyboard, operating tablet, hardcopy manipulation and document manipulation.

The operators' view was frequently directed to the screen, between 46% and 68% of the work time at the CAD terminal. This means that the graphic display was being observed during 2·5–3·5 hours per day per operator on average. In this aspect CAD-operating is comparable to conversational-terminal operating, as studied by Elias and Cail (1983). This means that from an ergonomic viewpoint a CAD terminal can be regarded as a conversational terminal, and, in fact, has even a higher rate of screen-viewing.

3.2. **Opinions on Graphic Displays**

The frequent viewing of the screen, together with the perception of small details in the drawings indicate the need for high quality displays. Of the questioned operators 20% thought that quality of the displayed images was rather poor, 12% criticized the display for being unsharp. It is necessary to differentiate between raster–scan displays and vector–storage displays, as the latter had better sharpness and stability of the displayed images. Table 3 shows the operators' opinion about the legibility of the display.

3.3. **Operating the Keyboard**

Operating the keyboard accounted for 14% to 22% of the work time, and seems therefore rather infrequent in comparison to other data-terminals. The reason was that frequently

TABLE 1. *Results of the work-sampling studies I.*
(pwb = printed wiring board)

			% of working hours working interactively on CAD		
		Department:	A	B	C
		Number of observations:	1681	1670	1177
Activity		Activity numbers	Mechanical design	pwb design	Electrical schematics
watching screen (without manipulation)		1	15	28	20
operating keyboard	watching screen	2	6	7	9
	not watching screen	3	8	7	13
operating tablet	watching screen	4	31	33	17
	not watching screen	5	17	10	9
hardcopy manipulation	on terminal table	6	< 1	< 1	1
	on deposit table	7	< 1	< 1	< 1
document manipulation	on terminal table	8	6	5	12
	on deposit table	9	8	4	3
various		10	9	6	16
sum of observations			100	100	100

TABLE 2. *Results of the work-sampling studies II.*
(pwb = printed wiring board)

		% of working hours working interactively on CAD		
	Department:	A	B	C
	Number of observations:	1681	1670	1177
Activity	Activity numbers	Mechanical design	pwb design	Electrical schematics
watching screen (incl. manipulation)	1 + 2 + 4	52	68	46
operating keyboard	2 + 3	14	14	22
operating tablet	4 + 5	48	43	26
hardcopy manipulation	6 + 7	< 1	< 1	1
document manipulation	8 + 9	14	9	15

used command strings were integrated in the tablet menu and activated by the digitizing pen, instead of the keyboard. In electrical schematic design (department C) the keyboard was used more than in the other applications, because of high quantities of text which were input to the drawings.

Table 3. *Satisfaction of 69 CAD operators on the legibility of the graphic display.*

	% 'Poor image quality'	% Characters unsharp'	% 'Poor resolution'
raster scan displays (number of users = 53)	25	13	30
vector storage displays (number of users = 16)	6	6	19
totals (number of users = 69)	20	12	28

The use of two input mediums, keyboard and tablet, gives rise to interference problems. Since the tablet was the primary input medium, the keyboard was generally placed eccentric, to the left or to the right of the tablet (see also Figure 1). The generally practised both-handed keying induced a twisted position of the trunk. Of the operators 41% said that they placed the keyboard on top of the tablet in the case of large keying quantities. The sum of the height of the tablet (70 mm middle of the tablet) plus the height of the keyboard (51 mm home-row) resulted in an unnatural position of hands and wrists while keying.

3.4. Operating the Tablet

Operating the tablet accounts for 26% to 48% of the work time. Of 38 workstations 13 were equipped with inflexible pen-cables, that were often caught in the corners of the tablets. Of the operators 15% regarded the flexibility of the pen as bad.

CAD-users adapted their tablet menus to their own desires. Of the operators 35% felt a need for improvement of the tablet-menus they used and 26% stated that the menu contained a number of elements they never actually used.

3.5. Hardcopy and Document Manipulation

Although 30 of 38 workstations were equipped with a hardcopy printer, hardcopy useage was a negligible percentage of work time. Of work time 9–15% was dedicated to the manipulation of other kinds of documents, including drawings, CAD-manuals, books and notes. In the case of department A, the greater part of the manipulation of large engine drawings was carried out on the additional reference table. In both other departments (B and C) the majority of drawings had size A3 ($42 \times 29{\cdot}7$ cm^2) or smaller and could therefore be handled on the terminal desk.

No less than 73% of the operators reported that insufficient terminal desk work area was provided.

3.6. Various Activities

The 'various'-category activities accounted for the remaining 6–16% of the work time. Eating, drinking coffee or tea, and talking with visitors or neighbours were the predominant activities in this category. These 'breaks' could not be regarded as a relaxation period from a physiological point of view, as the operator often stayed alert for a system prompt.

Moreover, 39% of the operators felt that CAD-work allowed less real pauses than did comparable drawing without CAD.

3.7. Illumination Levels

The illumination levels, measured on the tablet, on the keyboard and on the reference table are reported in Table 4.

TABLE 4. *Illumination levels.*

	Illumination levels on the workplace (lx)	
workplace element	Median	90% range
tablet	125	15–440
keyboard	125	15–505
reference table	118	15–500

It is concluded that the mean illumination levels are rather low. This is primarily due to the fact that all departments took measures to reduce ambient light in order to improve legibility of the display and avoid reflections. Of 9 departments 7 disconnected all fluorescent lamps and 6 of 9 kept blinds closed permanently.

However, if illumination is cut to levels under 100 lux, reading of documents is likely to suffer. Of the operators 22% thought that illumination levels on their workplaces were too low, especially for reading purposes.

The point must therefore be stressed that the studied CAD displays could hardly be used under normal office environmental lighting conditions. Special arrangements must be made, unless lighting characters of graphic displays improve.

3.8. Luminance and Luminance Contrasts

Results on luminance contrasts in CAD terminals are presented in Table 5.

Luminance levels were found to be generally low due to the relatively low illumination levels at the workstations. It is generally accepted that luminance contrasts of more than

TABLE 5. *Luminance contrasts.*

Comparison	Luminance contrasts < 3:1	3:1–10:1	> 10:1	Maximum
documents: screen (n = 38)	0	79	21	23:1
tablet: screen (n = 38)	5	77	18	18:1
window: screen (n = 26)	0	23	77	1450:1

10:1 should be avoided. Such contrasts were measured mainly between windows and screen background, in 77% of the workstations facing a window. Although luminance levels were generally low, 36% of the operators complained about annoying reflections on the screen which rendered reading more difficult.

3.9. **Health Aspects**

Figure 2 shows the incidence of physical impairments, involving the answers of 'almost daily' and of 'occasional' pain. To allow a comparison with other professional groups the results of a study on constrained postures of VDU operators by Hünting *et al.* (1980) were adopted.

Results are presented for pains in the neck, shoulders, right arm and right hand, for CAD operators (n = 69), data-entry-terminal operators (n = 53), conversational-terminal operators (n = 109) and traditional office workers, who do not use any terminal at all (n = 55).

The incidence of pains in CAD-operators is lower than in all other terminal operators. We assume the dynamic working methods, including frequent exercise, result in an absence of constrained postures in CAD operators.

Figure 3 shows the incidence of eye impairments, again involving the answers of 'almost daily' and of 'occasional' complaints. Together with the results of the CAD operators those of three comparison groups from a study by Läubli *et al.* (1981) are presented.

The occurrence of eye complaints in CAD-operators seems of the order of the complaints of other groups working with any kind of terminal. Like the other operators, they show more eye troubles than do traditional office workers. We assume that this is caused by the relatively long and intensive display watching.

FIGURE 2. *Incidence of almost daily (= ■) and of occasional (= ▩) pain, of four professional groups:*

1	=	CAD operators	(n = 69)
2	=	data-entry-terminal operators	(n = 53)
3	=	conversational-terminal operators	(n = 109)
4	=	traditional office workers	(n = 55)

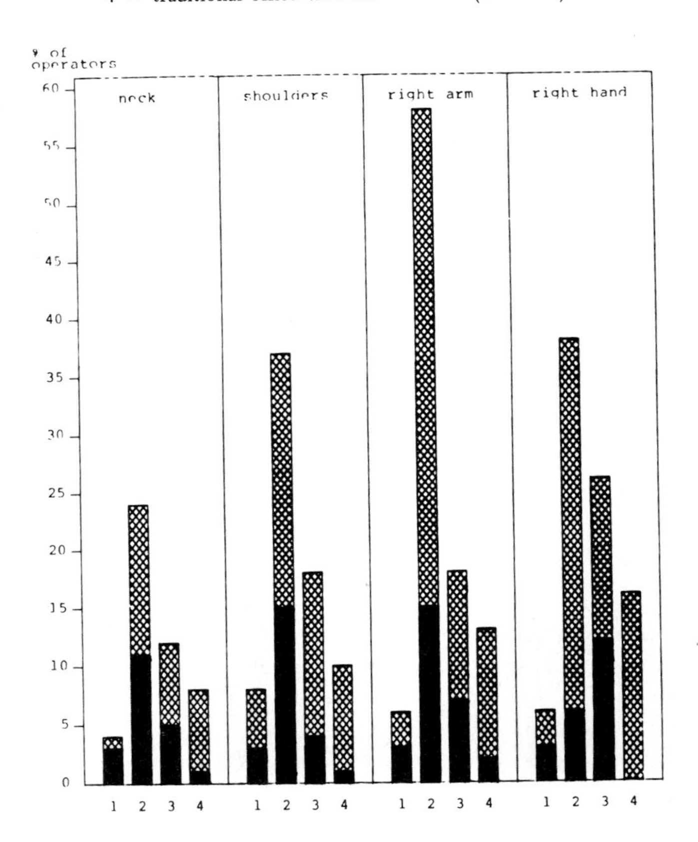

4. *Conclusions*

In the studied workstations the CAD-operators' view was directed on the screen about 50% of daily work time. Moreover, the perception of small details in drawings played an important role. This stresses the need for high quality displays.

In the opinion of their users, the studied display terminals could hardly be used under normal office-lighting conditions. Special environmental arrangements were made to avoid reflections on the screen. As ambient illumination levels were generally low, only the luminance contrasts screen to windows in the visual field were observed to exceed 10:1.

FIGURE 3. *Incidence of almost daily (= ▬) and of occasional (= ▩) eye complaints, of four professional groups:*

1 = CAD operators (*n* = 69)
2 = data-entry-terminal operators (*n* = 53)
3 = conversational-terminal operators (*n* = 109)
4 = traditional office workers (*n* = 55)

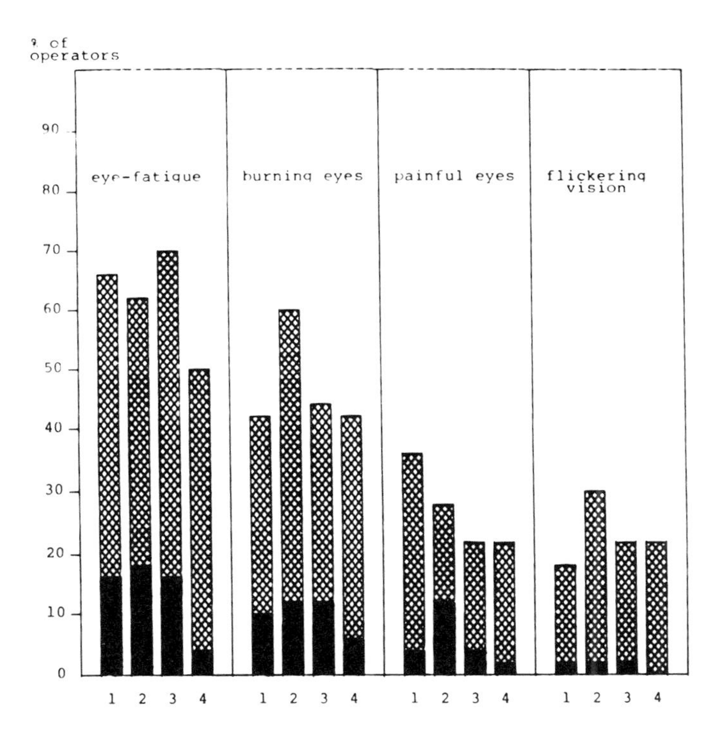

The studied CAD systems had a tablet and pen as primary input medium. The alphanumeric keyboard which served as a secondary input medium, was normally placed besides or on top of the tablet. Both positions resulted in unnatural body postures when keying. We conclude that the construction of a slim-line keyboard which can easily be put on and off the tablet is advisable.

Acknowledgement

The stimulus and funding which made this original study of the ergonomics of interactive graphics workstations possible was provided by a grant from Computervision Corporation of Bedford, Massachusetts, USA.

References

Elias, R. and Cail, F., 1983, Exigences visuelles et fatigue dans deux types de tâches informatisées. *Travail Humain*, **46**, 81–92.

Haller-Wedel, E., 1962, *Multimomentaufnahmen in Theorie und Praxis* (München: Carl Hanser-Verlag).

Hünting, W., Läubli, Th. and Grandjean, E., 1980, Constrained postures of VDU operators, in *Ergonomic Aspects of Visual Display Terminals*, edited by Grandjean, E. and Vigliani, E. (London: Taylor & Francis) pp.175–184.

Läubli, Th., Hünting, W., and Grandjean, E., 1981, Postural and visual loads at VDT workplaces. II. Lighting conditions and visual impairments. *Ergonomics,* **24**, No. 12, 933–944.

Predictors of Strain in VDT-Users and Traditional Office Workers

Steven L. Sauter
Department of Preventive Medicine, University of Wisconsin, Madison, Wisconsin, USA

Abstract

Survey and objective field data from 248 VDT-users and 85 traditional office workers were systematically analysed to separate and prioritize ways in which VDT-use might impact the jobs and the well-being of office workers.

Indices of both the general physical and psychosocial job environment were predictive of job dissatisfaction, mood disturbance, and general illness complaints. These attributes were more influential than individual characteristics, their effects were little different for VDT users than for non-users, and after adjusting for their effects, VDT-use contributed only to the prediction of mood disturbance—but the association was actually with improved moods. Despite the general lack of a direct or interactive impact on well-being, VDT-use was associated with significantly reduced job control and also less positive physical environmental conditions, an important predictor of all categories of strain.

Important predictors of specific somatic complaints among VDT-users included chair ratings for back–neck–shoulder discomfort, and corrective eyewear use and display quality ratings for eye-strain. Some commonly cited risk factors had little influence. Objectively assessed correlates of somatic complaints included workstation illumination for eye-strain, and gaze angle and keyboard placement for musculoskeletal strain.

1. *Introduction*

There is only limited empirical data which systematically links VDT-users' complaints to specific causal factors (Dainoff, 1982). In the present study, analyses are conducted to examine mechanisms suggested in the causal model shown in Figure 1. Paths A and B show, respectively, direct effects of individual attributes and job characteristics on strain. Path C shows an interaction of VDT use with job characteristics, and Path D depicts an additional *unique* effect of VDTs. The importance of these hypothesized effects on job dissatisfaction, mood disturbance, and illness complaints was explored through multiple regression

FIGURE I. *A theoretical model of stress–strain causation in VDT work.*

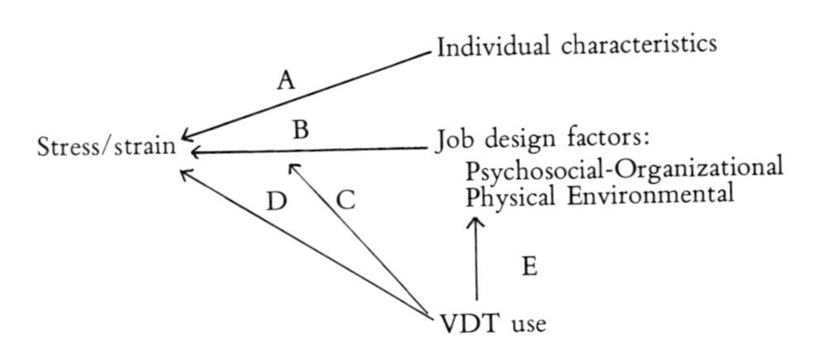

analyses performed upon survey data from a combined group of VDT users and their non-user counterparts. Path E shows a direct effect of VDT use on job characteristics themselves. This possibility was examined by a univariate comparison of job characteristics between VDT users and non-users.

Regression analyses were also undertaken to systematically examine the relationships of specific physical ergonomic and task design characteristics with eye and musculoskeletal complaints of VDT-users alone. Finally, objectively assessed physical workplace conditions and VDT-user anthropometric data were correlated (simple association only) with these somatic complaints.

Following is a synopsis of the results which are detailed more fully in the project final report (Sauter *et al.*, 1983).

2. *Methods*

Eighty-five traditional office workers and 248 VDT-users were administered (on-site) a survey seeking data on job psychosocial–organizational and physical environmental conditions, on individual biodemographic attributes, and on aspects of well-being. Objective data on working conditions were obtained for 25% of the VDT-users. Several of the research measures were ad hoc modifications of widely used scales of job characteristics and well-being (Caplan *et al.*, 1975; Insel and Moos, 1974; McNair *et al.*, 1971). Scale values for most multi-item measures were calculated from the first principal component of constituent items. Reliability coefficients are given parenthetically in all results tables.

The sample consisted of nearly all office workers with extensive office machine or VDT-use in six State of Wisconsin government agencies. The participation rate exceeded 90%. Breakdown by job activity was: data entry (34%), file maintenance–word processing (38%), clerical–secretarial (14%), supervisory–lead worker (7%), programmer (7%). The data entry and programmer subgroups were disproportionately VDT-users, and the clerical –secretarial subgroup disproportionately non-users. Thus, the sample reflects distributions in the current population of office workers. Mean hours of daily VDT and other office machine use in the two groups were 5·8 and 5·1 respectively.

3. *Results and Discussion*

3.1. **Effects of Individual Characteristics**

In separate regression analyses using the combined sample, measures of job dissatisfaction, mood disturbance, and general illness complaints were predicted from variables denoting whether or not workers: (1) resided in a family-type living arrangement, (2) were chief bread-winners for themselves or family, (3) were married, (4) used alcohol or (5) tobacco, and (6) received post-secondary education. Also included were variables denoting (7) number of people supported and (8) salary. Sex was not included because the sample was almost exclusively female. Age was not included because preliminary analyses already showed a strong age effect.

Results showed only marital status to be consistently predictive ($p < 0{\cdot}05$) of the three strain measures, but it accounted for less than 10% of the variance in each. Married workers reported reduced strain. The effect was the same for VDT-users and non-users alike (i.e., VDT use did not interact with marital status). As a control measure, age and marital status were therefore locked into analyses examining effects of working conditions.

3.2. **Effects of Working Conditions**

Table 1 describes the results of separate stepwise regression analyses to examine the influence of the job related variables shown to the left in Table 1 on the same three strain measures. Each column in Table 1 shows the results of the regression of the strain measure heading the column on all of the specified predictors.

Both psychosocial–organizational and physical environmental job attributes are seen to be predictive of all types of strain (path B, Figure 1). Curiously, cumulative daily time office machine use (VDTs for the VDT-users) predicts none. But perhaps the most notable aspect of these results is that when adjusting for other factors, VDT use/non-use (path D, Figure 1) is influential in predicting only mood disturbance (VDT-use is actually associated with improved moods). In regression models including interactions of the predictor variables, the variance explained in each strain measure was increased considerably, but the general effects shown in the top part of Table 1 were unchanged. Of particular interest, VDT use/non-use interacted significantly with only one variable in the prediction of all three strains (path C, Figure 1). Rising job demands were associated with increased mood disturbance for VDT-users, but not for non-users.

While VDT-use appears to have little unique or interactive impact on strain, it was associated with significantly less positive ratings ($p < 0{\cdot}01$) of the physical environment which in turn predict all three strains (path E, Figure 1). Job control, a predictor of dissatisfaction, was also reduced among VDT-users ($p < 0{\cdot}01$).

Finally, the combined effects of age and marital status (path A, Figure 1) contributed only 10–30% of the explained variance in these strain measures. Note that increasing age predicts reduced strain. This effect is commonly observed in occupational health research, and is usually attributed to a survival ('healthy worker') effect.

TABLE 1. *Results of the regression prediction of office worker strains.*

Predictors	Strain measures		
	Job dissatisfaction (0·92)	Mood disturbance (0·98)	Illness symptoms (0·95)
	Standardized regression coefficients		
Job control (0·84)[c]	−0·35	–[a]	–[a]
Social support (0·87)	−0·21	–[a]	–[a]
Workload demands (0·87)	–[a]	0·22	
Job future certainty (0·77)	−0·16	−0·28	−0·30
Physical environmental problem (0·84)	0·23	0·23	0·29
Job professional level	0·15	–[a]	–[a]
Hours daily office machine use	–[a]	–[a]	–[a]
VDT use (vs. non-use)	–[a]	−0·13	–[a]
Age	−0·25	−0·24	−0·19
Married (vs. not married)	−0·13[b]	–[a]	−0·20[b]
Model including interaction terms:			
VDT use × job demands	–[a]	1·38	–[a]
Variance explained (adjusted R^2)	51%	27%	34%

[a] indicates variable nonsignificant at $p \leq 0{\cdot}05$
[b] approximate value
[c] parenthetical values represent reliability coefficients (alpha, KR–20)

3.3. **Prediction of Eye and Musculoskeletal Complaints among VDT-users**

Table 2 shows the results of three separate stepwise regression analyses to examine the influence of the factors shown to the left in the table on eye and musculoskeletal strain. A number of variables commonly cited as potential risk factors in VDT-use do not contribute to the explanation of reported strain. A prime example is that cumulative hours of daily VDT-use is predictive of strain in only one case, but the effect is marginal ($p = 0{\cdot}046$) and contrary to intuition. Several effects, however, are consistent with conventional wisdom. Most notably, reports of display and ambient lighting problems are predictive of eye-strain, and chair comfort ratings and reports of problems with the workstation configuration are predictive of musculoskeletal strain. Note, however, the pronounced effect of job control, a more general psychosocial–organizational factor.

Of special interest is the corrective eyewear effect. Notice that the effect is significant after adjusting for age in prediction of eye complaints. Consistent with observations by other (Cakir *et al.*, 1978; Läubli *et al.*, 1981), VDT-users with corrective eyewear reported greater eye strain than those without. Consistent with Cakir *et al.* (1978), the effect was restricted mainly to users of monofocal lenses. These effects were much less evident in the control group.

TABLE 2. *Prediction of somatic complaints among VDT users.*

Predictors	Somatic complaints		
		Musculoskeletal	
	Visuo-ocular (0·87)	Upper torso (0·90)	Limbs/ extremities (0·78)
	Standardized regression coefficients		
Job control	−0·27	−0·26	−[a]
Job demands	0·18	0·19	−[a]
Display-image problems	0·25	NT	NT
Ambient lighting problems	$p = 0{\cdot}051$	NT	NT
Capacity for workstation lighting adjustability (0·54)[b]	−[a]	NT	NT
Extent of screen display viewing	−[a]	−[a]	−[a]
VDT-workstation physical configuration problems	NT	0·16	0·17
Capacity for VDT-workstation physical adjustability (0·66)	NT	−[a]	−[a]
Hours sitting/day	NT	−[a]	−[a]
Perceived chair comfort	NT	−0·21	−[a]
VDT response delays	−[a]	−[a]	−[a]
Discretionary breaks allowed	−[a]	−[a]	−[a]
Percent time interactive with VDT	−[a]	−[a]	−[a]
Data entry worker	−[a]	−[a]	0·20
File maintenance-word processing	−[a]	−[a]	−[a]
General clerical-secretarial worker	−[a]	−[a]	−[a]
Lead worker-supervisor	−[a]	−[a]	−[a]
Age	−0·21	−[a]	−[a]
Corrective eyewear used	0·25	0·17	0·19
Hours daily VDT use	−[a]	−0·15	−[a]
Variance explained (adjusted R^2)	21%	23%	10%

[a] indicates variable nonsignificant at $p \leq 0{\cdot}05$
NT indicates variable *not tested* in the regression analysis
[b] parenthetical values represent reliability coefficients (alpha, KR−20)

3.4. **Objective Correlates of Eye and Musculoskeletal Complaints**

Table 3 shows significant associations of musculoskeletal strain with both gaze angle and the keyboard adjustability variable. Strain scores increased with increasing gaze angle, and were reduced for VDT-users who had keyboards that were detached and adjusted away from the mainframe of the display unit. Note, too, the very strong association between the keyboard and gaze angle measures. In this case the gaze angle was much less severe (20° vs. 34°) for VDT users with detached−adjusted keyboards.

Table 4 shows that reported eye-strain is significantly associated with illumination at the workstation (variables 5 and 6). Workstation illumination (variables 4, 5, and 6) is positively

TABLE 3. *Anthropometric-workstation correlates of musculoskeletal complaints.*

Variables	1	2	3	4	5	6	7
1. Back posture	1·00						
2. Neck posture	0·50[b]	1·00					
3. Back support used	–[a]	–[a]	1·00				
4. Keyboard adjusted	–[a]	–[a]	–[a]	1·00			
5. Gaze angle	–[a]	–[a]	–[a]	0·72[c]	1·00		
6. Forearm angle	–[a]	–[a]	–[a]		–[a]	–[a]	1·00
7. Upper torso complaints	–[a]	–[a]	–[a]	–[a]	–[a]	–[a]	1·00
8. Limb/extremity complaints	–[a]	–[a]	–[a]	0·28[c]	0·39[d]	–[a]	0·50[d]

[a] indicates association nonsignificant at $p \leq 0{\cdot}05$
[b] Phi
[c] Eta
[d] Pearson *r*

TABLE 4. *Lighting related correlates of visuo-ocular complaints.*

[Tabled values = Pearson *r*]

Variables	1	2	3	4	5	6	7	8	9
1. Visuo-ocular complaints	1·00								
2. Screen background luminance	–[a]	1·00							
3. Keyboard luminance	–[a]	–[a]	1·00						
4. Illuminance at screen	–[a]	–[a]	–[a]	1·00					
5. Illuminance at keyboard	0·31	–[a]	–[a]	0·82	1·00				
6. Work surface illuminance	0·21	0·52	–[a]	0·73	0·87	1·00			
7. Screen-background luminance ratio	–[a]	0·71	–0·22	–[a]	–[a]	0·21	1·00		
8. Screen reflectance	–[a]	0·67	–[a]	–0·31	–0·39	–[a]	0·58	1·00	
9. Screen area with glare	–[a]	0·31	–[a]	–[a]	–[a]	–[a]	0·51	0·40	1·00
10. Average background luminance	–[a]	–[a]	–	0·44	0·46	0·42	–0·45	–0·31	–0·49

[a] indicates association nonsignificant at $p \leq 0{\cdot}05$

associated with background luminance levels. Thus, ambient lighting level is implicated as a cause of eye-strain. Interestingly, display related variables (luminance, screen–background contrast, reflectance, and glare) are not directly related to strain, but they tend to be related to ambient lighting indicators.

While the lighting, keyboard, and gaze angle effects seen in Tables 3 and 4 may come as no great surprise considering the importance generally attributed to these factors, a close reading of the VDT literature reveals little prior empirical demonstration of the relationships shown here. A word of caution is that unlike the previous regression results, Tables 3 and 4 show only simple relationships.

Acknowledgements

Other key members of the research team included Mark Gottlieb, Karen Jones, Kathryn Rohrer, and Vernon Dodson of the Department of Preventive Medicine, and William Raynor, Department of Statistics, University of Wisconsin. The study was funded entirely under contract from the US National Institute for Occupational Safety and Health (NIOSH), Michael J. Smith, project officer. However, the opinions expressed here are those of the author, and not necessarily of NIOSH.

References

Cakir, A., Reuter, J.H., Von Schmude, L., Armbruster, A. and Shultie, L.B., 1978, *Investigations of the Accommodation of Human Psychic and Physical Functions to Data Display Screens in the Workplace* (Bonn, West Germany: Der Bundesminister für Arbeit und Sozialordnung).

Caplan, R.D., Cobb, S., French, J.R.P., Van Harrison, R., and Pinneau, S.R., Jr, 1975, *Job Demands and Worker Health* (Cincinnati, Ohio: National Institute for Occupational Safety and Health).

Dainoff, M.J., 1982, Occupational stress factors in visual display terminal (VDT) operation: A review of empirical research. *Behavior and Information Technology*, **1**, 141–176.

Insel, P. and Moos, R., 1974, *Work Environment Scale—Form S* (Palo Alto, California: Consulting Psychologists Press, Inc.).

Laubli, Th., Hünting, W. and Grandjean E., 1981, Postural and visual loads at VDT workplaces: Lighting conditions and visual impairments. *Ergonomics*, **24**, 933–944.

McNair, D.M., Lorr, M. and Droppleman, L.F., 1971, *The Profile of Mood States* (San Diego, California: Educational and Industrial Testing Service).

Sauter, S., Gottlieb, M., Rohrer, K. and Dodson, V., 1981, *The Well-being of Video Display Terminal Users: An Exploratory Study* (Cincinnati, Ohio: National Institute for Occupational Safety and Health).

Experiences of Routine Technical Measurement Analysis of VDT Working Places in the Field of Occupational Health Service

R. von Kiparski
Institut für Arbeits- und Sozialhygiene (IAS), gemeinnützige Stiftung, Daimlerstraße 7, 7500 Karlsruhe, F.R. Germany

Abstract

Within the field of occupational health service covered by the IAS, which attends to more than 230 factories and public offices in the southern German area, light technical measurements are made routinely for the objective analysis of the visual strain at VDT working places.

In detail these measurements include the registration of illuminance, luminance ratios, luminance dispersements, oscillation coefficients, etc. In some cases, due to the request of the work council, the X-ray emissions were measured also.

The results of these cross-section testings are compared with the ergonomic recommendations. The most commonly voiced objections are discussed.

The possibilities to improve the light-situation, and therefore to reduce the working stress at minimal cost is explained with the aid of case examples.

1. *Introduction*

The Institut für Arbiets- und Sozialhygiene (IAS) is an inter-enterprise occupational health service at the moment servicing 230 companies in the south German area with more than 60 000 employees. A very important duty of this department is the technical measurement of VDT working places, and this is receiving more and more attention.

The aim of this study is to introduce a simple and ergonomic way to optimize the lighting situation at the VDT working place (von Kiparski, 1982).

2. *Methods*

Of the 620 VDT work places in companies serviced by the IAS 177 were included in the study. These places represent industry, banking and printing businesses. Measurement of luminance and illuminance was conducted with the help of a universal photometer. The error of measurement was below 5%, sufficient for practical examinations. The measurable angle to register the luminance is 1°; luminance of character as well as background luminance of the screen was recorded by a glass fibre optic with a diameter of 1 mm. We used a digital storage oscilloscope for the presentation of the oscillation coefficient of characters.

3. *Results*

The mean working times at the examined VDT working places show that 43% of the employees are occupied at the VDT for more than 4 hours. For the other 57% the total working time is not long enough to cause stress.

Figure 1 shows the measured horizontal illuminance of the work place in form of a histogram. No objections arose for about 50% of the working places. For 20% the illuminance was too low and for about 30% the illuminance was too high.

FIGURE 1. *Distribution of illuminance (Ordinate: % of work places).*

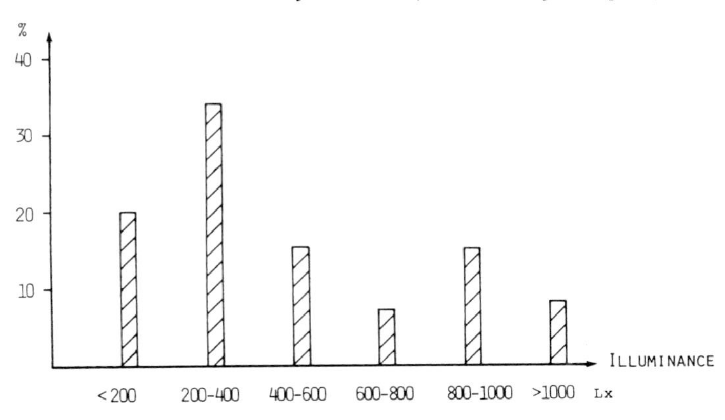

The requirement of the luminance ratio of < 5:1 of document to screen is an especially serious criterion using conventional light letters on a dark background (see Figure 2). This was accomplished if the mean luminance of the document was not too high. For already installed VDTs this is also the only possibility of improving the luminance ratio.

The measurement results of the luminance ratio of document to keyboard are rather

FIGURE 2. *Distribution of mean luminance ratio document/display.*

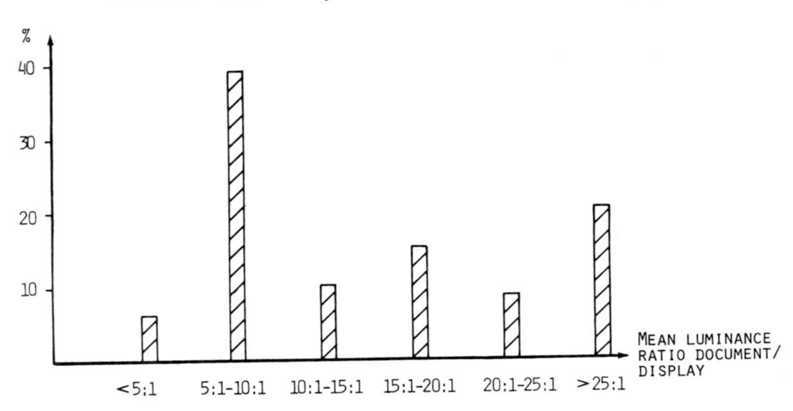

more favourable than those of document to screen. If black keyboards were used or the keyboards were placed in the shadow, the most unfavourable values resulted. Since data entry in most cases is done blindly (without looking at the keyboard), these measurement values did not receive much importance at our working place examinations.

FIGURE 3. *Distribution of luminance dispersement on screen by ambient light.*

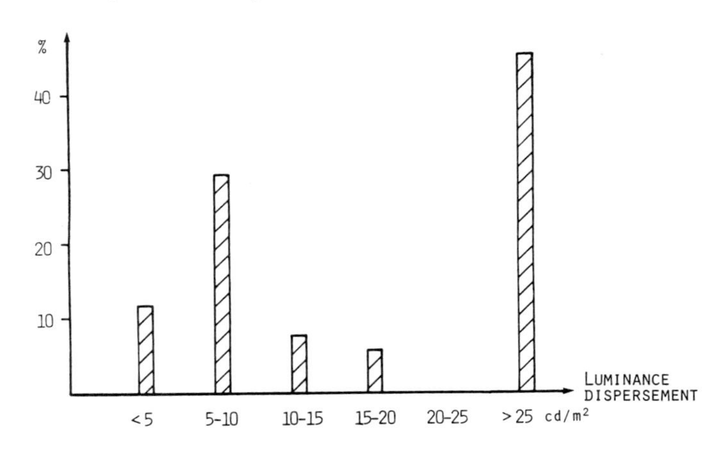

The distribution of luminance dispersement on the screens by the room illumination has two maxima (Figure 3). Because of great glare sources caused by windows the dispersement is above 25 cd/m^2 or because of not sufficient screened room illumination the dispersement is mainly in the range of 5 to 10 cd/m^2. In only 13% of the working places was the recommendation complied with.

The distribution of the luminances in the surrounding field of the VDT shows a similar characteristic (Figure 4). The great number of working places with luminances above 200 cd/m^2 indicates how even this especially disturbing influence is often ignored.

FIGURE 4. *Distribution of luminance in the surrounding field of the VDT.*

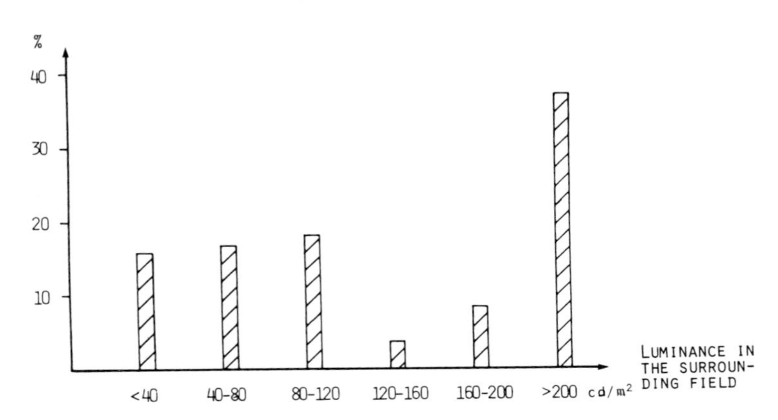

Our measurements of the oscillation coefficient in general resulted in uniformity figures between 0·5 and 0·6. Uniformity figures below 0·1 were not measured at all; therefore the criterion of Läubli *et al.* (1980) about oscillations can be called minor.

Because of fear expressed by members of the staff and the work council that VDTs would emit X-rays to a high extent, we frequently measured the X-ray radiation of the VDT screen at a distance of 5 cm from the surface of the screen. As expected the results stayed below 0·5 mR/h. This confirms the well-known fact that the X-ray radiation does not increase because of a functional disturbance or ageing of the apparatus.

4. *Discussion*

The illuminance measurements conducted by the IAS aimed to objectively study the strain and also provide and control possibilities for improvement of these working places.

While carrying out the measurements it is therefore necessary to document the marginal conditions such as arrangement of light fixtures, location of windows, position of VDTs at working places. If one intends to use the present measurement values for improvements one has to consider the relatively intense coupling of the measurement variables. For example the horizontal illuminance influences the ratio of direct to indirect illuminance, the ratio of document to keyboard as well as possibly the luminance dispersement on the screen. Luminance of characters and character contrast as well as luminance of the screen are closely coupled with each other.

References

Cakir, A., Hart, D.H. and Stewart, T.F.M., 1980, *Bildschirmarbeitsplätze* (Berlin: Springer).

Hoppe, K., Kiparski, R. von, Marschall, B. and Massmann, W., 1980, Die Arbeitsweise des Instituts für Arbeits- und Sozialhygiene (IAS) bei der arbeitsmedizinischen Betreuung von Klein- und Mittelbetrieben unter besonderer Berücksichtigung der arbeitshygienischen Meßtechnik. *Handbuch für Arbeitssicherheit*, Gruppe 10 105–156 Haufe, Freiburg/Br.

Kiparski, R. von, 1981, Beleuchtungstechnische Beurteilung von Bildschirmarbeitsplätzen. *Ergo-Med*, **5**, 2, 22–24.

Läubli, T., Hünting, W. and Grandjean, E., 1980, Visual impairments in VDT operators related to environmental conditions. *Ergonomic Aspects of Visual Display Terminals* (London: Taylor & Francis).

The Development of a Relevant Ergonomic Checklist for Designers of the New Technology Office

J. O'NEILL * AND R. BIRNBAUM
Ergonomics Unit, University College London, UK
*Now at: 17 Upper Culver Rd, St Albans, UK

Abstract

Recent developments are highlighting two major trends in office design, namely the increasing use of new technology and the trend towards flexible multifunctional office spaces. The subsequent specialized ergonomic considerations must be input to the design process, and those responsible for office design (architects, space planners and furniture designers) have indicated a preference for a practical design aid such as an annotated checklist. Existing ergonomic checklists were found inadequate for this purpose, and this study was set up to develop a new checklist to meet the need.

1. *Ergonomics in Offices*

For many years it has been recognized that the office, as a workplace, has as much potential need for ergonomic expertise as have the more traditional areas such as the production line. Fucigna (1967) points out the many interrelated factors that require optimization in the office environment, including the job procedures, the physical work environment, the location of work areas, and the equipment and furnishings. Preferably, ergonomics should be applied at the design stage of all aspects of the office work environment: there may otherwise be consequences in terms of implications for job performance, job satisfaction, and health and safety.

The traditional office environment has received some attention in the past, and much data is available for reference. Office designers are aware of functional recommendations for workplace design such as desk and chair height, illumination for writing tasks, postures for typing, and so on (Grandjean, 1980).

However, recent developments are highlighting two major trends in office design that, individually and interactively, are giving rise to concern. These trends are firstly, the increasing use of new technology in offices; and secondly, the increasing trend towards multifunctional, multi-unit, flexible office spaces. Traditional design approaches need to be revised as a result, and the ergonomic dataset enhanced alongside (Worthington, 1982).

Since the 1960s landscaped (or open-plan) offices have become more common, and today account for 40% of all office space in the US (Gaskie, 1980). The motivations that began the open-plan movement (increase in worker communication; economy in space, more democratic management style (Hedge and Travis, 1975)) still exist today, but added to them are needs for increasing flexibility in workspace usage; shared use of complex equipment; and open-plan areas with potential for office, storage or shop-floor use. Add to these the nature of the new technology equipment itself with its specialist environmental demands (Stewart, 1976), and the office designer obviously needs more support from the ergonomist than ever before.

2. *Ergonomics in Design*

Singleton *et al.* (1967) support the premise that the role of ergonomics in design must be constructive and participative and not simply a means by which to criticize the solutions of problems arrived at by engineers and designers. Moreover, the ergonomist cannot "recommend solutions which are more expensive than the traditional version unless he is prepared to provide evidence which justifies the additional expenditure".

Any ergonomic advice to designers, whether given by an ergonomist or through an ergonomic text, must therefore persuade the designer of the value and critical nature of ergonomic criteria. Additionally, some indication of the method of applying the information must be included. Since much office design work is the responsibility of architects, space planners, interior designers, or office furniture designers, and most ergonomics input tends to be from ergonomic texts and datasets, such design aids must also be practical, versatile and complimentary to the iterative design method.

Furthermore, the assumption must be made that complex, scientific, disruptive ergonomic research techniques are not likely to be used by office designers. Data about the design, or potential design, may nevertheless be gathered through interviews and observations of users, and through physical measurements of the working environment, such techniques being familiar.

3. *Use of Checklists*

The most efficient way to provide a comprehensive pertinent summary of ergonomic information for design problems is in the form of a checklist, which can guide, systemize and force the asking of pertinent questions (Easterby, 1967). The checklist questions can be phrased to include specific numeric data, or elucidate numeric data for comparison with datasets, and can also include interpretive statements that underline their critical nature.

There can be some disadvantages to checklists. Designers tend to have a preference for graphic or pictorial information (Meister and Farr, 1967) and by their nature checklists tend to be verbal, although there are examples of checklists backed up by diagrams, such as that of Cakir *et al.* (1979). Furthermore, the design process is an iterative one, and checklists work best when used in a formal and serial fashion (as in a machine maintenance check, for example). Certain aspects of the design process, such as the formulation of the design specification, may be missed altogether as the designer may limit his use of the checklist to certain stages in the design.

Notwithstanding these concerns, the checklist is a practical tool which is straightforward to use and as an *aide mémoire* or a data gathering tool it can be invaluable. Furthermore, designers are familiar with the use of checklists in other aspects of design such as manufacturing constraints. This study therefore concerns itself with the development of an ergonomic checklist for use by office architects and designers during workplace design, particularly for those multifunctional offices intended to include new technology installations.

4. *The Present Study*

The study involved the following exercises:

1. Review of existing relevant checklists.
2. Practical evaluation of these checklists in a multi-functional office environment.
3. Development of a new checklist combining valuable points from the published checklists* with new checkpoints.
4. Evaluation of the new checklist by architects and office designers.

Three particular checklists were selected for evaluation: 'Open-plan offices' (Kraemer *et al.*, 1977); 'Designing systems for people' (Damodaran *et al.*, 1980); and 'Visual display terminals' (Cakir *et al.*, 1979). Each covered some aspects of workplace design, either for open-plan design, or for new technology installations. No one checklist fulfilled all necessary aspects with the required approach or emphasis.

The practical part of the study took place in a large open-plan multifunctional office where computers and their peripherals were extensively used. A comprehensive assessment of the work environment was undertaken, using simple techniques familiar to non-ergonomists, to include factors such as air temperature, humidity and air movement; physical dimensions of furnishings and equipment; postural analysis; illuminance, reflectance and daylight factor; space and layout analysis; noise and acoustics. Interviews with the office staff and the office designers, and an opinion survey, gave information on the nature of the jobs and the tasks in each workstation, and highlighted design approaches and inadequacies.

The study showed up many key inadequacies in the office design, some due to the open-plan design (such as lack of privacy), some due to the new technology (such as worksurfaces too narrow to hold equipment) and some that may be attributed to an interaction between the two (such as the noisy printer causing irritation to all the office occupants). It was clear

* Copyright permission granted by publishers.

that there were limitations in ergonomic expertise and/or training among those responsible for the design of such a work environment. Moreover, maximum use had not been made of user participation and involvement in the design. As with Stewart's findings (1980) in a similar study, many ergonomic errors were basic in nature and could have been simply remedied with minor modifications.

This practical study simulated one intended use for the New Technology Office Checklist: the evaluation of existing office facilities in terms of their suitability for new technology and for multifunctional use. The other potential design objectives that could utilize the checklist, such as the design of a multifunctional workspace from scratch, or the design of flexible functional office furniture, or the conversion of cellular offices into open-plan, were borne in mind when the new checklist was constructed. Hopefully, the checklist design would be flexible enough to cater for a wide range of users. A variety of potential users were asked to comment on the new checklist and its utility, and various amendments made in response to their views. This evaluation is still continuing.

5. *Structure of the New Checklist*

The checklist does not presume any previous knowledge or expertise in ergonomics. Wherever possible, the checkpoints are quantitative, so as to be directly applicable in the design process. In many cases, they are phrased to imply the consequences to the user of the design if the ergonomic criteria are comprised. For example, the question 'Can each person view to infinity . . . so as to rest the eyes occasionally?' implies that the well-being of an office worker is enhanced by designing for a variety of viewing distances.

The checkpoints are phrased as questions, each biased towards a 'Yes' answer. The user is instructed that 'No' answers (including vague ones such as 'Maybe' or 'Sometimes') will alert him to the need to pay further attention to those points, perhaps with the aid of ergonomics texts or professional experts who can give advice.

The questions are to be directed at the most pertinent respondent, who may be the client or user of the design, or another member of the design team, or oneself: and where possible with reference to full-scale mockups or models. The checklist user himself determines the selection of topics to question at each stage of design, and the ordering of the questions, changing these according to the particular task in hand and the constraints acting upon it. Thus the checklist can be used in a flexible and iterative manner.

The checkpoints are stated at two levels, labelled 'critical considerations' and 'secondary considerations'. The idea here is that in selecting and using checkpoints, the checklist user must have some idea of how critical a particular checkpoint is, and the possible consequences to the workers if the checkpoint demands cannot be fulfilled. All design processes involve compromises, and where a person's well-being and job satisfaction are concerned, these compromises must be chosen carefully, on a scientific basis. The designer would know to give priority to those points under 'critical considerations' if compromise is required.

The checklist sections cover room layout, workstation design and seating, siting of equipment and job aids, and the physical work environment. The sections are not intended to be chronological, nor are the checkpoints within each section. About 50% of the checkpoints are drawn from the three published checklists mentioned above, and 50% are novel checkpoints developed by the present authors from the practical study and from exchange with design professionals.

6. *Conclusions*

The New Technology Office Checklist is an important addition to the ergonomic techniques available for use by non-ergonomists such as architects, space planners an furniture designers, in a fast-developing area where there is a dearth of ergonomic knowledge. It is flexible enough to compliment the iterative design process, comprehensive enough to cover the needs of a variety of workplace design objectives, and straightforward enough for use by non-scientists. It maximizes the 'self-help' approach to ergonomics, but does not discount user-involvement nor the rôle of the professional ergonomist when required.

Copies of the New Technology Office Checklist may be purchased from the authors.

Acknowledgements

This study was supported in part by Apple Computers Inc., Cork, and Westinghouse Inc., Dublin (Ireland).

References

Cakir, A., Hart, D.J. and Stewart, T.F.M., 1979, *Visual Display Terminals* (Chichester: John Wiley).

Damodaran, L., Simpson, A. and Wilson, P., 1980, *Designing Systems for People* (Manchester: National Computing Centre).

Easterby, R.S., 1967, Ergonomics checklists: an appraisal. *Ergonomics*, **10**, 5, 549–559.

Fucigna, J.T., 1967, The ergonomics of offices. *Ergonomics*, **10**, 5, 589–604.

Gaskie, M.P., 1980, Towards workability of the workplace. *Architectural Record*, August.

Grandjean, E., 1980, *Fitting the Task to the Man* (London: Taylor & Francis).

Hedge, A. and Travis, N., 1975, *User Evaluation of Open-Plan Offices* (Birmingham: University of Aston Applied Psychology Department publication No. 195).

Kraemer, Sieverts and Partners, 1977, *Open-Plan Offices. New Ideas, Experiences and Improvements* (London: McGraw-Hill).

Meister, D. and Farr, D.E., 1967, The utilisation of human factors information by designers. *Human Factors*, **9**, 71.

Singleton, W.T., Easterby, R.S., and Whitfield, D., 1967, *The Human Operator in Complex Systems* (London: Taylor & Francis).

Stewart, T.F.M., 1976, Human factors in the use of visual display units. *Visual Display Units and their Applications*, edited by D. Grover (Guildford: IPC Science and Technology Press), pp.153–176.

Stewart, T.F.M., 1980, Practical experiences in solving VDU ergonomics problems, in *Ergonomic Aspects of Visual Display Terminals*, edited by E. Grandjean and E. Vigliani (London: Taylor & Francis) pp.233–240.

Worthington, J., 1982, Workspace: loose-fit and specific. *Architects Journal*, August.

Health Hazards of VDTs

R. PINEAULT
Département de médicine sociale et préventive, Université de Montréal, Canada
AND D. BERTHELETTE
Institut de recherche en santé et en sécurité du travail du Québec, Canada

Abstract

This paper presents the conclusions arrived at by a multidisciplinary task force, set up by l'Institut de recherche en santé et en sécurité du travail du Québec. The conclusions concern visual, musculoskeletal, pregnancy, skin, neurological and stress problems suspected to affect VDT operators in relation to work organization, environment and equipment design hazards.

1. *Introduction*

L'Institut de recherche en santé et en sécurité du travail du Québec formed a task force with the mandate to investigate the relationships between the health problems suspected to result from VDT exposure and to formulate recommendations on prevention measures to be established in the workplace as well as studies that should be conducted in Quebec on specific subjects.

Two principles have guided the work of the task force:

1. The problems were analysed from different perspectives reflecting the multidisciplinarity of the group.
2. The problems were anlysed in a public health framework.

2. *Method*

The members of the task force had reviewed more than 200 publications dealing with the potential health hazards of video display terminals. The purpose of this work was to assess

* This paper is not an official position of the IRSST. It merely reflects the views of the authors.

the importance of health problems, both in terms of the frequency and severity, reported among VDT operators and to identify the health hazards associated with these problems.

In order to assess the degree of proof of the existence of a relationship between health problems and risk factors, the scientific value of the studies was evaluated according to the following scale:

2.1. Epidemiological study
2.2. Fundamental research
   1. On human subjects
   2. On laboratory animals
2.3. Clinical reporting of cases

The conformity of the investigations to methodological criteria has also been evaluated.

3. *Results*

3.1. **Visual Problems**

Only one researcher has ever reported clinical cases of cataracts among VDT operators. Scientific data show that the levels of radiation (infrared and microwave) required to produce cataracts are much higher than the levels emitted by VDTs. In addition, no study concludes that cataracts can be attributable to low radio frequencies or any other factor related to the use of VDTs. Consequently, the task force finds it highly unlikely that VDT work is associated with this problem.

Three studies have investigated the relationship of permanent eye damage to VDT work. They all conclude that occupational video viewing does not cause irreversible eye problems. However, these studies show several methodological weaknesses. First, the criteria used to select the subjects are not well defined. Second, very few confounding variables have been controlled. All this being considered, the task force concludes that even though the relationship between permanent eye damage and VDT work is highly unlikely, there is not sufficient proof to eliminate this possibility. At this point the task force did not have any recommendations to make.

Regarding visual fatigue, all the studies reviewed report a higher frequency of symptoms among VDT operators. Ergonomists have identified many factors that they suspect contribute highly to the visual discomfort of workers. The most important are:

1. Poor contrast between the characters and the screen background.
2. High luminance contrasts between the screen and the surfaces surrounding it (principally the source document).
3. Perception of glare and flicker on the screen.

Secondary factors:

1. Frequency and long duration of eye fixation on the screen.

2. Working more than 4 hours in front of the screen.
3. Existing oculovisual problems that have not been detected or that have not been appropriately corrected.

3.2. Musculoskeletal problems

No rigorous epidemiological study has tried to evaluate the frequency of musculoskeletal symptoms among VDT operators. Most efforts to estimate the occurrence of musculoskeletal problems have relied upon unvalidated questionnaires. These studies show serious methodological problems and thus it is not possible to conclude with certainty that VDT work is associated with a greater frequency of musculoskeletal problems than other types of clerical work.

On the other hand, ergonomic studies revealed workstation, task and work organization factors that might contribute to cause postural problems. The main factors incriminated are:

3.2.1. *Constrained Postures Related to:*

1. Incompatibility of visual tasks with manual activities requirements.
2. Unadjustable furniture.
3. Absence of forearm/hand supports.

3.2.2. *Static Positions Associated with:*

1. Task components, specially in parcellized work.
2. Fixed position of the screen.
3. Rapidity of task execution.

The use of ill-adapted bifocal lenses has also been implicated but the importance of this factor appears to be minimal.

3.3. Pregnancy Problems

A great concern has been raised in the public and particularly among VDT workers following the reporting by the Canadian Press of clustered cases of spontaneous abortions, congenital malformations and prematurity. Radiations emitted by VDTs, workload, postural immobility and stress were some of the factors indicated.

The first of the two epidemiological studies available reported a higher occurrence of spontaneous abortions and neonatal deaths among VDT operators. However, the researchers failed to identify factors associated with these pregnancy outcomes.

Several studies agree that the levels of X, UV, IR and high radio frequency radiations emitted by VDTs are much below the lowest levels of radiations for which effects on health are known. However, the task force considers that the lack of sufficient data on health effects of low radio-frequency radiations, static position, workload and stress does not disprove the hypothesis that these factors can constitute health hazards for pregnancy.

Considering the importance of pregnancy problems, the task force is supportive of the idea that the relationship between this problem and the VDT's occupational hazards has to be investigated in the most immediate future. At this point, the task force did not have any recommendations to make.

3.4. **Dermatitis**

Cases of facial dermatitis of VDT operators have also been reported.

The electrical potential difference between operators and VDTs, when the relative humidity is low and the floor is not carpeted with antistatic materials is suspected to be a hazard. Prevention of this health problem is relatively easy since the efficiency of the following preventive measures had been demonstrated:

1. Maintaining an adequate level of air humidity, given the climatic conditions.
2. Using antistatic materials to cover floor of VDTs offices.

3.5. **Photosensitive Epilepsy**

The hypothesis that occupational video viewing can produce convulsions among photosensitive epileptics has not been confirmed. It seems that the prevalence of epileptic conditions of all kinds in the population is about 0·5% and that among epileptics, only 4% show photosensitivity. Taking into consideration the absence of data on the occurrence of this problem among VDT operators and the low number of people susceptible that could be affected, the task force considers that this problem should constitute neither an intervention nor a research priority.

3.6. **Stress**

Several studies have revealed a higher frequency of self reported stress symptoms among VDT operators. The main factors suspected are:

1. Monotonous and repetitive tasks: they can require important psychomotor and psychic activities inductive of stress.
2. Mode of payment: stress would be more prevalent among operators paid on a piecework or production basis.
3. Frequent contacts with customers: there may be a discrepancy between customers' expectations and the administrative guidelines to which the operators have to conform.

4. Computer breakdown and delay: these problems affect above all operators paid by result and those who work in public services.
5. Lack of control on computer operations.
6. Long duration of work.

Considering the methodological problems faced by the studies on the one hand, and the absence of longitudinal study on the stress pathological issues on the other hand, the occurrence of stress and the relationship between tasks and work organization and incidence of health problems among VDT operators need to be evaluated further.

However, given the current status of our knowledge, it seems plausible that the stress felt by operators could make them prone to visual, musculoskeletal and pregnancy problems. This hypothesis should be tested in future studies.

4. *Conclusions*

Considering the number of workers involved, the rapidly growing introduction of computers in different activity areas and the worries of operators regarding the possible health hazards of VDTs, the task force stresses the importance of preventing the occurrence of the health problems identified.

However, it must be noted that most of the health problems reported are not specific to VDT work. In fact, task structuring, work organization and inappropriate layout of the workstation that can induce constrained and static postures, seem to be the main health hazards responsible for musculoskeletal, pregnancy and stress problems reported.

Regarding visual fatigue, some task and VDT characteristics (such as a visual object transmitting light) are suspected. Very few studies have concerned themselves with these aspects.

Finally, dermatitis can be related to electrical potential difference between the operator and the VDT. All these elements are also present in other working environments and could lead to similar health problems.

If a problem is suspected and prevention measures exist, they must be applied. In the absence of effective primary prevention measures, other actions can be taken, like reassignment of the pregnant worker and medical monitoring. These solutions must be temporary and limited. Studies must also be undertaken.

The task force recognizes the rapid development of new technologies; the task force's main lines of reasoning take into consideration choices that will be made with respect to technology development. Consequently, further studies on this matter will have to take these choices into consideration and be geared with either existing or developing technologies. The short and long term effectiveness and efficiency of the measures of intervention developed must be evaluated.

In addition, whatever the technological developments will be, it is important to prevent the problems associated with the work reorganization brought about by the introduction of VDTs.

Finally, considering the nature of the health hazards and the problems, namely the great diversity and complexity of interactions between these elements, studies on VDTs must be multidisciplinary in approach.

Software Ergonomics (Introductory paper)

T.F.M. Stewart
System Concepts Ltd, Museum House, Museum Street, London WC1A 1JT

1. *Introduction*

The aim of ergonomics is to improve the relationship between people, equipment, workplaces and the environment in which they interact. It does this by drawing on knowledge, techniques and theories from psychology, physiology, physics and any other relevant discipline. Increasingly, the equipment which now forms part of many people's working lives (and an increasing part in leisure also) contains a computer in some form.

The first area where poor computer ergonomics became recognized as a problem was in the design of terminal hardware. It may seem obvious now (and indeed should have been then) that the computer hardware should match human characteristics but a surprising amount of equipment appeared on the market where the keys were too small for accurate keying or where the characters on the display screen were poorly formed and illegible. For reviews of such problems see Cakir *et al.* (1980), Grandjean and Vigliani (1980) and Dainoff (1982).

Some of the early problems were caused by limitations in the technology for example of cathode ray tube design, but others were simply due to an inadequate appreciation of human characteristics among the manufacturers. In recent years, there has been substantial pressure from users, unions and advisory bodies on the suppliers to improve the hardware and there have been significant improvements in the technology itself. As a result, there are a growing number of computer products on the market in which the hardware really does exhibit good ergonomic design, although there are still some products which get the hardware interface hopelessly wrong.

Research and development work on hardware ergonomics is still continuing, especially as new products are developed, and as general standards rise. The International Standards Organization has an ergonomics sub-committee (ISO TC159 SC4) currently working on standards to eliminate the poor products and to increase the general level of computer hardware ergonomics (particularly VDTs).

Because most of the press publicity and indeed most of the obvious problems have been with hardware ergonomics, ergonomists have been criticized for regarding character legibility as the only issue (Crespy and Rey, 1983). This is clearly nonsense. From the early days of computing there has been a growing body of research focusing on the software interface—the interface between the users and the systems within the computer. Indeed, a number of the papers presented at the International Symposium on Man-machine Systems in September 1969 addressed just this topic. In the introductory paper of the special issue of the journal *Ergonomics*, Shackel discussed the importance of the software interface (Shackel, 1969). This interface has always been recognized by ergonomists as important, but now that many of the hardware problems have been solved, it assumes major importance.

Whereas a good hardware interface allows people to operate equipment, it is the software which determines what the system does and often what the user does also. Indeed, many systems have been introduced in such a way that their procedures have become full time jobs for certain of their users. These jobs were seldom designed to be rewarding or satisfying and problems of boredom and alienation have been reported (Eason *et al.*, 1974).

Another reason why the software interface is becoming more important is that advances in artificial intelligence techniques are leading to expert systems which perform tasks in ways that their users find it difficult to understand or follow (Feigenbaum and McCorduck, 1983).

Working in the software ergonomics area has not been without its problems. It proved difficult enough to provide solid cost justification for hardware ergonomics but it can be almost impossible for software ergonomics. Yet Gilb and Weinberg estimated that one design error in the IBM/360 operating system (using a space as a field delimiter) cost in the order of $100 million in lost computer time, corrupted files and debugging effort (Gilb and Weinberg, 1977). However one of the biggest problems in justifying software ergonomics is that people are even more adept at adjusting to bad software than they are to bad hardware. The costs of such adjustment, in terms of errors, delays, frustration and dissatisfaction are often well hidden.

Research in software ergonomics has been growing in recent years although it is not always labelled as such. Terms such as software psychology (Shneiderman, 1980) and cognitive ergonomics (Sime and Coombs, 1983) generally refer to the same area although the prime motivation of the researchers may be slightly different.

The literature is growing and there are the beginnings of theories being developed (Allen, 1982). In addition to the research evidence, there is a useful body of practical guidelines which have been derived over the years and which aim to provide specific interpretations of the research and advice to the system designer. For a particularly comprehensive set see Smith and Aucella (1983) and Gaines and Shaw (1983). However, there is a gap between theory and practice. Furthermore, the practising system designer often has enormous difficulty reconciling divergent views reported at conferences and in the literature (Macguire, 1982a). In some cases, the recommendations appear to conflict with each other as well as with the practical realities of system design.

In this paper, I would like to suggest that one reason for this apparent confusion is that usability and acceptability are not features of products or characteristics of systems but are judgements which people make. They therefore reflect the attitudes, experiences and expectations of the individual as well as the qualities of the product.

2. *Usability and Acceptability*

Attempts to improve usability and acceptability usually focus on minimizing the effort involved in using the equipment or system. For example, this may require special hardware design to make screens easier to read. Or it may involve reducing the complexity of the interface style or language. In addition, greater consistency, better training and conforming to well established population stereotypes can all help. It is even possible to build in suitable cues which encourage the user to adopt a suitable model in dealing with a novel system feature. A good example of this is the Office Technology Limited IMP workstation which uses a telephone handset to allow users to annotate text with speech. Without a recognizable handset, there would probably have to be rather careful instructions in how to leave and receive voice messages (Remington, 1981).

Another method of improving usability and acceptability is to reduce both the likelihood and, more significantly, the implications of errors. Gilb and Weinberg (1977) describe various means of anticipating input so that a certain amount of automatic checking can be done before too much damage is done. It is also possible to check out the most likely errors that users will make and ensure that their consequences are not too severe. For example, forgetting to change shift is a very common keying error yet some equipment puts such significant functions as 'clear the screen' on the same key as 'home the cursor' (Epson HX20 is an example). One of the most elegant exhortations made to designers to design interfaces capable of intelligent guesswork was made by Professor Al Chapanis when he said "To err is human, to forgive . . . design."

But much of the discussion on improving usability seems to overlook the fact that most users are willing to accept rather more effort and risk of error if the system is satisfying to use and performs some really useful function for them. I would therefore suggest that we can increase the subjective judgement of usability by improving the performance of the system for the user. For example, we might provide appropriate prompting which ensures that the user does not miss some potentially critical part of the task or we might tailor the service to fit his task more closely in terms of the structure of the dialogue or the contents of screens. We might even consider improving the friendliness of the messages or at least avoid being quite so rude and impolite as computer systems often are. Certainly the obscure and unintelligible language of much computing is unfriendly and may be intimidating to many ordinary users.

3. *User and Task Profiles*

In the previous section, we argued that usability and acceptability are essentially judgements which individuals make and so are highly user and task specific. At present we still know too little about the psychology of users and indeed the nature of man-computer tasks. These are both important areas for future research. However, there are three types of user/task combinations (which have been the subject of a number of studies) which we will consider further in this paper:

1. The computer professional developing a system for others to use.
2. The regular user of a clerical system.
3. The occasional user of an information system.

3.1. The Computer Professional Developing a System for Others to Use

The computer professional often has two quite distinct roles with respect to computer systems. As a professional his purpose is to act as an intermediary between computer technology and end users. This role may simply involve operating equipment of systems on behalf of others or it may require the creation or development of facilities for others to use directly.

In order to act in this professional capacity, he typically will use computer equipment and systems himself. In some cases this will be exactly the same facilities as the end users, but in other cases he will use specialized control or development facilities not normally available nor necessary to ordinary users.

Whilst developing systems for others, professionals often encounter problems with the hardware or software which they solve long before ordinary users ever get near the system. In doing this they often apply intuitive human factors knowledge and this can be remarkably effective. Many potentially fatal traps and bugs have been cleaned up during system development without the computer professionals getting any credit. Professionals, therefore, are interested in usability and acceptability both for users and for themselves.

For users, professionals need to be able to predict what the system will look like to the user and what the user might do to the system at certain key points. The good professional attempts to put himself in the user's position and anticipate both types of event. This can be rather difficult. Various studies have found striking differences in attitudes between computer professionals and ordinary users, for example see Hedberg and Mumford (1975).

It is too easy for the professional who has become highly proficient with a system to forget that the end users will not possess his knowledge of computing or of the system design. Of course, it is often necessary for the professional to create jobs which he himself would find totally unacceptable in terms of job satisfaction. The danger here is that he forgets that the end users may be more like him than he realizes and underestimate their requirements for intrinsic satisfaction and challenge in the job.

For themselves, professionals seem to be mainly interested in the performance of the equipment and systems they use. Research into their problems frequently focuses on such issues as the unpredictability of unstructured programs with an overabundance of GO TO statements (Jackson, 1980) or identifying effective problem solving strategies for improving programmer productivity (see Curtis (1982) for a thorough review of this area). The emphasis is on programming as a cognitive activity (Green, 1980) and on achieving good solutions to complex intellectual problems. Computer professionals are notoriously tolerant of unfriendly systems or difficult to use hardware and can often be seen working well into the night without additional financial reward, although it should be remembered that just because they do not complain does not mean that their efficiency has not been reduced by poor interfaces.

3.2. The Regular User of a Clerical System

The regular user of any system rapidly acquires the necessary skills for using the subset of facilities he uses regularly. Significant effort may be involved when he wishes to change from the well worn path of familiar options. The most important issue which has been widely researched is the impact the system has on the user's job and what effect that has on job satisfaction and motivation. One of the best known researchers in this area is Enid Mumford and she has pioneered a method of involving clerks and secretaries in the design of their own jobs as part of a sociotechnical systems design (Mumford, 1983). The development of office systems has made such direct user involvement all the more important and a substantial body of research is currently underway (Otway and Peltu, 1983).

3.3. The Occasional User of an Information System

The occasional user is the target for an increasing number of systems. Such users may well have a degree of choice over whether to use the system or not. It is therefore vital that the system presents an attractive and satisfying face to the occasional user with the minimum of effort and risk associated with the interaction. Videotex systems were designed specifically to serve such users and use simple failsafe menus and attractive colour graphics. Although videotex systems such as Prestel have been made very easy to use at the expense of real usefulness (Stewart, 1980), nonetheless they have represented a breakthrough in user interface design. The bulk of the research on casual users has focused on such topics as making computers tolerant of incorrectly typed input (Macguire, 1982b), making systems adapt to naive users (Thomas, 1981) and making user interfaces more enjoyable (Malone, 1982).

Of course, it is a mistake to imply that simply because the user only makes occasional use of the systems, its impact on his life is likely to be similarly insignificant. For example, the patient being interrogated by a computer diagnosis system is unlikely to be familiar with the interface but its potential impact and therefore its acceptability can quite literally be of life or death importance to him.

4. *Recommendations for Improving the User Software Interface*

In practice, each type of user and task combination has its own priority areas for improvement in the software interface. Indeed, the most appropriate targets for action vary quite significantly from one situation to another. What is necessary, feasible and cost effective in one system may not be in another. One of the reasons why it is difficult to reach agreement amongst researchers about the most effective actions may be that real users are prepared to trade off usability costs against benefits and this makes it difficult to establish the 'true' value of potential improvements.

I believe that all types of user may be concerned with all aspects of the interface. Which aspect benefits most from attention will depend not only on the type of user and on the task but also on the scope for improvement, the penalties involved in changing the system, the ease with which the improvements can be demonstrated, and the skills available in the organization to implement the changes.

As I pointed out above, research which has focused on computer professionals as users has concentrated on the performance of the system and in making sure that the facilities can be used effectively. Research into regular users of clerical systems has been mainly concerned with job design and satisfaction. The occasional user work has focused on making systems so easy to use that they require no new skills and often mimic paper or other familiar tools.

Although these foci have been useful in the past they may now inhibit a more creative approach to user interface design. Instead of just wondering how to improve the performance of systems for professionals and the satisfaction with systems for clerks and the effort in using systems for occasional users, we might benefit from turning our attention to the 'missing' aspects. How significant is job satisfaction to computer professionals or indeed to managers who may be occasional users? Weinberg's book *The Psychology of Computer Programming* (Weinberg, 1971) emphasized that programmers are people too but all too often we seem to regard them as mere extensions of the machine.

Surely clerks may also be concerned about the performance of systems. In an early study (Eason *et al.*, 1974), we found many clerks whose major source of dissatisfaction was the way the computer systems prevented them from doing a better job. Occasional users may not be solely concerned with fun and games, they may need to be able to do something really useful before they bother to use systems. Perhaps it has been the general uselessness of many systems which has prevented managers from using 'Integrated Management Information Systems' rather than just the unfriendly interfaces. The lack of enthusiasm shown by the general public in the UK for actually buying Prestel receivers and services may be more a reflection on the usefulness of the information than on the usability of the menus (although that too has been questioned (Stewart 1980)).

5. *Conclusion*

A systems approach is often advocated for the application of hardware ergonomics to complex man–machine systems. It seems to me that if we took a systems approach to the application of software ergonomics then our view of which aspects of the interface matter might be rather broader than current research would suggest. It may open our eyes to new possibilities for making software both more usable and more acceptable to a wider range of users performing a wider range of tasks.

References

Allen, R.B., 1982, Cognitive factors in human interaction with computer. *Behaviour and Information Technology*, **1**, 3, 217–236.

Cakir, A., Hart, D.J., and Stewart, T.F.M., 1980, *Visual Display Terminals* (Chichester: John Wiley).

Crespy, J., and Rey, P., 1983, *Work on Visual Display Units: Risks for Health*. World Health Organisation WHO/OCH/83.2.

Curtis, B., 1982, A review of human factors research on programming languages and specifications, in the *Proceedings of the Conference on Human Factors in Computer Systems*, Gaithersburg, Maryland, pp.212–218.

Dainoff, M.J., 1982, Occupational stress factors in visual display terminal (VDT) operation: a review of the empirical research. *Behaviour and Information Technology*, 1, 2, 141–176.

Eason, K.D., Damodaran, L., and Stewart, T.F.M., 1974, A survey of man-computer interaction in commercial applications. LUTERG Number 144, Department of Human Sciences, University of Loughborough.

Feigenbaum, E.A., and McCorduck, P., 1983, *The Fifth Generation* (Reading, Massachusetts: Addison-Wesley).

Gaines, B.R., and Shaw, M.L.G., 1983, Dialog engineering, in *Designing for Human–Computer Communication*, edited by M.E. Sime and M.J. Coombs (London: Academic Press).

Gilb, T., and Weinberg, G.M., 1977, *Humanised Input* (Cambridge, Massachusetts: Winthrop).

Grandjean, E., and Vigliani, E. (eds), 1980, *Ergonomic Aspects of Visual Display Terminals* (London: Taylor & Francis).

Green, T.R.G., 1980, Programming as a cognitive activity, in *Human Interaction with Computers*, edited by H.T. Smith and T.R.G. Green (London: Academic Press) pp.271–320.

Hedberg, B., and Mumford, E., 1975, The design of computer systems, in *Human Choice and Computers* (Amsterdam: North Holland/Elsevier) pp.31–59.

Jackson, M., 1980, The design and use of conventional programming languages, in *Human Interaction with Computers*, edited by H.T. Smith and T.R.G. Green (London: Academic Press) pp.321–347.

Macguire, M., 1982a, An evaluation of published recommendations on the design of man-computer dialogues. *International Journal of Machine Studies*, 16, 3, 279–292.

Macguire, M., 1982b, Computer recognition of textual keyboard input from naive users. *Behaviour and Information Technology*, 1, 1, 93–111.

Malone, T.W., 1982, Heuristics for designing enjoyable user interfaces: lessons from computer games, in the *Proceedings of the Conference on Human Factors in Computer Systems*, Gaithersburg, Maryland, pp.63–68.

Mumford, M., 1983, Successful system design, in *New Office Technology*, edited by H.J. Otway and M. Peltu (London: Frances Pinter) pp.68–85.

Otway, H.J., and Peltu, M., (eds), 1983, *New Office Technology* (London: Frances Pinter).

Remington, R.J., 1981, The transition from word processing to information processing: human factors considerations, in *Word Processing: Selection, Implementation and Usage in the 80's* (London: Online Publications) pp.261–270.

Shackel, B., 1969, Man-computer interaction—the contribution of the human sciences. *Ergonomics*, 12, 4, 485–500.

Shneiderman, B., 1980, *Software Psychology* (Boston: Little, Brown).

Sime, M.E., and Coombs, J.M., (eds), 1983, *Designing for Human–Computer Communication* (London: Academic Press).

Smith, S.L., and Aucella, A.F., 1983, *Design Guidelines for the User Interface to Computer-Based Information Systems*, prepared by The Mitre Corporation, Bedford, Massachusetts, Report number ESD–TR–83–122.

Stewart, T.F.M., 1980, Human factors in videotex, in the *Proceedings of the 4th International Online Information Meeting*, London, pp.87–96.

Thomas, R.C., 1981, The design of an adaptable terminal, in *Computing Skills and the User Interface*, edited by M.J. Coombs and J.L. Alty (London: Academic Press) pp.427–463.

Weinberg, G.M., 1971, *The Psychology of Computer Programming* (New York: Van Nostrand-Reinhold).

*Quality of Working Life and the Introduction of New Technology into the Office**

R.G. SELL
Work Research Unit, Department of Employment, Steel House, 11 Tothill Street, London SW1H 9NF

Abstract

This paper looks at the quality of working life and organizational aspects of introducing new technology into the office. It considers the options available in terms of restricting or widely disseminating information, having jobs with identity or broken down into small tasks, allocation of functions and the way in which the changes are introduced.

1. *Introduction*

The aim of this paper is to set in context the factors which need to be considered when looking at the organizational and quality of working life aspects of new technology in offices. It does not report to any extent on research which has been carried out because there is little to report. There is, as yet, little evidence of the major technological revolution in office technology which it is predicted will soon come about. Where there have been changes very few have been reported which have taken account of human needs and access to those cases which are not successful is difficult to obtain.

Whilst the majority of the research which has been carried out has been on the micro aspects of the man/machine interface, such as the design of the characters to be projected on a VDT screen, or the postural aspects of the work station, the acceptance and effective use of any new system is more likely to depend on wider aspects such as due account being taken of the user's needs and desires in the design of the individual job and the way in which they are put together to form a work organization (Eason and Gower, personal communication).

 The views expressed are those of the author and may not represent the views of the Department of Employment.

The degree to which the user's needs and desires are considered depends in turn on the values of those who are responsible for laying down the system specification. All too often lowest possible capital cost and the need to get the system operating by a specified date are the declared priorities rather than the long term effectiveness in terms of operator satisfaction and health and system efficiency.

This paper will look at some of the options which there are and the consequences of the various choices.

2. *Modern Information Technology*

The main recent development in information technology is the ease and relatively cheap cost of providing information as compared with previously. The main choice is whether this information is restricted to a relatively small group and used as a management control device or whether it is distributed widely as a means of opening up the organization and allowing a wider degree of participation. Information is power.

One way in which information can be used as a management control device is by measuring the number of key depressions which each operator makes. This is obviously only important if the emphasis is on differentiated jobs where some people are employed entirely on data entry tasks.

Another aspect of restricting access to information has occurred where more senior people have had their own terminals to interrogate the computer when people at operating levels, who need the information to carry out their job, do not. This particular problem is becoming less serious as equipment costs come down and more terminals get installed.

The problem still remains, however, of a senior person in an organization being able to get hold of detailed information direct from the data base such as the details of deliveries to individual customers without necessarily knowing the logic on which delivery decisions have been taken. If he confronts the order handler with this information when there has been a complaint from a customer it may lead to more defensive and perhaps less effective behaviour in the future.

Where information does become widely distributed such that senior people have access to detailed information and junior people are able to know better overall trends in performance it is necessary to look at the level of trust which exists in the organization. Organizational development activities may be necessary to help build up trust such that the people at the one level understand the way in which those at the other levels are working and agree the basis on which decisions are being made.

3. *Organization and Job Design*

There is a large risk that office jobs will become more like factory jobs with many people having to do repetitive, short cycle, paced tasks such as data entry. A more satisfactory job design situation is to have people doing a complete job such as all aspects of order processing; receiving the order, entering it on the computer, allocating priorities, arranging deliveries, progressing the order, dealing with enquiries, for a group of customers.

If an organization's main concern is with the capital cost of equipment it is very likely to move towards increased specialization with the emphasis on data processing clerks, centralized word processing pools, etc. This is likely to lead to employees with little intrinsic interest in jobs. One much publicized system in the UK which led to widespread customer dissatisfaction because of the time taken to process enquiries was based on this principle. Without close links between the originator and the processor of data errors are more likely. These links are difficult to establish with centralized special functions.

If some degree of specialization is to be retained it is likely to bring better results in terms of both quality of working life and efficiency if it is within a working team. If secretarial/word processing work is carried out within a working team rather than via a central function the improved understanding between organization and processor will both reduce errors and increase commitment. It also allows more opportunity for the specialist to be involved in other activities of the team when the demand for the specialist skill is temporarily low.

If jobs are designed such that each person performs a number of related but different tasks and if automation equipment continues to get cheaper it is likely that most people will only be using a VDT for a small part of their working day. This may make the micro-ergonomic aspects less important than the organizational aspects although more time may be spent by many more people in using terminals. It is important therefore to determine the main ergonomic principles to be applied.

Job design problems are likely to arise in the future much more in regard to managerial jobs. One aspect is that people at senior levels may be required to have keyboard skills which they may see as a lowering of the skill level of their job. On the other hand they may, if they have developed an interest in computers, see keyboard operations as a natural development of their job.

A more important problem is likely to be the effect on the data collecting and processing aspects of the manager's job. To what extent will he be able to exercise his own judgement? Will there be moves towards making his decision making skill unnecessary? Will the system assume that all the information necessary for a decision will be collected and manipulated by the computer making the manager's job unnecessary?

These issues need to be faced up to and arrangements made to ensure that the computer is used as a tool to aid the manager in his job and not as a replacement which, in actual practice, can only carry out those parts of his job for which there are hard figures. Computers cannot exercise subjective judgements nor deal with unquantifiable factors.

4. *Job Loss*

There has been more concern expressed about job loss than on the need for good job design. This concern was initially about clerical level jobs but it has now spread to middle management level jobs as well.

In the author's view much of the current concern by trades unions and employees about the ergonomic and health aspects of VDTs is a displacement from the fear of unemployment, leading to a hope that by attacking these aspects overall introduction times will be slowed down. British trades unionists have traditionally paid little attention to ergonomic and health problems such as noise. With VDTs, however, they have been much concerned with less tangible problems such as the possible radiation hazards of the CRTs.

There seems little doubt that as office automation develops with electronic mailing and filing systems and voice entry devices become more common many simple clerical jobs may become redundant. On the other hand the ease with which data can be obtained will make it economic to do many things which cannot be contemplated today. Where the balance will come between job loss and job creation is therefore difficult to predict.

5. *Management of Change*

In the experience of the Work Research Unit change is more likely to be carried out successfully if those whose jobs will be affected by the change are actively involved in the design of the system and the decisions which have to be made. No one likes work changes imposed on him unilaterally without his agreement as this is a breaking of their psychological contract with the employing organization.

Whilst it is easy to say that participation is necessary it is more difficult in practice to carry it out.

All participation activities have to take account of the institutional environment in which they have to work. Not only do the managers and the managed have to be involved but also account has to be taken of the views of the trades unions at both company and district level. Their attitude needs to be considered and their cooperation or at least their acquiescence obtained.

A major investment of time is required and the willingness to make this depends very much on the values and disposition of the top management. If their values encourage the necessary time investment they are less likely to be looking for a return on that investment in hard financial terms although they will, of course, expect improvements in overall organizational performance which can be assessed in less tangible ways.

If there are people in key positions who actively disagree with participation a difficult problem of principle arises. Can you say 'you will participate'? If the disagreement persists decisions have to be made as to what to do with that person. If they have been long service, loyal employees can you ignore that and remove them from their job? Perhaps the fairest solution is to help them to find an off-line job where participation is less important.

A particular problem arises with the introduction of new powerful computer systems whose potential it may be difficult for the non-expert to understand until the system is actually installed. The best recommendation that can be made here is that the system designer ensures that the system is flexible enough to accommodate difficult patterns of working (Eason and Sell, 1981).

6. *Conclusion*

Although the topics considered in this paper are less amenable to laboratory study and precise measurement than are the more detailed aspects of the man/machine interface, failure to consider them in applying new systems is likely to lead to considerable mental strain and failure to take account of them in experimental studies is likely to lead to misinterpretation of results.

Although they are less tangible, their overall effects are, if anything, more likely to influence both system performance and effectiveness.

Reference

Eason, K.D., and Sell, R.G., 1981, Case studies in job design for information processing tasks, in *Stress, Work Design and Productivity*, edited by E.N. Corlett and J. Richardson (Chichester: John Wiley) pp.195–208.

From Work Analysis to System Design

L. Pinsky
Laboratoire de Physiologie du Travail et d'Ergonomie du Conservatoire National des Arts et Métiers, 41, Rue Gay-Lussac, 75007 Paris, France

Abstract

An approach for research on cognitive processes in computer use and for improvement of system design is set up from a study on an interactive coding system. It is based on 'ergonomic experimentation'. Its results show the prominent part taken by a 'cooperative principle' in the operator's interpretation of computer's responses and the characteristics of his 'natural' reasoning for action. The diagnosis allows to define two effective notions for design: problem solving aid and cooperative competence for the system.

1. *Introduction*

Although cognitive aspects of work were present in the operators' complaints about nervous fatigue and in the system designers' practice, only recently have they come into the focus of ergonomics, beside the considerations of hardware and environmental problems (Shackel, 1980).

This relatively new situation calls for the definition of appropriate strategies both for knowing the actual operator's cognitive processes and for providing an ergonomic support to system design (Barnard *et al.*, 1981; Allen, 1982).

In this paper we will illustrate the notions and the methods of an ergonomic research carried on for the designing of an interactive system.

2. *The Interactive Coding System*

In an investigation on occupational information, people had filled out a printed form. The matter is to code this information. There is a classification made of pre-established occupational categories. One of these categories must be assigned to the respondent's occupation.

The operator transmits to the system an occupation designation in ordinary language and several precoded variables (wage earning status, duty, . . .). An algorithm for automatic coding is started. If it succeeds, a category is assigned and a 'category description' is sent back to the operator for his approval. If it fails a 'message' proposing different possibilities is displayed.

In order to be more flexible in accepting occupation designations, the system may not take into account all the words of the transmitted designation. Otherwise the algorithm does not use systematically all the coded variables; it depends upon the transmitted occupation designation and the occupational area.

3. *The Ergonomic Experimentation*

In order to be able to make a diagnosis of the future working situation we used the following procedure:

1. We worked with a group of 10 employees similar to the intended operators (the same population).
2. At each step of the design we built up an experimental situation in such a way that the operator achieved a cognitive activity as close as possible to that of the future working situation.
3. We analysed this activity.
4. We defined a diagnosis corresponding to the future situation by interpreting the results of the analysis on the basis of what we knew of the future situation at this step.

We call this procedure 'an ergonomic experimentation'. It is simultaneously an effective heuristic to produce hypotheses, a way to validate them and a means to test proposed ergonomic transformations.

The work analysis dealt with two types of record:

1. Observable behaviour: video tapes were made of the operator's actions (filling in or modifying the screen-form) and the computer's response, as they occur.
2. Verbalization: tape recordings were made of the operator's comments or accounts about his reasoning both while working with the system and afterwards, on the basis of the videotapes.

The interpretation of the data so collected required extensive discussion with the operators.

4. *Work Analysis: Notions and Results (Pinsky and Theureau, 1982)*

The general framework of our work analysis is a theory of action. It is a requisite of the critical features of the operator's behaviour: an intentional, conscious, planned and goal-directed behaviour (von Cranach, 1982).

The operator's reasoning designs and guides his actions. This reasoning for action is based on his interpretation of the computer's responses.

4.1. Interpretation and Dialogue

When interpreting the computer's responses the operator is not only concerned with the meaning of words and sentences, the syntax and the context (linguistic and extralinguistic) he also expects the responses to comply with a 'cooperative principle'.

This hypothesis is drawn from the Grice's approach of the human conversation (Grice, 1975) and is confirmed by the examination of the 'infelicities of the dialogue'. As a matter of fact we can describe all these infelicities as caused by the violation of some cooperative rules (Pinsky, 1983):

1. The system takes account of all the information transmitted.
2. The system is pertinent.
3. The system is coherent.
4. The system gives all the information necessary to the exchange.
5. The system is clear in the expressions it uses.

We distinguish two levels of violation:

1. 'Pathological dialogue': the dialogue is interrupted, the operator leaves off.
2. 'Irregular dialogue': the operator is merely surprised.

These failures are not only anecdotal, they will have very noticeable adverse effects in the working situation.

4.2. Reasoning for Action

Standard logic, even adapted, succeeds to grasp only very partial aspects of this kind of natural thinking. We are led to construct a 'natural logic' on quite new bases. J.B. Grize has opened a way for such a construction (1982). We have tried to follow this way.

While reasoning the operator deals with specific objects: the occupation to be coded and the category description or the message terms. These are discursive wholes, set up respectively by the respondent and the taxonomist. The operator may consider only some parts of them.

The operator carries on two main kinds of operations on these objects.

1. 'The Bringing Together': the operator establishes a relation of proximity between two objects. To do so he may use different types of bringing together: he may consider that the two objects have a common property; that some parts of them belong to the same semantic field or one of these parts is semantically included in the other; that the designations of the objects have a common word or that they are located at the same 'generality level'.

2. 'The Differentiation': the operator constructs a difference between the two objects. Here too, several modes may be used.

While reasoning, the operator combines these different operations: for instance, a differentiation may weaken or eliminate a previous bringing together. Differentiations may be reduced by other higher level operations which, for instance, take account of the system's functioning.

5. *Ergonomic Contributions to System Design*

This work analysis allows us to define two main general notions for the improvement of the system design:

1. 'The problem solving aid competence' of the system. It means for instance to make more complete and clearer the nomination of the category descriptions and the message terms or to mark on the screen the words of the occupation designation and the coded variables taken into account by the automatic algorithm.
2. 'The cooperative competence', a necessary condition for the previous one, aims to make the computer's responses cooperative. For instance, in order to design the messages we gave a set of specific rules such as:
   (a) A message must not ask for an information already transmitted.
   (b) Before sending a message, all the information already transmitted must be used.
   (c) The coding problem must be explicitly set up.
   (d) And so on . . .

These notions and criteria are more accurate and effective than usual concepts like 'flexibility, friendliness, naturalness, usability and acceptability'.

6. *Conclusion*

We have showed that work analysis can provide powerful means to account for actual cognitive processes, that it can be achieved during a design process through a series of ergonomic experimentations. On the practical standpoint this method is effective by giving criteria for design. The necessary developments would concern: the deepening of theoretical notions, the analysis of other types of interactive systems, the testing of the generality of the design rules.

References

Allen, R.B., 1982, Cognitive factors in human interaction with computers, in *Directions in Human Computer Interaction*, edited by A. Badre and B. Shneiderman (Norwood, N.J.: Ablex Publishing Corp.).

Barnard, P.J., Hammond, N.V., Morton, J., and Long, J.B., 1981, Consistency and compatibility in human-computer dialogue. *International Journal of Man-Machine Studies*, **15**, 87–134.
Grice, H.P., 1975, Logic and conversation, in *Syntax and Semantic: Speech Acts*, Vol. 2, edited by P. Cole and J.L. Morgan (New York: Academic Press) pp.41–58.
Grize, J.B., 1982, *De la logique à l'argumentation* (Geneva: Librairie Droz).
Pinsky, L., 1983, What kind of 'dialogue' is it when working with a computer?, in *The Psychology of Computer Use*, edited by T.R.G. Green, S.J. Payne and G.C. van der Veer (London: Academic Press) pp.29–40.
Pinsky, L., and Theureau, J., 1982, *Activité cognitive et action dans le travail* (Paris: Conservatoire National des Arts et Métiers, Collection de Physiologie du Travail et d'Ergonomie, no. 73).
Shackel, B., 1980, Dialogues and language—can computer ergonomics help? *Ergonomics*, **23**, 857–880.
von Cranach, M., 1982, The psychological study of goal-directed action: basis issues, in *The Analysis of Action*, edited by M. von Cranach and R. Harre (Paris: Cambridge University Press, Editions de la Maison des Sciences de l'Homme).

Task Analysis in Applying Software Design Principles

K.L. Kessel
The Koffler Group, 3029 Wilshire Blvd., 200, Santa Monica, California, USA

Abstract

Because the usefulness of software is determined by user perception and cognition, task analysis must incorporate relevant user mental attributes in comparing user needs with task requirements. Problems studying user behaviour have prevented descriptions of precise recommendations for software design in task analysis. These difficulties include complexities inherent in the subject matter and the approach many investigators have taken in their studies. A model-building approach based on research in human perception and cognition is suggested for a more accurate assessment of user behaviour under various task conditions. The appropriate level of detail for a task analysis is determined by user cognitive models and the necessity of fine level descriptions and predictions.

1. *Introduction*

1.1. Task Analysis and Models of User Behaviour

Task analysis for software design describes operations required to obtain a goal and compares them with user abilities (Drury, 1983). Different task operators will act as a function of various strategies used to obtain the goal. Thus, a line on a CRT screen may be accessed with continuous depressions of a 'line feed' key, with the movement of a mouse, or with a command to move a cursor up or down *n* lines. Each operator has user mental and physical requirements associated with it. A user must know how to use a mouse or a command and must understand how the operator can accomplish the goal. Consequently, the usefulness of a user-system interface (USI) is determined by human perceptual and cognitive attributes. These then become a part of task analysis.

Depending on user capability, different operators should result in various error rates and error types, and task completion times. If user behaviour were well understood then it might be possible to make predictions about which operators would yield the best task completion times and the lowest error rates. User cognitive models would also allow predictions about newly developed operators. Unfortunately, models of user behaviour normally lack the precision that is necessary for making analytic (quantitative) predictions (Moran, 1981). Theories rarely progress beyond a 'box and arrow' psychology (Reisner, 1981; Sheil, 1982), and little more than the most general recommendations, such as, 'make the interface consistent', are made (Hammer *et al.*, 1983; Heckel, 1982; Maguire, 1982). This lack of rigour in applying and developing guidelines has its origin in the subject matter and in the approach investigators take in their studies.

1.1.1. *Definitions and Measurements*

Studies of user behaviour frequently suffer from methodological problems. Definitions of mental characteristics are often consequences of experimental procedures. As the testing method changes, so do the results. For example, Badre (1982) used two different measures to define cognitive information 'chunks'. These measures were uncorrelated and confused the definition of a chunk. Malone (1982) attempted to provide USI guidelines based on preferences in a few incomplete studies of computer game playing. Considering only one game, Malone's results were complicated by a large difference between the preferences of male and female subjects. Studies of mental workload likewise present difficult and often disparate definitions of workload based on physiological, subjective survey or performance data.

Definition and measurement of complex human mental attributes requires a more robust approach than has been applied. Different measures of the same theoretical concept must converge on the same conclusions if the concept is to have validity (Garner *et al.*, 1956). Constraints on the applicability of mental concepts must additionally be delineated.

1.1.2. *Research Approach*

The models of user behaviour needed for implementation in a task analysis have not been forthcoming partially as a result of the direction software ergonomics has taken in its approach to research questions. Rather than constructing models of user behaviour, particularly quantitative models, ergonomists have focused on statistically significant group differences and give little indication of a parameter's relative importance in modifying user behaviour. Ergonomists need to know how much of the variance in user activity can be accounted for with a given variable, and not simply that the variable may affect behaviour.

2. *Task Analysis and Model-Building for an Applied Science*

2.1. **Cognitive Science in Task Analysis**

Ergonomists must develop timely and applicable recommendations and methods for designing ergonomic software. This cannot be done by attempting to discover fundamental processes of mental activity (Meister, 1982). Rather, this project is relegated to basic researchers. For ergonomics, basic research is a template for constructing hypotheses and models of user behaviour. Card, Moran and Newell (1983) provide the most explicit demonstration of this approach in performing task analysis. Values for parameters describing user behaviour are obtained from basic research literatures (for example, reaction-time data) and from several of their own experiments. Models of user behaviour are based on these value parameters and predictions are calculated from the resulting models. Thus, the time to complete a task might be calculated from finger-to-keyboard keystroke time, mental decision time and document scanning time. Although Card *et al.* did not fully investigate user problem solving and motivation, their approach could be expanded to incorporate these into their models. For example, a user may choose a command that requires a longer keystroke and execution time than another command performing the same function through different means if the latter command is more likely to produce errors. Predicting user behaviour might be modelled as the frequency of such errors and the time to restore the original state as an inverse relationship of the probability of using that command.

Reisner (1981, 1982) has taken a similar approach by developing a formal grammar for describing tasks and predicting user behaviour in those tasks. Actions that complete a task, such as 'Move cursor' or 'Retrieve from long term memory', are described as a sequence of consecutive events. Likely sequences and their execution times are derived from the basic research literature (similar to Card *et al.*'s parameter values), common sense or ergonomic studies. These predicted values are then substituted for the formal grammar. Different devices can be assessed by comparing their resulting formal grammars.

The necessity of basic cognitive and perceptual research for modelling user activity in a task analysis is demonstrated in Reisner's formal grammar and in Card *et al.*'s 'calculational' methodology (Moran, 1981). Both are model-building approaches and are consistent with the need for making precise predictions of user behaviour in task analysis. Different levels of analysis from entire tasks to single keystrokes can be described with user models.

2.1.1. *Molar and Molecular Analyses*

Tasks such as text-editing and monitoring may be broken down into subtasks including adding, deleting or copying text (Smith, 1981). Subtasks can be broken down further into simple transactions such as single keystrokes followed by system responses. Reisner's formal grammar is an analysis at this level of transaction, as is Card *et al.*'s keystroke-level model. The appropriate level of analysis is dependent upon task requirements and the ability to model behaviour at finer levels. The time it takes for a neurological signal to be sent from brain to muscle in typing tasks would be overly precise for most tasks. Knowing that a typist can type *x* number of words per minute might be sufficient knowledge. Indeed, a molar level analysis such as 'Edit text, close file' that requires *y* hours to complete might be sufficient. Determining the appropriate level of analysis depends on the amount of precision necessary for software design decisions. The decision to 'edit file' versus 'use command Z', versus 'keystroke "ctrl" "A"' and the importance of specifying those actions will determine the appropriate level of analysis.

More molecular analyses will not necessarily have greater predictive power. For example, the keystroke-level model (Card *et al.*, 1983) did not predict user behaviour as well as more molar analyses for text-editing tasks. Molecular measurements are more difficult than molar measurements. To record eye movements of people looking for errors on a CRT or the rate at which they key in commands and the accompanying mental processes used to choose those commands, is much more difficult than recording the mean time needed to edit a simple 'typo'. Each measurement brings possible sources of variability. At the finest level steps may be missed, added inappropriately, or placed out of sequence. Consequently, the decision for more molecular analyses must be considered in light of its relative benefits and costs.

3. *Conclusion*

In the context of this paper, software design guidelines are roughly equivalent to models of user perceptual and cognitive behaviour. The models are derived from basic research and are intrinsic to task analysis. Predictive capabilities should increase as factors such as skill and motivation are included in user models. The level of detail required for task analysis is determined by the accuracy of user models and measurements, and by the level at which software design decisions are to be made. The advantages of a model-building approach in an applied science include greater accuracy in prediction, determination of behaviour magnitudes and the application of basic research findings in task analysis.

References

Badre, A.N., 1982a, Selecting and representing information structures for visual presentation. *IEEE Transactions on Systems, Man, and Cybernetics, SMC–12*, pp.495–504.

Badre, A.N., 1982b, Designing chunks for sequentially displayed information, in *Directions in Human/Computer Interaction*, edited by A.N. Badre and B. Shneiderman (Norwood: Ablex) pp.179–193.

Card, S.K., Moran, T.P., and Newell, A., 1983, *The Psychology of Human-Computer Interaction* (Hillsdale: Erlbaum).

Garner, W.R., Hake, H.W., and Eriksen, C.W., 1956, Operationism and the concept of perception. *The Psychological Review*, **63**, 149–159.

Hammer, M., Kunin, J.S., and Schoichet, S., 1983, What makes a good user interface? *OAC Conference Digest* (Philadelphia: AFIPS) pp.121–130.

Heckel, P., 1982, *The Elements of Friendly Software* (QuickView Systems, Los Altos: Interactive Systems Consultants).

Maguire, M., 1982, The development of dialogue design guidelines for a computer based local information system to be used by the general public. *Proceedings of Human Factors in Computer Systems*, pp.350–354.

Malone, T.W., 1982, Heuristics for designing enjoyable user interfaces: lessons from computer games. *Proceedings of Human Factors in Computer Systems*, pp.63–68.

Meister, D., 1982, Where and what are the data in human factors? *Proceedings of the 26th Annual Meeting of the Human Factors Society*, pp.722–726.

Moran, T., 1981, An applied psychology of the user. *ACM Computing Surveys*, **13**, 1–11.

Reisner, P., 1981, Formal grammar and human factors design of an interactive graphics system. *IEEE Transactions on Software Engineering, SE*–7, pp.229–240.

Reisner, P., 1982, Further developments toward using a formal grammar as a design tool. *Proceedings of Human Factors in Computer Systems*, pp.304–308.

Sheil, B.A., 1982, The psychological study of programming. *ACM Computing Surveys*, **13**, 101–120.

Smith, S., 1981, Design guidelines for the user-system interface of on-line computer system: a survey report. *Proceedings of the 25th Annual Meeting of the Human Factors Society*, pp.509–512.

Cognitive Complexity Related to Image Polarity in the Aetiology of Visual Fatigue

S.E. Taylor, B.W. McVey, and W.H. Emmons
Human Factors Centre, IBM Corporation, San Jose, California, USA

Abstract

The mismatch in image contrast polarity typically found between source documents and CRT displays has been suggested in the aetiology of visual fatigue. The greater complexity involved in the encoded comparison strategy required by mixed polarity (compared with the visual comparison strategy available with matched polarity) is hypothesized to be fatiguing. Subjects compared alphabetic characters presented sequentially on a CRT in matched or mixed polarity with or without a delay between them. With no delay, the reaction time was greater for mixed polarity, suggesting that polarity does indeed contribute to the determination of which comparison strategy must be used. However, the real world task of changing visual fixation between a source document and a display imposes a minimum of 500 ms delay. With a delay of 500 ms, no difference in reaction time was observed. Thus it is concluded that in this situation, the encoded comparison strategy is necessary for matched as well as mixed polarities. There is no apparent cognitive advantage for matched polarities between source material and display.

1. *Introduction*

It is a very common situation for a visual display terminal (VDT)-operator to be confronted with a dark character source document and a light character display. Though this situation is common, it has been deemed by some to be a major contributor to visual fatigue.

The hypothesis that visual fatigue is due to polarity differences came about partly because of the results of studies (Bauer and Cavonius, 1980; Radl, 1980) which suggest that operators performed better with dark character displays. It has been suggested that visual fatigue occurs when the operator must compare information in one polarity with another since this is believed to involve greater complexity and perhaps even different or competing processing mechanisms (Krueger, 1981).

Research shows that comparison of information that is visually the same is faster than comparison of information that is visually dissimilar. The former information is presumed to be compared on a visually encoded basis while the latter is presumed to be compared on a basis involving some encoding of the visually presented information into some other form, such as a name-encoding process (Posner, 1978).

There are two parts to the necessary hypothesis. One part is that mismatched polarity is more complex. The other part is that this complexity leads to visual fatigue. Thus the hypothesis is rejected if the first part is not true. The first step in the research project was to determine whether or not mismatched polarity is more complex.

The next logical step was to determine whether or not such a difference would be of practical concern. Research has shown that the visual versus name-encoding differences in information processing, reflected in reaction time, disappear when sufficient time intervenes between the two items being compared. An operator must expend a minimum 500 ms in eye movement latency and duration when changing fixation between some item on a source document and the same item on a display (Taylor, 1979). It was hypothesized that any advantage for matched polarity such as facilitation of a visual comparison would be absent with typical task delays and result in some other comparison strategy such as name encoding.

It should be mentioned that any factor which makes two stimuli visually dissimilar negates the advantage of making a visual comparison and, typically, forces a name-encoding strategy, in proportion to the dissimilarity. Many source documents are handwritten. They are at least in a different font, of a different size, and in a different part of the visual field. All of these differences, in addition to the eye movements upon which the present research was predicated, would be expected to eliminate any advantage of matched polarity because the name-encoding strategy must be used. Thus any or all of these factors could have been used to evaluate the practical significance of the relationship between polarity and comparison strategies.

2. *Method*

2.1. Subjects and Apparatus

The study consisted of two experiments. In the first, 16 subjects participated; in the second, 18 subjects. The stimuli were presented on a Tektronix 4023 CRT display terminal with a white P4 phosphor presentation on a dark grey background. Under control of an IBM System/7 program, stimuli were presented on the screen in dark character or light character format. The stimuli appeared in two adjacent fields in the approximate middle of the screen. The light and dark backgrounds were 151 cd/m^2 and 24 cd/m^2 respectively. The strokewidths for dark characters were 16% greater than for light characters.

The character set was restricted to ten letters of low confusability (A, C, P, T, L, M, S, V, E, J). Both the small size of the set and the low confusability allowed for easy visual matching of characters. Two function keys were assigned as response keys: one key was used to indicate that the second stimulus was the same character as the first, the other key indicated that the characters were different.

The computer program presented stimuli with equal frequency of same and different characters and recorded the response made by the subject and the reaction time to make the response. Data were collected on tape for later analysis.

2.2. **Procedure—Experiment 1**

The subjects sat at the CRT terminal and were told to respond 'Same' if the characters were the same letter of the alphabet or 'Different' if they were not. The pairs of characters were presented in all possible polarity combinations, that is, dark–dark, dark–light, light–dark, and light–light characters for first and second characters, respectively. The order in which they received the four polarity combinations was counterbalanced in a balanced Latin Square design with four replications. All stimulus presentations were in sessions of 30 trials. All of the trials in one session were of a single polarity combination. Thus subjects received 120 test trials at each polarity combination for a total of 480 test trials. After initiating a session, a character appeared in the left presentation zone and remained for 250 ms. The first character then disappeared and was immediately followed by the second character appearing in the right presentation zone. Coincident with the onset of the second character the reaction timer was started. The second character remained present until the subject had depressed the 'Same' or 'Different' key, at which time it disappeared and the reaction time was recorded.

2.3. **Procedure—Experiment 2**

Character presentations were as described above, except that a 500 ms delay was introduced between the offset of the first character and the onset of the second character. Only the dark–light character pairs were used for a mixed polarity condition and only the dark–dark character pairs were used for a matched polarity condition. Subjects were given two sessions of 30 trials at each of the two polarity combinations.

3. *Results and Discussion*

Means were computed for the blocks of four sessions, totalling 120 trials for Experiment 1, and two sessions, totalling 60 trials, for Experiment 2. The block means were the data used in analyses of variance.

3.1. **Results and Discussion—Experiment 1**

Analysis of correct and incorrect responses revealed very low error rates not differing between polarity conditions. Table 1 presents the basic reaction time (RT) data for

Experiment 1. Clearly subjects responded considerably more quickly to pairs that were the same character. This finding is consistent with the literature involving such same-different judgements (Posner, 1978).

TABLE 1. *Mean RT* for various character pairs: Experiment 1*

pairs	Character polarity			
	dark–dark	light–light	dark–light	light–dark
All	503	502	518	513
Same	487	489	507	501
Different	518	516	530	525
Same, correct	489	489	508	502
Different, correct	519	517	531	525

*Reaction time in ms

The other observation to be made about the data in Table 1 is that the means for both kinds of matched polarity (dark–dark and light–light) are not different from each other nor are the means for both kinds of mixed polarity (dark–light and light–dark) different from each other. The finding of no difference between dark–dark and light–light is particularly meaningful. It suggests that dark character polarity is not more quickly processed than light character polarity. Insofar as RT is an index to the complexity of processing information according to its polarity, there is no difference between the two polarities.

Table 2 shows for each pair type the mean RT for matched polarity pairs and for mixed polarity pairs, along with the *F*-ratio for the particular orthogonal contrast of matched vs. mixed polarity and the Type I error probability. In all cases, the difference in mean RT between matched and mixed polarity pairs is statistically significant. The mean RT for matched polarity pairs is less than for mixed polarity pairs. The difference is small, about 13 ms averaged across the different pair types.

TABLE 2. *Comparison of matched vs. mixed polarity: Experiment 1*

pairs	Mean RT* by polarity		Results of orthogonal contrasts	
	matched	mixed	$F(1, 42)$	p
All	503	516	8·831	0·005
Same	488	504	7·698	0·008
Different	517	528	4·773	0·035
Same, correct	489	505	8·107	0·007
Different, correct	518	528	4·195	0·047

*Reaction time in ms

The results of Experiment 1 seem to be unambiguous. The data themselves clearly show that subjects respond more quickly to matched polarity pairs. In terms of the theories given to explain these phenomena, it appears to be the case that when the characters are in matched polarity, a rapid, visual comparison can be made. However, when the characters are in mixed polarity, a more time consuming comparison must be made. The subject must encode both characters in some form which can be compared, perhaps by giving them each a name (the name of the letter), a more time consuming comparison method.

Thus the original hypothesis that information comparison strategies differ for matched vs. mixed polarity pairs appears to have received support from the results of Experiment 1. However, Experiment 1 does not reflect the operator's real task. The operators do not have information from a source document located within a visual fixation of displayed information. The operators must reposition their eyes back and forth between the two sources of information. Such eye movements consume time. This and many other differences between source documents and displays might eliminate the advantage of matched polarity. Experiment 2 was undertaken to explore that possibility.

3.2. **Results and Discussion—Experiment 2**

The results of Experiment 2 were analysed by analysis of variance with a repeated measures design. The mean RT and the *F*-ratios for the comparisons of mixed vs. matched polarity for each response type are shown in Table 3. None of the measures showed a difference at the $p < 0{\cdot}05$ level of significance. The results of the comparisons with a delay of 500 ms give support to the hypothesis that the advantage of matched polarity does not exist in the real world situation, in which eye movement latencies introduce delays.

TABLE 3. *Comparison of matched vs. mixed polarity with 500 ms delay: Experiment 2*

	Mean RT* by polarity		Results of ANOVA	
pairs	matched	mixed	$F(1, 42)$	p
All	486	483	0·077	N.S.**
Same	469	476	0·303	N.S.
Different	502	490	1·876	N.S.
Same, correct	472	478	0·259	N.S.
Different, correct	503	490	2·204	N.S.

*Reaction time in ms

**Not significant ($p > 0{\cdot}05$)

4. *Conclusions*

An attempt can be made, now, to understand the role of polarity in making such comparisons. The role of polarity in the aetiology of visual fatigue based on hypothesized increased information-processing complexity with mixed polarity can then be assessed.

The first question to be answered was whether or not image polarity was an attribute of a stimulus which would, when the same, facilitate the comparison of two items. If it were not such an attribute, then the two-part hypothesis could not be true that mixed polarity is cognitively more complex and that this complexity leads to visual fatigue.

Experiment 1 subjected the first part of that hypothesis to the test. The results clearly show that image polarity is an attribute which, when the same, facilitates the comparison of two stimuli. The standard interpretation is that when the image polarity of two items to be compared is the same, a visual comparison is facilitated. When the image polarity of the two items is not the same, they must first be encoded on some basis that does allow comparison, such as name-encoding. The visual comparison strategy, when possible, is performed more quickly than the name-encoded comparison strategy. Presumably, the longer latency of the latter indicates that it is cognitively more complex.

Having established that image polarity is related to cognitive complexity, the question remained about whether or not the advantage of matched polarity exists in a real situation. Any factor which makes two stimuli visually dissimilar contributes to eliminating the advantage of making a visual comparison. Experiment 2 shows that when a delay is present, the subjects cannot use a visual comparison strategy. They must use the name-encoding comparison strategy. It makes no difference whether or not the image polarity of the source document and the display are the same. Whether such complexity as is involved with the name-encoding strategy is fatiguing, making the image polarity of source documents and displays the same cannot affect the strategy that the operator uses, and hence cannot affect any postulated fatigue.

References

Bauer, D., and Cavonius, C.R., 1980, Improving the legibility of visual display units through contrast reversal, in *Ergonomic Aspects of Visual Display Units*, edited by E. Grandjean and E. Vigliani (London: Taylor & Francis) pp.137–142.

Cavonius, C., 1982, Changes in contrast and sensitivity from VDT usage. Paper presented at the IX International Ergophthalmological Symposium, San Francisco.

Krueger, H., 1981, Personal Communication, October, Munich.

McVey, B.W., Clauer, C.K., and Taylor, S.E., 1984, A comparison of anti-glare contrast-enhancement filters for positive and negative image displays under adverse lighting conditions. This vol., pp.405–409.

Posner, M.I., 1978, *Chronometric Explorations of Mind* (Hillsdale, NJ: Erlbaum).

Radl, G.W., 1980, Experimental investigations for optimal presentation-mode and colours of symbols on the CRT-screen, in *Ergonomic Aspects of Visual Display Terminals*, edited by E. Grandjean and E. Vigliani (London: Taylor & Francis) pp.127–136.

Taylor, S.E., 1980, Visual fixation: the influences of retinal and extraretinal information, doctoral Dissertation, Northern Illinois University, 1979, *Dissertation Abstracts International*, 40, 5446B, University Microfilms No. KHN80–11179.

Taylor, S.E., McVey, B.W., and Emmons, W.H., 1983, *Cognitive Complexity Related to Image Polarity in the Etiology of Visual Fatigue* (San Jose, CA: Human Factors Centre, IBM Corporation, HFC–45).

Stress as a Function of Increased Cognitive Load at a VDT

W. Barfield
School of Industrial Engineering, Purdue University, West Lafayette,
Indiana 47907, USA

Abstract

An experiment is reported which investigated the effects of computer induced cognitive load on performance. Cognitive load was varied by manipulating scrolling rates and density levels of characters on a video display terminal. Subjective reports of stress were elicited by a stress questionnaire. Results showed a significant performance decrement for scrolling rates but no differences for density levels. The subjective stress questionnaire indicated that subjects reported more stress as scrolling rates increased.

1. *Introduction*

Currently computers are being used for a variety of purposes in offices and other work environments. Tasks such as word processing, data entry, and text and data manipulation are frequently performed by office personnel with the aid of a computer. These tasks often require workers to visually monitor information which is moving across the video display terminal (VDT) screen. The result of moving information across the VDT screen is referred to as scrolling (Bury *et al.*, 1982; see also Brooke and Duncan, 1983).

Increased rates of scrolling information or text on a VDT may lead to increased cognitive load, decreased performance, and also stress (Salvendy, 1982).

In the research study which follows, the speed of scrolling information and the density of information on the VDT screen were varied to manipulate cognitive load. The following hypothesis was investigated: increased rates of scrolling and increased density levels (characters per line) will lead to performance decrements, increased errors and stress.

2. *Methods*

2.1. Experimental Design

Scrolling rates and density levels were manipulated as independent variables in the experiment. The four scrolling rates were: 20 (low), 30 (medium 1), 45 (medium 2), and 67.5 (high) lines/minute (all derived from a pilot study). In addition, two density levels were used, either four or seven characters/line. The dependent variables were three performance measures and subject's answers to a questionnaire, the Feeling Tone Checklist, which was used to obtain subjective reports of stress (Pearson and Byars, 1956). This questionnaire contains questions which allow the subject to relate his or her physical and psychological attitude at that moment. Also, six questions relating specifically to the scrolling task were included in the questionnaire (heavy eye strain, have to work too fast, etc.).

In addition a repeated-measures design was used. The eight treatment combinations (4 scrolling rates by 2 densities) were randomized for each subject.

2.2. Subjects

Ten university students participated as volunteers or for class credit in the experiment (4 males, 6 females; $\bar{X}$ = 22 years). All subjects had previous experience with computers.

FIGURE 1. *Example of characters scrolled across VDT screen. Target appears along left margin. Instructions to identify target are also shown.*

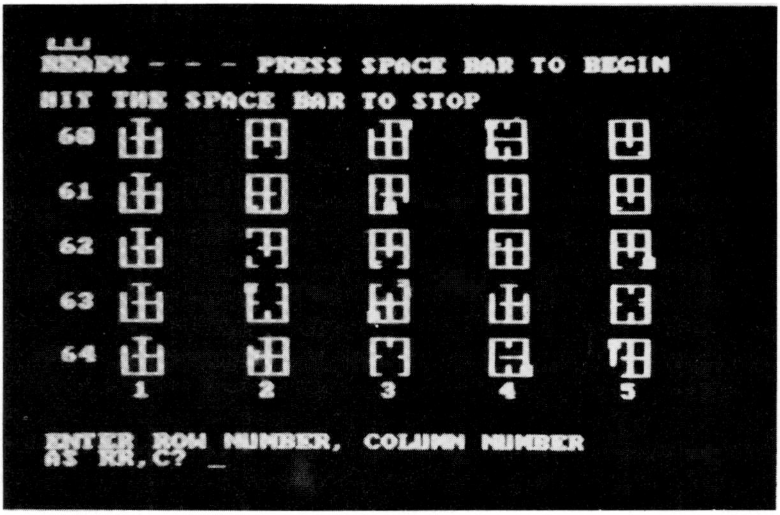

2.3. Apparatus

An IBM personal computer was used to generate graphic characters which were randomly presented to each subject (Figure 1).

2.4. Procedure

Subjects searched for a visual target which appeared randomly among graphic characters on the VDT screen for each trial. The target was scrolled along the left screen margin for comparison purposes. When subjects detected a target they pressed the 'skip bar' and then were prompted by the software to type the row and column numbers which allowed the identification of the target. Each subject completed at least 31 visual search trials for each session. Each session lasted 30 minutes followed by a ten minute rest period. During the rest period subjects completed the stress questionnaire.

3. *Results*

3.1. Performance Measures

Performance measures were based on the following criteria. A correct target decision occurred when a subject stopped the scrolling mode of the VDT and typed in the correct row and column numbers for the character which matched the target. If the subject stopped the scrolling mode when the target was not on the VDT screen, a false identification was recorded. Finally, a miss occurred if the subject did not respond during a trial (every trial had one target).

Table 1 shows the results of the ANOVA computed for correct target detection.

As revealed by Table 1 the interaction between scrolling rates and density levels was not significant. In addition density levels were not significantly related to correct target detection.

TABLE I. *ANOVA for correct target detection.*

Source	*df*	*ss*	*MS*	*F*
A (subjects)	9	197·61	21·96	1·58
B (density)	1	25·31	25·31	1·8
AB	9	125·31	13·92	
C (scrolling)	3	861·73	287·24	35·55[a]
AC	27	218·13	8·08	
BC	3	74·23	24·74	< 1
ABC	27	868·63	32·17	

[a] $p < 0{\cdot}001$

Scrolling rates were found to significantly affect the subjects' ability to detect the target. Moreover, Neuman–Keuls multiple comparisons were computed for the four levels of scrolling. The results showed that all levels of scrolling rates were significantly different from each other with the exception of scrolling 20 and 30 lines/minute.

Figure 2 shows the frequency of correct target detections for scrolling rates for each density level. A high rate of scrolling is associated with a decreased level of target detection. The frequency of correct target detection is not affected by increasing scrolling rates from 20 to 30 lines/minute, nor by different density levels.

FIGURE 2. *Frequency of correct target detection for scrolling rates for each density level.*

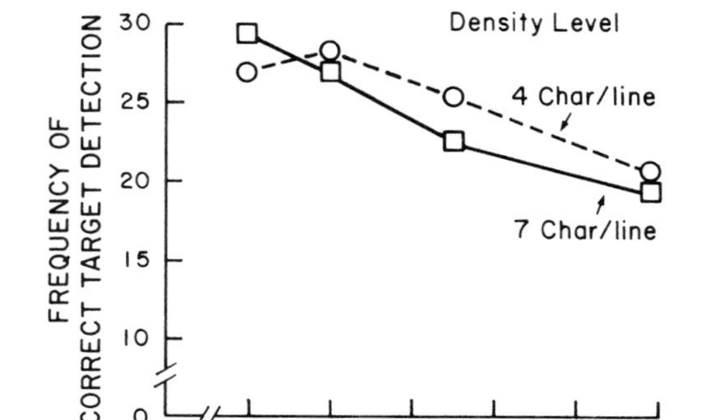

An ANOVA for errors revealed that scrolling rates had a significant effect on missing the target ($F(3, 27) = 17{\cdot}61$, $p < 0{\cdot}001$) but no effect on the false identification of the target. Neuman–Keuls comparisons showed that missing the target was different for every level of scrolling with the exception of 20 and 30 lines/minute. Figure 3 graphically shows this relationship.

The results of the stress questionnaire are shown in Table 2.

Examination of the frequencies of responses in Table 2 indicate that subjects responded differently to the three possible alternatives of 'better than', 'same as', and 'worse than' for each level of scrolling. In general, subjects responded 'better than' less often and 'worse than' more often as scrolling rates increased for each density level. This trend suggests an increased level of stress for subjects as scrolling rates increased.

4. *Discussion*

Increased scrolling rates lead to decreased target detection, increased errors and subjective reports of stress. Density levels had no effect on these measures. Thus, the speed at which information is presented on a VDT is critical to overall task performance.

FIGURE 3. *Frequency of errors for scrolling rates.*

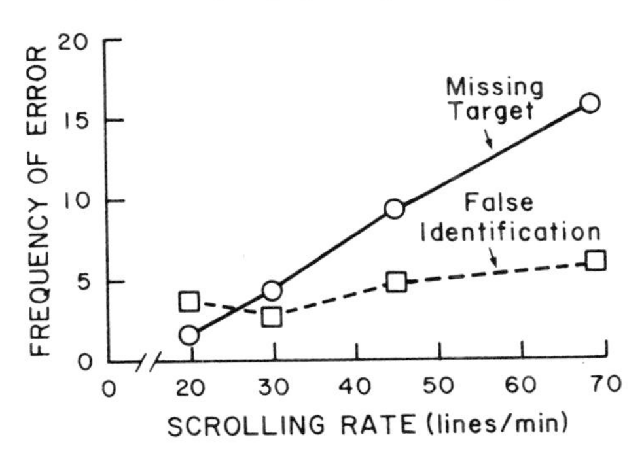

TABLE 2. *Frequency of responses to stress questionnaire.*

	Scrolling rate											
	Low			Medium 1			Medium 2			High		
Density	BT	SA	WT	BT	SA	WT	BT	SA	WT	BT	SA	WT
4	92	42	36	76	54	40	76	35	59	62	51	57
7	97	41	32	63	62	46	64	60	46	49	52	69

BT = Better than SA = Same as WT = Worse than

There were no significant differences in performance between scrolling rates '20' and '30' for either density level. This suggests that we can scroll more information on a line without significantly increasing errors or decreasing correct target detection for these rates.

An examination of Figure 2 shows that the predominant search strategy as scrolling rates increased was to refrain from responding unless reasonably sure that the target was on the screen. This is indicated by a relatively stable 'false identification of the target error' versus 'missing the target' error.

Lastly, the subjective stress questionnaire revealed a general trend for increased reports of stress as scrolling rates were increased.

Acknowledgements

I thank Bill West for development of the software, F. Lynne Taylor, Nelson Tamplin, and Mica Robertson for their assistance in running subjects. I am also indebted to Dr G. Salvendy, and Sheue-Ling Hwang for their valuable contributions. Part of this study has been published in the Proceedings of the Human Factors Society 27th Annual Meeting, Norfolk, Virginia, 1983.

References

Brooke, J.B., and Duncan, K.D., 1983, A comparison of hierarchically paged and scrolling displays for fault finding. *Ergonomics*, **26**, 465–477.

Bury, K.F., Boyle, J.M., Evey, R.J., and Neal, A.S., 1982, Windowing vs scrolling on a video display terminal. *Proceedings: Human Factors in Computer Systems*, 15–17 March, Gaithersburg, Maryland, pp.41–44.

Pearson, R.G., and Byars, G.E., Jr, 1956, *The Development and Validation of a Checklist for Measuring Subjective Fatigue* (Randolph Field, Texas: School of Aviation Medicine, USAF, Report 56–115).

Salvendy, G., 1982, Human–computer communication with special reference to technological developments, occupational stress and educational needs. *Ergonomics*, **25**, 435–447.

Efficiency of Data Entry by VDUs—A Comparison Between Different Softwares

C. Romano and A. Sonnino
Institute of Occupational Health, University of Turin, Italy

Abstract

A group of VDU-users were asked to enter the same alpha-numeric data into a computer using the same VDU but three different softwares. One implies the use of a mask. The others are of an interactive type (each data being requested by a specific question on the screen) though differing in that only one of them gives an acoustic signal when ready; random delays are introduced in order to simulate the variable work speed of a heavily loaded computer.

A great advantage in terms of performance and of reduction of stress is obtained when using a mask.

When performing without a mask the use of the beep to signal that the computer is ready to accept the data significantly lowers input times and reduces the stress of the operators.

1. *Introduction*

There is a growing awareness of the key role exerted by the design of the software regarding both the quality of the performance and the well-being of the operator. Recent papers (Huchingson *et al.*, 1981; Tullis, 1981) deal with the improvement of the human/computer interface, in terms of graphic type of information, format and load of the message.

Our recent field experience (unpublished data) dealt with complaints from VDU-operators about the inconvenient way the data were to be keyed in.

The operators' main complaint was the uneven pace at which the requests for each data appeared on the screen (sequentially) depending on the variable computer load. According to these operators, this led to frequent errors and a relevant stress.

On the basis of this experience, we compared some different softwares devoted to introducing data.

2. *Methods*

We have simulated the task of a hospital administrative clerk consisting in the keying-in of patients' names and code numbers of the clinical examinations performed.

Patients were characterized by a fixed number of identifying items and a variable number (ranging from 1 to 10) of items referring to the exams performed.

Data had to be introduced into the computer from a print-out source in three different ways. Two out of the three consisted in a series of requests which appeared on the screen with random delay from 0 to 2 seconds in order to reproduce those typical of a busy computer (Figure 1).

FIGURE 1. *Display format in Test 1 and 2.*

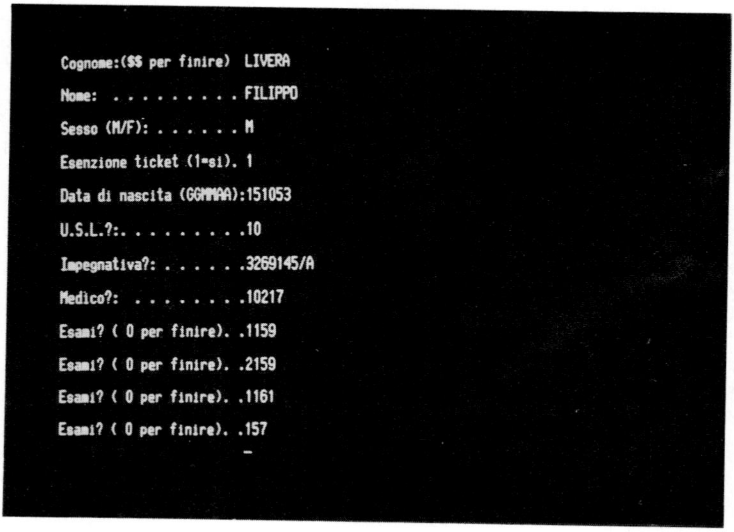

```
Cognome:($$ per finire) LIVERA
Nome: . . . . . . . . . FILIPPO
Sesso (M/F): . . . . . . M
Esenzione ticket (1=si). 1
Data di nascita (GGMMAA):151053
U.S.L.?:. . . . . . . . .10
Impegnativa?: . . . . . .3269145/A
Medico?: . . . . . . . .10217
Esami? ( 0 per finire). .1159
Esami? ( 0 per finire). .2159
Esami? ( 0 per finire). .1161
Esami? ( 0 per finire). .157
```

The appearance of each request on the screen signified that the computer was ready to receive the answer. In the first software (Test 1) each request was indicated by means of an acoustic signal. In both cases data were not accepted if keyed in during the delay time. The third software (Test 3) was based upon the use of a mask (Figure 2) in which the beginning of each field was automatically reached by the use of a tabulator key. In this case a delay time was present only at the end of each group of data (i.e. of a single patient); the filling out of the mask was actually independent from the response of the computer, being a task of the VDU itself.

Twenty subjects (15 females and 5 males) entered the study. Ages ranged between 25 and 40. All the subjects were familiar with the VDUs being programmers or data entry operators. None of them had, however, previously used the programmes tested. Each subject was requested to enter data in three different consecutive days, starting at the same

FIGURE 2. *Display format in Test 3.*

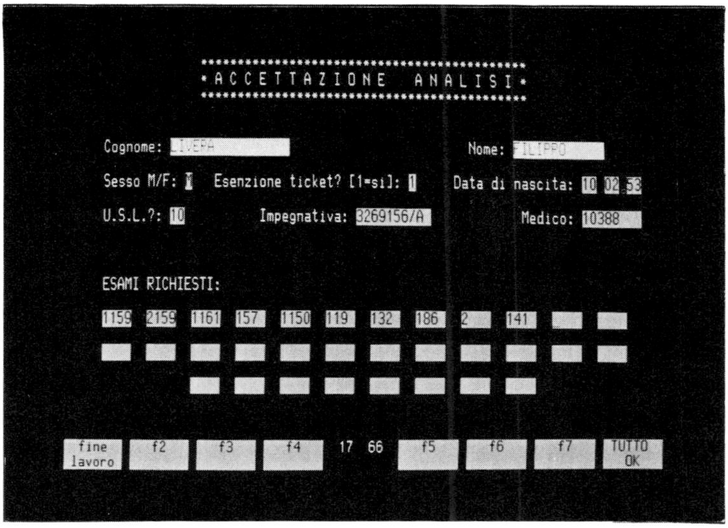

time of day; the first day software No. 1 was used, the second day No. 2 and the third day No. 3. Each test implies two hours of work, including spontaneous pauses, beginning after at least 15 minutes of training.

Data to be introduced were exactly the same in each test and for each subject.

At the end of the third test a questionnaire was administered in order to obtain information about likes and dislikes, subjective feelings of efficiency and stress.

We compared the data for all three tests in terms of time and errors. When data of a single patient took more than 150 seconds to be keyed-in, they were eliminated from time analysis. Each difference between the reference and the keyed in items was defined as an error.

3. *Results*

Figure 3 shows the mean time for input of a single patient's data. As predictable, the shortest keying-in time was observed in Test 3; the acoustic signal improved the speed in Test 1 when compared with Test 2.

Figure 4 shows the input time for a single patient, with 10 items referring to exams; the mean value of three patients is reported, measured at start time and after 30, 60, and 90 minutes of work. The time shows a clear decrease in Test 3 and to a lesser extent in Test 1. The trend is less evident in Test 2. This finding is presumably due to the benefits of rising confidence with the test, which seem to exceed the effects of fatigue, also due to the short duration of each test.

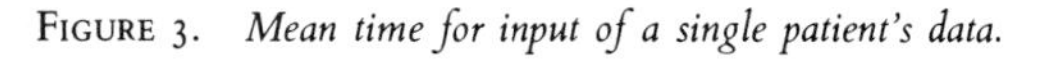

FIGURE 3. *Mean time for input of a single patient's data.*

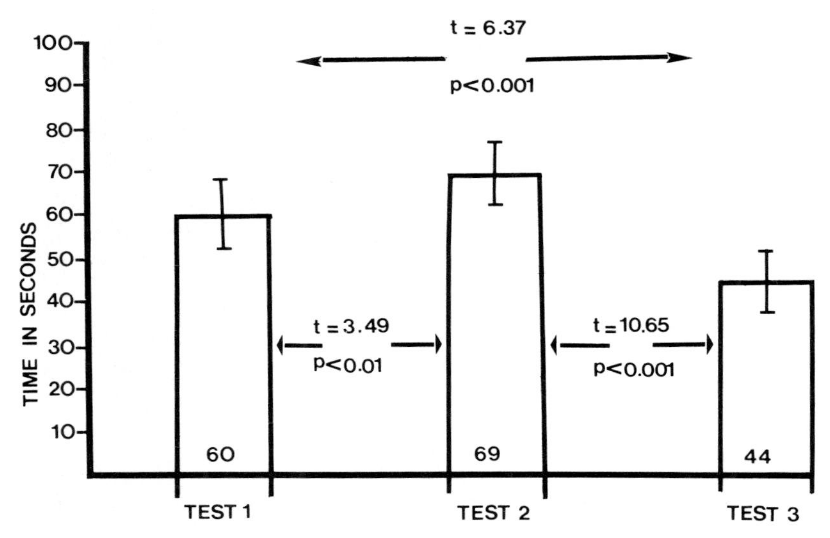

FIGURE 4. *Time of input for a single standardized set of data (see text) measured at fixed delays from beginning of the test.*

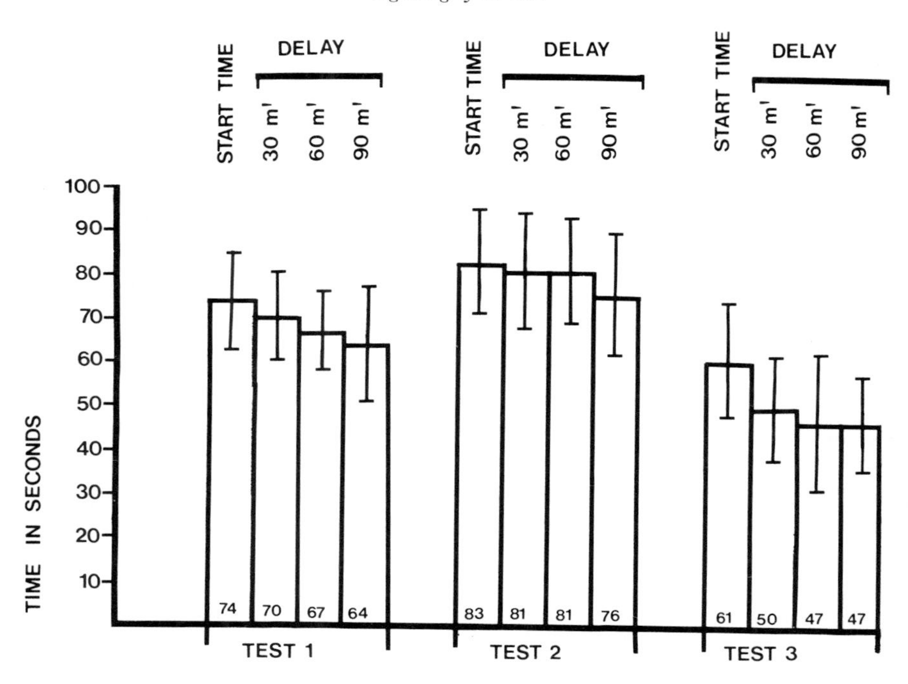

Errors do not globally vary significantly amongst the 3 tests. We found no reliable trend related to time. The high value of the calculated standard deviations does in fact reduce the significance of the different number of errors found in each test during the 30 minute periods of observation and in particular the significance of the rising number of errors in the last period of tests 2 and 3.

Regarding the subjective evaluation of the three tests, all the subjects declared they preferred the third one. The reasons for this preference were attributed essentially to its high speed and to a lesser extent to its simplicity, to the possibility of correcting the wrong inputs and to the lack of delays. In fact all but one subject classified the 3 tests, from the stress point of view, in the following order: $2 > 1 > 3$. When dealing with keying-in precision, 15 subjects reported to have made the highest number of errors in Test 2, 5 in Test 1 and none in Test 3.

4. *Discussion*

The analysis of our data clearly shows that a great advantage in terms of performance and reduction of stress is obtained when the input of data is made through a mask.

When performing without a mask the use of the 'beep' to signal that the computer is ready to accept the data significantly lowers input times and reduces the stress of the operators. This could be due to the lower strain on visual functions deriving from the possibility of entering data without having to continuously look at the screen.

None of the three data-entry programmes used in this study showed any difference in terms of errors.

However, this finding is biased by the relevant scattering of data, which may be due to the small number of subjects studied and the short time characterizing our tests. Furthermore, this result differs from the subjective evaluation of the operators.

In conclusion the present study confirms the importance of the software when introducing data; it points out the usefulness of using a mask, whenever possible, or at least of giving acoustic signals to inform the operator that the computer is ready.

At any rate the study itself should be seen as a preliminary one, needing further confirmation through tests applied to larger groups of people working for a longer period of time.

References

Huchingson, R.D, Williams, R.D., Reid, T.G., and Dudeck, C.L., 1981, Formatting, message load, sequencing method, and presentation rate for computer-generated displays. *Human Factors*, **23** (5), 551–559.

Tullis, T.S., 1981, An evaluation of alphanumeric, graphic and color information displays. *Human Factors*, **23** (5), 541–550.

Reorganization of the Telephone Information Service from Telephone Books to VDTs

A. SONNINO
Institute of Occupational Health, University of Turin, Italy
AND G. MORUZZI
SIP, Società Italiana per l'esercizio telefonica, Rome, Italy

1. *Introduction*

The remarkable increase of telephone users, which has taken place in Italy in recent years, has made it necessary to reorganize not only the installations, materials used and main services but also auxiliary services, as these were proving progressively more and more inadequate to meet the users' demand. Among many others the subscriber telephone service was reorganized by means of an electronic processing system using a series of video terminals.

The introduction of this equipment caused a series of reactions in the personnel involved, in part due to the natural suspicion of new technology and in part caused by the inadequacy of the premises where the video terminals were to be installed. With this change age-old problems were brought up, such as job organization and the serious problem of trying to work efficiently in over-crowded conditions with as many as 40 to 60 people all talking, out of necessity, at the same time.

2. *Methods*

Although a series of changes were made in various branches to meet the personnel's requirements, it was soon obvious that by trying to resolve individual problems one at a time, other, and at times more serious, problems were created.

In trying to solve this problem, the management carried out surveys both in Milan and Turin branches, giving the local university occupational medicine department the job of identifying the harmful and discomforting factors. They were also asked to find suitable solutions which might be used as guidelines in the construction and re-structuring of similar work places. In order to look deeper into the problem surveys were also carried out in telephone information rooms of some provincial areas in Turin.

The following aspects were taken into consideration:

1. The indoor climate, the lighting, the background noise and the voice through headphones.
2. Ionizing radiation emitted by the video terminals.
3. The workplace and postures.
4. The state of health and instrumental examinations of sight and hearing.
5. The personnel's work behaviour based on a series of individual and group interviews.

The group of telephonists was compared with a control group, made up of clerks employed by the firm who had never used VDTs, in order to verify the last two points.

3. *Results*

At the end of this investigation, it was shown that:

1. The indoor climate conditions, apart from the fact that the air conditioning was inadequate, worsen when 40 to 60 VDT-users work together, as in this section. In fact the terminals themselves together with the operator's body heat produce about 200–500 W, which changes thermal conditions very rapidly.
2. The lighting system, which was insufficient and a cause of complaint even when telephone books were used, was found to be completely inadequate when VDTs were in use, not being fitted with appropriate reflectors. Also the plastic covers reflected light onto the screens causing interference. The operators' natural reaction was to turn off part of the lighting leaving the room badly and unevenly lit, which is absolutely unacceptable (3 to 150 lux).
3. The background noise of the room at peak hours, is about the same as that coming from the voices in the headphones consequently causing interference.
4. Previous work spaces proved inadequate when occupied by terminals, as these had never been ergonomically planned.
5. The 'social' part of the investigation included, as already mentioned, an epidemiologic (medical examinations, psychosomatic survey, sight and hearing instrumental examinations) and a behavioural (individual and group interviews) survey.
6. The lack of technical knowledge on the part of the users created an atmosphere of suspicion, which led to 'false problems' beyond all real dimensions.

During the survey a series of medical examination were carried out in Turin: 465 medical check-ups, 507 sight tests, 507 audiometric tests subdivided into the telephonists and control group. The results showed no remarkable difference from a clinical point of view between the 2 groups.

The instrumental examination showed no optical strain was caused by VDTs. The average visual acuity appeared to be normal and the same for both groups: no change was found in the phoria, which would be a demonstration of eye fatigue.

The visual symptomatology investigation showed a subjective difference only in the 'eye burning' symptom (49·5% in the telephonists and 36·8% in the clerks). This problem though real from a subjective point of view, did not show any conjunctiva disorders.

The audiometric tests showed normal results for the telephonists. The graph curve was almost equal to that of the clerks.

The psychological effort required by this task can be outlined as follows:

1. Understanding the request (the subscriber speaks badly, is lengthy, gives useless information; interference due to the background noise, etc.).
2. Giving the requested information in a simple and comprehensive form for the subscriber.
3. Retention of information already seen on the screen.

Other interpersonal problems have to be added to this psychological load, which include keeping one's 'self control' when speaking to the subscriber.

As all these factors go towards building up the emotional and psychological load, in theory a foreseeable consequence might be the occurrence of emotional disorders more frequently than in other situations. Although some literature does speak about the existence of the so-called 'telephonist's neurosis', when comparing the control group with the telephonists, no such effect was observed. In fact the emotional balance measurement, obtained on the basis of the psychosomatic symptoms observed at various levels (cardiovascular, gastrointestinal, neuromuscular, etc.) is exactly the same in both groups.

It is interesting to note that video terminal operators complain of troubles such as: headaches, shoulder and backaches, neck pains and insomnia. In our research the presence of such disorders, from the anamnesis was shown to be no higher than in the control group.

We may conclude that no specific harmful factors emerged both at organic and psycho-emotional levels.

It is worthwhile noting, however, that fatigue, both at the end of a day's work and after a holiday, is greater in the group of telephonists than in the clerks (respectively 10·5% and 4·6% after a holiday and 76% and 59% at the end of a day's work.

Our survey also underlined the importance of adequate lighting for VDT premises; in fact 79% of the operators have a bad opinion of the plant, whereas the clerks (who work in rooms with similar electric equipment) are far less critical (less than 50%). It is meaningful that fatigue is present in the telephonists after only two and a half hours work and in the clerks after five and a half hours.

Telephonists claim that 99% of their arguments are due to the aforementioned indoor climate problems.

The ergonomics analysis showed a series of parameters to be modified:

1. Rooms must be smaller for better temperature control according to variation in heat sources, so as to reduce background interference.
2. A better distribution and homogeneity of natural and artificial lighting together with the right colour temperature must be guaranteed.

3. CRT must have no reflective surface and the VDT must not be higher than the operator's eye level.

4. *Conclusion*

Particular attention was paid to the construction of the newly installed VDT terminals, produced by Selenia.

Technological modifications are constantly made to improve working conditions. A recently experimented device to reduce line disturbance in the headphone has been fitted. This device also enables the telephonists to adjust the volume of the incoming message.

The company, in cooperation with trade unions, have agreed a system whereby indoor climate and lighting conditions can be systematically checked.

Nationally controlled environment commissions—made up of trade union members and management representatives working in single provinces and regions—will evaluate this information and make environmental changes when and where necessary.

In 1984, 5500 telephone workers will be asked to take part in a survey which includes a series of medical check-ups and tests. These will be carried out by the various local University Institutes of Occupational Medicine, and will include specialized vision tests, hearing and osteoarticular apparatus.

These results will be compared with the previous survey and will bring to light any eventual relationships between health conditions and the activity performed.

Small work groups (20–30 units) have been created in an effort to solve the work organization problem; the old method of supervisors has been changed to a system which makes each individual operator responsible and leaves them a self-management as far as work load distribution is concerned. At the same time the range of tasks performed by each operator was broadened so as to reduce monotony and de-responsibilization, which are amongst the prime causes for lack of work motivation.

Further Reading

Brouha, L., 1967, *Physiology in Industry* (Oxford: Pergamon Press).
McIntyre, D.A., 1980, *Indoor Climate* (London: Applied Science Publishers).
Philips, 1980, *Lighting Manual* (Eindhoven: Philips).

Implementation of an ADP-System to Calculate Salaries: Evaluation of the Implementation Process and Changes in Job Content and Work Load

P. HUUHTANEN
Institute of Occupational Health, Department of Psychology, Laajaniityntie 1, 01620 Vantaa 62, Finland

Abstract

A sample of 24 payroll clerks in nine public institutions were followed through different stages of the implementation of an ADP-system. Two other groups were used as controls. Both the two-year implementation process and the changes in job content and strain were evaluated. Many laborious manual tasks had been computerized, and the work involved more qualification demands than manual salary calculation. The study also revealed problems similar to those encountered in other studies: time pressure, dependence on the computer system; and the amount of new kinds of routine tasks. Both physical and mental strain were slightly reduced, but the transition period itself was experienced as stressful. The study indicated that the technical process of computerization was well planned. More attention should be paid to social and psychological aspects. Suggestions on how to improve the training, information, and planning of such changes are discussed.

1. *Introduction*

The computerization of administrative work has gained much attention among social researchers during the last few years. The scope of the behavioural research extends from stress reactions of terminal workers (Johannsson and Aronsson, 1979; Salvendy, 1982) to societal phenomena like unemployment, new values in the post-industrial societies and effects of the automation on the way of living in general. Much research has been devoted to changes in the occupations and qualifications of the employees (Järvinen, Kirjonen, Tyllilä and Vihmalo, 1982). Of growing interest has been the implementation process of ADP systems in the workplaces (Eason, 1982).

The aim of this study was to investigate how the computerization of the calculation of salaries affected job content, the demands of the work, mental load, and work organization

(Huuhtanen, 1983). The sample comprised three groups of payroll clerks working at 22 public institutions: one group of payroll clerks calculated salaries manually (n = 17); one group used computerized systems (n = 25); and the follow-up group (n = 32 initially), which was followed through the two-year transition period. The other two groups were reference groups.

The computerized system used to calculate and pay civil servants' monthly salaries is based on the phase distribution model developed by the State Computer Center. The collection and coding of necessary information as well as data entry are done separately within each government institution. The calculations are carried out centrally at the State Computer Center. All the material is checked and corrected before the salaries are paid through bank accounts. The State Treasury carries the main responsibility for the implementation, maintenance, and development of the system.

The theoretical framework was based on psychological stress theories supplemented with data on work organization (Figure 1). It was assumed that the effects of automation on the work and the payroll clerks would be mediated via work organization and the distribution of tasks. Personal characteristics, motives and work experience were also thought to affect the degree of perceived stress experienced with the implementation of the new technology.

FIGURE 1. *The framework of the study.*

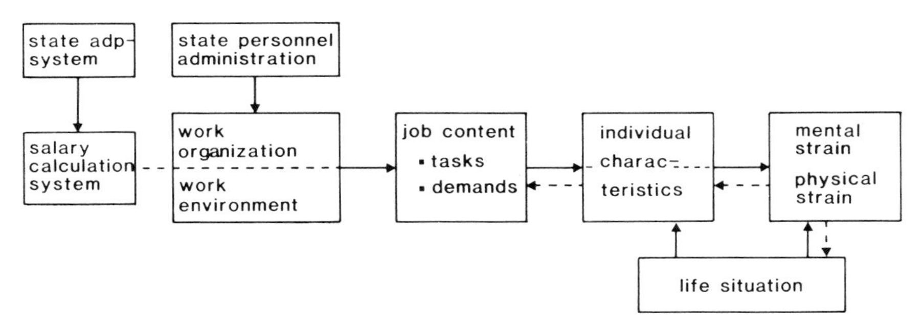

2. *Methods*

Information was collected by interviews, questionnaires, and self-ratings. The follow-up group was interviewed a total of three times: before the transition, immediately after the transition; and 8 to 10 months after the transition (Figure 2).

The topics on which data were gathered through interviews included job contents, the demands of the work, work organization, individual needs and motives, evaluation of the new computerized system, the preparation and implementation of the transition, perceived health, and attitudes about computerized office work in general. The questionnaires were used to explore the dimensions of job contents, job satisfaction, social contacts, and long-term stress reactions.

FIGURE 2. *The collection of data.*

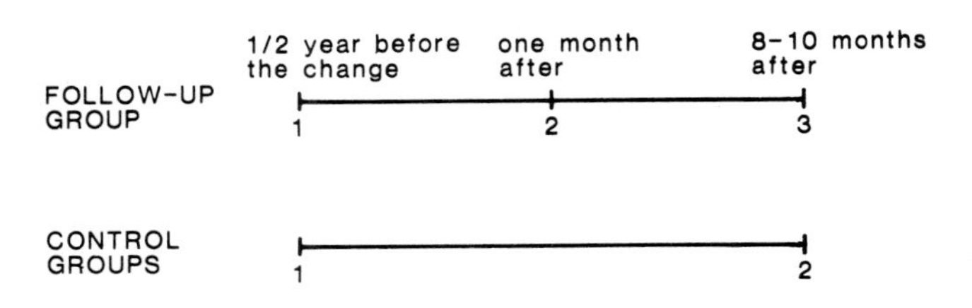

3. *Results*

This article summarizes the initial results on the change in the work and the transition process itself. The report relies heavily on the payroll clerks' subjective evaluations and reactions.

For the most part, the payroll clerks felt that the changes in job contents were positive. Many laborious manual tasks (calculations, collecting and typing out payroll lists for different registers, etc.) had been computerized, and the work now required more knowledge and skills (Figure 3). Fears of increased monotony and simplified job demands had not proved to be true, or at least these drawbacks had not yet materialized 8 to 10 months after the transition.

Despite these positive effects, the study also revealed problems similar to the problems encountered in other studies. Time pressure and dependence on the computer system were considered problematic. The amount of checking the lists and codes had increased, and difficulties were experienced in the correction of errors.

Both the physical and the mental strain of the work were slightly reduced over time (Figure 4). Automation did not lead to excessive use of visual display terminals, as data entry normally took two to four days a month. Table 1 illustrates the decrease of the physical strain of the follow-up group.

The transition period itself (one to two months) was experienced as stressful. Work-related problems could not be forgotten at home, and the amount of energy available for leisure activities decreased (Figure 5).

The payroll clerks proposed many suggestions on how to improve the training, the preparatory information, and the instructions distributed in conjunction with such transitions (Table 2). They also had ideas about the follow-up actions necessary after the implementation of such transitions.

4. *Discussion*

The study clearly indicated that the technical, social, and psychological aspects of the whole implementation process have not been mastered equally. Much more attention should be

FIGURE 3. *Evaluations of the quantitative or qualitative changes caused by the ADP-system (follow-up and ADP-based groups together, n = 45).*

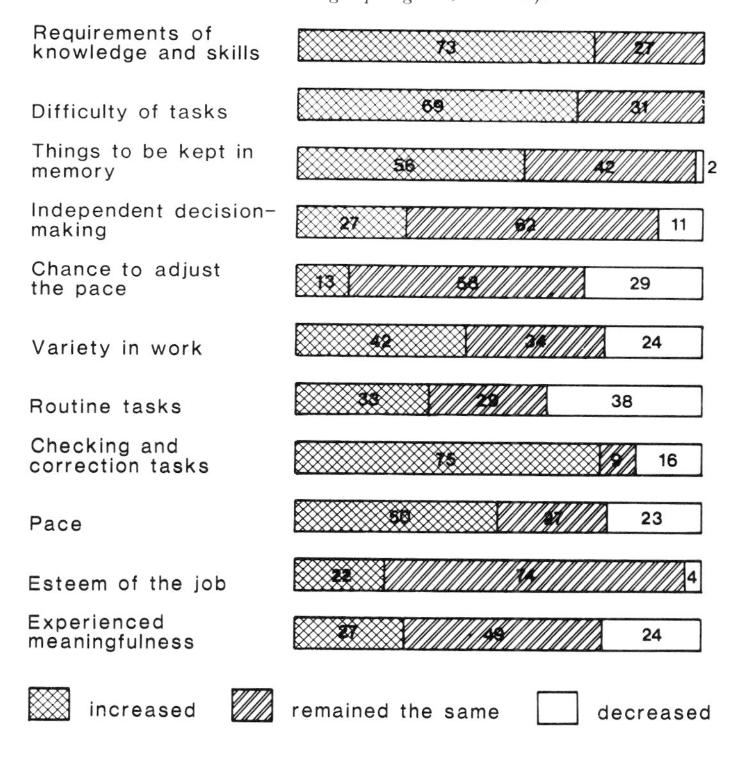

FIGURE 4. *Frequency of the experiencing of overload before and after the change (follow-up group,* n = *23).*

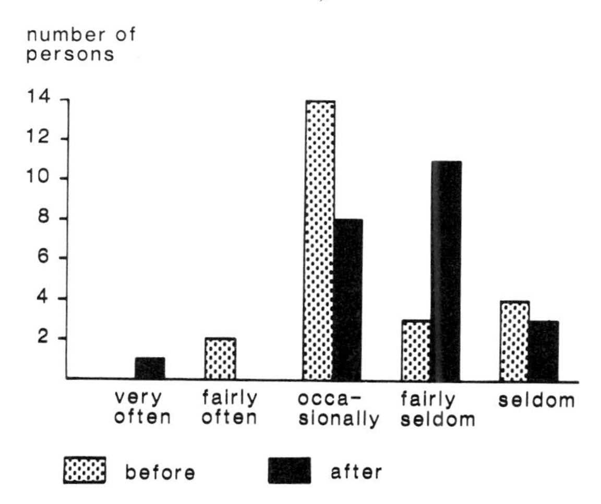

TABLE 1. *Recurring pains in different parts of the body before and after the change in the work of the follow-up group (number of persons).*

Part of the body	Follow-up group (n = 23) before	Follow-up group (n = 23) after	Control group (n = 29) before	Control group (n = 29) after
Neck	15	10	19	16
Shoulder	6	1	6	9
Elbow	2	0	5	5
Wrist	6	2	5	8
Fingers	5	2	3	4
Back	8	7	13	11

TABLE 2. *Problem areas experienced by the payroll clerks during the implementation of the ADP system.*

1. Information distributed beforehand
2. Preparations for the change inside the institutions
3. Scope of the ADP system
4. Timing of education and training
5. Giving a coherent picture of the ADP system
6. Keeping the instructions up to date during and after implementation of the change
8. Labour resources during the transition period
9. Time of year of the change

FIGURE 5. *Frequency that occupational pressures followed the clerks home before and after the change (follow-up group,* n = *24).*

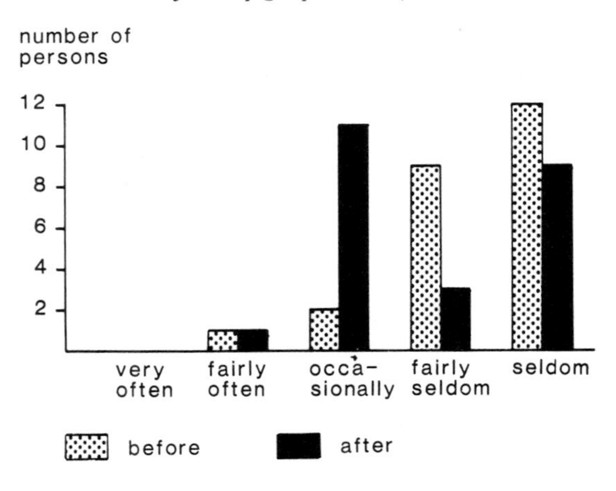

paid to the social and psychological aspects and effects of computerization. One way to soften the transition would be to increase user participation in the planning and implementation stages. More information and training are needed for both employees and supervisors.

Deeper analyses of this material will be done as case studies that concentrate on individual reactions and on the relation between an objective analysis of the work and the employees' own subjective evaluations.

References

Eason, K., 1982, The process of introduction information technology. *Behaviour and Information Technology* **2**, 197–213.

Huuhtanen, P.T.J., 1983, Implementation of computerized system and its efects on job content, work load, and work organization in office work. *Työterveyslaitoksen tutkimuksia* 201 (Helsinki Institute of Occupational Health). [In Finnish with English and Swedish summaries.]

Johansson, G., and Aronsson, G., 1979, Stressreaktioner i arbete vid bildskärmterminal. Stockholms universitet, Psykologiska Institutionen, *Rapporter*, nr 27, Stockholm.

Järvinen, P., Kirjojen, J., Tyllilä, P. and Vihmalo, A., 1982, Analysis and designing of jobs in a man-computer system. Department of mathematical sciences, University of Tampere, *Report* A74, Tampere.

Salvendy, G., 1982, Human-computer communications with special reference to technological developments, occupational stress and educational needs. *Ergonomics*, **25**, 6, 435–447.

The Perception of Display Delays During Single and Multiple Keystroking

J.M. Boyle and T.M. Lanzetta
Human Factors Center, IBM Corporation, San Jose, California 95193, USA

Abstract

Two separate experiments were conducted to determine the perceptual threshold for time delays between key entry and the appearance of a character, or characters, on a VDT. For the single keystroke case, in which one character was entered and subjects indicated whether or not a delay occurred between key depression and character display, the threshold was 165 ms. In the multiple keystroke experiment, strings of characters were entered while typists visually attended to the display. For this case, the perceptual threshold was 100 ms.

1. *Introduction*

A well-designed user/computer interface should match the capabilities of the computer with the perceptual and cognitive capabilities of the human operator. The effects of system response time on user performance is an aspect of this symbiosis that is of particular interest to both system designers and the human factors professional. At the most elemental level, we have to understand the circumstances under which a user will begin to notice system response delays. In interactive computing systems, a delay occurring between a user input such as a single keystroke or a command, and a response from the computer is likely to have an adverse effect on human performance if the delay is sufficiently long. Studies have demonstrated that in some instances where system response delays occur, there is a corresponding decline in task performance as well as a tendency for users to develop negative attitudes about the general usefulness of the system (Barber and Tiernan, 1978; Thadhani, 1981). If a system is designed such that system response time is always less than the perceptual threshold of the user, user performance problems may be reduced. The objective of the two studies reported here is to provide input to system designers about human perception of duration during execution of a highly automatic task, namely text typing.

As the objective interval between the presentation of two stimuli to the same or different modalities is increased, a perception of simultaneity gives way to a perception of succession. The point at which this occurs is called the threshold of succession or threshold of discontinuity. The threshold depends on the modality of presentation, the intensity of the stimuli, and attentional factors (Fraisse, 1963). For intramodal stimulus presentation, approximate perceptual threshold values for each modality are: vision—100 ms, hearing—50 ms, and touch—50 ms (Fraisse, 1963). Intermodal stimulus presentation typically yields values ranging from 50 to 100 ms (Tinker, 1953).

While the above values may provide general guidelines, they pertain to situations in which the subject is the passive recipient of the stimuli. That is, both stimuli are exteroceptive. In the case of typing, however, both stimuli are not exteroceptive. The subject's keystroke, a motor response, provides an interoceptive stimulus. Only one previous study has attempted to determine the threshold of succession for such stimuli, reporting a threshold of 50 ms for a task in which the release of a key activated a neon bulb (Biel and Warrick, 1949). No previous research has attempted to determine the threshold of succession for key entry and display update in the context of VDT usage.

The following two sections describe studies of perceptual thresholds for time delays between key entry and display of the keyed character, first when a person keys only a single keystroke, and then for the case where a stream of continuous keystrokes are entered.

2. *Experiment 1: Single Keystroke Case*

2.1. **Subjects**

Forty-six subjects, 22 technical professionals and 24 clerical workers, participated in the testing.

2.2. **Apparatus**

The experimental hardware consisted of an IBM 3278 keyboard and a VDT. The display had a refresh rate of 30 frames per second with interlace. The delay from the time the refresh buffer was updated until the first dots of the character were presented on the screen varied from 0 to 16·7 ms. A minimum delay of 16·7 ms and a maximum delay of 33·3 ms could occur before the entire character was displayed. Delays between a subject's keystroke and the presentation of the character on the VDT were controlled by a computer program.

2.3. **Procedure**

Fifteen equal interval delays ranging from 0 to 400 ms were selected on the results of pilot research. Each delay was randomly selected ten times for a total of 150 trials. On each trial, a cursor was displayed on the centre of the twelfth line on the VDT. The subject was instructed to depress the 'H' key while visually attending to the cursor. The subject's keystroke resulted in an 'H' being displayed above the cursor which remained stationary. The subject then reported whether or not a delay occurred between the keystroke and the appearance of the character on the VDT by keying 'Y' or 'N'. The entire session lasted ten minutes.

2.4. **Results**

A total of 460 samples were collected for each delay interval. For each interval, the percentage of 'Yes' responses (i.e. delay perceived) for all subjects was computed. Figure 1 shows the cumulative percent of delays reported for each of the intervals.

FIGURE 1. *The percentage of delays perceived by 46 typists in the single keystroke case.*

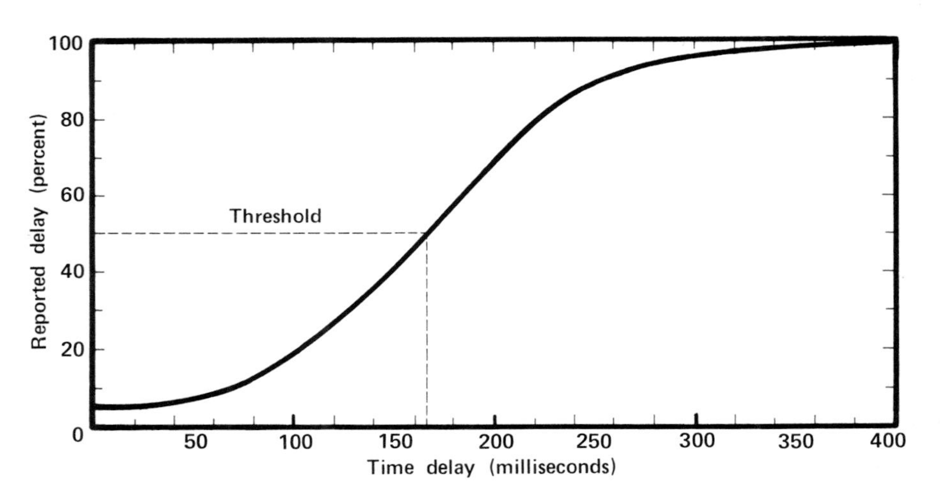

The group threshold of succession is 165 ms. That is, this value evoked a positive response on 50% of the trials. For delays less than 165 ms, a keystroke and the appearance of the character on the VDT were perceived, on the average, as occurring simultaneously. Conversely, a keystroke and the appearance of the character on the VDT were perceived as occurring in succession for delays greater than 165 ms.

Other percentile points can be determined from Figure 1. For example, a 70 ms delay was perceived only 10% of the time. Delays of 240 ms or longer were detected nearly all the time.

Although delays occurred on 93·9% of the trials, a delay was perceived on only 44·7% of the trials. Also, on approximately 6% of the zero-delay trials, subjects reported a delay.

This reporting error can be attributed to errors in keying responses, or more likely, to a tendency to guess. The relatively low magnitude of this value suggests that subjects did not resort to guessing often.

2.5. **Discussion**

The threshold value of 165 ms obtained in the present study is substantially larger than the value of 50 ms reported by Biel and Warrick (1949). This discrepancy may be due, in part, to context effects resulting from the range of delay intervals studied. Biel and Warrick used five delay intervals ranging from 0 to 120 ms. In the present study, 15 delays ranging from 0 to 400 ms were used. In that the perceived value of a particular stimulus is known to be influenced by the value of the other stimuli presented (Dember and Warm, 1979), it is likely that subjects adjusted their response criterion upward in the present study, resulting in a higher threshold.

Another factor that may have contributed to the magnitude of this discrepancy is the experimental hardware. The minimum delay before the entire character was displayed on the VDT was 16·7 ms. Although the subject may have made a decision regarding the presence or absence of a delay on less than the full matrix of dots comprising the character, this seems unlikely. Thus, the obtained threshold should probably be adjusted downward by 16·7 ms, yielding an adjusted threshold value of approximately 145 ms. An even more conservative adjustment would be to decrease the obtained value by the average amount of time it takes to display the entire character (i.e. 25 ms), yielding an adjusted threshold value of 140 ms.

3. *Experiment 2: Multiple Keystroke Case*

The results of Experiment 1 provide a precise value for the threshold of succession between a single keystroke and the appearance of a character on a VDT. The information is of limited value, however, since the task itself is not representative of most real-world tasks. Thus, the second experiment sought to determine the threshold of succession while typing strings of text material.

3.1. **Subjects**

Twenty-four experienced typists from a temporary help agency participated in Experiment 2. A minimum typing speed of 60 words per minute was required.

3.2. **Apparatus**

Same as in Experiment 1.

3.3. Procedure

The delays used in Experiment 2 were the same as for Experiment 1. Thus, there were fifteen equal interval delays ranging from 0 to 400 ms. Each delay was randomly selected 10 times for a total of 150 trials. The onset of the delay between each keystroke and the appearance of the corresponding character was coincident with the keystroke. On each trial, a cursor was displayed in column 1 of the twelfth line of the VDT. The typists were instructed to key continuously while attending to the screen. The stimulus material consisted of sentences and phrases that were short and easily memorized (e.g. 'There's a lot more to golf than having the equipment and the intent.'). Only alphabetic, space and punctuation keys were permitted to be typed. Typing continued until a judgement regarding the presence or absence of a delay could be made. At that point, the subject responded by keying either a 'Y' or 'N'. A maximum of 200 keystrokes could be typed before a response was required.

3.4. Results

For each delay interval, the percentage of 'Yes' responses was computed. Figure 2 shows the psychometric function obtained for continuous typing. The percentage of reported delays is plotted for each delay interval. For this experiment, the absolute threshold is 100 ms. For delays below 100 ms, subjects on the average perceived their keystrokes and the appearance of the corresponding characters on the VDT as occurring simultaneously. Delays longer than 100 ms resulted in perceptions of succession, the appearance of the character following a keystroke by a noticeable delay.

Delays were reported on 62·1% of the trials, and on approximately 19% of the zero-delay trials, a delay was reportedly observed. As in Experiment 1, this reporting error largely reflects a subject's tendency to guess. It appears that most of this type of error is due to the difficulty in discriminating short delays (20–40 ms) from no delay.

3.5. Discussion

The threshold obtained in the case of continuous typing is 65 ms lower than the threshold determined for a single keystroke. Clearly, temporal acuity is greater for this task. That is, subjects in the multiple keystroke study had a larger sample of keystrokes upon which to base a decision. If the subject failed to detect a delay on the first keystroke, multiple opportunities to do so still remained. Given such an explanation, however, one might expect a reduced tendency to guess, a finding which was not obtained. Delays were reported on 17% of the zero-delay trials in the present experiment, as compared to 6% of the trials in the single keystroke study.

Another explanation which accounts for the lower threshold obtained in the present experiment is rooted in the nature of the task. It is unlikely that subjects type at a constant rate while entering text. Highly familiar and well-practised words such as 'the' are typed in a rapid burst. Thus, when the keystroke interval between two successive keystrokes is less than the experimental delay the prior keystroke will not be displayed until after the second keystroke has occurred, making the delay more noticeable.

FIGURE 2. *The percentage of delays perceived by 24 typists in the multiple keystroke case.*

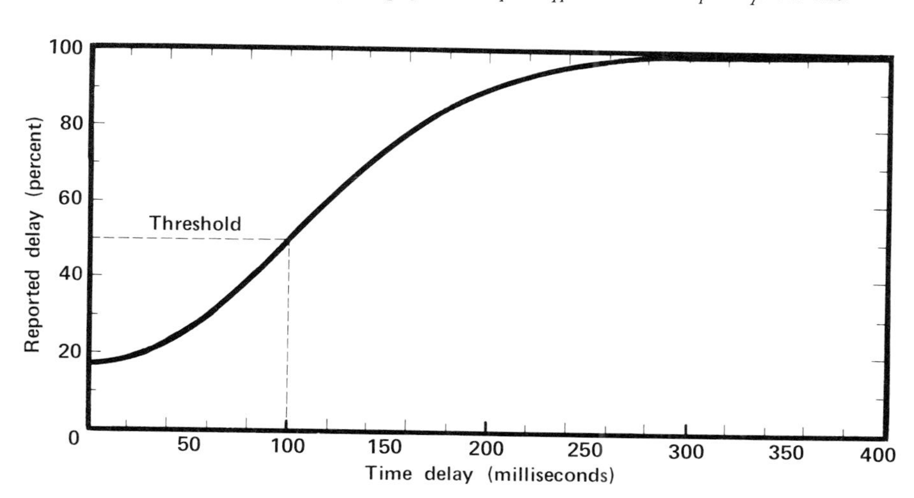

4. *General Discussion*

Overall, the results of these studies indicate that for optimal user performance, delays between key entry and character display should not exceed 165 ms in the unlikely case of a single keystroke, or 100 ms for tasks requiring multiple keystrokes. In that perceptual thresholds are not fixed either within or between individuals, these values are approximate. However, it is likely that they represent the minimum perceptual threshold, as the tasks employed were relatively simple, thus allowing users to devote most of their processing resources to the task of delay detection. Future studies should attempt to determine threshold values for tasks requiring concurrent processing, such as text composition and editing. For tasks requiring greater cognitive load, it is hypothesized that perceptual resources ordinarily available for delay detection will be used by the task, thus increasing threshold values.

References

Barber, R.E. and Tiernan, B.B., 1978, Response time and operator perception. *Proceedings of The Human Factors Society 22nd Annual Meeting*, Santa Monica, California.

Biel, W.C. and Warrick, M.J., 1949, Studies in perception of time delay. *American Psychologist*, No. 4, 303.

Dember, W.P. and Warm, J.S., 1979, *The Psychology of Perception* (New York: Holt, Rinehart & Winston).
Fraisse, P., 1963, *The Psychology of Time* (New York: Harper & Row).
Thadhani, A.J., 1981, Interactive user productivity. *IBM Systems Journal*, **20**, 4, 407–423.
Tinker, M.A., 1953, Temporal perception, in *Psychology*, edited by E.G. Boring, H.S. Langfield and H.P. Weld (New York: John Wiley).

Use of Magnitude Estimation for Evaluating Product Ease-of-Use

R.E. Cordes
IBM Corporation, Tucson, Arizona 85744, USA

Abstract

User-performance measures may not be sensitive indicators of product ease-of-use and should be supplemented by subjective estimates. This paper briefly examines a problem using subjective estimates and discusses a solution using magnitude estimation. Also, an applied software study using magnitude estimation (line production) is presented.

1. *Introduction*

There appear to be relatively few reported experimental studies that examine the ease-of-use of software products. This may be due to the lack of an acceptable methodology for conducting usability evaluations (Frankosky, 1983). It is generally assumed that by examining user-performance measures (such as training time, number of errors, error–recovery time, number of requests for assistance, number of attempts, and percent successful), a product's ease-of-use can be abstracted (Miller, 1971). While some investigators equate performance decrements to a lack of usability (typically through an inability to pass so-called 'usability requirements'—see Bennett, 1979), their operationally defined criteria lack any empirical basis or foundations. Mental work load, effort, vigilance, and other subjective variables may be more of a determining factor in the perception of ease-of-use than actual user-performance measures.

The purpose of this paper is to examine the use of subjective estimates (using magnitude estimation) as a supplement to performance measures for evaluating the usability of software products. A brief review will be given, followed by a discussion of an actual application of the approach.

2. *Subjective Estimates*

Subjective estimates appear to be quite promising for measuring mental work load (e.g. Williges and Wierwille, 1979; Wierwille and Connor, 1983, and see Moray, 1982 for a review) and should have similar success in evaluating product ease-of-use. Also, subjective estimates of 'perceived ease-of-use' (see Bethke, 1982 for use of this term) are easy to obtain and do not interfere (if requested post hoc) with tasks to be evaluated.

Obtaining subjective estimates can present a challenge in the design of an experiment. Using subjective estimates requires that each subject use each product in order to directly compare its ease-of-use (i.e. a repeated-measures design). However, unless the user interface with the products is quite different, you cannot have the same subject use and make comparisons between products. Often, it is the case that using one product will provide most of the information necessary to use a comparison product (e.g. when evaluating the documentation of a product and its re-design). Obviously, with such strong transfer effects, counterbalancing the order of presentation would be ineffective. The only practical choice available is to use a different group of subjects for each product (i.e. a between-subjects design). However, when using this design, a person cannot make direct ease-of-use comparisons between products.

The method used in this paper to circumvent this problem is to have each group rate its respective products (or tasks) against a common reference or standard that does not interfere with the tasks under evaluation. Having at least one common reference task or product for each group allows comparisons to be made between the subjective estimates of each group. In this manner, subjective comparisons can be obtained without having each person directly use and compare each product. This indirect-comparison approach provides a baseline calibration for the subjective estimates.

3. *Magnitude Estimation and Task Difficulty*

Magnitude-estimation procedures (Stevens, 1956) may be helpful in evaluating a product's ease-of-use. In magnitude estimation, subjects are typically asked to assign a number (of their own choosing) that will match the magnitude of a perceived stimulus. The procedure appears to be successful in identifying the relationship between most sensory continua and their stimulus magnitude (e.g. loudness, vibration, brightness, taste, heaviness, and warmth—see Stevens, 1975 for a more exhaustive list). Its success in psychophysics has led others to use the procedure in other areas where the physical stimulus is not readily measurable. For instance, magnitude estimation has been applied to such diverse measures as political dissatisfaction, perceptions of national power, pleasantness of odours, and aesthetic value of drawings (see Stevens, 1975 for sources).

There is little work reported using magnitude estimation to evaluate ease-of-use. A reason for this may be the lack of a true zero-point for the establishment of an ease-of-use ratio scale. On the other hand, studies examining perceived difficulty are reported in the literature (e.g. Borg, Bratfisch and Dornic, 1971, and Haverland, 1979). Since ease-of-use and difficulty are presumably inversely related and difficulty has a natural zero-point (i.e. the difficulty of doing nothing), perceived difficulty is the preferred measure. For this reason, perceived difficulty was also the subjective dependent measure of choice for the following study.

4. *Experimental Illustration*

4.1. **Background**

Both the direct- and indirect-comparison approach were applied in a usability evaluation comparing the command interface and a prototype panel interface of an IBM product called HSM (Hierarchical Storage Manager). HSM is a software product that provides space management (such as backup, recovery, migration, and recall) of TSO data sets. While HSM can and usually does provide these functions automatically, they are also under user control. This control is normally available via the HSM command language. The prototype was developed on ISPF (Interactive System Productivity Facility—see Joslin, 1981, or Maurer, 1983, for a review) to enhance the usability of HSM release 3.2 (announced in October 1983) by providing an alternative panel-entry interface to the HSM functions.

4.2. **Method**

4.2.1. *Subjects*

Sixteen people served as subjects in this study, all of whom had some experience with TSO/ISPF but had never used HSM directly before.

4.2.2. *Procedure*

The participants were divided into two groups of eight people each. One group used available on-line support only (TSO help for HSM and panel tutorials). The other group, in addition, were given the HSM User's Guide and HSM Command Reference Summary Card. Half of the participants in each group used the prototype panels first and the other half used the HSM commands first to accomplish eight comparison tasks (invoking the HLIST, HALTERDS, HBACKDS, HDELETE, HRECOVER, HBDELETE, HMIGRATE, and HRECALL functions). The order of task presentation was also counterbalanced across subjects.

The three dependent measures collected were task-completion time, number of errors per task, and perceived task difficulty. A measure of task difficulty was obtained after each task by having subjects draw a horizontal line on sheet(s) of graph paper (¼-in square blocks) matched in length to their perceived difficulty. A short line (two blocks in length), representing the difficulty of simply logging on to TSO, served as the reference task for making comparisons.

4.3. **Results**

A distinction in the analysis was made as to whether the tested HSM tasks were usable (i.e. capable of being performed within the time constraints of ½ hour per task) and the ease with which they could be learned. The results revealed that the product was usable on these tasks. According to a binomial test, the tasks' failure rate was quite low, allowing the acceptance of a usability claim of at least 80% success rate ($p < 0{\cdot}05$).

The ease-of-learning of the HSM tasks was evaluated by comparing the results of using the HSM commands with the prototype interface. A 2 (support) × 2 (interface) × 8 (tasks) MANOVA was conducted on the comparison data. The difference between the HSM command and the prototype panels was significant ($F3,\ 12 = 9{\cdot}21$; $p < 0{\cdot}002$). With all dependent measures, the HSM commands proved superior. The task variable was also significant ($F21,\ 276 = 3{\cdot}14$; $p < 0{\cdot}0001$). There was no significant benefit in having hard-copy documentation available (subjects tended to rely on the one-line help) and no interactive terms were significant ($p > 0{\cdot}05$).

4.4. **Discussion**

Further analysis revealed that the primary source of user difficulty was associated with two of the prototype panels. Analysis of the data omitting these sources of difficulty eliminated the significant differences between the two interfaces. As a result of this study a number of the panels (including the two mentioned above) were redesigned to improve their ease-of-learning.

Of special interest in this study was the application of magnitude estimation (line production) in the evaluation of perceived task difficulty. The participants in the study found it quite natural to draw lines to match their perception of task difficulty. Also, the use of a standard reference task (logging on to TSO) was successful in permitting the between-subjects magnitude estimates to be included in the multivariate analysis. The common reference task served to calibrate the subjective estimates of both groups (with manuals and without manuals) and thus allowed meaningful between-subjects comparisons.

References

Bennett, J.L., 1979, Incorporating usability into system design. Paper presented at Design '79 Symposium, Monterey, Californa, April.

Bethke, F.J., 1982, Measuring the usability of publications. Santa Theresa Laboratory, San Jose, California; IBM Technical Report TR03.183, February.

Borg, G., Bratfisch, O. and Dornic, S., 1971, On the problems of perceived difficulty. *Scandinavian Journal of Psychology*, **12**, 249–260.

Frankosky, R., 1983, Software interface ease of use: Metrics and Methods. *Proceedings of the 2nd Annual Phoenix Conference on Computers and Communications, IEEE Cat. No. 83CH1864–8*, 534–537.

Haverland, E.M., 1979, Magnitude Estimation: a new tool for measuring subjective test variables. US Army Tropic Test Center Report, June, No. 790601, 1–58.

Joslin, P.H., 1981, Systems productivity facility. *IBM Systems Journal*, **20**, 388–406.

Maurer, M.E., 1983, Full-screen testing of interactive applications. *IBM Systems Journal*, **22**, 246–261.

Miller, L.A., 1971, Human ease of use criteria and their trade-offs. Poughkeepsie, N.Y.; IBM Technical Report TR00.2185, April.

Moray, N., 1982, Subjective mental workload. *Human Factors*, **24**, 25–40.

Stevens, S.S., 1956, The direct estimation of sensory magnitudes—loudness. *American Journal of Psychology*, **69**, 1–25.

Stevens, S.S., 1975, *Psychophysics: Introduction to Its Perceptual, Neural, and Social Prospects* (New York: John Wiley).

Wierwille, W.W. and Connor, S.A., 1983, Evaluation of 20 workload measures using a psychomotor task in a moving-base aircraft simulator. *Human Factors*, **25**, 1–16.

Williges, R.C. and Wierwille, W.W., 1979, Behavioral measures of aircrew mental workload. *Human Factors*, **21**, 549–574.

A Comparison of Cursor-Key Arrangements (Box Versus Cross) for VDUs

W.H. EMMONS
Human Factors Center, IBM Corporation, San Jose, California 95193, USA

Abstract

On video display terminals, cursor-positioning keys (usually four in number) are generally arranged in a box (rectangular) arrangement, or in a cross (cruciform) arrangement. Performance of 32 inexperienced subjects and 10 experienced subjects on the box arrangement was measured. Measures included time to perform the task, number of extra moves taken, the number of non-cursor keys struck, and the time to locate the first cursor key in a move sequence. The inexperienced group performed better on the cross arrangement for all measures except the number of noncursor keys struck. The experienced group showed no differences in the performance measures with the exception of the greater number of extra moves on the box arrangements. Test results showed that subjects performed generally better with the cross cursor arrangement.

1. *Introduction*

Several layouts of cursor-positioning keys have been used by manufacturers of video display terminals. The most popular layouts are the box arrangement and the cross or cruciform arrangement.

An informal survey of 23 European manufacturers of video display terminals found that 14 used the cross, 5 used box, 2 used a straight row layout and 2 used contextual cursors. Intuitively, the cross arrangement appears to be the better layout since the location of the keys is a cue to the direction of movement.

The purposes of the present study were to determine:

1. Do people show a difference in performance and/or preference between the two arrangements.
2. Would terminal users experienced on the box arrangement have difficulty in transferring to the cross layout?

1.1. Keyboards

The arrangements of the cursor keys on the keyboards used in this study are shown in Figures 1 and 2. The keyboards were otherwise mechanically and electronically identical. The cursor keys were 'Typamatic', with a 0·5-second delay before onset of Typamatic mode and a repetition rate of 10 characters per second.

FIGURE 1. *Box arrangement.*

FIGURE 2. *Cross arrangement.*

2. *Procedure*

Thirty-two subjects with little or no experience in cursor-positioning tasks participated in the first phase of this experiment. The second phase employed ten subjects experienced in the use of the box arrangement.

A subject's task began when the system displayed an 'S' (Start) and a '#' (target) on the terminal screen. A sample display can be seen in Figure 3.

The cursor was prepositioned at the start location. The subject would then key in single strokes, or use Typamatic (continuous) mode, to move the cursor toward the target. When the cursor reached the target, the trial was concluded.

To begin the next trial, the operator keyed ENTER, which brought up a new display with S and # in new positions. Each session consisted of 120 such trials. The trials presented to the operator were selected from a 120-trial pool in a random sequence without duplication.

The length of cursor movement was divided into three categories as shown in Table 1. Each trial required one of the following eight types of cursor movement.

right only	right and up
left only	right and down
up only	left and up
down only	left and down

The 120 trials in a session consisted of 5 trials from each category of length for each of the 8 types of movements. During each session, the minimum number of cursor moves (characters or lines) required to complete the 120 trials was 3422.

FIGURE 3. *Sample display screen of cursor-positioning tasks.*

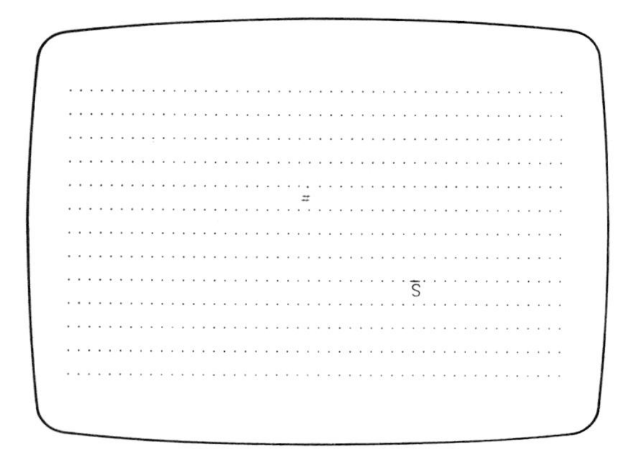

TABLE I. *Intervals of cursor travel.*

Direction of cursor movement	Intervals of cursor travel		
	Short	Intermediate	Long
Right or left (characters)	11–15	21–25	36–40
up or down (lines)	4–8	12–16	18–22

The mean number of moves for short, intermediate, and long movements were 14·28, 27·75 and 43·50 respectively. The mean number of moves for all trials was 28·51.

3. *Experimental Design*

Each subject completed four sessions (two sessions on each keyboard) which were presented in a 2 × 2 Latin-square design.

The stimulus presentation, recording of keystrokes, and their associated time of occurrence were controlled by a computer program.

4. *Results*

The four dependent variables were time to complete a trial, the time from the presentation of the stimulus screen until the first keystroke, the number of extra or superfluous moves made while manoeuvering the cursor to the target, and the percentage of noncursor keys accidentally struck (number of noncursor keys divided by the number of required direction changes multiplied by 100.)

4.1. **Inexperienced Subjects**

An analysis of variance was computed for the four dependent variables for each of the 24 combinations of lengths and directions (Simons and Clauer, 1978). Since the results were essentially the same for all of the combinations, only the statistics for the combined data are presented here. Table 2 shows the means, the 95% confidence interval of the differences between the means, and the probabilities of the differences between the means being due to chance.

TABLE 2. *Performance results using inexperienced subjects.*

Dependent variables	Cursor arrangement		95% interval	Probability
	Box	Cross		
Time per trial (seconds)	6·15	5·91	±0·14	0·002
ENTER to 1st move (seconds)	1·51	1·37	±0·08	0·008
Extra moves per trial	0·83	0·70	±0·09	0·003
Percent of noncursor keys	0·25	0·49	±0·16	0·005

All of the measures show significance beyond the 0·05 level of confidence. The performance differences are all in favour of the cross arrangement with the exception of accidental depression of noncursor keys.

A total of 19 subjects preferred the cross arrangement, 12 preferred the box arrangement, while one subject expressed no preference.

A chi-square test of hypothesized population proportions (Bradley, 1976) was applied to the preference data and was not significant at the 0·05 level of probability.

4.2. **Experienced Subjects**

Table 3 shows the results for the experienced group. No significant differences were found for any of the dependent variables with the exception of extra moves. There were a greater number of extra moves with the box arrangement.

TABLE 3. *Performance results using experienced subjects.*

Dependent variables	Cursor arrangement		95% confidence interval	Probability
	Box	Cross		
Time per trial (seconds)	5·30	5·27	±0·14	N.S.*
ENTER to 1st move (seconds)	1·26	1·22	±0·11	N.S.
Extra moves per trial	0·45	0·33	±0·10	0·03
Percent of noncursor keys	0·20	0·27	±0·20	N.S.

*Not significant, $p > 0.05$

Four subjects preferred the box arrangement, three preferred the cross layout, and three expressed no preference. The chi-square test was not significant at the 0·05 level of probability. Nine of the ten experienced subjects felt they would have no difficulty in switching to the cross pattern.

5. *Discussion*

The results indicate that the cross cursor arrangement is superior in performance measurements for inexperienced subjects. Even subjects experienced on the box arrangement performed just as well on the cross layout.

Much of the advantage for the cross cursor arrangement for the inexperienced group appears to be due to the fact that the subjects could locate the first key to be struck faster (1·37 seconds for the cross and 1·51 for the box). These findings are in agreement with the findings of Fitts *et al.* (1953) concerning stimulus–response compatibility. This advantage disappeared in the case of the experienced subjects, which might be expected from their familiarity with the box arrangement.

The inadvertent striking of noncursor keys for both the box and cross layouts was less frequent in the experienced group, reflecting their superior keying ability.

6. *Conclusions*

1. The cross cursor arrangement is superior in all performance measures except for the percentage of inadvertently struck noncursor keys for subjects with little or no experience in a cursor-moving task.
2. Subjects with experience on the box layout perform at least as well on the cross pattern.
3. The preference data is inconclusive.

References

Bradley, J.V., 1976, *Probability, Decision, Statistics* (Englewood Cliffs, New Jersey: Prentice Hall).

Fitts, P.M., Seeger, C.M. and Stewart, C.M., 1953, S-R compatibility: Spatial characteristic of stimulus and response codes. *Journal of Experimental Psychology*, **46**, 199–210.

Simons, R.M. and Clauer, C.K., December 1978, *LATIN: A PL/1 program for the analysis of replicated latin square experimental design* (San Jose, CA: IBM Human Factors Center Report HFC-28).

Effect of the Amount and Format of Displayed Text on Text Editing Performance

M.J. DARNELL AND A.S. NEAL
Human Factors Center, IBM Corporation, San Jose, California 95193, USA

Abstract

An experiment was conducted in which 28 typists used both a 20 line, 80 character display and a single line, 32 character display connected to the same screen-oriented text editing system. The typists completed 384 editing tasks using each display. Each task required typists to make a revision which was marked on a printed copy of a document. The results showed no practical productivity difference between the two displays for common editing revisions after a moderate amount of practice.

1. *Introduction*

Screen-oriented text editors can vary in the amount of text displayed. Typical displays show 20 lines of 80 characters of text. Expert opinion concerning the amount of text shown on text editor displays is that more is better (Finseth, 1982). For common editing tasks, however, it is not clear that the use of a typical display would result in greater productivity than a display showing a much smaller amount of text. In the manuscript editing task, where a typist makes a series of revisions marked on a printed copy of a manuscript, the typist's actions are focused on a small part of the manuscript at a time (Card, Moran and Newell, 1980). The typist needs to view only that small part of the manuscript containing the current revision. By one estimate, 89% of revisions are within single lines of text (Allen, 1982). Thus, a single line display would seem adequate with respect to the amount of text displayed for purposes of common editing revisions. On the other hand, a larger display may facilitate a typist's ability to visually locate a revision point on the display or printed page because of the amount of text displayed and the similarity of format between the displayed and printed manuscript.

To address these issues, an experiment was conducted in which typists used a screen-oriented text editing system with both a 20 line, 80 character display and a single line, 32 character display. This report focuses on the typists' productivity while performing common editing revisions after a moderate amount of practice.

2. *Method*

2.1. **Subjects**

Twenty-eight skilled typists, recruited from a temporary employment agency, participated as subjects in the experiment.

2.2. **Equipment**

2.2.1. *Display and Keyboard*

The same video display system was used for both display conditions. The large display which showed documents through a 20 line by 80 character window, was equipped with a movable cursor. The small display, which showed documents as a continuous string of text through a single line, 32 character window, was equipped with a fixed position cursor in the centre of the display. The text editing system keyboard used the standard IBM typewriter layout with the addition of some outboard function keys.

2.2.2. *Editor*

The keyboard and display system were driven by a screen-oriented editor designed for a series of editor–display experiments.

On the large display, the cursor could be moved line-by-line vertically and word-by-word horizontally by using four cursor keys. The character keys inserted characters at the cursor position, pushing the cursor to the right and dynamically reformatting the remaining text in the document. On the small display, the text could be scrolled line-by-line and word-by-word using the four cursor keys. The character keys inserted characters at the cursor position and scrolled them to the left.

A 'Find' function could be used by pressing the 'Find' key, typing the target text string and pressing a cursor key to indicate the direction of search in the document.

The large display was equipped with back and forward paging functions which showed the previous or following 18 lines of text, respectively.

The editing system was also equipped with a 'Move' function that was used to transfer strings of words from one part of a document to another, and a 'Delete' function that was used to erase lines, words or characters.

2.3. **Tasks**

A task set consisted of a fixed sequence of 128 types of editing tasks composed of the factorial combination of 4 types of document format, 4 vertical cursor movement distances, 2 lengths of revision, and 4 types of revision.

The four types of document format were continuous narrative text, business letters, outlines with two levels of indentation, and tables composed of three columns. Eight different documents corresponding to each type of document format were used in the experiment.

The four vertical cursor movement distances corresponded to the number of lines between the initial cursor position and the revision point. The four distances were 0 lines (but more than 12 characters), 1 to 3 lines, 7 to 12 lines, and 21 to 40 lines.

The two lengths of revision corresponded to the amount of text affected by a revision. Small revisions affected word strings with less than 25 characters, all on the same line. Large revisions affected word strings with more than 72 characters on not more than 3 adjacent lines.

The four types of revisions were insert, delete, replace, and move. An insert required only typing words into a document. A delete required only erasing words from a document. A replace required deleting words from a document and inserting words in place of those deleted (or vice versa). A move required transferring a word string from one part of the document to another.

Each document was printed unjustified in 12-point 12-pitch type on a single page of 8½ by 11 paper with a maximum of 60 lines of text and 80 characters per line.

Each printed document was marked with a single revision indicating either an insert, delete, replace, or move. The position of the cursor in the initial display of a document was also marked on the printed document. Special symbols for 'Carrier Return' and 'Tab', which were shown on displayed documents, were not shown on printed documents. Each document was initially presented on the large display so that the cursor was positioned within the top three lines of the display.

2.4. **Procedure**

Typists participated in both large and small display conditions in a counterbalanced order. For each display condition, the typist was given a demonstration, and required to complete a tutorial, two practice task sets and one test task set.

During a task set, the typist was seated before the keyboard and display system with a binder of printed documents to the left. For each task, the typist turned to the next page in the binder to see a printed copy of the next document. Meanwhile, the document was presented on the display. The typist was required to move the cursor to the revision point in the document and make a single revision. Upon completing the revision, the typist pressed a special key labelled 'DONE' which automatically loaded the next document into the editing system.

3. *Results*

Confidence limits were drawn on the time and error data from the test task set. Each set of reported confidence limits is based on a 95% confidence that the treatment mean difference is within the reported limits.

3.1. **Editing Errors**

Each task was classified as to whether the final result of the revision contained any non-typographical errors. These were errors that the typist had failed to correct. Errors that the typist made and corrected during the task were not identified but are reflected in the task editing time.

The mean proportion of tasks with errors was 0·035 and 0·041 for the large and small displays, respectively. The confidence limits for the mean difference are $-0{\cdot}018 \leq 0{\cdot}006 \leq 0{\cdot}130$. (The middle term (0·006) is the absolute value of the mean difference. The left and right terms are the lower and upper confidence limits, respectively. Using this convention, if the lower limit is positive then the difference between the means is statistically significant.)

3.2. **Editing Time**

The mean editing time reflects the average amount of time from the first keystroke of each task to the Done keystroke. The mean editing time for the large and small displays was 23·95 s and 23·43 s, respectively. The confidence limits for the mean difference are $-1{\cdot}52 \leq 0{\cdot}52 \leq 2{\cdot}56$.

The editing time for each task was partitioned into two additive components termed locating time and revising time. Locating time, the time required to position the cursor at the revision point, included the interval between the first keystroke of the task and the first keystroke of a revising action. Revising time included the interval between the first keystroke of the revising action, and the Done keystroke.

3.2.1. *Locating time*

The mean locating time for the large and small displays was 7·25 s and 7·75 s, respectively. The confidence limits for the mean difference are $-0{\cdot}35 \leq 0{\cdot}23 \leq 0{\cdot}82$. However, there was a difference between display conditions in mean locating time that depended on the vertical cursor movement distance (see Figure 1). The locating time was significantly less for the large display when the vertical cursor movement distance was 0 and 7–12 lines. (The confidence limits for the mean difference between the large and small displays for the 0 and 7 to 12 line vertical cursor movement distances are $0{\cdot}03 \leq 0{\cdot}57 \leq 1{\cdot}10$ and $0{\cdot}04 \leq 0{\cdot}98 \leq 1{\cdot}91$, respectively.)

3.2.2. *Revising Time*

The mean revising time for the large and small display conditions was 16·44 s and 15·69 s, respectively. The confidence limits for the mean difference are $-0{\cdot}86 \leq 0{\cdot}75 \leq 2{\cdot}36$. However, there was a difference in the revising time between the large and small displays that depended on the type of revision task (see Figure 2). The revising time for the large display was significantly greater for insert tasks, but significantly less for move tasks. (The confidence limits for the mean difference between the large and small displays for insert and move tasks are $0{\cdot}76 \leq 2{\cdot}35 \leq 3{\cdot}93$ and $0{\cdot}04 \leq 2{\cdot}47 \leq 5{\cdot}44$, respectively.)

FIGURE 1. *Effect of the vertical cursor movement distance on mean locating time for the large and small displays.*

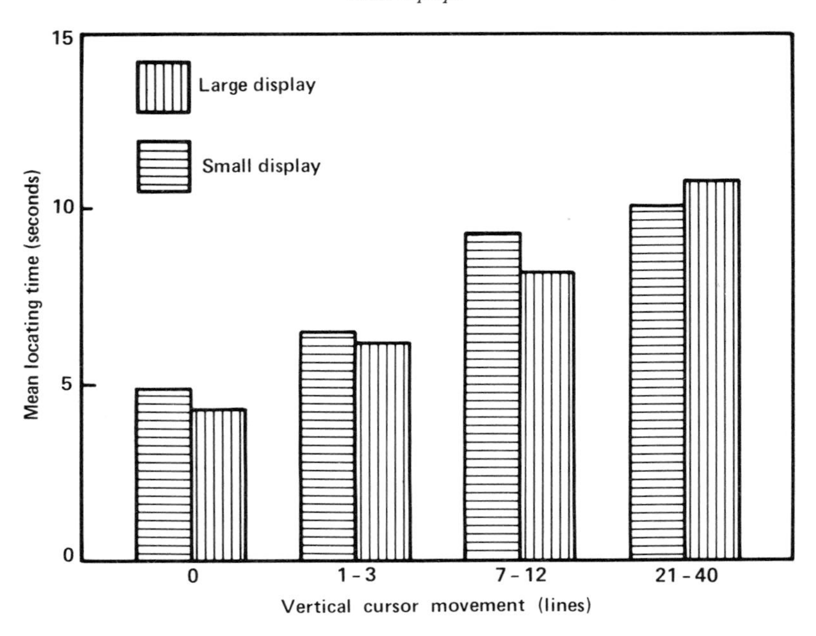

4. *Discussion*

To inspect the locating time advantage for the large display for tasks in which the vertical cursor movement distance was 0 and 7 to 12 lines, the mean locating time was partitioned into its additive components (i.e. proportions of time spent using the cursor keys, 'Find' function, and 'Paging' function). For both displays the frequency of use of the 'Find' function to position the cursor depended on the vertical cursor movement distance. At the 0 line distance, the frequency of 'Find' function use per task was 0·08 and 0·09 for the large and small displays, respectively. Thus, the cursor keys were almost exclusively used to position the cursor on both displays. On inspection, it was apparent that typists consumed more time between cursor keystrokes on the small display than on the large display. Cursor positioning on the different displays probably encouraged different strategies. On the large display, typists could visually locate the revision point, note its position, and then move the cursor to that position. On the small display, the revision point was not initially shown. Thus, typists were required to scan the text as they scrolled the display to position the

FIGURE 2. *Effect of the type of revision on mean revising time for the large and small displays.*

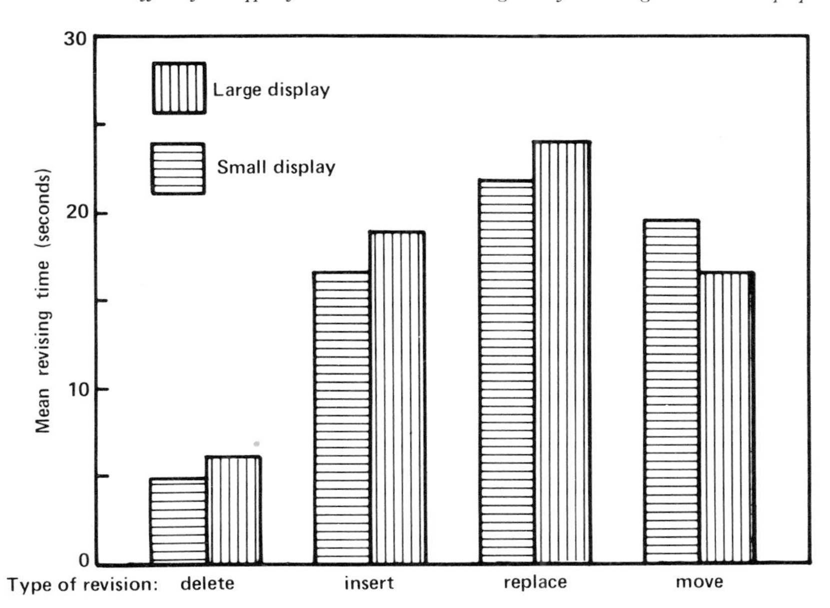

revision point at the cursor, particularly as they approached the revision point. The time to scan is reflected in time intervals between cursor keystrokes.

At the 7 to 12 line distance, the frequency of 'Find' function use per task was 0·92 and 1·07 for the large and small displays, respectively. Thus, the Find function was used more than once per task on the small display. Part of the locating time disadvantage for the small display was caused by time spent cursoring, probably to visually locate the position of the cursor in a document following the unsuccessful usage of the Find function.

In the present experiment there was only a single revision mark to locate on each page. One might reasonably argue that if revision marks were distributed over a printed document page, locating a particular revision mark would be facilitated using the large display because of the greater amount of text shown and the similarity in format between the displayed and printed document. However, in one study of editing task characteristics, revisions were found to be clustered together instead of being evenly distributed throughout documents (Allen, 1982). So searching for a revision point on a typical printed page may usually involve searching for the cluster of revision marks without much reference to the position of the cluster on the page.

To inspect the differences in revising time between the displays that depended on the type of revision, the mean revising time was partitioned into its additive components (i.e. proportions of time spent inserting, deleting, and moving text, and repositioning the cursor between revising actions.) The large display had an advantage over the small display in move tasks. The major reason for this advantage is that the use of the 'Move' function required marking the beginning, ending and destination points of the to-be-moved word string, and the cursor keys were most frequently used to position the cursor at these points. Since cursoring was more time consuming on the small display, use of the 'Move' function was also more time consuming.

The advantage for the small display in insert tasks was associated with a higher frequency of the use of the 'Delete' function on the large display to delete words or larger units of text, even though insert tasks did not require the use of the 'Delete' function. The text inserting rates for the two displays were the same. Apparently, there was a higher probability of inserting text in the wrong position on the large display.

To conclude, the use of the large display resulted in a small advantage in locating time for certain vertical cursor movement distances and in revising time for move tasks but resulted in a small disadvantage for insert tasks. Overall, there was no difference between the displays in editing time or editing errors. Thus, the use of a small display (a single line of 32 characters) resulted in the same productivity as a standard 20 line, 80 character display for the mix of manuscript editing tasks examined in the experiment.

References

Allen, R.B., 1982, Patterns of manuscript revisions. *Behavior and Information Technology*, **2**, 177–184.

Card, S.K., Moran, T.P. and Newell, A., 1980, Computer text-editing: an information processing analysis of a routine cognitive skill. *Cognitive Psychology*, **12**, 32–74.

Finseth, C.A., 1982, Managing words: what capabilities should you have with a text editor? *Byte*, 7 (4), 302–310.

Unexpected Consequences of Participative Methods in the Development of Information Systems: the Case of Office Automation

M. Diani
Centre d'Etudes Sociologiques, CNRS, 82 rue Cardinet, 75017 Paris, France
and S. Bagnara
Istituto di Psicologia, CNR, Via dei Monti Tuburtini 501, 00157 Rome, Italy

Abstract

In a large unionized bureaucracy, where decisions were made in a participatory fashion, the introduction of computer based information systems produced unanticipated consequences: competitive work relations, redundancies, inflexible task allocation, and new forms of social control. Our study suggests the need to rethink the ways in which these systems are introduced with particular reference to:

1. Organizational work load (Bagnara and Diani, 1983).
2. A major question concerning the technological nature of office automation (OA).
3. The limits of participative methods of work redesign.

1. *Introduction*

The study of the development and generalization of information technology, and more specifically of computer based information systems points at fundamental modifications in the meaning and contents of human work. These changes primarily affect operators (Grandjean and Vigliani, 1980); but they are also relevant for unions, management and organizations (Zuboff, 1982; Diani, 1982). The office automation (OA) process is very often followed by a series of social phenomena. Currently these phenomena are viewed as 'unexpected consequences', but in fact, a better definition would be that of 'predictable unexpected consequences' of office automation.

The most common consequences are the following (Mowshowitz, 1980; Mumford and Sackman 1975):

1. Resistance to adopt and to use the new technologies.
2. Underutilization of the machines.
3. Lack of organizational effectiveness.
4. Development of conflicts and organizational frictions.
5. Reduced economic performances.

In the last ten years, an increasing number of methods have been developed in order to help prevent these conflicts (Otway and Peltu, 1983); and all methods are also, more or less, defined as participative and democratic to facilitate the acceptance of the new technology. The aims and characteristics of the participative method can be summarized as follows:

1. Democratic decision of implementation of OA, concerning the time and mode of implementation, providing information, education and training.
2. Involvement of users in the design process of the new organization.
3. Quality of working life, in particular by establishing physical ergonomic standards and agreements related to the time spent operating on VDT.
4. A less hierarchical and centralized structure associated with an increase in productivity and job satisfaction.

Establishing a strong consensus regarding the new office technology and the new organizational setting is the most important constraint and aim of participative methods.

2. *Results*

In this paper we deal with the participative introduction of office automation in a French national organization, after a long series of studies, conducted by a multidisciplinary team.

The implementation of OA took place in a large national bureaucracy characterized to some extent by a democratic structure.

The National Agency for Production of Energy (OPEN) has its own organization to administrate social security and corporate activities for the personnel: the Mutual Society. This is a structure with the following characteristics:

1. Highly unionized at every level of the hierarchy.
2. Managers elected on union lists.
3. Cooperative organization of work up to a certain point.

We visited several regional branches at different stages of the process of office automation. The majority of the personnel are women, grouped by age, under or over 35. In this organization the tasks are of two types:

1. Structured, formalized and foreseeable: updating and processing files, payments, settlements.
2. Unstructured, difficult to formalize: communications, relations with other institutions, special cases.

Before the introduction of OA, the Mutual Society used the centre for data processing (EDEN). Created almost 15 years ago this is a centralized structure, whose speed, accuracy, reliability, and QWL declined very fast, both for technological reasons and because of the expanded complexity of tasks. The availability of new information systems on the market appeared as a suitable means to attain three ends:

1. Better quality of services offered to the members of OPEN.
2. Increase of QWL through the suppression of EDEN and the recomposition of tasks performed there in offices of the Mutual Society.
3. Decentralized management allowing more autonomy to some decision-making processes to regional branches of Mutual Society.

Procedures of participative design were followed, but under two constraints produced by the characteristics of the democratic organization of the Mutual Society:

1. Preservation of employment and qualification, despite all the foreseeable modifications of the organizational structure introduced by the OA.
2. A limitation to 4 hours/day of work on VDT, by creating a 'bynome': two operators alternating on the same VDT during the day.

3. *Real Unexpected Consequences of OA*

After the introduction of OA in several territorial branches of the democratic organization, despite the participative design, there were some really unexpected consequences. It is likely that, in order to establish a very broad consensus and given the particular structure of our organization, the necessity to maintain the agreements mentioned above became an organizational constraint, raising unexpected consequences of OA:

1. Complexity of the cultural conflicts.
2. A paradoxical division of work.

3.1. **Cultural Conflicts**

The EDEN and the Mutual Society personnel resisted the introduction of office automation and information technology. The first groups were motivated by their technical knowledge, and by past experience of computer mediated work: rigidity, failures and psychological isolation. The latter were worried both by the newcomers, considered as dangerous competitors and by a radical, and mostly unknown, change in their working conditions. But it was only the actual combination of these two sub-population's culture in the workplace that produced an unintended effect given that the goals of the two groups were largely the same, at least before the participative design process (Sainsaulieu, 1977; Crozier, 1983).

3.2. The Paradoxical Division of Work

In general, it has been observed that the introduction of OA produced important changes in job profiles. But one of the pledges of the democratic organization is to maintain the qualifications and employment. First, most of supervisor's tasks were integrated into the software or decentralized to the operators: supervisors and operators are hierarchically differentiated while in fact having exactly the same job. All the supervisory levels, including the office managers of the Mutual Society, see their task altered by the new computer-based system: in some cases their tasks are completely eliminated. In this case the creation of the new 'professional figures' is more a linguistic solution than an organizational feature and reality.

There is also a second and more unexpected consequence affecting the operators working on VDT 'as bynome'. In this case, the operators literally do not know what to do during the second half of the working day, because the most important operations are performed through the VDT. Moreover, individual differences or special operations can require more or less time and are difficult to plan in advance, creating from the beginning problems among members of the bynome, that have increased over the first months of experience.

The computer-based system worried the members of the Mutual Society in part because of the risks of impersonal control and the introduction of standardized criteria of efficiency and individual performance by the machine: the solution proposed in the democratic organization added the pressure of the 'peer-group control'. All these factors tend to increase a series of phenomena related to a cognitive mediation of work in organized systems.

4. *The Cognitive-Mediation of Work*

With computer-based technologies, a component of mental work-load that has been insufficiently studied becomes visible: it is represented by all the variables that define the social organization of aims and the division of work. For instance, given the same conditions, the number of errors varies as a function of the control, directly exerted both by the hierarchy or the peer-group: an operator devotes more attention to the 'controller', is less concentrated on the task and the likelihood of errors increases. Furthermore, the importance of the error itself is immediately enlarged by the organizational dimension created by the social comparison of the two members of the 'bynome'.

This organizational work-load is relevant for two main reasons (Bagnara and Diani, 1983). First of all, the market of office automation provides hardware and software that present very little flexibility, based upon analysis of work procedures according importance almost uniquely to structured activities. This technological rigidity of OA devices makes it difficult for operators to respond to uncertainty, to perform non-structured tasks and to consolidate every type of informal and learning by tasks expertise. Formal negotiations and informal arrangements with the hierarchy and the managers became much more difficult: a number of mental activities shift from accomplishing the tasks to the standardized rules, trying to adapt behaviour to the goals and tasks of the computer-based system.

Once work-goals and rules are standardized, less space is left to the bargaining process:

the integration of procedures does not increase the richness of tasks and job satisfaction, but leads instead to an increase of attention, concentrated on control, perceived as direct and non-negotiable. Errors are perceived as organizational failures (Bagnara and Diani, 1983).

In this technological phase of development of OA, only a low percentage of the organization of work can be standardized and automated. The rest is composed of coordination, control and integration between the task procedures and the office goals and function no longer performed by supervisors and managers: an increasing part of individual's attention is then devoted, at the cognitive level, to these organizational tasks.

Within the democratic organization three sources of control and organization seem to be contradictory with one another:

1. A standardized control (software–hardware).
2. The peer-group control ('bynome').
3. The cognitively mediated set of organizational goals to be performed by individuals (Bagnara *et al.*, 1983).

5. *Conclusion*

In all cases observed a totally unexpected consequence was observed: the increase of mental work-load related to the cognitive mediation of organizational factors. This organizational work-load is very difficult to manage, measure and prevent. On the one hand, the incompatibility between organizational models and computerized procedures can lead to a paradoxical and forced development of forms of social competition, comparison and 'conflictual cooperation'. On the other hand, even with participative methods of introduction of OA, the democratic division of work, at least in the type of organization that we have observed, instead of establishing a mutual confidence among operators seems to reinforce the elements of rigidity that are produced by OA devices. It is likely that the definitions of participative methods should be improved by paying less attention to formal aspects of work redesign, while establishing a more systematic relation between the technological nature of computer-based information systems and its reflect on cognitive-mediation of work and organizational goals.

References

Bagnara, S. and Diani, M., 1983, Carico di lavoro organizzativo derivante dall'innovazione tecnologica di tipo informatico. *Studi Organizzativi*, XV, II.

Bagnara, S., Diani, M. and Salmaso, D., 1983, Lavoro umano e automazione d'ufficio. *Atti dell' 84e Congresso della Società Ingegneri Eletrici*, C3, 23–27.

Bagnara, S., 1983, *Il lavoro umano nell'ufficio automatizzato*. Working paper (Rome: Istituto di Psicologia, Consiglio Nazionale delle Ricerche).

Crozier, M., 1983, Implications for the organizations, in *New Office Technology: Human and Organizational Aspects* edited by H.J. Otway and M. Peltu (London: Frances Pinter).

Diani, M., 1982, *Conséquences organisationelles des nouvelles technologies de l'information*. Working paper (Paris: Centre d'Etudes Sociologiques, Centre National de la Recherche Scientifique).
Grandjean, E. and Vigliani, E., (eds), 1980, *Ergonomic Aspects of Visual Display Terminals* (London: Taylor & Francis).
Mowshowitz, A. (ed), 1980, *Human Choice and Computers*, 2 (Amsterdam: North Holland).
Mumford, E. and Sackman, H. (eds), 1975, *Human Choice and Computers* (Amsterdam: North Holland)
Otway, H.J. and Peltu, M. (eds), 1983, *New Office Technology: Human and Organisational Aspects* (London: Frances Pinter).
Sainsaulieu, R., 1977, *L'identité au travail* (Paris: Presses de la Fondation Nationale des Sciences Politiques).
Zuboff, S., 1982, Computer mediated work: the emerging managerial challenge. *Office: Technology and People*, **1**, 66–70.

Visual Functions in Office—Including VDUs (Introductory paper)

H. KRUEGER
Department of Hygiene and Ergonomics, ETH-Center, CH–8092 Zurich, Switzerland

1. *Introduction*

Work in today's offices requires good vision whereas many traditional jobs, i.e. mining, heavy industry, farm work or underground and surface building, require only poor visual acuity. The number of employees in offices increases each year and, additionally, the number of VDUs in respect of text-computers installed in offices increases. So on the one hand more and more employees must have good visual functions and on the other hand the organizational stress is growing. High investments are followed consequently by a necessary increase in productivity.

Visual work in offices may be characterized as close-work requiring good visual functions. In most cases artificial light is used, perhaps combined with daylight. Now the special requirements of VDUs must be considered.

There is no possibility of studying all visual functions but the following parameters should be mentioned: visual acuity; adaptation; viewing-distance; flicker of VDU and lighting; and individual factors and impairments. More details will be given in the special papers.

2. *Common Remarks*

In ergonomics an investigation normally starts with a field study. An instrument often used is a questionnaire about complaints. But this self-diagnosis of employees themselves using a questionnaire should only be a first step. The next step should be an evaluation on the basis of physiology or psychophysics. A third step may be another evaluation following field study after intervention. So the task of ergonomists is deducing causally founded connections from correlations. In the area of vision, this means finding the common rules of vision or the individual factors that result in complaints.

A complication is, that additional factors not considered may introduce the correlations. Working men are not finite, limited physical systems but open to many influences from environment and organization of their job as well.

1. Non-visual, unspecific parameters may project on to the symptoms of visual stress via the autonomic nervous system by means of central nervous stress.
2. Investigations in physiological sciences can only give a positive answer or no answer. A negative result may belong to a wrong question, a wrong method or a wrong sample of subjects.
3. Sometimes human behaviour looks chaotic in a mathematic meaning. Without any reason it may change from one moment to another.

There are many results in the area of vision. But we have to consider that vision research is mainly interested in the principles of information processing of the visual system. The aim is normally to find out how the system works and not how easily it works. Thresholds are of limited profit for practical use if the aspects of comfort and discomfort are neglected. Many data are evaluated with students in their twenties, their height of performance. As long as we do not differentiate between different groups, we shall not find any significant differences between different viewing conditions but only tendencies towards different complaints. An indicator could be the variance of means. We should always take into account that there is no such animal as the 'average observer'.

Furthermore it is important to correlate results of specific sensory recognition with those outside the area of specific recognition. Specific cells of the visual system project on to specific centres of the visual cortex, but also via unspecific pathways onto unspecific cortical areas like the reticular formation in the brain stem. A lack of specific recognition does not mean that there is no unspecific reaction in the organism.

Finally there are no absolute limits where comfort ends and discomfort begins.

3. *Visual Acuity*

Many authors have found a positive correlation between work and visual acuity, both in the field and in the laboratory: ophthalmologists require different values of visual acuity for different jobs, as summarized by Pape *et al.* (1976). The value for office work is a visus of 0·8 for a distance of 50 cm. This value is suitable for a conventional office but not for a modern one with VDUs. As Schmidtke and Schohe (1967) have shown for a sample of more than 9400 industrial workers, there is a high percentage with lower acuity.

With respect to these people we have to remember that otherwise VDUs are used as an aid for persons with visual impairments. We should not only ask for a sufficient visual acuity, but also for an adjustable character size as we do with respect to brightness and contrast.

4. *Adaptation*

Good visual acuity is based on sufficient contrast, brightness and adaptation in the central and peripheral visual field as well. In comparison with traditional office work, a new aspect was introduced by VDUs. The low refresh rate of many screens made it necessary to choose the negative presentation with respect to the flicker sensitivity of the eye. At the same time the question arose whether the positive or the negative contrast would be the most suitable one. Many studies failed in the search for differences of performance data. On the other hand significant personal preferences for the positive one were reported. The difference between objective and subjective data is still unsolved. The effect of positive or negative presentation of areas with alphanumerics on long-term and short-term adaptation is not completely known yet.

5. *Viewing distance*

Vision in today's offices is vision of near distances. A short viewing distance needs a well-tuned concordance of accommodation and convergence. Figure 1 gives an overall view of accommodation and convergence according to the classical results of Fincham *et al.* (1957). The best conditions are when each value of accommodation leads to a corresponding value of convergence with no convergence stimulus. The result of different viewing distances is the line of phoria and that of Donders, respectively. In addition to that line there is an area within which the discrepancy between accommodation and convergence can be compensated by fusional convergence or divergence. Within this area heterophoria induced by chromatic aberrations of the optical system of the eye, by drugs like alcohol or analgesics or by spectacles may be compensated by the visual nervous system. It is well known that this compensation often leads to headaches.

The area of compensation is smaller if the person has lost compensation and tries to achieve a single, binocular image by fusion. Then the border line of fusion is given by the

FIGURE 1. *Correlation between accommodation and convergence (diagrammatic), see text.*

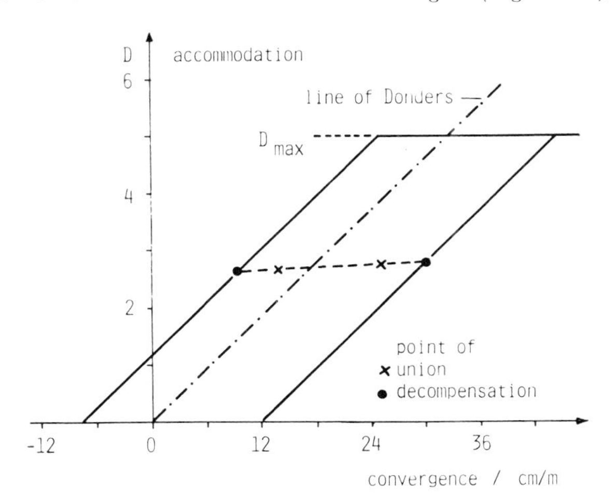

points of fusion. Table 1 summarizes the data of 10 students for 2 targets at a distance of 50 cm. The area of compensation decreases significantly ($p < 0{\cdot}01$) if the target is not a Maltese cross but a text. Therefore, in real situations occurring during office work, the usable area of compensation for single spaced, unstructured texts is smaller than the one measured in the normal screening tests with simple targets.

TABLE 1. *Area of fusion for differing measuring conditions.*

	Fusional convergence				Fusional divergence			
	Point of decomposition		Point of union		Point of decomposition		Point of union	
	Mean	Variance	Mean	Variance	Mean	Variance	Mean	Variance
Maltese cross	8·4°	0·19°	5·2°	1·5°	5·9°	0·42°	2·6°	1·2°
Text	7·9°	0·79°	4·6°	2·6°	5·1°	0·76°	0·4°	0·14°

In practice, the question of comfort or discomfort of fusion along the line of phoria is not yet answered. Figure 2 tries to give an answer with a sample of 10 emmetropic students. The subjects read a normal typewritten text of high contrast with different values of induced heterophoria which they had to compensate through fusion. The subjective feeling of stress was measured in an intermodal arrangement by means of a correlated hand force. Already small values of fusion increased the feeling of stress. Fusional divergence seemed to be little more fatiguing than convergence. The complaints of fusion lead to a third dimension of Figure 1.

FIGURE 2. *Intermodal measured effort of experimentally induced fusional vergence.*

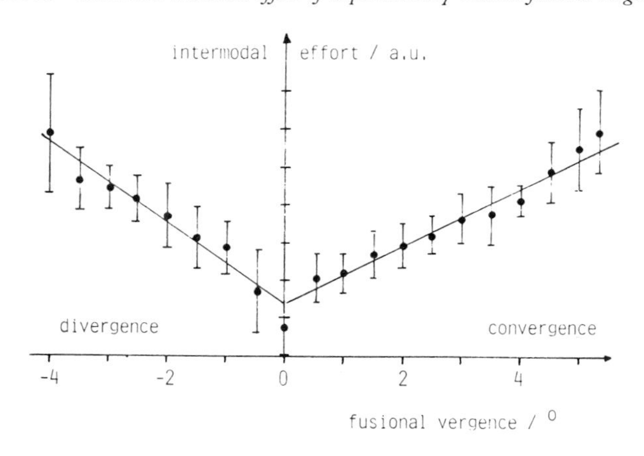

As already postulated, first by Weber in the 19th century, the resting position of accommodation lies in front of the eyes between near and far points. The value of this resting position is not fixed either intra-individually or inter-individually. It relies on the dynamic equilibrium of the ergotropic sympathetic and the trophotropic parasympathetic nervous system. Any excursion from the resting position means an increased load for the ocular muscle system. The mean of young subjects lies around 1·25 D (Leibowitz and Owens, 1975; Östberg, 1980; Zuelch and Krueger, 1982) and shifts to the far point with increasing age (Figure 3). Only at this distance does the depth of field spread symmetrically around the distance of fixation (Zuelch and Krueger, 1982). Since accommodation to a nearer distance means muscular effort, the depth of field is shifted asymmetrically from the distance of the stimulus towards the resting position. The assumption of symmetry often mentioned seems to be a wrong extrapolation. For a short distance the field of sharp vision normally lies behind the target and not in front of it.

FIGURE 3. *Resting position of accommodation measured by means of an IR-optometer.*

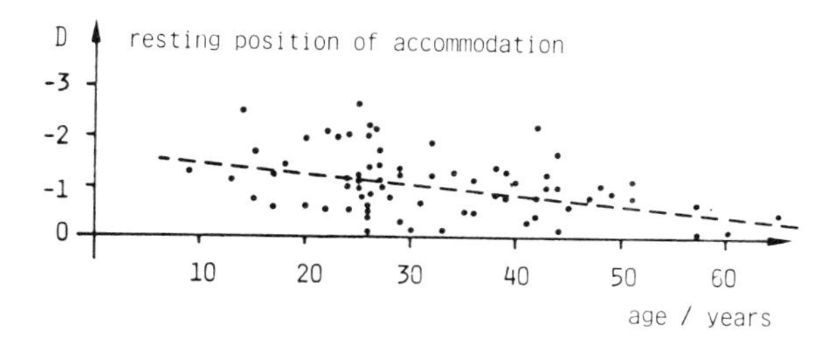

The viewing distance of 50 cm usually recommended is shorter than that of the resting point and should mean less visual comfort. In fact, Grandjean *et al.* (1983) reported a preferred distance of about 60 to 90 cm ($\bar{x}$ = 76 cm).

In order to get the right viewing distance the nervous system has to decide about sharpness of the retinal image. This is not troublesome if we use sharp targets. But in practice the visual system is often exposed to characters of considerable blur. Characters are blurred by etched surfaces in front of the VDUs. We have bad paper copies. As we have found, blur shifts accommodation (Korge and Krueger, 1983) toward the resting position. One more reason which should propagate this value is the best distance of vision. The inability of the accommodation servo loop to use fluctuations of accommodation to get a sufficient input signal for the loop adds a functional, physiological blur to that of the unsharpness of the target.

The velocity of accommodation decreases parallel to the decrease of the ability to accommodate. In a field study we have found a decrease, beginning at the age of 40, for different changes of accommodation within the region of 0·3 to 3·0 D by means of an IR-optometer (Krueger and Hessen, 1982). The ability to accommodate within a short time ends at an age between 50 and 55. To a small extent, this decrease in accommodation may be compensated by a decrease of the pupil width throughout life, but the decreasing ability to accommodate should show some effects on the behaviour at the working place with VDUs. Figure 4 demonstrates the effect of age on the viewing distance. For the presented results two groups of different age (20–30 and 45–55 years old) did the same task. They looked for typing errors of synthetic words on a VDU. The two registrations are representative for both groups. While the young subjects moved back and forth, the older ones were tied to the display because of loss of accommodation. This hints at the variable 'age'.

FIGURE 4. *Viewing distance during a reading and correcting task using VDU, keyboard and paper for a young and an older subject.*

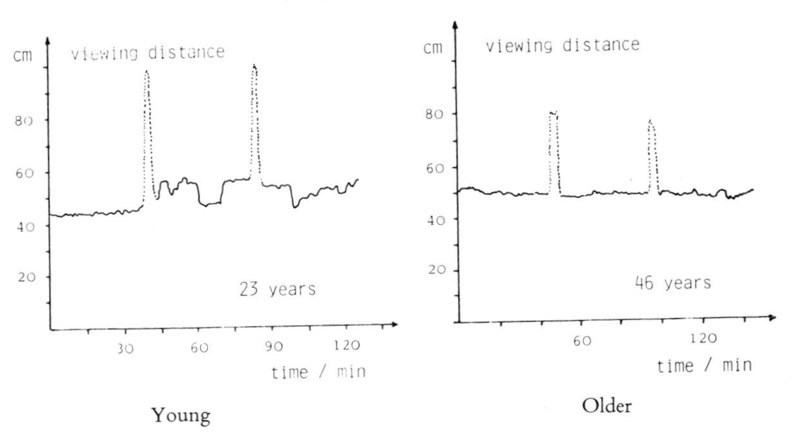

Chromatic aberration of the optics of the eye is often mentioned together with VDUs. Hyperopia for red and myopia for blue on a dark background are well known. If the colours are desaturated the effects on accommodation can be neglected (Krueger and Mader, 1982).

6. *Flicker*

Our knowledge about flicker sensitivity of the eye is mainly derived from the basic investigatations of de Lange and Kelly. A systematic study based on areas with characters is still needed. Additionally, the distribution of the flicker fusion frequency is not yet established.

7. *Individual Factors and Impairments*

Insufficient visual acuity was already mentioned above. Another individual factor is colour deficiency. This will gain more importance in the future if not only CADs use multicoloured displays but also text systems. Usually those subjects with colour deficiencies have difficulties if the colours lie on lines of equal or similar colour recognition, for example a line between red and green. Therefore colour codes should be used together with other redundant codes.

8. *Conclusions*

On the one hand there are many data from basic vision research. However, on the other hand there are remarkable gaps if we consider unspecific reactions correlated with information processing in the visual channel. The effect of impaired visual functions should be taken into greater consideration.

References

Fincham, E.F. and Walton, J., 1957, The reciprocal actions of accommodation and convergence. *Journal of Physiology*, **137**, 488–508.

Grandjean, E., Huenting, W. and Piderman, M., 1983, VDT workstation design: preferred settings and their effects. *Human Factors*, **25**, 161–175.

Korge, A. and Krueger, H., Influence of edge sharpness on the accommodation of the human eye. *Graefes Archive*, to be published.

Krueger, H. and Hessen, J., 1982, Objective, continuous measurement of accommodation in the eye. *Biomedical Technology*, **27**, 142–147.

Krueger, H. and Mader, R., 1982, The influence of colour saturation on the chromatic error of accommodation of the human eye. *Fortschritte der Ophthalmologie*, **79**, 171–173.

Leibowitz, H.W. and Owens, D.A., 1975, Anomalous myopias and the intermediate dark focus of accommodation. *Science*, **189**, 646–648, *Journal of the Optical Society of America*, **65**, 1121–1128.

Östberg, O., 1980, Accommodation and visual fatigue in display work, in *Ergonomic Aspects of Visual Display Terminals*, edited by E. Grandjean and E. Vigliani (London: Taylor & Francis) pp.41–52.

Pape, R., Blankenagel, A. and Kaiser, J., 1976, *Berufswahl und Auge*, 4, Auflage (Stuttgart: Ferdinand Enke).

Schmidtke, H. and Schober, H., 1967, *Sehanforderungen bei der Arbeit* (Stuttgart: Gentner Verlag).

Zuelch, J. and Krueger, H., 1982, Bedeutung der Schärfentiefe und Wahrnehmungstiefe der Augen für das Sehen am Arbeitsplatz. *Arbeits-, Sozial-, Präventivmedizin*, **17**, 1–4.

A Mechanism of Mental Stress Response on VDT Performance

M. Kumashiro
Department of Human Factors Engineering, School of Medicine, University of Occupational and Environmental Health, Japan, 1–1 Iseigaoka, Yahata Nishi-ku, Kitakyushu, 807, Japan

Abstract

The most significant finding from this project was that a perceived stress response as measured by CFF-test occurred after 30 min of VDT performance and continued to increase throughout the remaining 60 min on the VDT task. On the other hand, visual strain and a disrupted feed back system (which because of more complex functions involve the whole cortex) as measured by near-point accommodation and TAF-test reached a significant level after 60 min on the VDT task. The visual and perceived stress responses did not show recovery after 60 min of recess, whereas central and peripheral nervous functions including motor performance showed recovery within that same rest period.

1. *Introduction*

Several field studies have suggested that a major complaint of VDT operators is visual fatigue, and that this symptom is one of the most important problems with the use of VDT performance (Cole, 1979; Rey and Meyer, 1980; Läubli *et al.*, 1980; Dainoff *et al.*, 1981). Östberg (1980) reported that the accommodation response was significantly reduced. However, Happ and Beaver (1981) found there was no correlation between the workers' subjective complaints of visual and general fatigue and their performance. According to Stammerjohn *et al.* (1981) the factors that triggered off most visual function complaints were screen glare, angle, brightness and flicker. From the above evidence it is generally agreed that the use of CRT screen is one of the major causes of operators' occupational stress. However, when researching workers' performance it should be remembered that occupational stress is caused not only by visual fatigue from using a CRT screen but also by the lowering of psychophysiological functions due to the monotony of the repetitive work. For example, according to Gunnarson and Östberg (1979) repetition caused increased complaints of monotony from the VDT operators.

An experimental study was, consequently, undertaken to evaluate the effect of CRT-displayed presentation of visual information on psychophysiological functions in a man–computer repetitive task. Changes in the visuosensory function system and in cerebrocortical activity levels with time were studied in a limited continuous task in which recognition, judgement and output decision elements needed to be completed within a very limited cycle time.

In addition, a mechanism of the onset of visual and mental stress responses induced by VDT operation was suggested on the basis of the results obtained.

2. *Methods*

2.1. **Task**

For the VDT task the subjects had to carry out mental multiplication in 60 mathematical problems presented for 1 min on a 12 in character display (6 columns × 10 lines) and then to select the answers from 3 sets of digits. For a control task without the use of a CRT display (non-VDT task), a box of the same dimensions as the CRT display in the VDT task was used and problems were presented on a sheet (printed in black on a white paper) inserted in and removed from the front of the box (corresponding to the display area) each minute. Other conditions were matched as closely as possible to those used in the VDT task.

Mathematical stimuli presented consisted of one-figure numerals excluding 0 and 5, and they were randomly combined. A response device had 3 keyswitches 2 cm in diameter positioned at distances of 5 cm. The switch on the left was to indicate numerals 1, 2 or 3 for the tail numeral of the answers. The switch in the middle was for numerals 4 or 6 and on the right for numerals 7, 8 or 9.

2.2. **Subjects**

The subjects consisted of 7 healthy adult male students of 21 and 22 years of age in both the types of task. An eyesight of 1·5 for both eyes was confirmed in all of them by the optometry.

3. *Part I. Preliminary Experiment*

3.1. **Procedures**

The VDT and non-VDT tasks were performed for 1, 10, 20, 30 and 60 min to measure variations in heart rate (HR) during each performance.

3.2. Results

The pre-load HR (78 beats/min) was higher in the VDT as well as non-VDT tasks than the resting HR (71 beats/min). The HR during performance further exceeded the pre-load HR. On comparison of the HR variation between the VDT and non-VDT task, the HR during load tended to increase, though very slightly, in the VDT task. The mean HR of the 20 min load was lowest (85 beats/min) in all VDT task conditions.

4. *Part II. The Experiment*

4.1. Experimental Procedures

In both the VDT and non-VDT tasks, the entire course of the work load applied was 90 min, split into 3 courses of 30 min continuous task. Immediately after each course, the task was suspended for 8 min for examination of psychophysiological functions. This procedure was repeated three times. Items examined were the function of maintaining concentration (TAF) (Takakuwa, 1971; Kumashiro and Saito, 1978), critical fusion frequency (CFF) and near-point accommodation. Such examination was performed 6 times, i.e. before and immediately after the 30, 60 and 90 min task load, and again 30 and 60 min after the task period. In addition, EEG changes with eyes open and eyes closed were measured at each time of above examination only in the VDT task.

4.2. Results

4.2.1. *Near-Point Accommodation*

The variation of near-point accommodation with the passage of work time was compared between the VDT and non-VDT tasks in Figure 1. In the VDT task, as compared with the pre-load value, the near-point distance was somewhat prolonged at the 60th min of the task onwards ($p < 0{\cdot}05$). Furthermore, this prolongation phenomenon of the near-point distance was also noted at each time of recess after completion of the VDT task ($p < 0{\cdot}05$ and $p < 0{\cdot}10$ after 30 and 60 min respectively from the termination of the 90 min VDT task). Thus significant recovery of the accommodation function was not observed. In the non-VDT task, on the contrary, significant prolongation of the near-point distance was not noted at any point of measurement compared with the pre-load level.

Comparing the VDT and non-VDT tasks, the difference in prolongation of the near-point distance was fairly large ($p < 0{\cdot}10$) already at the 30th minute of the task onwards and it increased further with time showing significant difference at 60 min after load, at the termination of the 90 min task ($p < 0{\cdot}01$) and also at 30 min of recess after the task period ($p < 0{\cdot}05$). The difference between the 2 types of tasks still remained large ($p < 0{\cdot}10$) even at 60 min after the termination of the task period.

FIGURE 1. *Changes in near-point accommodation in a VDT performance (solid line) and in a non-VDT performance (dashed line) as a control with working hours.*
● Significant lowering compared to pre-value ($p < 0{\cdot}10$).
◐ Significant lowering compared to pre-value ($p < 0{\cdot}05$).
* Indicate the significant differences between a VDT-performance and a non-VDT performance
($^{*}\ p < 0{\cdot}10$, $^{**}\ p < 0{\cdot}05$, $^{***}\ p < 0{\cdot}01$).

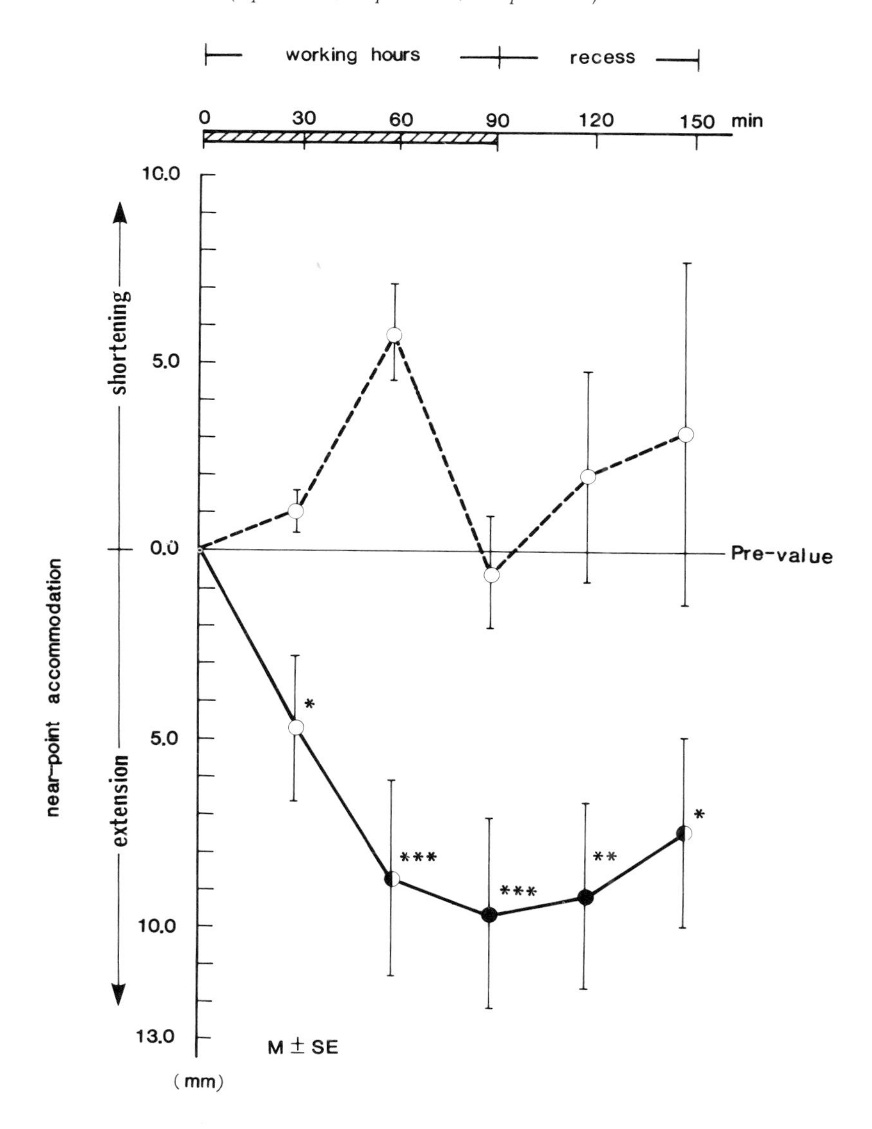

4.2.2. *Critical Fusion Frequency (CFF)*

In the VDT task, CFF was significantly reduced ($p < 0{\cdot}10$) at 30 min after load, as compared with the pre-load value, which was subsequently followed by further reduction ($p < 0{\cdot}05$). On the other hand, in the non-VDT task, a significant reduction of CFF occurred only after 90 min of cumulative work load ($p < 0{\cdot}05$). On comparing the VDT and non-VDT tasks, significant differences were noted at each measurement point. Furthermore, even at 60 min of recess after the task period there still existed a considerable difference between the 2 types of tasks ($p < 0{\cdot}05$; see Figure 2).

FIGURE 2. *Variations of critical fusion frequency (CFF) with working hours.*
● Significant lowering compared to pre-value ($p < 0{\cdot}10$).
◐ Significant lowering compared to pre-value ($p < 0{\cdot}05$).
* Indicate the significant differences between a VDT-performance and a non-VDT performance
($^{*}\ p < 0{\cdot}10$, $^{**}\ p < 0{\cdot}05$, $^{***}\ p < 0{\cdot}01$).

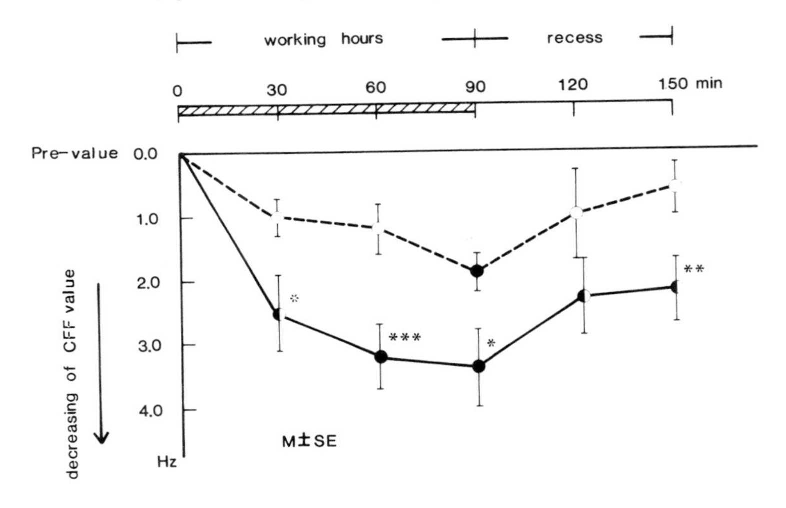

4.2.3. *Function of Maintaining Concentration (TAF)*

The level of TAF·L in the VDT task was reduced considerably after 60 min of cumulative task load ($p < 0{\cdot}10$) and the reduction became significant on termination of the task load ($p < 0{\cdot}05$). On studying the process of spontaneous recovery after the 90 min task period, TAF·L gained a rapid recovery after 30 min of recess and was significantly restored 60 min after the recess ($p < 0{\cdot}05$), as compared with the value immediately after 90 min on the task. On comparison between the VDT and non-VDT tasks, the difference between them grew large only after 60 min of cumulative task load ($p < 0{\cdot}10$).

4.2.4. *Variations of EEG and Peaks in the Alpha Wave Band*

During the VDT task period the power spectrum of the alpha waves recorded from the parietal regions shifted to the lower end of the spectrum (8·54 Hz) but recovered 30 min after the work period to the initial state (10·00 Hz).

4.3. **Discussion**

As indicated by the results of HR, CFF, near-point accommodation, TAF and EEG variations in this study, the following mechanism may be postulated in association with the onset of visual and mental stress responses induced by VDT operation (Figure 3).

FIGURE 3. *Schematic view of a development of visual and mental stress response on VDT performance.*

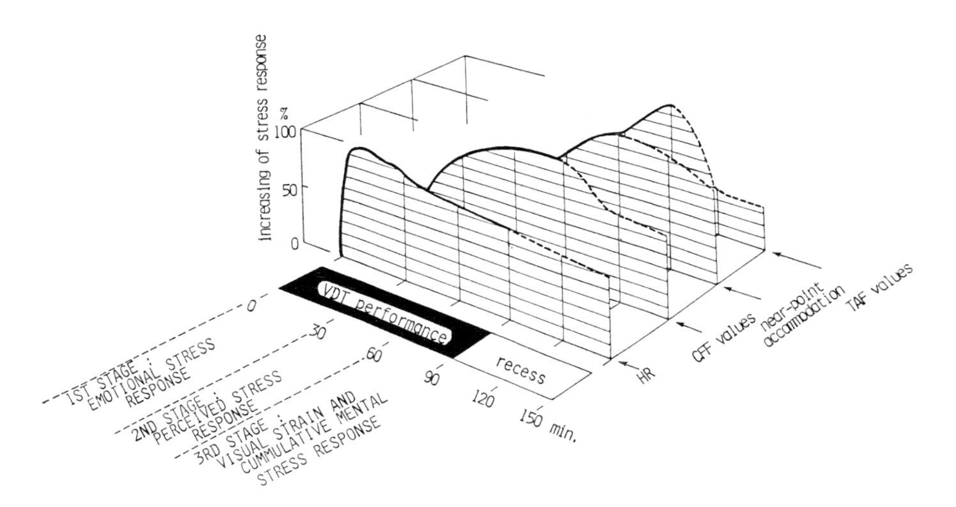

4.3.1. *First Stage*

In a repetitive computer-paced work within a very short cycle time, elevation of HR occurred almost equally regardless of the use of a CRT screen or the span of a task time. However, the use of CRT screen produced a tendency to HR acceleration, as compared with non-use of CRT screen. Leplat (1978) stated that the HR variation can be an index to measure the work load represented by emotional stress and tension. It was suggested that emotional stress response that appears in the form of such cardiac acceleration is a phenomenon that precedes the onset of other responses.

4.3.2. *Second Stage*

The significant reduction of CFF value that occurred at 30 min after load application, only in the VDT task, may be regarded as a synergistic effect of the stressor evoked by the computer-paced work itself and the stressor evoked by the use of CRT screen. Simonson and

Enzer (1941) and other studies suggested that the reduction of CFF may be interpreted as representing the loss of handling capacity of cortical visuosensory information which reflects the loss of sensory and cognitive functions due to a decreased level of arousal. Therefore, low CFF value caused by the 30th min of task onwards in VDT operation could be described as a disruption of an open loop system comprising visual input, to cortex, to motor response. In other words, it may be interpreted as an appearance of perceived stress response represented by malfunction of the visuosensory information input system.

Moreover, perceived stress response in terms of CFF decrease makes a most dramatic appearance when the task time exceeds 60 min.

4.3.3. *Third Stage*

1. A significant correlation was noted between the reduction of near-point accommodation and the duration of the VDT task. Also significant prolongation of the near-point distance was noted especially from the 60th min of the VDT task onwards. These suggested that the critical point at which objective visual strain developed as a typical symptom in the VDT task was found around 60 min after commencement of the task.
2. In addition, at 60 min after task load, conspicuous reduction of the TAF·L value was observed in the VDT task.

In contrast to the CFF-test, the TAF-test is said to reflect the state of the cerebral cortex, and close relationship existing between autonomic nervous balance and TAF (Takakuwa, 1982). Therefore, the low TAF·L value could be described as a disrupted feed back system which because of its more complex functions involves the whole cortex.

Moreover, throughout the VDT task there is a noticeable but insignificant shift of the peak alpha wave level to slower alpha waves, indicating a decrease in arousal level. A VDT operation that lasts for 60 min may begin to produce cumulative mental stress responses; these phenomena are shaped into distinct symptoms at 90 min after the task load.

On the other hand, recovery from these stress responses is preceded by a change in the arousal level of the cerebral cortex and activation of cortical activity, followed by gradual remission of a visual strain and a perceived stress response.

References

Cole, B.L., 1979, Visual problems associated with visual display units, in *Ergonomics and Visual Display Units*, edited by B. McPhee and H. Howie. (Sydney: Ergonomics Society of Australia and New Zealand), pp.29–42.

Dainoff, M.J., Happ, A. and Crane, P.C., 1981, Visual fatigue and occupational stress in VDT operators. *Human Factors*, **23**, 421–438.

Gunnarsson, E. and Östberg, O., 1979, *The Physical and Psychological Working Environment in a Terminal–Based Computer Storage and Retrieval System*, Stockholm, Swedish National Board of Occupational Safety and Health, Department of Occupational Medicine, Report 35.

Happ, A.J. and Beaver, C.W., 1981, Effects of work at a VDT-intensive laboratory task on performance, mood, and fatigue symptoms. *Proceedings of the Human Factors Society 25th Annual Meeting*, pp.142–144.

Kumashiro, M. and Saito, K., 1978, Studies on the learning effects of target-aiming performance, *Japanese Journal of Industrial Health*, **20**, 212–217.

Läubli, T., Hünting, W. and Grandjean, E., 1980, Visual impairments in VDU operators related to environmental conditions, in *Ergonomic Aspects of Visual Display Terminals*, edited by E. Grandjean and E. Vigliani (London: Taylor and Francis), pp.85–94.

Leplat, J., 1978, Factors determining work load. *Ergonomics*, **21**, 143–149.

Östberg, O., 1980, Accommodation and visual fatigue in display work. *Displays*, July, 81–85.

Rey, P. and Meyer, J.J., 1980, Visual impairments and their objective correlates, in *Ergonomic Aspects of Visual Display Terminals*, edited by E. Grandjean and E. Vigliani (London: Taylor & Francis), pp.77–83.

Simonson, E. and Enzer, N., 1941, Measurements of fusion frequency of flicker as a test of fatigue of the central nervous system. *Journal of Industrial Hygiene and Toxicology*, **23**, 83–89.

Stammerjohn, L.W., Jr., Smith, M.J. and Cohen, B.G.F., 1981, Evaluation of work station design factors in VDT operations. *Human Factors*, **23**, 401–412.

Takakuwa, E., 1971, Maintaining concentration (TAF) as a measure of mental stress. *Ergonomics*, **14**, 145–158.

Takakuwa, E., 1982, *Evaluation of Fatigue and the Function of Maintaining Concentration (TAF)*, Hokkaido University Medical Library series, Vol. 14, Hokkaido University School of Medicine, Sapporo.

The Dynamics of Dark Focus and Accommodation to Dark and Light Character CRT Displays

S.E. Taylor and B.W. McVey
Human Factors Center, IBM Corporation, San Jose, California 95193, USA

Abstract

Accommodation to dark and light character CRT displays was measured. No differences in level or variability of accommodation were found (the latter was 0·1 diopters). The utility of dark focus as a possible index of stress in the accommodative system was evaluated. With 8 subjects continuously monitored over 20 min in the dark on 3 days by an infrared servo optometer, dark focus was found to be highly unstable, typically having a range of values over 20 min in excess of 1 diopter. Its utility in indexing accommodative stress is, thus, questionable.

1. *Introduction*

It has been hypothesized that the accommodation system of the eye is stressed by the nature of a CRT display. In support of that hypothesis, evidence has been put forth suggesting that accommodation deviates from optimum as revealed by shifts towards an idiosyncratic dark focus. In this paper, both the 'idiosyncratic dark focus' construct and the accommodative activity of the eye in the presence of dark and light character displays are examined.

There have been suggestions in the literature that dark character displays constitute a better stimulus for the eye than light character displays. This confusing issue has involved many hypotheses and conflicting findings (see McVey *et al.*, 1984; Taylor *et al.*, 1984).

Another important line of research, which bears indirectly on the same issue, is that of dark focus. The topic of dark focus or intermediate resting accommodation arose in connection with understanding the effects of VDT viewing on accommodation (see Dainoff, 1982). In general, evidence of a deficiency in VDTs is offered in the form of a shift away from accommodation to a target at a known distance in the direction of dark focus (Murch, 1982; Östberg, 1982).

The issue of dark focus is somewhat controversial. Whether it exists, whether it is idiosyncratic to individuals, and, most particularly, the exact value of it are controversial

(Simonelli, 1983; Toates, 1972). But several authors have suggested that dark focus is implicated in many other relationships between some variable and accommodation (e.g. Leibowitz and Owens, 1978). Simonelli concludes that accommodation can be seen as a compromise between the position of the stimulus (required focal effort) and an individual's dark focus. The total evidence seems to suggest that the compromise is proportional to the accommodative 'stress' which results in the compromise.

This research attempted first to discover whether or not any differences in accommodation level or variability (which we call accommodative stability) could be found as a function of image polarity. Second, the question of dark focus was looked at with the view that it might reveal accommodative stress as a function of image polarity. We were also interested in testing the utility of dark focus by measuring it with an infrared optometer continuously over time with a subject absolutely in the dark. The goal was to discover whether or not dark focus was a sufficiently stable attribute of an observer to allow it to be used as an index of accommodative stress.

2. *Method*

2.1. **Subjects**

Seven subjects, 4 males and 3 females, participated in the experiment. Subjects were selected for their compatibility with the visual monitoring equipment to be described below. Compatibility required that the subjects be able to maintain a 2·5 to 3·0 mm pupil diameter while accommodating to a target at 3 diopters (3 D). In addition, subjects were given an eye examination and were required to demonstrate 20/20 or better vision at near and far point without correction.

2.2. **Equipment**

Accommodation in the lens of the eye was measured using an SRI International Infrared Servo Optometer incorporated in an SRI International Dual Purkinje Eyetracker (5th generation). For a description of this equipment and its capabilities, the reader should see Crane and Steele (1978). The optometer accurately measures accommodation to within about 0·1 diopters. Accommodation was measured in the right eye with the left eye occluded. The subjects viewed stimulus material through a dichroic mirror and were not aware of the infrared signal from the optometer.

The material to which the subject accommodated was either hardcopy or text on the CRT. To prevent difficulties with the optometer caused by pupillary constriction below 2·5 mm, it was necessary to maintain relatively low light levels. The hardcopy was text printed on paper which was illuminated with an incandescent source at 1·48 cd/m^2. The bright character display had a background luminance of 0·07 cd/m^2 and a character luminance of 8·5 cd/m^2, resulting in contrast modulation of 0·98. The dark character display had a background luminance of 10 cd/m^2 and a character luminance of 1·12 cd/m^2, resulting in contrast modulation of 0·8. The text displayed on the CRT was identical with that on the hardcopy and consisted of a series of stories. An IBM Series/1 computer collected all data from the optometer and also presented the stimuli on the CRT which was an IBM 3101 Display Terminal.

2.3. **Procedure**

2.3.1. *Phase I*

The procedure in Phase I of this experiment was part of a larger study also discussed in Rupp *et al.* (1984) in these same proceedings. Each subject received 4 treatments on each of 2 days: hardcopy (paper), CRT in focus with light characters, CRT in focus with dark characters, and CRT defocused with light characters. In each treatment, the subject's accommodation was monitored for 10 s while viewing the target both before and after a 5 min period during which the subject read text. The CRT display or hardcopy was located at 53 cm (1·88 D).

2.3.2. *Phase II*

In this phase, subjects underwent three sessions of dark focus monitoring. The subject was positioned at the optometer and enshrouded in black cloth until all light was eliminated. Subjects' accommodation was monitored for 10 s at 0, 5, 10, 15, and 20 min. Subjects were tested on 3 days in order to test for temporary changes in dark focus related to prior accommodative history. On the first day, subjects maintained essentially relaxed accommodation (did not read or view objects closer than 2 m for at least one-half hour before the dark focus session); on the second day, they read at 2 D for 5 min before the dark focus session; and on the third day, they read at 2 D for 30 min before the dark focus session.

3. *Results*

3.1. **Phase I Results**

In this paper, we will only report on the comparisons of accommodative activity between dark and focused light character displays. The data were, however, statistically analysed by analysis of variance with all four conditions included. Mean accommodation to the display was 1·61 D for both light and dark characters. The apparent underaccommodation is the result of the difference in accommodation to the green phosphor compared with paper illuminated by the incandescent source. See Rupp *et al.* (1984) for further explanation.

The more important comparison between light and dark character displays, which reflects an increase or decrease in stress on the accommodative system, is accommodative stability, defined as the r.m.s. accommodation within the 10 s monitoring periods (smaller values indicate greater stability). For dark character displays, averaged over days, period, and subjects, the accommodative stability value is 0·10 D; for light characters, it is 0·09 D. The difference is small and not statistically significant. To allow some precision in identifying their equivalence, a 95% confidence interval was constructed from the analysis of variance for the difference in accommodative stability between dark and light character displays. Adding the resulting +/− 0·04 D to the observed difference allows us to assert that statistically the largest difference that could be expected between the two is 0·05 D.

Thus neither the mean accommodation nor accommodation stability differs between dark and light character displays.

3.2. **Phase II Results**

Averaging across both days, all five 10 s periods, and all seven subjects, the mean dark accommodation is 1·36 D and the mean accommodation stability is 0·25 D. Note that local accommodation stability is two-and-one-half times as great as was found with a visible target.

However, these values do not demonstrate the central finding. Dark focus is highly variable over time. The construct of a static and idiosyncratic dark focus (which in some papers has been specified to two decimal places) seems inappropriate when we characterize the variability of dark focus over time. Figure 1 shows the mean dark focus values across subjects for each day and period. For Day 1, the variability of these estimates is 0·42 (s.d.); for Day 2, 0·43; and for Day 3, 0·60. The data, by the way, do not show any significant residual accommodation on Days 2 and 3.

Perhaps the most revealing indication of the variability of dark focus is to be found in Table 1. Table 1 shows for each subject and day, the range of dark focus values observed. The last row in the table shows the mean range for the seven subjects on each day. Dark focus, on the average, varies over a range in excess of 1 D. It is clear that focus is changing greatly and continuously in the dark!

FIGURE 1. *Dark focus means over 20 min in the dark on 3 days (expressed in diopters).*

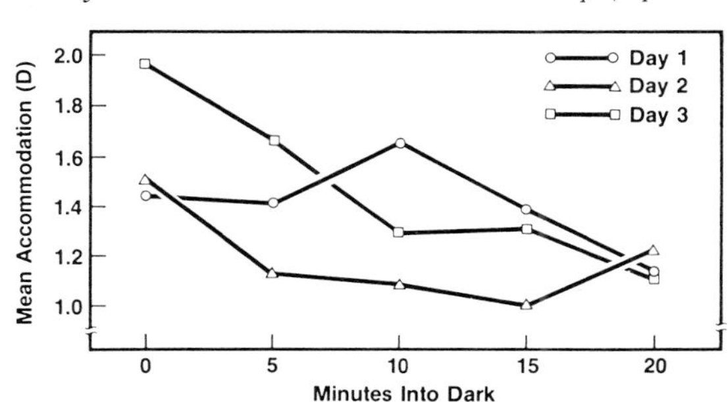

TABLE I. *Dark focus range*[a] *for each subject on each day.*

Subject	Day 1	Day 2	Day 3
1	1·05	1·93	2·46
2	0·83	0·28	1·05
3	2·06	1·46	1·20
4	0·07	1·20	1·17
5	1·74	0·79	1·90
6	0·31	1·09	1·23
7	1·13	0·69	1·85
Mean	1·03	1·06	1·55

[a] in Diopters.

4. *Discussion*

Phase I of this research clearly demonstrates that there is no accommodative stress relating to the choice of image polarity in VDT viewing. All of the subjects easily accommodated to the display with the same level of accommodation and with remarkable stability. One must conclude that these data do not support a recommendation for one polarity over another.

Phase II of this research reveals results which certainly create problems for the use of the dark focus construct in evaluating VDTs. Accommodative stability was greatly reduced in the 10 s samples. More importantly, the range of dark focus values obtained for a given subject was typically quite large. Although the average of these dark focus measurements was in the neighbourhood of 1·35 D, the variability was so great (as great within subjects as between) that the mean values are meaningless.

Our values are not atypical. Dark focus means reported in the literature have ranged from 0·5 D to 2·5 D! It appears that most of our subjects were accommodating in the dark, but not consistently at any particular value. Thus dark focus does not appear to be a sufficiently stable attribute that it can serve as a construct on which to base estimates of the accommodative stress associated with VDT tasks.

One can only hesitantly comment on the data in the literature. Perhaps the visible stimuli required for the laser optometer or polarized vernier optometer are stimuli to accommodation. Or perhaps the inability of those devices to continuously monitor accommodation masked the variability of dark focus.

However, our research clearly calls into question the utility of the dark focus concept at least in VDT research.

Acknowledgement

The authors are indebted to Maryellen Çiak for her extraordinary effort in learning the SRI equipment and collecting data and to Vic M. Ramos whose programming and electronic skills proves absolutely and literally indispensable.

References

Crane, H.D. and Steele, C.M., 1978, Accurate three-dimensional eyetracker. *Applied Optics*, **17**, 691–705.

Dainoff, M.J., 1982, Occupational stress factors in visual display terminal (VDT) operation: a review of empirical research. *Behaviour and Information Technology*, **1**, 141–176.

Leibowitz, H.W. and Owens, D.A., 1978, New evidence for the intermediate position of relaxed accommodation. *Documenta Ophthalmologica*, **46**, 133–147.

McVey, B.W., Clauer, C.K. and Taylor, S.E., 1984, A comparison of anti-glare contrast-enhancement filters for positive and negative image displays under adverse lighting conditions. This vol., pp.405–409.

Murch, G.M., 1982, Visual fatigue with prolonged display use. *SID 82 Digest of Technical Papers (Los Angeles: Society for Information Display)*, pp.200–201.

Östberg, O., 1982, Accommodation and visual fatigue in display work, in *Ergonomic Aspects of Visual Display Terminals*, edited by E. Grandjean and E. Vigliani (London: Taylor & Francis) pp.41–52.

Rupp, B.A., McVey, B.W. and Taylor, S.E., 1984, Image quality and the accommodation response. This vol., pp.254–259.

Simonelli, N.M., 1983, The dark focus of the human eye and its relationship to age and visual defect. *Human Factors*, **25**, 85–92.

Taylor, S.E., McVey, B.W. and Emmons, W.H., 1984, Cognitive complexity related to image polarity in the aetiology of visual fatigue. This vol., pp.175–180.

Toates, F.M., 1972, Accommodation function of the human eye. *Physiological Review*, **52**, 828–863.

Image Quality and the Accommodation Response

B.A. Rupp, B.W. McVey and S.E. Taylor
IBM Corporation, San Jose, California 95193, USA

Abstract

The accommodation response to blur produced by defocusing the distal image and blur produced by the chromatic aberration of the eye was measured. Differences in the spectral composition of the image did produce differences in the mean level of accommodation. However, in all cases the stability of accommodation was not significantly different. Blur did not produce any increased activity of the accommodation mechanism.

1. *Introduction*

There have been concerns that viewing VDTs may unduly stress the accommodation mechanism. Mourant *et al.* (1979), for example, speculated that the lower resolution of a VDT, as compared to hard copy, would cause the accommodation mechanism to continually hunt in an attempt to bring the image into sharp focus. The concern expressed by Rey (1977) was that, because of chromatic aberration of the eye, a bimodal spectral luminance distribution such as that found in the P4 family of phosphors would cause the accommodation mechanism to continually refocus between the peaks of the spectral distribution, thus potentially stressing the system.

Fincham (1951) has shown that defocusing the distal image will not cause a change in the accommodation response. One should not expect any increase in the normal oscillations of the accommodated lens to be produced by lowering the resolution of a display. However, there is a voluntary component in the accommodation response that it would be interesting to compare the accommodation response as well as accommodation stability to hard copy with displays of different levels of resolution. Toates' (1972) review of the accommodation function does include chromatic aberration as a stimulus for accommodation. Although the initial concern was over a monochromatic display that used a particular phosphor, the concern, if valid, would be of particular importance for multicolour VDTs. The combinations of the primaries of a multicolour VDT inherently produce multimodal spectral distributions.

The first part of this study will deal with the accommodation response to hard copy and two levels of display resolution. The second part will deal with the accommodation response to unimodal and bimodal spectral distributions. Accommodation level and stability will be measured to determine if there are differences in the accommodation response due to image blur or spectral composition.

2. *Part I. The Accommodation Response to Blur and Mode of Presentation*

2.1. **Method**

2.1.1. *Subjects*

Five subjects ranging from 17 to 21 years of age were used in this part of the study. The subjects all had uncorrected 20/20 or better vision.

2.1.2. *Apparatus*

An IBM 3101 display terminal with an added neutral density filter with a thin film antireflection coating on its first surface was used to present the focused and defocused light character CRT images. The images presented on the display terminal were continuous text. The hard copy was of similar continuous text printed on white paper. Viewing distance was 53 cm. Diffuse incandescent light was used to illuminate the task area at a level of about 7 lux. The photometric characteristics of the three presentations may be seen in Table 1.

A photographic reproduction of the two levels of display resolution may be seen in Figure 1. The luminance distributions, as measured across a vertical stroke, for the focused and defocused characters are shown in Figure 2.

Accommodation level was measured by the use of an SRI infrared optometer. That device was described in the preceding paper (Taylor and McVey).

TABLE 1. *Photometric characteristics of the three presentations.*

Presentation	Background (cd/m^2)	Character (cd/m^2)	Contrast (Modulation)	Contrast (Ratio)
Focused display	0·07	8·5	0·98	121:1
Defocused display	0·07	5·6	0·97	80:1
Hard copy	1·5		*c.* 0·8	*c.* 10:1

FIGURE 1. *Photographic reproduction of two CRT displays, one focused and one defocused.*

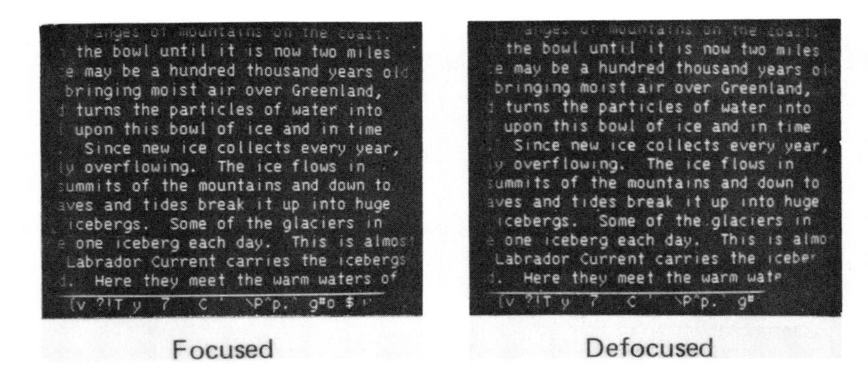

FIGURE 2. *Luminance distribution across a vertical stroke of a focused and defocused character on a CRT. At a viewing distance of 50 cm, one minute of arc equals 0·145 mm.*

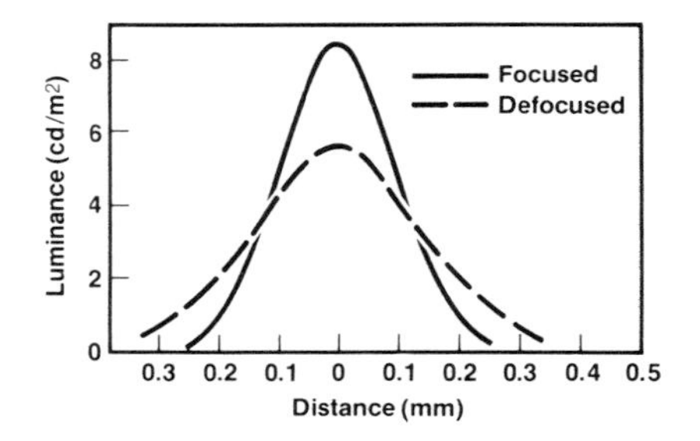

2.1.3. *Procedure*

Accommodation was measured continuously for a 10 s period at the beginning and end of a 5 min reading task. Each subject was tested under each of the 3 conditions on 2 separate days. A 5-min rest period was given between each condition.

2.2. Results/Discussion

The level of accommodation was calculated at intervals of 0·05 s over each of the 10-s test period. Those values were used to determine mean accommodation and accommodation stability. Accommodation stability is defined as the root mean square of the data points. The results are given in Table 2.

The difference in mean accommodation between hard copy and displayed images was significant at the 0·001 level, $F(3, 12) = 14{\cdot}55$. Other differences, focused/defocused, beginning/end of the reading task, day 1/day 2, for both mean levels and stability of accommodation were not statistically significant.

TABLE 2. *Accommodation to various presentations of text.*

Condition	Mean accommodation (diopters)	Accommodation stability (r.m.s.)
Focused display	1·61	0·085
Defocused display	1·61	0·098
Hard copy	1·82	0·101

After the data were analysed it was assumed that the additional 0·2 diopters of accommodation made for the hard copy as compared to the CRT presentations was due to the chromatic difference between the two presentations. The P39 phosphor of the display produces a relatively narrow band of green while the incandescent light that illuminated the hard copy was broad band and biased towards the longer wavelengths. To test this assumption one of the subjects was recalled and the accommodation responses to the hard copy presentation were measured when the material was illuminated by the incandescent source and by light from the display. The subject showed a 0·23 diopter increase in accommodation for the incandescent source. The difference in accommodation between the display and hard copy may be attributed to chromatic differences in the presentation.

3. *Part II. Accommodation Response to Uni- and Bimodal Spectral Distributions*

3.1. **Method**

3.1.1. *Subjects*

Four subjects ranging from 17 to 23 years of age were used in this part of the study. The subjects all had uncorrected 20/20 or better vision and normal colour perception.

3.1.2. *Apparatus*

A high contrast negative image transparency of a schematic of an optical device was mounted on the front surface of a piece of opal glass. The opal glass was rear illuminated by two independent light sources. One light source was filtered by a No. 55 Wratten filter (green) and the other by a No. 22 Wratten filter (orange). The relative spectral luminance of the two sources may be seen in Figure 3. The green source produced a line or symbol luminance of 10·6 cd/m^2 and the orange source produced a line or symbol luminance of 8·8 cd/m^2. Viewing distance was 50 cm and all room lighting was turned off. An SRI infrared optometer was used to measure accommodation.

FIGURE 3. *Weighted spectral luminance for Wratten filters No. 22 and No. 55. The weighting is a product of transmission and relative spectral sensitivity.*

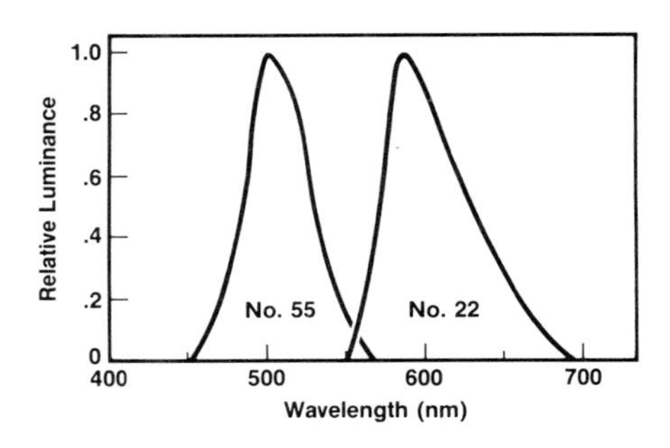

3.1.3. *Procedure*

Accommodation was measured continuously for 10-s periods during each of the three conditions. In the first condition half of the subjects viewed the figure illuminated by the green source and half viewed the figure illuminated by the orange source. The sources were reversed for the second condition and in the third condition the figure was illuminated by both sources at the same time.

3.2. **Results/Discussion**

The accommodation response and stability were determined the same way as in Part 1. The results are given in Table 3.

The accommodation stability, a r.m.s. of about 0·1 diopters is about that expected for the accommodated eye (Campbell, 1959). The multimodal spectral distribution, the green plus orange, did not produce any additional accommodation activity; accommodation did not shift between the two peaks. There was a 0·27 diopter difference in the accommodation response between the green and orange conditions. That is about the dioptric difference between those two colours with respect to the chromatic aberration of the eye. It is interesting to note that the accommodation response to the combined colours was essentially an averaging of the responses to the individual colours.

TABLE 3. *Accommodation response with different spectral distributions.*

Illuminant	Mean accommodation (diopters)	Accommodation stability (r.m.s.)
Green (Wratten No. 55)	1·60	0·078
Orange (Wratten No. 22)	1·87	0·085
Green + orange	1·74	0·083

References

Campbell, F.W., Robson, J.G. and Westheimer, G., 1959, Fluctuations of accommodation under steady viewing conditions. *Journal of Physiology* (London), **145**, 579–594.

Fincham, E.F., 1951, The accommodation reflex and its stimulus. *British Journal of Ophthalmology*, **25**, 381–393.

Mourant, R.R., Lakshmanan, R. and Herman, M., 1979, Hard copy and cathode ray tube visual performance—are there differences? *Proceedings of the Human Factors Society 23rd Annual Meeting*, 367–368.

Rey, P., 1977, Problèmes visuels aux écrans de visualisation. *Seminar Proceedings*, 16 December 1977 (University of Geneva).

Toates, F.M., 1972, Accommodation function of the human eye. *Physiological Review*, **52**, 828–863.

Far Point of VDU Operators Measured in situ

A. Serra
Istituto di Clinica Oculistica, Cattedra di Ottica Fisiopatologica dell'Università di Cagliari, I–09100 Cagliari, Italy

Abstract

Refraction at the far point of a number of clerks is measured *in situ* using a modified Badal's optometer, the targets being the digits on coloured displays. Subtle changes (within about 1 dioptre) are found as a function of screen dominant wavelength and/or environmental illumination level.

1. *Introduction*

In modern offices coloured displays are widely used. In some cases the eye–screen distance is standardized, in other cases it may vary unpredictably within wide limits. The implications of the chromatic aberration of the eye on the estimates of refraction, in relation to the phosphor spectral emission, are tackled in the recent ergo-ophthalmological literature.

In the present report we support the suggestion (Östberg, 1980) that sight should be measured *in situ*, at the work-place. We measured the far point of a number of clerks by using as targets the digits themselves appearing on the display. We wonder whether deviations are met, suggesting that as the colour varies, the wavelength focused on the retina also varies.

2. *Method*

We used a modified version of Badal's optometer to measure (monocularly) the refraction at the far point. Briefly, a 6 dioptre lens is fastened on to a ruler. The eye's nodal point is kept fixed at lens focus, in the object space. The lens–display distance is varied until the blur-to-sharp transition is attained. First measured in millimetres, this liminal distance is converted into dioptres. The lens is frameless, so that the illumination at the eye from the surround is not screened out.

Three groups of subjects (clerks in different offices, scientists) were tested. The number of subjects in each group was 14, 14 and 8. Their ages ranged from 26 to 52 years.

3. *Experimental Findings and Discussion*

The data shown in Figure 1 were obtained from 14 subjects (28 eyes) by using as a target which was first a digit appearing on a white screen (W), and then on a green screen (G). The two displays were flanked, and their luminances were set at the values currently adopted for routine work. Office lighting was a mixture of natural light and light from daylight fluorescent tubes.

FIGURE 1. *Abscissae: refraction at the far point for a white target, in dioptres. Ordinates: the same for a green target. Each point refers to a different subject.*

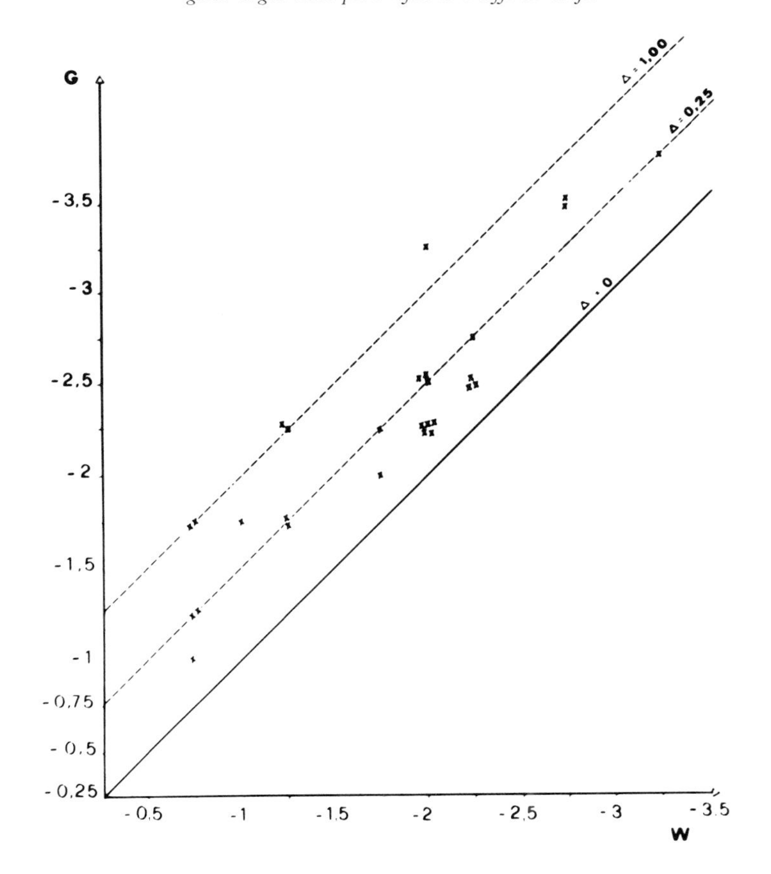

FIGURE 2A. *Ordinates: difference (D) in the far point estimates for the pairs of target colours denoted on the abscissae: w, white; r, red; y, yellow; g, green.*

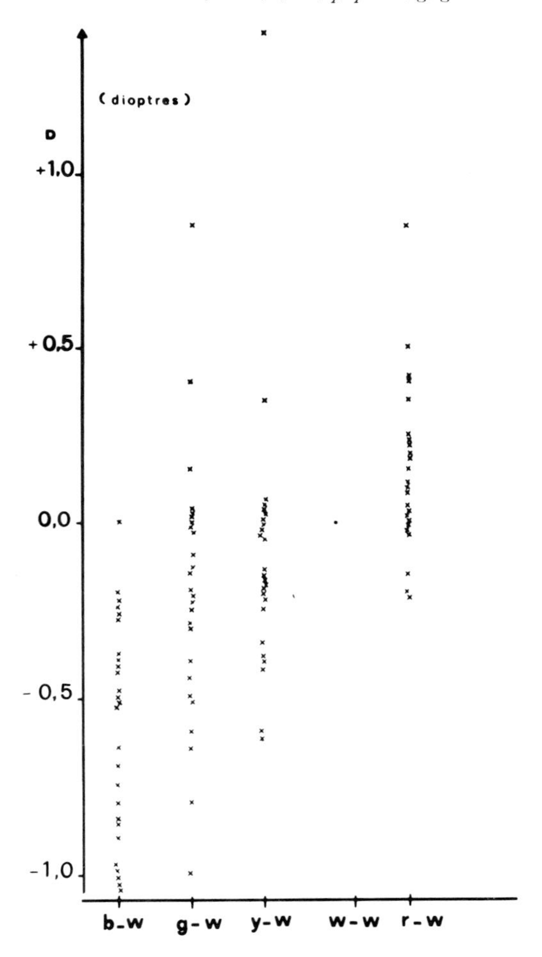

Our findings show that a sort of myopization occurs when passing from the white to the green display. Its magnitude exceeds one dioptre in one case only.

The data shown in Figure 2a were obtained from another group of 14 subjects (28 eyes) by using as targets the digits appearing on a multicolour display. Let us take as a reference the far point measured, for every subject, by the use of a white (w) target.

Our findings show that when the blue (b) target is shown, the eye becomes slightly myopic; when the red (r) target is shown, there is a trend toward hyperopia. This mirrors the wavelength dependence of eye chromatic aberration.

Now, the spectral emission of the phosphors covers a relatively broad band. In spite of this, it seems that when a solitary digit of a given colour is seen, there is a shift of the wavelength focused on the retina.

FIGURE 2B. *Chromatic aberration of human eye, in dioptres, d (ordinates) versus wavelength (abscissae). Circles: Wald and Griffini's data; crosses: Ivanoff's data. Full line: theoretic aberration, dashed line: real aberration.*

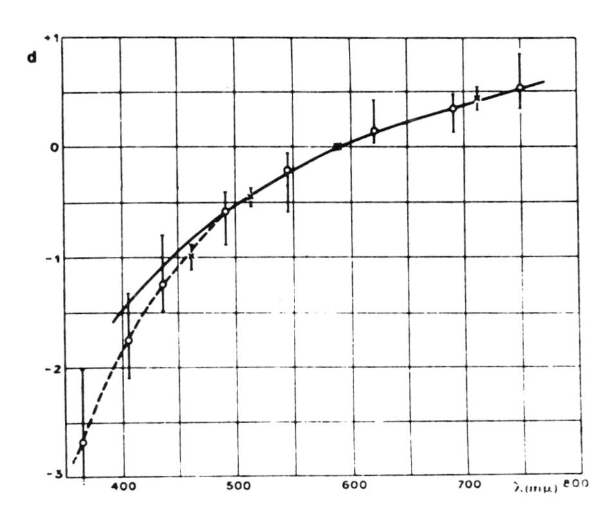

Lastly, let us consider the influence of room lighting. We measured the refraction at the far point of 8 subjects (16 eyes) under 2 different environmental conditions to which the eyes were well adapted. The design luminance was kept at the level usually adopted by the operators. In the first case a luxmeter placed close to the eye recorded 40 lux, in the other case 8 lux.

FIGURE 3. *Frequency distribution of the differences (in dioptres) in far points under lower and higher environmental illuminations. Open circles, green target; crosses, white target.*

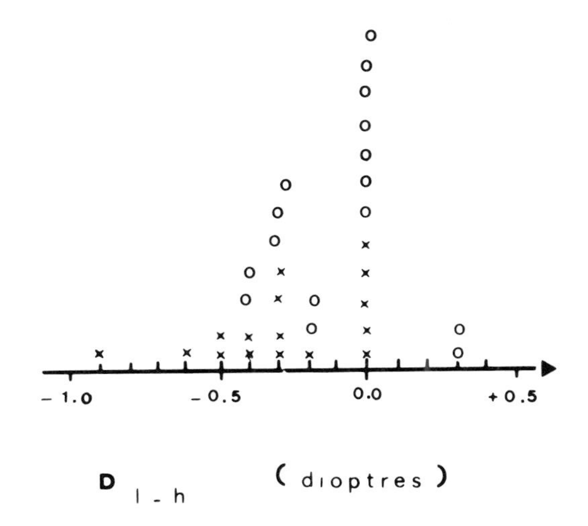

As is shown in Figure 3, it is noted that the dimmer environment involved a myopization, the so-called twilight-myopia.

All these data seem to indicate that photometric conditions may be responsible for subclinical effects on the operator's refraction.

References

Krüger, H., 1980, Ophthalmological aspects of work with display workstations, in *Ergonomic Aspects of Visual Display Terminals*, edited by E. Grandjean and E. Vigliani (London: Taylor & Francis) pp.31–40.

Östberg, O., 1980, Accommodation and visual fatigue in display work, in *Ergonomic Aspects of Visual Display Terminals*, edited by E. Grandjean and E. Vigliani (London: Taylor & Francis) pp.41–52.

Display Parameters for Improved Performance and Reduced Fatigue: an Experimental Study

M.J. Schmidt
Digital Equipment Corporation, Andover, Massachusetts, USA
and J.M. Camisa
Rockefeller University, New York, New York, USA

Abstract

Evidence suggests that VDT users may improve performance and reduce fatigue if CRT refresh rates exceed line frequency, if screen and surround approximate field luminance, and if contrast is moderate. VDT fatigue and performance depend on factors other than display parameters, however, so that existing field study data are not very helpful in making precise deductions about optimal display characteristics. We describe here a large scale experimental study in progess to assess the effects of display characteristics on visual fatigue (changes in spatial and temporal contrast sensitivity) and VDT performance. Emerging data show that both kinds of contrast sensitivity change during 3 hour VDT sessions. Contrast sensitivity should prove to be a powerful, sensitive tool for helping find optimal display parameter values.

1. *Introduction*

A growing body of scientific evidence supports the conclusion that VDT usage, *per se*, causes no harmful physiological effects and is not a health hazard (e.g. Grandjean, 1980). It is also clear, however, that many office VDT users find their work tiring and uncomfortable. VDT users tend to report eye-strain, headache, and other stress symptoms more often than non-VDT workers. A Canadian study, for instance, found during a 3-month period that 77% of VDT workers report eye-strain, compared to 56% of non-VDT users (Canadian Labour Congress, 1982). A government study in the United States found VDT users reporting significantly more discomfort than non-VDT clerical workers, in terms of blurred vision, burning eyes, back pain, and 20 other classes of complaint (Smith *et al.*, 1981; NIOSH, 1981). Concern over VDT use has led to regulations on VDT design in Europe (e.g. DIN standard 66/234) and in requests from labour organizations in the United States and Canada for government regulation of VDT work conditions.

Which aspects of the VDT job setting are most responsible for these complaints? And what can VDT designers do to improve user comfort and performance? We are in the midst of a large-scale study at the Digital Equipment Corporation to address these questions to display parameters of the VDT. The large number of field study data are not very helpful in making precise deductions about optimal display characteristics, because performance and fatigue depend on several interacting factors apart from the display itself. In our study, other major sources of difficulty such as the physical environment, computer system performance, software design, and psychosocial aspects of the job setting are held constant, insofar as possible.

Other research suggests that fatigue and performance are related to such display parameters as CRT luminance and flicker (Isensee and Bennett, 1983), contrast polarity (Bauer and Cavonius, 1980), character sharpness, and presence and absence of high-contrast screen surrounds (Camisa and Schmidt, 1984). These factors are independent variables in our study. Dependent variables are spatial and temporal contrast sensitivity, and eight measures of VDT performance (typing speed, error rate, etc.). Spatial and temporal contrast sensitivity are sensitive and relatively complete measures of visual capability. We wish to show in this paper that they are useful measures of visual fatigue in a study designed to find optimal display screen characteristics.

Our approach is to measure a subject's spatial and temporal contrast sensitivity, conduct a VDT work session (while capturing VDT performance data), and then measure both kinds of contrast sensitivity again. First results of this study show that both types of contrast sensitivity change after a three hour VDT session.

2. *Methods*

2.1. **Subjects**

To date, 95 Digital employees (ages 20–65) have passed health, visual, and keyboard-skills screening criteria, to participate as subjects.

2.2. **Apparatus**

Each subject is tested with three displays: one monitor for each type of contrast sensitivity test and a work session VDT. Display screens are 30 cm (diagonal) and anti-glare coated.

2.2.1. *Contrast Sensitivity Monitors*

Spatial contrast thresholds are obtained from a monitor displaying sinusoidally modulated vertical grating patterns at 0·5, 1·0, 2·0, 4·0, 7·9, and 15·9 cycles per degree (cpd). A measure of temporal contrast sensitivity is obtained for spatially uniform fields at refresh rates of 30, 40, 50, 60, 72 and 90 Hz. Screen luminance is a constant 50 cd/m^2; these monitors are viewed from 1·0 m. Stimulus generation and subject data recording are controlled by a dedicated PDP-11/03 computer.

2.2.2. *VDT Work Session Monitors*

Six Digital VT-100 type VDTs have been modified to serve as fully functional terminals, while scanning at 30 Hz (interlaced) and 50, 60, 80, 90, and 120 Hz (non-interlaced), in the first experiment in progress to examine effects of faster scan rate. Work session VDTs are viewed from 0·5–0·75 m; in the first experiment, mean screen luminance (while filled with text) is 50 cd/m^2. For each subject, worktable, chair, and VDT are arranged to the recommended configuration (Grandjean, 1980). VDT work session tasks are controlled by a time-shared VAX 11-780 computer.

2.3. **Experimental Procedure**

The 3·5 h experimental session has three components:

2.3.1. *Contrast Sensitivity Pre-Tests (15 min)*

Binocular spatial and temporal contrast sensitivity functions are obtained using a self-tracking method: contrast thresholds are taken as the mean contrast of six ascending and six descending contrast reversal points.

2.3.2. *VDT Work Session (3 h)*

A VDT work session immediately follows the pre-tests. The subject's task is to retype lines of text (from magazine and newspaper articles) as they appear on the screen. Each line must be completed before the next line appears. The subject thus fixates the screen throughout the work session. Feedback on VDT performance is given throughout the work session; subjects are encouraged to 'score' as high as possible (type as fast and as accurately as possible). After each hour there are 5-minute rest breaks.

2.3.3. *Contrast Sensitivity Post-Tests (15 min)*

The pre-test procedure is repeated immediately after the VDT work session.

3. *Results and Discussion*

Data obtained so far support several preliminary conclusions:

1. Baseline differences between subjects, on contrast thresholds and VDT performance

FIGURE 1. *Spatial contrast sensitivity functions for a typical subject, obtained before and after a 3-h VDT work session. The horizontal axis represents width of bars in the grating test pattern in terms of spatial frequency—light/dark bar cycles per degree of visual angle. The vertical axis represents visual sensitivity as the reciprocal of contrast threshold. Decrements in contrast sensitivity following VDT work are most pronounced at the middle spatial frequencies.*

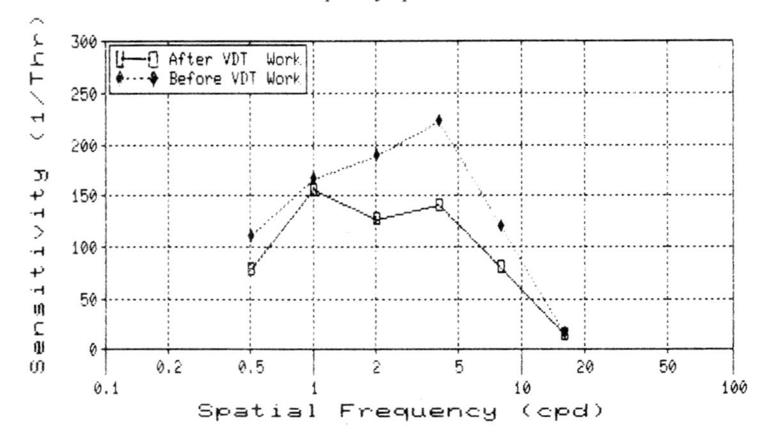

measures, tend to be at least as large as the experimental effects. Thus, if we are to obtain results reliable enough, and clear enough to have any impact on VDT design, either large numbers of subjects or repeated-measures experimental designs must be used.

2. Spatial contrast sensitivity generally decreases after a 3-h VDT session, with sensitivity decrements most pronounced at middle spatial frequencies (2·0–7·9 cpd). Temporal contrast sensitivity, however, may change in either direction and more data are needed before conclusions are appropriate. Figure 1 shows typical spatial contrast sensitivity functions before and after a 3-h work session, for one subject.
3 VDT performance (typing rate, error rate, and so on) fluctuates throughout the 3-h work session; subjects generally perform best during the first 30 min of each hour, after which error rate increases.

Acknowledgements

We wish to thank Digital's Tom Stockebrand and the Southwest Engineering Group for continued support in executing this study. Dar Berhoozi, Claude Sigel, John Gilstrap and Pat Lauffenberger are contributing significantly to its progress.

References

Bauer, D. and Cavonius, C.R., 1980, Improving the legibility of visual display units through contrast reversal, in *Ergonomic Aspects of Visual Display Terminals*, edited by E. Grandjean and E. Vigliani. Proceedings of the International Workshop, Milan, March, 1980 (London: Taylor & Francis) pp.137–142.

Camisa, J.M. and Schmidt, M.J., 1984, Performance, fatigue and stress for the older VDT user. This vol., pp. 276–279.

Canadian Labour Congress, 1982, Towards a more humanized technology: Exploring the impact of video display terminals on the health and working conditions of Canadian office workers (Ottawa: CLC Labour and Education and Studies Centre).

Grandjean, E., 1980, Ergonomics of VDUs: review of present knowledge, in *Ergonomic Aspects of Visual Display Terminals*, edited by E. Grandjean and E. Vigliani (London: Taylor & Francis) pp. 1–12.

Isensee, S.H. and Bennett, C.A., 1983, The perception of flicker and glare on computer CRT displays. *Human Factors*, **25**, 177–184.

NIOSH, 1981, Potential health hazards of video display terminals (Washington: Department of Health and Human Services, NIOSH publication number 81–129).

Saito, M., Tanaka, T. and Oshima, M., 1981, Eyestrain in inspection and clerical workers. *Ergonomics*, **24**, 161–173.

Smith, M.J., Cohen, L. and Stammerjohn, L.W., Jr., 1981, An investigation of health complaints and job stress in video display operations. *Human Factors*, **23**, 387–400.

Umbach, F.W., Kalsbeek, J.W.H. and Bosman, D., 1982, A device for the measurement of contrast resolution, spatial and temporal resolution by means of a VDU screen (Enschede: Twente University of Technology).

Working at Visual Displays: the Influence of Age

H. JIRANEK, W. KUGELMANN AND H. KRUEGER
Institute of Work Physiology, Technical University Munich, Barbarastr. 16,
D–8000 Munich 40, FR Germany

Abstract

By means of a search- and comparison-task (pronounceable nonsense words of two different lengths were used) the influence of three codes (brightness-regulation, reverse-contrast and underlining) on speed and quality of work was investigated. Two age groups took part in the experiments—Group 1: 11 subjects ranging from 20 to 30 years of age and Group 2: 8 subjects 45 to 55 years of age. The evaluation of the results was based on the hypothesis that different codes evoke different cognitive effort, but younger subjects are more capable of compensating possible disadvantages. Two three-factor analyses of variance show that the older subjects actually work worse in a code-specific way. A theoretical approach concerning the evaluation of codes was developed.

1. *Introduction*

Software ergonomics has to aim at the evaluation of software-configurations in order to improve software. Nowadays it is generally accepted that concepts that are good for one user-group may be bad for another, because users differ, for instance, in terms of experience (Benz and Haubner, 1983) or using-frequency (Eason, 1980). The idea of a fictional average user should be abolished.

The difficulties certain users have compared to others, might indicate not only that the software features are unsuitable, but also that the 'high performers' are more likely to be capable of compensating for the software's shortcomings. The latter is particularly valid when the users resemble each other in the important aspects mentioned above.

This is what we thought of when we compared the performance data of two age-groups over different codes and different word lengths in a search and comparison task; and we have tried to base the evaluation upon these ideas.

The mere interpretation of statistical means does not seem to be sufficient, as there are sometimes contradictions between performance data and the subjective assessment of software configurations, which have not been interpreted (see Benz and Haubner, 1983). What is needed is a theory-guided interpretation. We have tried to develop a theoretical approach to make it possible to evaluate the results of the experiment.

2. *Method and Subjects*

2.1. **Subjects**

Two groups of subjects took part—Group 1: 11 students from various faculties, ranging from 22 to 27 years of age (mean age = 24·6, s = 1·75) and Group 2: 8 subjects 45 to 55 years of age (mean age = 47, s = 3·74). Five subjects were graduates, one a precision mechanic, one commercial employee and an office worker.

2.2. **Configuration of the Task**

A monochromatic display (Phosphor P39) was filled with pronounceable nonsense words (background test). On this background twelve areas were marked (five words per area, four areas per code). The codes were brightness-regulation, contrast-reversal and underlining. The position and shape of the marked areas was chosen randomly with one restriction: areas of the same code should not touch each other. The areas of each code were composed of words of two different lengths (short words: mean = 5·25 letters, long words: mean = 7·25 letters); the same word was distributed ten times on the four areas of the respective code. That means: each of the six different words appeared ten times in one single picture. Of these words 30% were written in the wrong way. The correct writing of the six task words for one image, with their respective code, was given on a printed document.

The subjects had to count the deviant words and to put the result in a keyboard connected to a microcomputer. Altogether 60 images were to be worked through, that means that 3600 nonsense words had to be checked. One learning session (20 images) with immediate sound-feedback preceded the main test.

2.3. **Collected Data**

'Time' and 'accuracy' (i.e. the subject's error rate) were measured for any single task and so could be referred to code and length of the task words. At the end of each main session the subjects had to 'evaluate' the different codes in a questionnaire regarding readability, stress and subjective preference.

3. *Results*

The computation of the data was done separately for working time and accuracy with a three-factor analysis of variance (group × word length × code) with repeated measures on two factors (word length and code) (See Bortz, 1979).

3.1. The Influence of Age

The age has a significant impact ($p < 0{\cdot}05$) on the overall working time and the average time used for one single task (see Figure 1), the influence on the accuracy is still inconclusive ($p < 0{\cdot}10$, Figure 1).

FIGURE 1. *Working time per task (s) (a); and frequency of errors (%) (b) for younger and older subjects.*

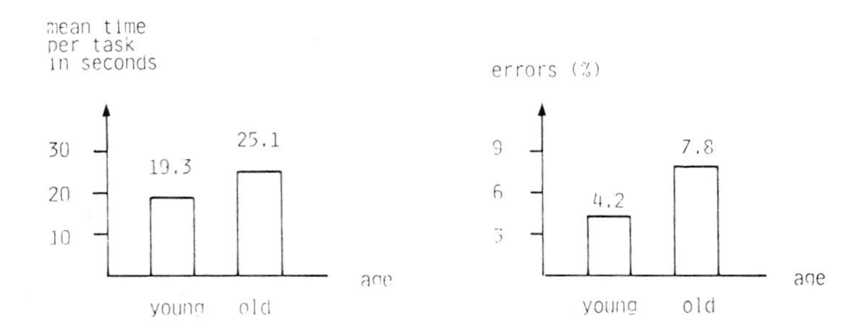

The accuracy of work within the codes is due to age, that means that there is an interaction ($p < 0{\cdot}05$) between the factors age and code. Figure 2 illustrates this connection. This interaction, however, comes about as a result of the code underlining. There is an

FIGURE 2. *Frequency of errors (%) of the two age groups with the different codes.*
young subjects o — o, old subjects o – – o
b = brightness-regulation, c = contrast-reversal, u = underlining

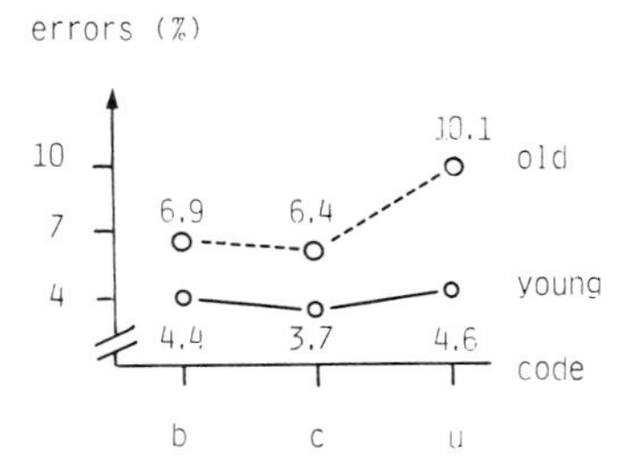

interaction code × age group ($p < 0{\cdot}05$) for 'working time' too, if the code underlining is eliminated from the calculations (see Figure 3).

FIGURE 3. *Working time per task (s) of the two age groups with the used codes.*
young subjects o — o, old subjects o - - o
b = brightness-regulation, c = contrast-reversal, u = underlining

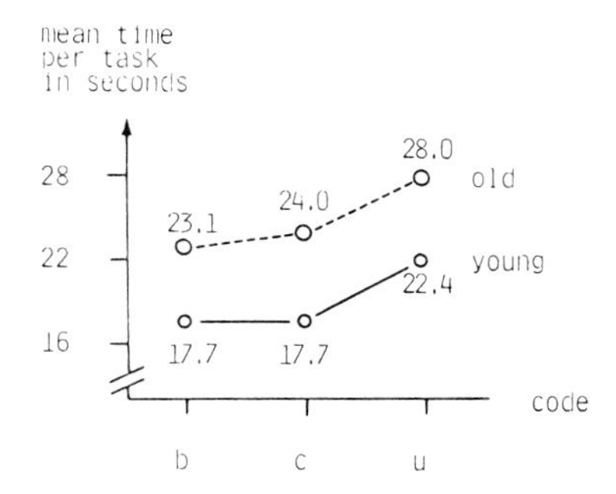

3.2. The Influence of the Different Codes and Word Lengths

Because this article focuses on the influences of age, we give here only a very brief description of the results, which are based upon the data of both groups together ($n = 19$).*

1. The word length influences the working time ($p < 0{\cdot}01$), but not the working accuracy.
2. A significant interaction ($p < 0{\cdot}05$; if underlining left out $p < 0{\cdot}01$) can be found between code and word length: Working time between short and long words differs most with brightness-regulation least with underlining.
3. The factor coding significantly ($p < 0{\cdot}01$) influences working times. This influence is due to the extremely long working time used for the areas marked with underlining. Contrast reversal and brightness-regulation are similar in the use of time.
4. The same effect was found for the accuracy of work also being influenced by the factor code. Again that result is caused by underlining.

3.3. Subjective Data

The subjects had to assess the codes used with different questions (see Section 4.1.). The answers to 'all questions' indicate that brightness-regulation is valuated rather better than contrast-reversal, while underlining is very clearly put onto last place by the subjects. The older subjects however definitely prefer brightness-regulation more than the younger ones.

* Final Report by H. Jiranek, W. Kugelmann and H. Krueger in the press.

4. *Discussion*

We do not aim at making statements on age-processes but we are interested in improving software. The involvement of the two age groups makes it possible to evaluate the codes against a theoretical background.

4.1. **Model for Code-Evaluation**

If we suppose that the form of information processing does not differ fundamentally between subjects that are similar in terms of experience, using-frequency, intelligence and so on, there is no reason to assume that older subjects use other strategies to solve the task of which the experiment consists.

On the other hand the 'cognitive effort' or the 'perceptual effort' need not be the same with different codes, because they may be different in terms of strain. But human beings are capable of compensating many of the shortcomings a certain code may have. These compensational processes probably run unconsciously. The less a certain code facilitates the information processing with reading and analysing the respective Gestalts, the more it will elicit compensational effort by the one who is working with it and the more it will be prone to disruption. We suppose that this disturbance-proneness increases when people get older.

Let us illustrate these statements by Figure 2. If we had taken only young people (lower line), we would have come to the conclusion that there is no important difference between the codes used, and they do not influence the working quality. By means of the second group, however, following the assumptions we made it is to be supposed that underlining is most prone to disturbances and the younger subjects are merely more capable of compensating. In mathematical terms: it is exactly the interactions which the interest has to focus on!

Figure 3 shows the average working times and can be interpreted in the same way. As we stated, there is a significant interaction between the codes brightness-regulation and contrast-reversal on the one hand, and age on the other. According to our theory, that means contrast-reversal is evoking more compensational processes, probably not during the search process but during the reading processes.

Of course we must not ignore the overall means. It can clearly be seen that underlining requires more time.

If the phenomenon we called 'cognitive effort' does exist it must cause cognitive stress. This stress should be reflected in the results of questionnaires. The questionnaire data we received correspond exactly to our interpretation of the working data. This is why we suggest treating questionnaires not as appendages to research but as instruments of great value.

In the same way which we made use of the dichotomy of the sample for interpretation, we based further findings on the dichotomy of the task itself by using long and short words. We found that this enabled us to make statements about the amount of *a priori* attention different codes seem to elicit.

The problem of software evaluation can be summarized in the following way: it is necessary to take more than merely the statistical means of performance measures, because it must be decided whether a certain result comes about by means of the quality of the software or by means of compensational processes which are combined with stress and strain. This requires a theoretical background of research. Obviously the psychometrical assessment gives relevant hints.

References

Balzert, H. (ed.), 1983, *Software-Ergonomie*. Reports of German Chapter of the ACM 14 (Stuttgart: Teubner).

Benz, C. and Haubner, P., 1983, Codierungswirksamkeit bei Informationsdarstellungen in Bildschirmmasken, in *Software-Ergonomie*, edited by H. Balzert, pp.124–134.

Bortz, J., 1979, *Lehrbuch der Statistik für Sozialwissenschaftler* (Berlin: Springer).

Eason, K.D., 1980, Dialog design implications of task allocation between man and computer. *Ergonomics*, **23**(9), 881–891.

Performance, Fatigue and Stress for the Older VDT User

J.M. Camisa
The New York Association for the Blind, The Rockefeller University, New York, New York, USA
and M.J. Schmidt
Digital Equipment Corporation, Andover, Massachusetts, USA

Abstract

The human visual system decomposes the retinal image and relays the spatio-temporal information via parallel channels. Neural pathways carry the information in the form of both excitatory and inhibitory mechanisms. Understanding the effect of VDT use on the visual system requires an understanding of how the sub-channels are affected by the spatio-temporal characteristics of the video display. In this paper, the effects of neighbouring contours, level of adaptation and age are discussed. Results from the visual evoked potential analysis of these factors indicate that:

1. Surrounding contours have an effect on the processing of spatio-temporal characteristics of the video image.
2. There are characteristic changes in inhibitory and excitatory influences during prolonged VDT use.
3. The cycle of depletion for these influences is different for older versus younger observers.

The conclusion is that creation of an optimal VDT will have to balance these influences in such a way as to reduce the fatigue in the neural system for observers of different ages and visual needs.

1. *Introduction*

Video display terminal (VDT) operation involves primarily use of the visual system and related eye/hand interactions. It is generally accepted however, that the human visual system was not designed primarily for the purpose of reading. In fact, the notion of 'a' system is rather misleading. Actually, it is composed of specialized sub-systems that have evolved which carry very specific kinds of visual information. Information about different aspects of the retinal image is analysed by parallel sub-systems projecting to the visual cortex (Regan, 1982).

Research has demonstrated that functional sub-channels exist for visual information dealing with contrast, size, movement, orientation and colour. If we wish to assess the effects of VDT use on visual function, we have to then assess its effects on each of these channels. In terms of the older VDT user, we also have to take into account the differential effects that the ageing process has on the performance of these channels (Sekuler *et al.*, 1982).

With age comes a host of changes in the optical and neural hardware of the human visual system (Carter, 1982). There are the effects due to the progressive yellowing of the crystalline lens causing changes in colour perception and a reduction in the light reaching the retina. There is also a decline in the ability to accommodate that allows for sharp focus of retinal images over a range of distances. There are, however, changes with age that indicate a more central or neural decay in function. The adjustment of the visual system to a change from light to dark environments known as dark adaptation declines significantly after age 60. Some of this reduction in dark adaptation is due to changes in the neural system (Pitts, 1982). It has been suggested that the increase in threshold is due to interruption of normal metabolic processes within nerve cells in the brain and retina (McFarland and Fisher, 1955).

The basic unit of the human nervous system is the nerve cell. Information carried by the cell is usually digital; that is it fires or it does not fire. The message carried from one cell to another can signal either excitation, that is causing the next cell to trigger or inhibition preventing the cell from triggering. In the human nervous system interacting neural networks of inhibition and excitation are largely responsible for most operations. In the visual system inhibition is used to help define details in the retinal image. The visibility of a spot of light on the retina is influenced by the proximity of other light or dark areas nearby. This lateral inhibition is a basic operation supplying information about the retinal image.

What has this to do with VDT use? Remember that the visual system is really a group of sub-systems each tuned to different aspects of the retinal image. All of these systems have inhibitory networks. A video image unlike the printed page is not a static image. There are elements of the image which change from micro-second to micro-second. Many of these changes are generally believed to be below threshold. But how do they affect the basic operation of the visual nervous system? What effect does age have on these inhibitory and excitatory mechanisms which are extremely sensitive to changes in the retinal image?

The data we present comes from two ongoing studies. The first is basic research on the process of spatial patterns as measured by the visual evoked potential technique, whereby the bioelectrical activity of the brain is recorded from scalp electrodes and is presented in this paper. This research is being conducted at the Rockefeller University in collaboration with Drs Vance Zemon and Floyd Ratliff of the Department of Biophysics. The second is a carefully controlled study on the effects of VDT use on vision and visual test performance. This is being conducted at Digital's Southwest Engineering in Albuquerque, New Mexico, in collaboration with Tom Stokebrand.

2. *Method*

Visual evoked potential (VEP) recording is a method of analysis whereby the event related electrical activity of the brain is recorded via scalp electrodes. The harmonic components of the averaged brain wave response to the visual stimulus gives us information about how the visual system processes the retinal image. VEP recording is a diagnostic tool in neurology and neuro-ophthalmology, providing a very sensitive analysis of neural transmission (Camisa *et al.*, 1981). Using Fourier analysis allows a decomposition of the brain waves into the contributions from channels conveying information about the spatial and temporal characteristics of the stimulus (Zemon and Ratliff, 1981).

The subjects were 8 young subjects (aged 15–18) and 4 older subjects (aged 50–53). The subjects viewed a Tektronix 608 monitor at a distance of 1 m. Patterns were generated by a computer controlled visual stimulator. All subjects were male and had a visual acuity of 20/20 with correction. Evoked potentials were recorded from the scalp over the primary visual cortex from three gold EEG electrodes.

The stimulus consisted of a central disk surrounded by three contiguous annuli, and radially divided into light and dark segments. By modulating the light and dark segments at a combination of temporal frequencies we were able to measure lateral interactions between neighbouring neurons and the visual system. We were specifically interested in:

1. How surrounding contours affected the processing of the image in the centre.
2. How these interactions were affected by visual adaptation and age.

3. *Results*

It had already been reported that the presence of a static surround either lighter or darker than the test field caused significant suppression of all components of the human VEP (Zemon and Ratliff, 1981). We found that a dynamic surround also reduced most components of the VEP. The results showed that with increasing viewing time the suppression changed in a cyclic fashion. There was an initial sharp decline in VEP amplitude with a gradual recovery over a one-hour period. There were some individual differences within each age group. However, there was a marked difference between the cycle of suppression for the old v. young group. The young group showed many more cycles during a given recording period than the older group. In addition, the older group took much longer to recover amplitude than the younger group.

4. *Discussion*

The results of these studies demonstrate that processing of visual information as displayed on a VDT screen is significantly influenced by surrounding areas that are not the same mean luminance as the test field. Also that this influence on visual processing changes with viewing time and has different characteristics depending on the age of the observer. These changes in the components of the VEP are considered to reflect fluctuating levels of neurotransmitters in the brain. The primary inhibitory neurotransmitter in the visual cortex is gamma-aminobutyric acid (GABA). These findings may indicate higher concentrations of GABA in the cortices of the older subjects.

These results indicate that there may be a different set of optimal display parameters for older v. younger VDT users. Individual differences even within the age groups might necessitate creating VDTs with adjustable screen characteristics to allow the user to fine tune the display for personal use.

References

Camisa, J.M., Bodis-Wollner, I. and Mylin, L.H., 1981, The effect of stimulus orientation on the visual evoked potential in multiple sclerosis. *Annals of Neurology*, **10**, 532–539.

Carter, J.H., 1982, The effects of aging on selected visual functions: color vision, glare sensitivity, field of vision, and accommodation, in *An Aging and Human Function*, edited by R. Sekuler, D. Kline and K. Dismukes (New York: Alan R. Liss Inc.) pp.121–130.

McFarland, R.A. and Fisher, M.B., 1955, Alternations in dark adaptation as a function of age.*Journal of Gerontology*, **10**, 424–428.

Pitts, D., 1982, The effects of aging on selected visual functions: dark adaptation, visual acuity, stereopsis, and brightness contrast, in *An Aging and Human Visual Function*, edited by R. Sekuler, D. Kline and K. Dismukes (New York: Alan R. Liss Inc.) pp.131–159.

Regan, D., 1982, Visual information channeling in normal and disordered vision. *Psychological Review*, **89**, 407–444.

Saito, M., Tanaka, T. and Oshima, M., 1981, Eyestrain in inspection and clerical workers. *Ergonomics*, **24**, 161–173.

Sekuler, R., Kline, D. and Dismukes, K. (eds), 1982, *An Aging and Human Function* (New York: Alan R. Liss Inc.).

Shahnavoz, H., 1982, Lighting conditions and workplace dimensions of VDT operators. *Ergonomics*, **25**, 1165–1173.

Zemon, V., Kaplan, E. and Ratliff, F., 1980, Bicuculline enhances a negative component and diminishes a positive component of the visual evoked cortical potential in the cat. *Proceedings of the National Academy of Sciences*, **77**, 7476–7478.

Zemon, V. and Ratliff, F., 1981, Intermodulation responses. *Journal of the Optical Society of America*, **71**.

Focusing Accuracy of VDT Operators as a Function of Age and Task

L. HEDMAN AND V. BRIEM
Högskolan Vaxjö, Sweden

Abstract

Twenty-nine operators at a Swedish Telecom Enquiry Centre participated in tests of eye-strain, including measurements of accommodation and convergence near-points and visual focus of accommodation using a laser optometer. The main study, which was divided into job-rotation and no-job-rotation phases, lasted for 6 months, during which the operators were tested at least twice a day in connection with work at a VDT or in a control condition. The results show that, given the specific job environment in the study, there was no evidence of negative visual effects connected with the use of VDTs. It is concluded that the most important variable in this type of study, apart from the operator's visual characteristics, is the age of the operator, and that particular attention must be paid to the differential effects of this factor.

1. *Introduction*

Several studies have presented evidence that VDT work may cause visual problems among office workers (Grandjean and Vigliani, 1980; Dainoff *et al.*, 1981). These problems are usually assumed to be temporary, and are often attributed to the high degree of visual processing required in this type of work. However, neither the exact nature nor the principal causes of visual discomfort and strain that VDT workers sometimes complain of have as yet been fully established.

Among the indicators of eye-strain are the accommodation near-point, convergence near-point and the refractive power of the eye. The last is an indicator of accommodative strain, and is frequently obtained with the 'laser-optometer technique' (c.f. Leibowitz and Hennessy, 1975).

According to some theorists, intensive visual tasks result in a temporary reduction in the ability of the eye to accommodate to different distances. However, even if this statement seems a reasonable one to make, the consequences of it have not been adequately tested experimentally. For instance, how does job rotation influence visual task performance, and what role does an operator's age play in this context?

The field study described below was carried out at the Swedish Telecom Enquiry Centre in Malmö (South of Sweden Telecom Area), and was intended to reveal whether computer supported catalogue work alone (VDT) was associated with more signs of eye-strain than mixed VDT and manual catalogue work (MAN). (For further information about the study, its design, methodology, details of results, and discussion, see Hedman and Briem, 1983.)

2. *Method*

2.1. **Subjects**

There were 29 telephone operators, 27 female and 2 male, aged between 17 and 54 years of age (median age = 27) and they were selected according to the criteria of age, sex, work experience, work schedule, and eye status to form a representative sample from the population of operators at the Centre.

2.2. **Design**

The main study lasted 6 months and was divided into two phases: job-rotation phase (R-phase: first 3 months) and no-job-rotation phase (NR-phase: second 3 months). During the R-phase, the subjects worked for the first 2 hours of a test day at a 'relevant' (test) task (VDT or MAN), the second 2 hours at a 'non-relevant' task (the other task, MAN or VDT), and the last 2 hours at the relevant task again. The rest of the test day was taken up with meals, rests, and three visual test-sessions: immediately before the beginning of work in the morning and after each of the two work sessions at the relevant task. In the NR-phase, the subjects spent all 6 working hours on the day of testing doing the relevant task, and the subjects were tested only twice, before and after the work session. Subjects who worked at the VDT were allowed 15 minutes rest per hour. All subjects took part in every experimental condition in a random order within a fully balanced factorial design.

2.3. **Visual Tests**

1. Accommodation near-point (ACCN).
2. Convergence near-point (CNVN) tests with a conventional RAF near-point rule.
3. Visual accommodation relative to targets at 0·25, 0·5, 1·0 and 6 m (corresponding to 4, 2, 1, and 0·17 diopters), as measured with a modified Field Laser Optometer (FLO) technique (cf. Hedman and Briem, 1983).

2.4. Analyses

The BMDP ANOVA program P2V (Dixon and Brown, 1979) was used for the analyses of the data reported here. Main independent variables were 'age' of the observer (3 groups: < 21, 21–40, > 40 years); relevant 'task' and 'test-session' on the day of testing; and 'distance' of the target in the FLO-tests (4 light-focus targets: standard test plaques subtending equal visual angles at 0·25, 0·5, 1·0, 6 m; 1 dark-focus target at 6 m). Independent analyses were carried out for the two phases of the study. Accommodation was given as deviation in diopters of visual fixation from expected value for each visual target. This transformation of the FLO measurements:

1. Makes the measurements for different visual targets directly comparable.
2. Highlights the importance of the deviations from 'expected' measures in the discussion of the results.

When needed, the data were normalized before statistical analyses.

3. *Results and Discussion*

No clear main factor differences were noted either between the eye-strain indicators obtained relative to the two tasks or between indicator values obtained before and after

FIGURE 1. *Average near-points of accommodation and convergence (ACCN and CNVN) for the 3 age groups in the NR-phase, measured at test-session 1 (morning) and 2 (evening).*

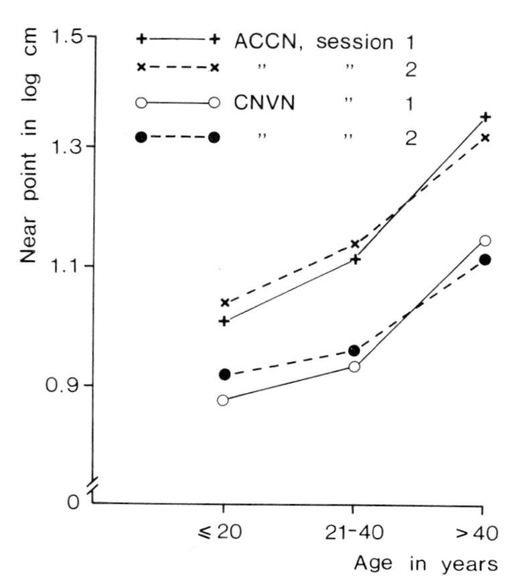

work. However, some interactions between factors were obtained, and the most important of these are discussed below.

In the NR-phase, an interaction exists between 'age' and 'session' for the near-point measures (see Figure 1: for ACCN, $p = 0{\cdot}015$ (top), and for CNVN, $p = 0{\cdot}059$ (bottom). Several authors have reported increased ACCN and CNVN values at the end of a work-day (e.g. Brown *et al.*, 1982), and this has been taken as an indication of eye-strain. In our study, age has a differentiating influence on the near-point measures. Thus, the effect shown in the youngest age group is in accordance with that shown by previous investigators, while the opposite is true of the oldest age group. This result may be explained as occurring because of the effects of two different factors, i.e. eye-strain and practice, which have opposite effects on the near-point measures (see Hedman and Briem, 1983, for an extended discussion).

FIGURE 2. *Light-focus (LF) and dark-focus (DF) at 0·17 diopters for the 3 age groups measured at test-sessions 1 (morning), 2 (midday), and 3 (evening) in the R-phase. (a) VDU, (b) MAN.*

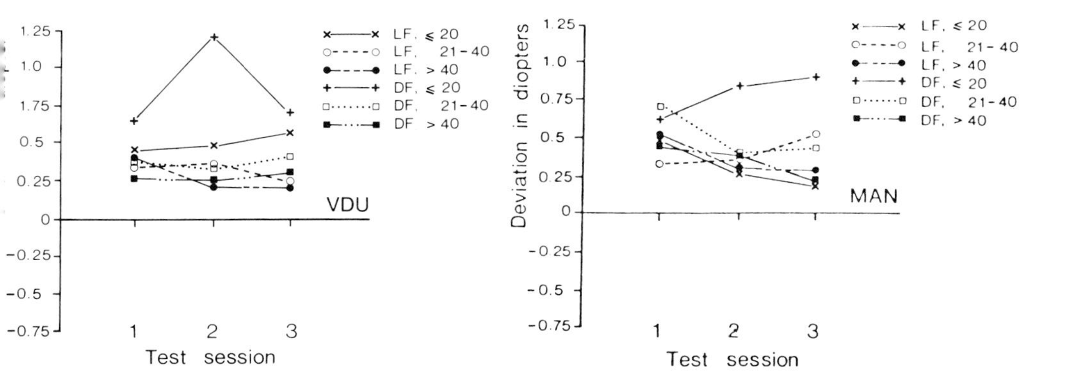

The ACC measurements did show up differences between VDT and MAN in one single case: dark-focus ACC of the youngest subjects in the VDT condition for visual targets at 0·17 diopters in the R-phase was greater at the midday than the evening test (Figure 2a). The overall trend is for the ACC measures for targets at 6 m to decrease from a positive deviation in the first test of the day towards a null deviation in later tests, except for the dark-focus of the youngest group which shows the opposite trend (Figure 2a and 2b). These results were not replicated in detail in the NR-phase.

According to Östberg (1980), 'temporary work-induced myopia' occurs after a period of relatively intensive work at a VDT. Results supporting this hypothesis have been presented by Östberg *et al.* (1980) who used subjects in the age range 25–40, and by Haider *et al.* (1980) who found decrements in their subjects' visual function after work at VDTs. In our study, the middle-age group showed no reliable signs of temporary myopia which only appeared in the youngest group. In general, our results do not support those of the authors cited above. However, there are several important differences between our study and other studies of these accommodation effects, such as that:

1. We used subjects in a wide age-range, in three age-groups.
2. Our study spanned a relatively long period and included repeated tests under different experimental conditions.

What our results allow us to conclude is that age is a critical variable which must be taken into careful consideration in the design of job environments that contain a large proportion of visually intensive work. Furthermore, as long as the work-load for the operator does not exceed some, still unspecified, critical point, nor extend to overly long working hours with few rest periods, then it seems that neither the inclusion of VDTs in the job environment nor the number of hours worked at a VDT has any effects of practical significance on the operator's visual function.

References

Brown, B.S., Dismukes, K. and Rinalducci, E.J., 1982, Video display terminals and vision of workers: summary and overview of a symposium. *Behavioural and Information Technology*, **1**, 121–140.

Dainoff, M.J., Happ, A. and Crane, P., 1981, Visual fatigue and occupational stress in VDT operators. *Human Factors*, **23**, 421–438.

Dixon, W.J. and Brown, M.B. (eds), 1979, *BMDP–79: Biomedical Computer Programs, P-Series* (Berkeley: University of California Press).

Grandjean, E. and Vigliani, E., (eds), 1980, *Ergonomic Aspects of Visual Display Terminals* (London: Taylor & Francis).

Haider, M., Kundi, M. and Weissenböck, M., 1980, Worker strain related to VDUs with differently coloured characters, in *Ergonomic Aspects of Visual Display Terminals*, edited by E. Grandjean and E. Vigliani (London: Taylor & Francis), pp.53–64.

Hedman, L.R. and Briem, V., 1983, Changes in focusing accuracy of VDU operators as a function of age, hours worked, and task. Paper submitted for publication.

Leibowitz, H.W. and Hennessy, R.T., 1975, The laser optometer and some implications for behavioural research. *American Psychologist*, **30**, 349–352.

Östberg, O., 1980, Accommodation and visual fatigue in display work, in *Ergonomic Aspects of Visual Display Terminals*, edited by E. Grandjean and E. Vigliani (London: Taylor & Francis), pp.41–52.

Östberg, O., Powell, J. and Blomkvist, A.C., 1980, Laser optometry in assessment of visual fatigue. University of Lulea, Sweden, Technical Report, No. 1:T.

Measuring Perceived Flicker on Visual Displays

B.E. Rogowitz
IBM T.J. Watson Research Center, Yorktown Heights, NY 10598, USA

Abstract

We have developed a procedure for measuring the perceived flicker on a display terminal. The measurements are taken at three levels of display contrast, and for each level, the percentage of the general population which will perceive flicker can be estimated. This technique can be used for displays of various sizes, refresh rates, luminance levels, polarities and phosphors.

1. *Introduction*

The refresh CRT (cathode ray tube) is currently the most commonly used technology for displaying information on a visual display terminal (VDT). The inner surface of an evacuated glass bottle is coated with phosphor, and symbols and characters are 'written' on the screen by a beam of electrons which scans the phosphor surface. In this technology, therefore, the characters on the screen are not constantly illuminated. The light output depends on the rate at which the information is scanned (the 'refresh rate') and on the time course of the phosphor's glow (the 'persistence'). The amount of flicker perceived depends critically on the temporal characteristics of this light; the faster the refresh rate and the longer the persistence, the less likely we are to see flicker. But, this physical measurement is not enough to predict whether or not flicker will be seen.

As shown in Figure 1, flicker sensitivity depends on more than the temporal waveform of the light. In this experiment, subjects were asked to look at a light with a very simple temporal waveshape, a 50% duty cycle square wave in time. A psychophysical procedure was used to measure that rate where the intermittent light appeared to be 'flicker-free'. The temporal frequency where a light is flicker-free is called the critical flicker frequency (CFF). Perceived flicker depends on luminance level. For all the conditions measured, the greater the luminance, the faster the light has to be modulated in order for it to appear steady. While 30 Hz eliminates the perception of flicker at low luminance levels, 50 or 60 Hz may be required at higher luminance levels. Furthermore, at all luminance levels, for lights

FIGURE 1. *Critical flicker frequency related to target characteristics.*

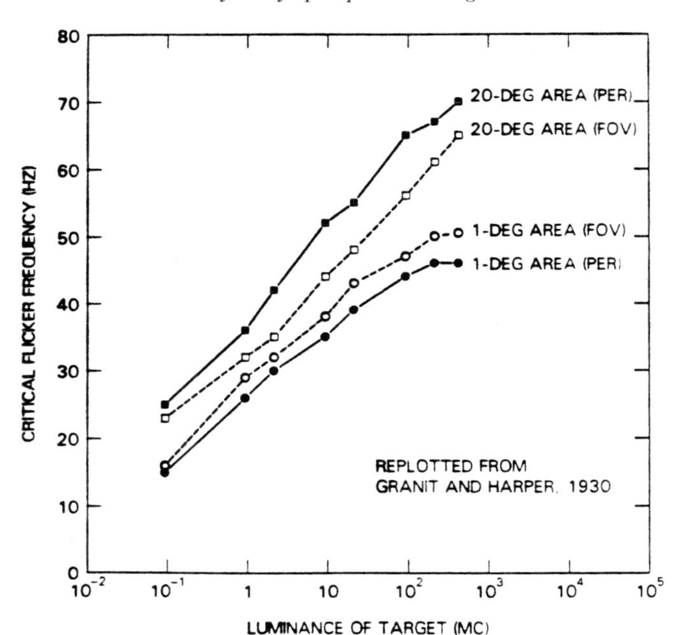

which are physically identical in their temporal characteristics, the greater the angular subtense of the light, the more it appears to flicker.

The perception of flicker also varies from individual to individual. Figure 2 shows the critical flicker frequency measured for 100 subjects looking at a 2-degree square-wave light. These data show that flicker sensitivity is quite variable and that the older subjects tend to be less sensitive to flicker than the younger subjects. This last point is of particular concern to us, since VDT users tend to be younger than the average office worker (Canter and Davies, 1982; US Bureau of the Census, 1981).

In our flicker procedure, we first measured the distribution of flicker sensitivity for 100 subjects, using a 2-degree square-wave light. The next step was to measure the amount of flicker perceived on various displays at three levels of contrast. We used a matching procedure wherein the temporal frequency of the standard 2-degree light was varied until the amount of flicker perceived on the display and the amount of flicker perceived on the standard light was equal. In this procedure, we measure the amount of flicker perceived on the display in terms of the amount of flicker perceived on the standard light. Since each subject's flicker sensitivity is known with respect to the general population, we can estimate the proportion of the population likely to perceive flicker in that display. Since all displays are calibrated under a constant illumination, and each subject's sensitivity is referenced to the general population, data for different displays tested in different laboratories with different subjects can be compared.

FIGURE 2. *Critical flicker frequency related to age.*

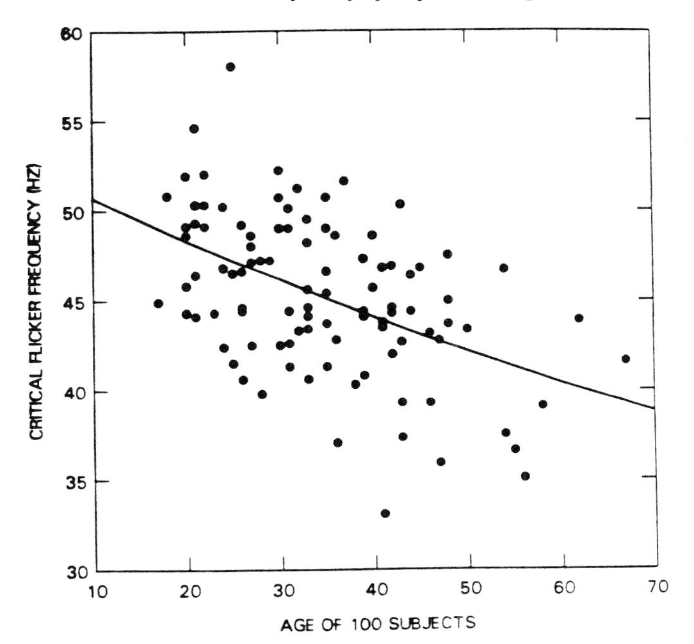

2. *Calibrating the Room, the Displays, and the Standard Light*

All experiments were conducted in a room of known ambient illumination, approximating a 600 Lux office. The light from the well-adjusted overhead fluorescent light was filtered through white paper until the illumination falling on the display screen and on the standard light was 284·5 Lux.

All displays were tested under this standard illumination. The luminance reflected from each screen, however, varied from display to display depending upon the transmissivity of the glass and the anti-reflective coating of the screen. We calculated, for each screen, the amount of character luminance necessary to produce characters of 55%, 70%, and 85% contrast. By convention, contrast is defined as the difference between the background luminance and the average pel luminance divided by the sum of these two values. In running the experiment, the display screen was filled with 5-letter word-like clusters, filling 70% of the character boxes.

The standard flicker light was simply a square-wave current drive to a glow modulator tube, viewed through two separate pieces of diffusing glass. Under the standard room illumination, the trough luminance of the flicker light measured 8·5 cd/m^2; the peak luminance measured 178·5 cd/m^2. This corresponds to a luminance modulation (over time) of 89·9%.

FIGURE 3. *Cumulative distribution of critical flicker frequency. Data replotted from Figure 2.*

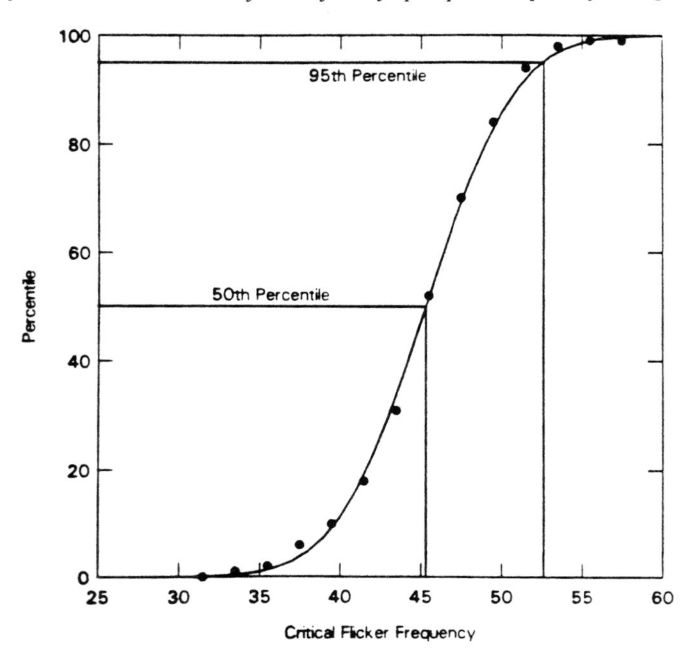

3. *Population Data for Flicker Sensitivity*

Figure 3 shows the cumulative distribution of flicker sensitivity in the population of 100 subjects tested. These data are fit by a cumulative gaussian with a mean of 45·3 Hz and a standard deviation of 4·38. This means that for the standard light we used, the mean critical flicker frequency was roughly 45 Hz, and subjects with a CFF of 53 Hz were in the 95th percentile. This distribution allows us to 'calibrate' our observers. If someone in the 95th percentile sees no flicker, it is likely that at most 5% of the population will see flicker on that display. If someone in the 50th percentile sees no flicker, up to half the population might still see that display flicker.

4. *Measuring Display Flicker*

Figure 4 shows the physical arrangement of the standard light and the screen used in the flicker matching procedure. The subject sits, with his chin in a chin rest, 20 in (50·8 cm) from the standard light and 20 in from the display screen. First, the observer's critical flicker frequency is measured as before, using a method of limits. Next, the amount of

FIGURE 4. *Arrangement used for the flicker matching procedure.*

flicker perceived on each display is measured in terms of the amount of flicker perceived on the standard light. For each contrast level of each display tested, the experimenter adjusts the temporal frequency of the standard light until the subject judges the display and the light to flicker equally. If the display does not flicker, the observer matches it to a standard light which also does not flicker. That is, he matches the display to a standard light modulated at his CFF. If the display does appear to flicker, the subject matches it to a standard light which also flickers. That is, the temporal frequency of the standard light is adjusted to a value beneath the observer's CFF. The greater the perceived flicker on the display, the lower the temporal frequency of the matching light.

The match is accomplished through a bracketing procedure. First, the temporal frequency of the standard light is turned to a rate above the observer's CFF and the subject is asked to judge which flickers more, the light or the screen. The subject is asked to look directly at the centre of the flicker light and directly at the centre of the VDT, and is free to look back and forth between the display and the light as often as he chooses before making his decision. Next the light is turned to a frequency well below the CFF, and the subject again judges which flickers more, the light or the screen. This procedure is then repeated with each successive high temporal frequency a little less high and each successive low temporal frequency a little less low. The result of this process is the convergence upon that temporal frequency of standard light for which the subject can no longer decide which flickers more. It takes roughly 15 repetitions (30 judgements) to reach a convergence point. The mean of three such convergence points is called the 'matching frequency'.

Two subjects' matching data for one display are shown in Figure 5, where the matching frequency is plotted as a function of display contrast. The horizontal dotted line indicates the subject's critical flicker frequency. The lower the temporal frequency of the matching light relative to the subject's CFF, the more that screen appears to flicker. The subject whose data are represented in the left-hand panel has a CFF of 52·25 Hz. For this subject, the display does not appear to flicker at the low contrast level (55%); the standard light is

FIGURE 5. *Flicker matching data of two subjects on a display which is flicker-free.*

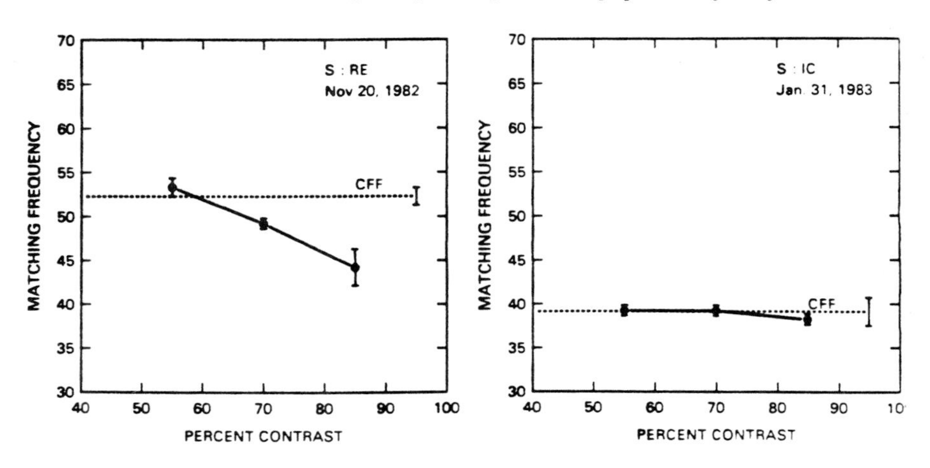

adjusted to a frequency equal to his CFF. At a medium level, 70% contrast, however, the subject matches the flicker on the screen to a lamp frequency beneath his CFF. At this contrast, the display appears to flicker. The perceived flicker is greater still in the high-contrast condition. A

FIGURE 6. *Flicker matching data for five subjects for low, medium and high contrast.*

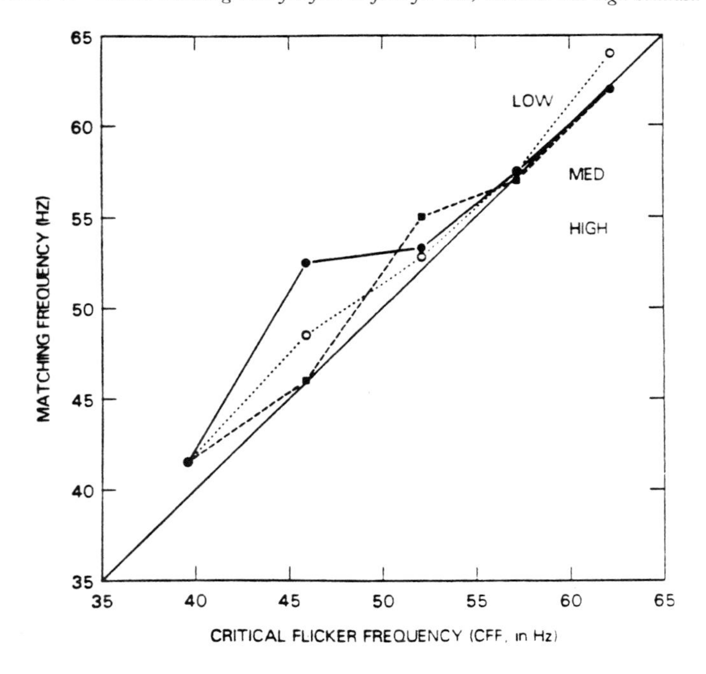

threshold of 52·25 puts this subject in the top 10% of the population tested with the standard lamp. Since this subject does not see flicker at the lowest contrast level, it is likely that this display, at this contrast, will be flicker-free to 90% of the general population. At the higher contrast levels, however, flicker is clearly detected. It is likely that *at least* 10% of the population will detect flicker at these contrast levels.

For the low-sensitivity subject (right-hand panel) this same display does not appear to flicker. Independent of display contrast, the subject never matches the display flicker to a rate below his CFF. The fact that he does not see flicker, however, does not provide us with very much information. Since, with a CFF of 39, this subject is in the bottom 10% in flicker sensitivity, all we know is that at least 10% of the population will perceive that display as flicker free.

Figure 6 shows how the data from many subjects can be combined. Here, the matching frequency, the temporal frequency of the standard light for which the light and the screen flicker equally, is plotted as a function of the observers' critical flicker frequency. There are three values for each observer, showing the matching frequency for low, medium, and high contrast displays. Data from five observers are replotted here with each observer's data represented as three points aligned vertically over his CFF. The extent to which the display is 'flicker-free' is reflected in the extent to which all points lie on or above the vertical diagonal where the matching frequency and the CFF are equal. This display, for these observers, is essentially flicker free.

FIGURE 7. *Flicker matching data for the same subjects on another display.*

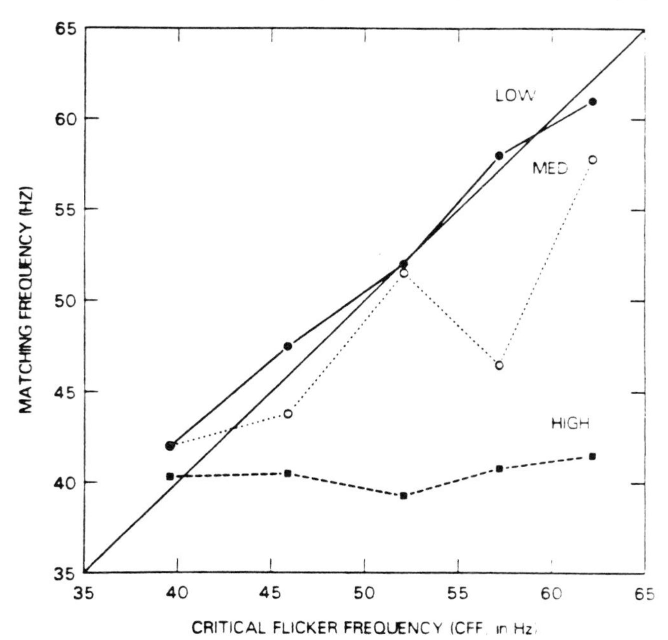

The next figure, Figure 7, characterizes a display which is flicker-free to these same observers only at the low contrast level. At 70% contrast, the three least sensitive observers match the display to a light modulated at their threshold values, whereas the two high sensitivity observers match the display to a standard light modulated at a rate beneath their CFF. In the high contrast condition, this display flickers so badly that even the low-sensitivity subjects see flicker. Only a small proportion of the general population will be spared the sensation of flicker on this display when the contrast is turned up to 85%.

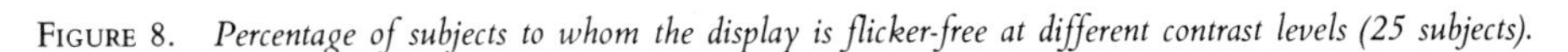

FIGURE 8. *Percentage of subjects to whom the display is flicker-free at different contrast levels (25 subjects).*

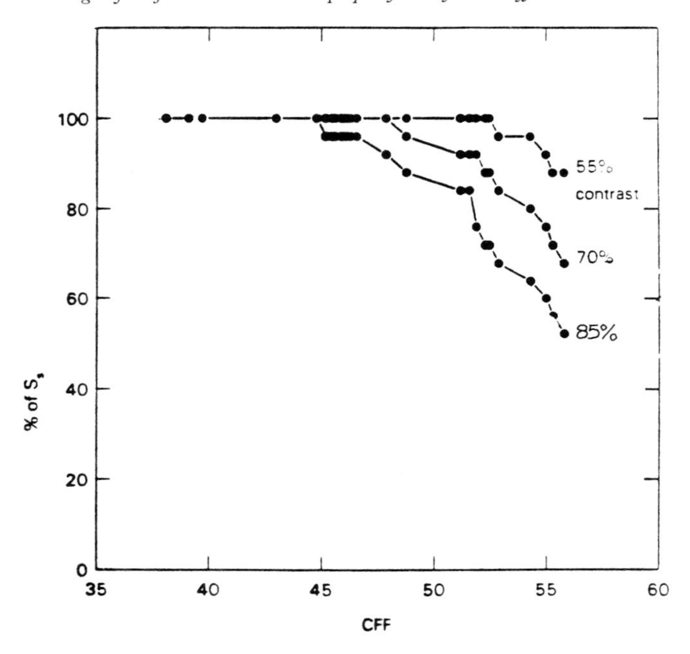

Figure 8 replots the flicker matching data in a different form. Here, instead of depicting the *degree* to which the display flickers to each of the subjects, it depicts the *number* (proportion) of subjects for whom the display is flicker-free. By convention, a display 'flickers' to a given subject when the matching frequency is at least 1 s.d. lower than the subject's CFF. The display is 'flicker-free' whenever the matching frequency is not more than 1 s.d. below the subject's CFF. To construct this graph involves:

1. Counting the number of subjects for whom the display is flicker-free at each CFF represented.
2. Accumulating the proportion of subjects for whom the display is flicker-free across CFF.

So, the graph plots the proportion of subjects for whom the display is flicker-free at each successively higher CFF. If a display were perfectly flicker-free at all levels of contrast for all subjects, the data points would all fall on a horizontal line at 100%. If at each successively high value of CFF the observer sees flicker, the height of the data point drops a notch from the previous position. If no additional observers are added to the number perceiving flicker, the height remains constant, and the segment connecting this point to the last is horizontal.

5. *A Flicker Standard*

Although there is no evidence that flicker has ill effects on the health or vision of VDT users, perceived flicker is annoying. Various standards committees have decided, thus, that display terminals should not flicker. One possible technique for deciding whether a display is 'flicker free' would be to use the flicker-matching technique under controlled testing conditions with a few high-sensitivity observers. The more stringent the criterion for accepting a display, the higher the CFFs of the observers. For example, if displays should be flicker-free to 95% of the population, then, when tested using the standard light, all subjects should have CFFs equal to or exceeding 53 Hz. These subjects would then perform the flicker-matching test. If these observers see no flicker at the criterion contrast level, then the display would be declared 'flicker-free'.

The main question remaining is where to set the various criteria. In our opinion the criteria should be quite high. We suggest that visual display terminals be flicker-free to 95% of the population up to 85% contrast. That is, users should be able to have a 12:1 contrast ratio without perceiving flicker.

6. *Conclusions*

The flicker-matching procedure provides a foundation upon which we can evaluate the perceived flicker on visual display terminals. Unlike methods previously devised, it references each observer's judgements of flicker to the flicker sensitivity of the general population. This, in turn, allows us to predict the response of the general population to each particular display from the data of relatively few subjects.

References

Canter, D. and Davies, I. (1982) Ergonomic parameters of word processors in use. *Displays*, **3**(2), 81–88.

Granit, R., and Harper, P. (1930) Comparative studies on the peripheral and central retina II: Synaptic reactions in the eye. *American Journal of Physiology*, **95**, 211–227.

US Bureau of the Census, *Statistical Abstract of the United States: 1981* 102nd ed, Washington DC.

Lighting, Glare Measurement and Legibility of VDTs (Introductory paper)

H.L. SNYDER
Department of Industrial Engineering and Operations Research, Virginia Polytechnic Institute and State University, Blacksburg, Virginia 24061, USA

1. *Introduction*

As the 'paper office' has been transformed into the 'electronic office', several functional and aesthetic incompatibilities have become apparent. The paper office has been typified by high contrast, positive (black-on-white) images viewed under relatively high illumination (800 to 1500 lux) and in rooms with windows which let in sunlight and provide views to look at. With the introduction of the VDT, efforts have been made, often to a Herculean extent, to retain these aesthetically desirable and interesting visual characteristics while at the same time achieving comfortable and efficient use of the electronic displays. In many situations, visual discomfort has resulted from the indirect glare of the natural or artificial light sources from the VDT screen. Efforts to reduce the glare by the application of anti-reflective (AR) surfaces ('filters') have met with mixed success; unfortunately, many of these AR coatings or devices are little more than poor optical quality coloured sheets of plastic. Recent measurements, examples of which are indicated below, indicate that a large majority of these AR filters significantly reduce image quality rather than improve it.

2. *Display Requirements*

The visual requirements of the display must be met *in situ*; that is, while the various display parameters of a 'good' display can be stipulated in the abstract, the usefulness and suitability of the display depend greatly on its interaction with the surrounding illumination, the layout of the workplace, the eyesight of the worker and other variables. Poor environmental control can severely compromise the displayed image in spite of every good intention by the manufacturer. An appropriate place for evaluation is the displayed image in the working environment, not in the manufacturer's laboratory.

While there are functional interactions among the display variables that contribute jointly to legibility, readability and comfort, it is helpful to take a simplistic view of display requirements by stipulating the following conditions *in situ*: contrast at least 9:1 (modulation > 75%); character size at least 16 arcminutes; raster modulation < 15%; and no visible dot or line structure to the character matrix. Research (Howell and Kraft, 1959; Snyder and Maddox, 1978) has demonstrated that character legibility is significantly reduced as the contrast (modulation) of the character against its background decreases below 90% for single characters in a non-contextual display and below 75% in a contextual (e.g. text) display. Similarly, these researchers have demonstrated that the optimum angular subtense for legibility for a non-contextual display is about 1 arcdegree for a non-contextual display and between 16 and 25 arcminutes for a contextual display. The 16–25 arcminute range agrees with some current national standards. Selection within this range depends largely upon edge sharpness and contrast.

It is important that the character structure and the raster line structure of the display should not be clearly visible to the user. When the raster structure is visible, performance degradations occur. For example, Beamon and Snyder (1975) demonstrated that reductions in raster modulation to 10% by introducing spot wobble to the scanning CRT spot improved performance significantly. While observers tend to prefer a 'sharp' display in which the raster lines are distinctly visible (raster modulation in excess of 20%), performance degradation results from this fixed position visual 'noise'.

Even more important than the raster modulation is the visibility of the matrix structure of the individual characters, caused by dot spacing being greater than dot diameter. From a design standpoint, this is the result of a mismatch between line rate and dot size or resolution. The effect of this mismatch is an increase in reading time for text passages, as illustrated in Figure 1. As the space between character dots increases relative to the dot size, the

FIGURE 1. *Effect of dot spacing on character legibility.*

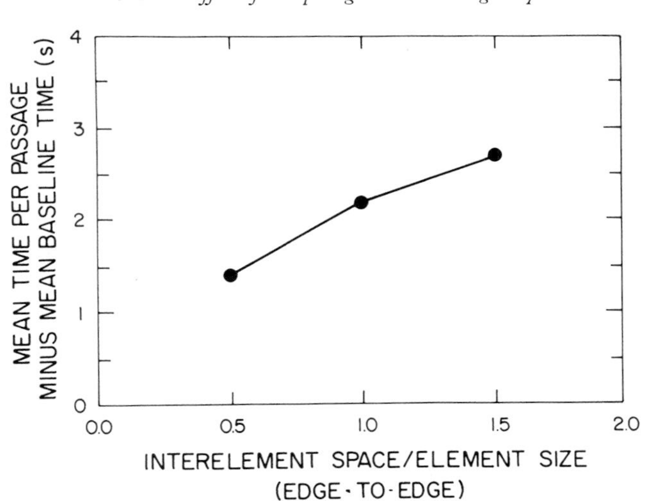

reading time also increases above the baseline reading time for stroke characters (Snyder and Maddox, 1978). Extrapolation of this function to a zero spacing suggests no reading time increase for stroke characters; that is, the closer a dot matrix character approaches the structure of a stroke character, the more readable is the text.

The character size, raster modulation, and dot modulation requirements are relatively easy to measure and verify in a manufacturer's laboratory. Also easy to verify is the relationship between legibility and character design or font. Unfortunately, manufacturers rarely perform this type of research and few make use of the existing data on the subject.

It has been demonstrated that character font and matrix size can have a significant effect upon character legibility and readability. For example, a 5 × 7 dot matrix font is less legible than is a 7 × 9 matrix font, which is in turn less legible than a 9 × 11 font (Snyder and Maddox, 1978). In addition, there is a substantial effect on legibility and readability of the interaction between font and matrix size. As illustrated in Figure 2, the Huddleston font is more legible than three other fonts in a 5 × 7 matrix size, but the Huddleston and Lincoln/Mitre fonts are equally legible in 7 × 9 or 9 × 11 matrix sizes.

FIGURE 2. *Effect of matrix size and font on character legibility.*

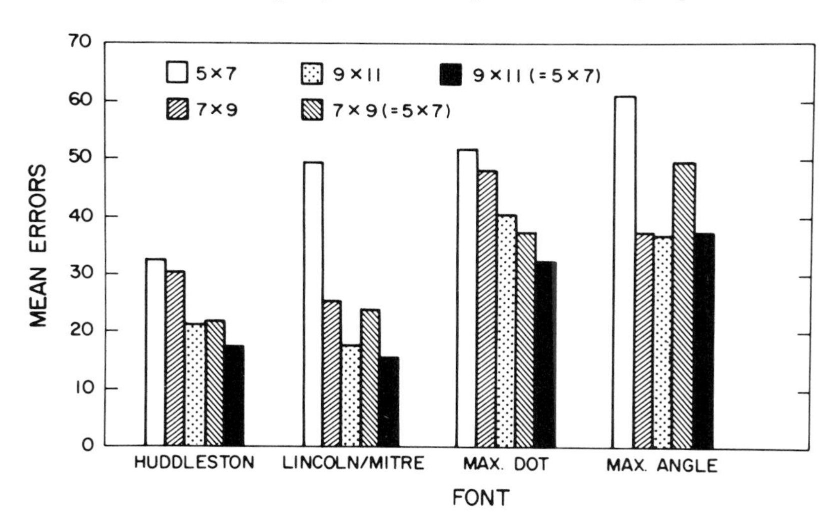

Very recent research in our laboratory has further proved the superiority of the Huddleston font, particularly for flat panel displays which are subject to failures of single picture elements (pixels) or entire lines of addressable pixels. In this research (Abramson, Mason and Snyder, 1983), it was demonstrated that the Huddleston font produced shorter text reading times for all levels of per cent pixel failures (Figure 3) and for all types of display failure (Figure 4). Thus, the selection of a display font should be made carefully and should be based upon research results, not only upon aesthetic desires of the manufacturer. Too often this is not the case as there is no current standardization of fonts, either informally or nationally.

Figure 3. *Effect of font and per cent cell failures on text readability.*

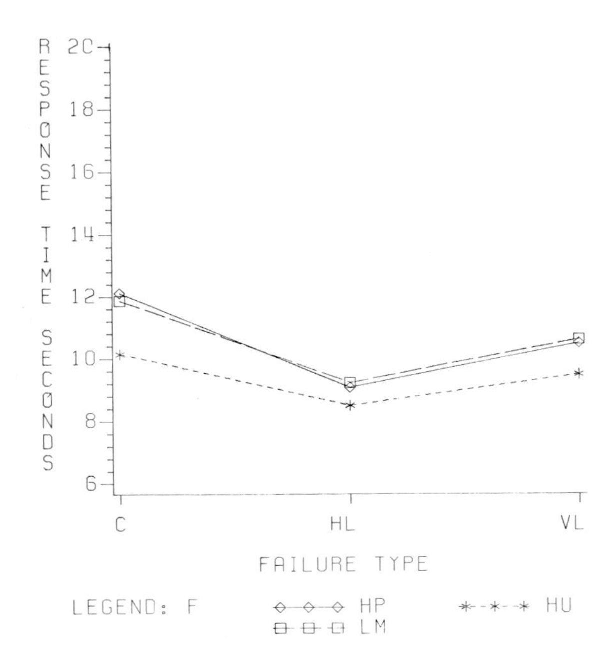

Figure 4. *Effect of font and failure type on text readability.*

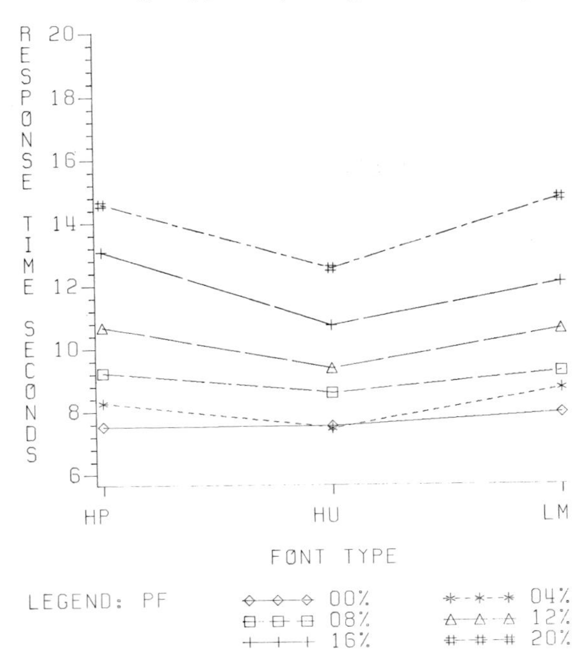

3. *Glare and its Measurement*

The existence of ambient illumination introduces the possibility of reflected glare from the display and the resultant reduction in displayed contrast. Glare sources contribute to reduced display effectiveness by decreasing the contrast of the displayed image. In addition, focused glare sources cause visual difficulties if the glare source contains texture of patterning that competes with the displayed information. Although numerous AR coatings and attachable AR filters are available, the effectiveness of most of these has yet to be assessed quantitatively. A recommended approach to this measurement problem follows.

One means of quantifying the quality of the displayed image is through the modulation transfer function, or MTF (Snyder, 1980). This function is simply the contrast (modulation) expressed as a function of the size of the bars on a sine-wave grating, with increasing spatial frequency (e.g. cycles per unit visual angle) denoting decreasing bar width. More modulation per unit spatial frequency indicates greater contrast and perceived sharpness to the displayed image. Studies have shown that increases in the area under the MTF correlate highly with perceived image quality and with performance of the person using the display (Snyder, 1980). Thus, the measurement of the MTF of a display in the presence and absence of a glare source can provide a quantitative index of the reduction of image quality caused by the glare source. In addition, the modulation at the lowest spatial frequency is a direct measure of the displayed contrast of the characters.

Figure 5 illustrates the line spread function (LSF) measured from a typical VDT in the absence of a glare source. The LSF is Fourier transformed to yield the MTF illustrated in the lower part of Figure 5, which is followed by some image statistics that define the conditions of the measurement. Of particular interest is the integrated area under the MTF, which serves as an index of image quality.

By comparison, Figure 6 illustrates the same type of measurements for this VDT when a veiling light source delivers surrounding illumination on the display surface sufficient to produce a reflected luminance of 75 cd/m^2. The displayed image MTF is accordingly reduced and the image appears lower in contrast. Figure 7 is of the same form and indicates the resulting measurements obtained when a typical amber filter is placed over the display. It should be noted that not only does the amber filter not improve the contrast or image quality above that obtained without the filter, but also it does not restore it to the no-ambient-illumination image quality level. Thus, the amber filter has the effect of producing no improvement in contrast or glare suppression; in fact, it reduces image quality further than would be the case if no filter were applied under the existing ambient illumination conditions! Clearly, the image quality is higher without the amber filter. Similar data and results have been obtained for other types of filters.

A patterned glare source also produces an image seen by the observer at an optical distance different from that of the display. Some researchers have suggested that this additional image from the focused glare source causes changes in accommodation of the eye, leading to experiences of visual fatigue. While such relationships have yet to be supported experimentally, there is no question that focused glare sources are at best annoying and distracting, and at worst may cause visual fatigue. Therefore, their measurement and quantification are needed to evaluate the effects of various AR coatings and filters on glare source suppression.

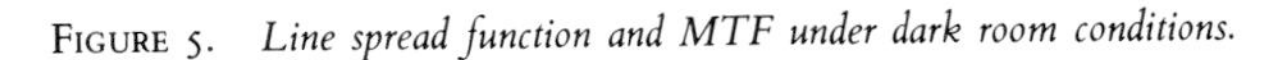

FIGURE 5. *Line spread function and MTF under dark room conditions.*

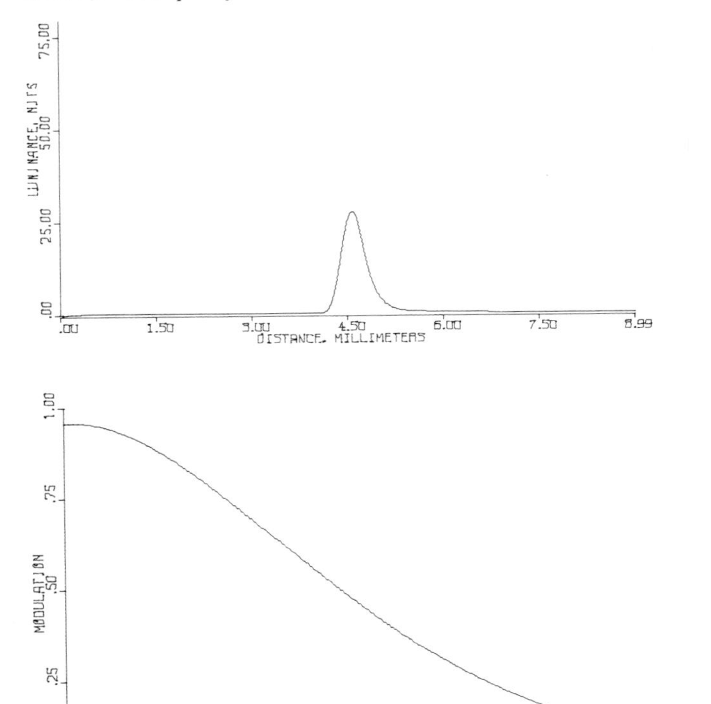

DATA FILE	: H22439	PHYSICAL APERTURE	.025 MM	MTF	=	1.0549
MONITOR #	: HP2622	MAGNIFICATION	: 1.100	EP	=	.7410
FILTER	: NONE	TOTAL # POINTS	: 399.	SSF	=	1.5186
GLARE	: DARK	POINT SPACING	: .0248 MM	M(0)	=	.9576
DIRECTION	: H	SAMPLES / POINT	: 150.			
		SAMPLING RATE	: 1000.00 HZ			

The evaluation of changes in the 'image quality' of a glare source can be accomplished in a manner similar to the evaluation of the display image quality in the presence of the glare source. Specifically, the MTF of the glare source image is measured by aiming the microphotometer at the display and focusing on the image of the glare source itself. An LSF of the glare source is obtained, which is then transformed into a glare source MTF. This MTF can be compared with and without ambient illumination for various AR coatings and filters.

Measurements made to date on various AR coatings and glare sources indicate that few of the popular attachable devices improve image quality in the presence of ambient glare. In

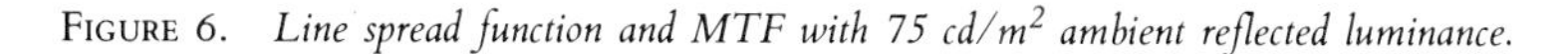

FIGURE 6. *Line spread function and MTF with 75 cd/m² ambient reflected luminance.*

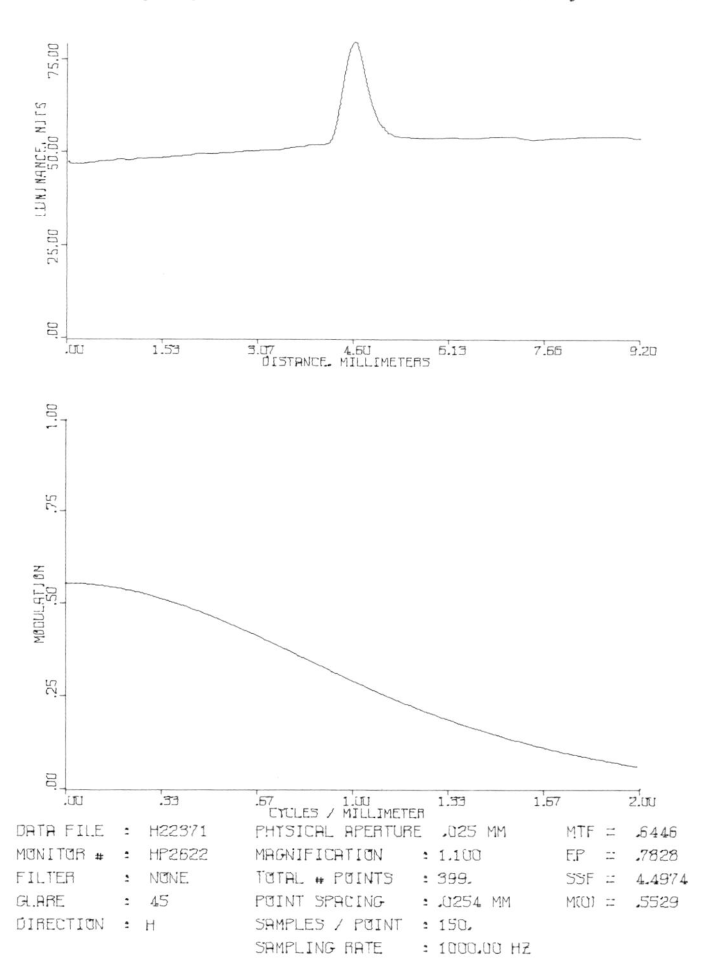

many cases, the addition of these devices serves to reduce further image quality rather than to improve it. Measurements of the focusing of the glare source by the attached filter or coating indicate similar results. While several filters or AR attachments significantly diffuse the glare source images, as indicated by a reduction in the MTF, these devices also tend to diffuse the displayed image; i.e. they result in reduced sharpness of the displayed image. These results, made for a number of AR coating and filters under a variety of ambient conditions, suggest that some manufacturers of such devices have either failed to understand the problem from the user's point of view or have failed to develop a technology adequate for many of the environmental conditions in which VDTs are currently used.

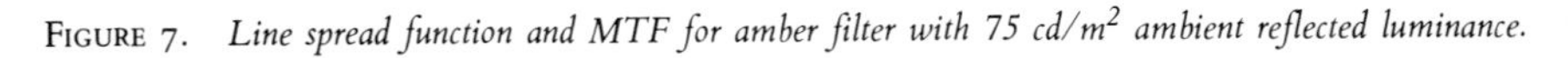
FIGURE 7. *Line spread function and MTF for amber filter with 75 cd/m² ambient reflected luminance.*

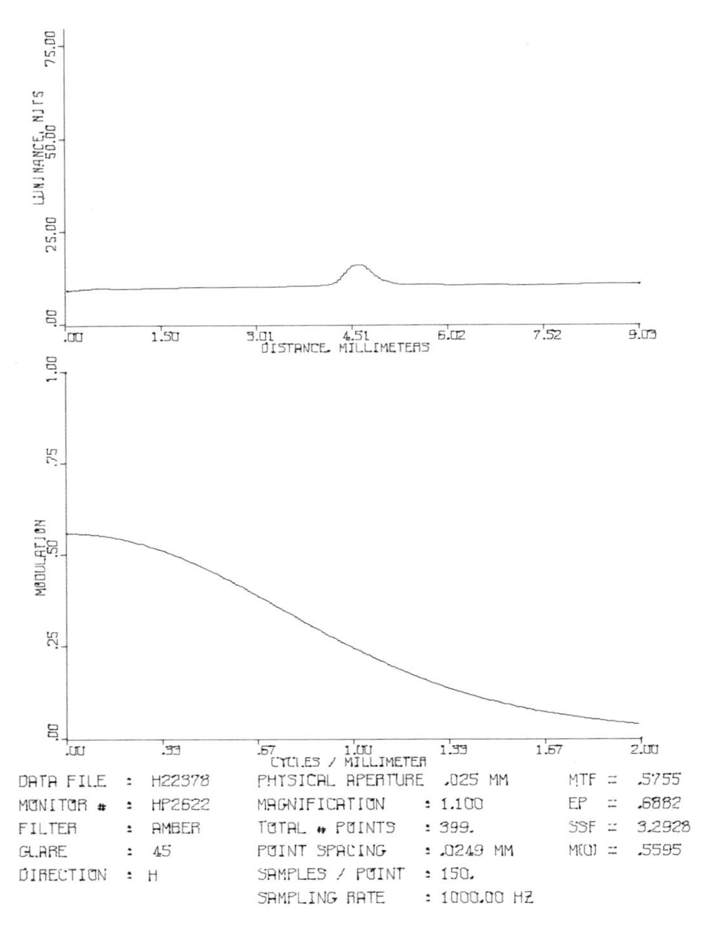

4. *Implications for Recommended Illumination Levels*

Field experience, discussions with numerous professional colleagues, laboratory experimental results, and the above described glare measurements all lead to a similar conclusion. Specifically, any appreciable amount of ambient illumination, from either artificial or natural sources, can significantly reduce the contrast of displayed information and result in a focused glare source pattern. While it is desirable to retain natural lighting sources and artificial lighting levels in the order of 700 lux or more for non-VDT activities in the typical office environment, current display technology is often incompatible with this need.

Several solutions are available. First, ambient office illumination can be reduced to a level totally compatible with the unfiltered VDT, that is, with the VDT not dependent upon any glare suppression from an AR coating or filter. Such levels are in the order of 200 lux and generally cause the office to appear dimly illuminated. These levels are inadequate for viewing hard-copy (paper) documents except by those with good vision using high-contrast documents. For persons having ocular media opacities, low vision, or substantial correction and reduced accommodative range, this ambient level is inadequate. In such cases, local task lighting is necessary and must be designed to avoid reflected or direct glare problems for other workers.

Second, it is possible to use higher illumination levels, with either natural or artificial lighting, and to baffle illumination sources adequately to avoid glare conditions. Such an approach requires very careful attention to the office architecture and is frequently beyond the capabilities of many office designers. Further, the need for such care is usually underestimated, resulting in unfavourable glare conditions for some workers. Because this approach is difficult and often incompatible with offices undergoing rearrangement or redesign, other solutions are more generally useful.

Third, and perhaps of greatest importance, is the need for manufacturers and designers of equipment to understand the nature of glare, its measurement, and the requirements for displays to reduce the glare problem without reducing image quality. This appears to be a fertile field for physical and behavioural research, a field that should have lucrative results for the successful manufacturer.

5. *Conclusions*

Until AR coatings and glare filters reach the level of technology needed to filter glare adequately without reducing image quality, problems will continue in the lighted office environment. These problems are amplified by the persistent design of VDTs having character fonts and matrix sizes with reduced legibility and readability, raster and dot spacing with too great a modulation, and inadequate contrast under ambient illumination conditions. At the present time, reduced ambient illumination can improve conditions for many workers, but the final solution may lie in an improved design for AR coatings and filters. It is reasonable to expect that a catalogue of filter applications may be the ultimate solution, with the manufacturer recommending a particular field-installable AR filter for each type of glare condition, so that the user can select the AR filter or coating most suitable for the environment in which the VDT is used. Only in this manner does it appear that natural lighting, high ambient illumination, and adequate display image quality will coexist with visual comfort.

References

Abramson, S.R., Mason, L.H. and Snyder, H.L., 1983, The effects of display errors and font styles upon operator performance with a plasma panel. *Proceedings of the Human Factors Society*, 28–32.

Beamon, W.S. and Snyder, H.L., 1975, *An Experimental Evaluation of the Spot Wobble Method of Suppressing Raster Structure Visibility*, Technical Report AMRL–TR–75–63, Wright-Patterson Air Force Base, Ohio.

Howell, W.C. and Kraft, C.L., 1959, *Size, Blur, and Contrast as Variables Affecting the Legibility of Alphanumerics on Radar-type Displays*. Technical Report 59–536, Wright-Patterson Air Development Center, Ohio.

Snyder, H.L., 1980, *Human Visual Performance and Flat Panel Display Image Quality*. Technical Report HFL–80–1, Virginia Polytechnic Institute and State University, Virginia.

Snyder, H.L. and Maddox, M.E., 1978, *Information Transfer from Computer-generated Dot-matrix Displays*. Technical Report HFL–78–3, Virginia Polytechnic Institute and State University, Virginia.

The Effect of Variation of Saccadic Eye Movement on VDU Operation

S. Yamamoto and K. Noro
University of Occupational and Environmental Health, School of Medicine,
1–1 Iseigaoka, Yahatanishiku, Kitakyushu 80, Japan

Abstract

In order to characterize VDU work, a visual search experiment was carried out. The experiment continued for two hours and the subjects' eye movements and performance were measured. We found a variation in saccadic eye or levelling eye movement and this was analysed and its effect studied.
The results were as follows:

1. The intervals of occurrences of the levelling eye movement shortened and the number of occurrences of the levelling increased.
2. The levelling eye movement lowered the performance level.

1. *Introduction*

When searching regularly displayed word lists, we usually read those lists column-wise or row-wise. In these cases, the regulated saccadic eye movement and fixation are observed like a step function by recorded eye movements. On loading such a task for a long period of time, a change of eye movement was observed; that is, the levelling eye movement. Variations of saccadic eye movement are usually found in the clinical ophthalmological examination. This examination was done by tracking an indicator with the eye.

In VDU work, operators move their eyes with their mind. This is not the same as reading the word list and tracking the indicator. Essentially there are differences in the eye movement between reading and tracking. The reading eye movement is reflected by the controlled mind. Any change in that mind influences the information processing. A change in performance means a change in visual information processing. The performance is one measurement of the output of the work. Now we will investigate the relationship between the performance and the levelling eye movement.

2. *Levelling Eye Movement*

We observed the regulated saccadic eye movement and fixation in a visual search experiment with VDU.

We sometimes observed the variation of eye movement; that is, the flattened saccadic eye movement. This variation is illustrated in Figure 1.

We call this variation the levelled eye movement. Bahill and Stark (1975) reported the relationship between visual fatigue and glissade eye movement in eye tracking. In VDU work, however, the operator did not perform eye tracking. The operator moved his eyes to read the words listed on VDU.

3. *Method*

The experiment using VDU, called 'name search test', was carried out.

In the first stage, an instruction for the subjects was displayed on the VDU. This instruction was the name of groups; for example, the names of countries, names of flowers,

FIGURE 1. *The levelling eye movement. In the first half the saccadic eye movement is performed in the right to left eye movement but this saccadic eye movement was flattened, showing levelling eye movement. Upper line: 1 second per scale. Middle line: Saccadic movements right–left. Lower line: Up-down movements.*

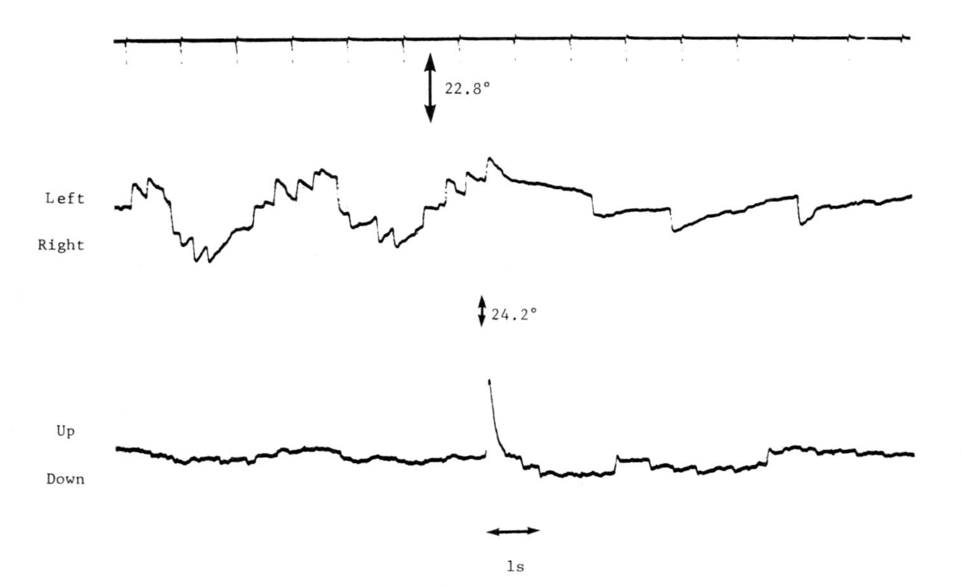

names of animals, etc. We prepared 10 groups, and this instruction was shown for 5 seconds.

In the second stage, the word list was displayed. The list was made up of 60 words and names of various kinds, and some indicated words were hidden in the list. We called this indication a target. The subjects searched for the target from the list displayed on the VDU. Whenever the subjects found the target, they pressed down the zero key of the 10-key numerics in the VDU. When they completed the task, they pressed down the key number 3 of the 10-key numerics, and the target and list changed at random. This task was repeated for two hours.

4. *Results*

The levelling eye movement began to appear 30 minutes after the start of the experiment. These occurrences of the levelling eye movement increase from the middle to a later period. This subject performed 205 tasks in 2 hours' work and the correct numbers were 102 (49·0%) tasks. The number of targets was 445 in 205 tasks and the found targets were 290 (65·2%) and the overlooked targets were 155 (34·8%).

The levelling eye movement firstly appeared at the 65th task at 38 minutes after the start of the experiment. The number of the levelling were 47 tasks and the rate of the levelling per task was 22·9%. The targets of the tasks in the levelling were 50, the found targets were 18 (36·0%) and the overlooked targets were 32 (64·0%).

5. *Discussion*

5.1. **The Regularity of the Occurrence of Levelling Eye Movement**

We will investigate the regularity of the occurrence of the levelling eye movement. Firstly the trend of the occurrence of this levelling was studied. We assumed that this eye movement uniformly occurred in the experimental period. If this assumption was correct, the following statistics U_0 would follow the normal distribution, where

$$U_0 = \frac{\dfrac{\Sigma t_i}{n} - \dfrac{1}{2} t_0}{t_0 \sqrt{\dfrac{1}{12} n}}$$

The t_i indicates the time the levelling occurred, where $0 < t_i < t_0$ and t_0 is 120 minutes and n is the number of occurrences of the levelling.

The number of levellings were 47 and S t_i = 230174·5 s. So we have U_0 = 4·7291 and

$U_0 > 1{\cdot}96$. The assumption was rejected with high significance. It could prove the existence of a trend in the levelling. So we estimated the linear regression and the result was

$$Y = 109{\cdot}87 - 0{\cdot}069\ t$$

where t is the time. The duration time of the levelling increased and the following regression was required

$$Y = 6{\cdot}07 + 0{\cdot}08\ x$$

where x indicated the passing time.

From above, passing the time of VDU work, the number of occurrences of the levelling was increased and the interval of the occurrence decreased. The duration time of the levelling also increased.

5.2. **Levelling Eye Movement and Performance**

Focusing on the target, we investigated the relation between levelling eye movements and performances. We considered performances in the following two cases; one in the case of the levelling and the other in the normal saccadic eye movement. Table 1 shows the number of found targets and overlooked targets in those two cases.

TABLE I. *The relation between eye movement and target.*

	Found target	Overlooked target
Levelling eye movement	18	32
Saccadic eye movement	272	132

From this table, we calculated the chi-square value and obtained the fact that $\chi^2 = 21{\cdot}2$ and the value of chi-square distribution $\chi^2\ (0{\cdot}05)$ was $3{\cdot}84$. From this result, this contingency table was shown to be highly significant. From this, we can see that the overlooked targets in the levelling eye movement are not the same as in the saccadic eye movement. It can be considered that the levelling eye movement influences the overlooked target. We rewrote that table at percentage and the result is shown in Table 2.

From Table 2 in the levelling eye movement, we can see there are twice as many overlooked errors as in the normal saccadic eye movement. Table 3 shows the results of the four subjects.

From these results we can say there are no differences between the saccadic eye movement in the normal and in the levelling eye movement.

TABLE 2. *Relative ratio between eye movements and targets.*

	Found target	Overlooked target
Levelling eye movement	36·0%	64·0%
Saccadic eye movement	68·9%	31·1%

TABLE 3. *The results of the relation between performance and eye movement (%).*

Subject		A	B	C	D	mean ± s.d.
Saccadic eye movement (Total)	found	68·9	85·7	92·4	77·7	81·2 ± 10·1
	overlooked	31·1	14·3	7·6	22·3	18·8 ± 10·2
Saccadic eye movement (in normal eye movement)	found	68·7	86·1	92·9	78·3	81·5 ± 10·4
	overlooked	31·3	13·9	7·1	21·7	18·5 ± 10·4
Saccadic eye movement (in the levelling eye movement)	found	70·5	75·0	89·2	76·8	77·1 ± 8·0
	overlooked	29·5	25·5	10·8	23·2	22·1 ± 8·0
Levelling eye movement	found	36·9	12·5	20·0	20·0	22·1 ± 9·9
	overlooked	64·0	87·5	80·0	80·0	77·9 ± 9·9

This levelling produced the occurrences of overlooked targets and lowered performances. This levelling eye movement is considered to be the effect of a loaded eye muscle. From this, we cannot get the visual information in the levelling which results in overlooking thus decreasing the performance.

6. *Conclusion*

The levelling eye movement was observed in the visual work. This levelling eye movement produced overlooked errors and decreased the performance. It is important to observe this eye movement in order to evaluate the visual load.

References

Bahill, T.W. and Stark, L., 1975, Overlapping saccades and glissades are produced by fatigue in the saccadic eye movement system. *Experimental Neurology*, **48**, pp.95–106.

Bellamy, L.J. and Courtney, A.J., 1981, Development of a search task for the measurement of peripheral visual acuity. *Ergonomics*, **24** (7), pp.497–509.

Yarbus, A.L., 1967, *Eye Movements and Vision* (London: Plenum Press) 222 pages.

Analysis of the Relationship Between Saccadic Movements and Reaction Times of VDU Operators

K. NORO AND S. YAMAMOTO
University of Occupational and Environmental Health, School of Medicine, 1–1 Iseigaoka, Yahatanishiku, Kitakyushu 807, Japan

Abstract

Operators' information processing in VDU work was investigated. Eye movement was measured, the response process was characterized, and the relationships of the response process to the VDU discussed. From this, the effect of VDU software was described.

1. *Description of VDU Work by Saccadic Movements*

The subject is instructed to search for and read specified words in a word list displayed on a VDU. The subject's saccadic movements during the search reading are recorded. The meaning of eye fixation is clarified by analysis of the recorded waveform. The relationship between the subject's reaction time and search efficiency and the relationship between the subject's eye movement regularity ratio and work performance are made clear. The subject's adaptability to VDU work is evaluated according to the obtained reaction time and saccadic movement regularity. From the derived findings we have determined whether the amount of information displayed on the VDU screen is correct.

2. *Method*

2.1. **Procedure**

An experiment using VDU, called 'name search test', was made.

In the first stage, instruction for the subjects was displayed on the VDU. This instruction was the name of groups; for example, the names of countries, names of flowers, names of animals, etc. We prepared 10 groups, and this instruction was shown for 5 seconds.

In the second stage, a word list was displayed. The list was made up of 60 objects and names of various kinds, and some indicators were hidden in the list. We called this indicator a target. The subjects searched for the target from the list displayed on the VDU. Whenever the subjects found the target, they pressed down the zero key of the 10-key numerics in the VDU. When they completed the task, they pressed down the key number 3 of the 10-key numerics, and the target list changed at random. This task was repeated for two hours. It required the operator's ability to search out quickly objective words or symbols without referring to each sentence or word.

3. *Results*

3.1. **Role of Eye Fixation**

In the present experiment, the saccadic movements of the subject were recorded on a data recorder, and the recorded saccadic movements were reproduced on a digital storage oscilloscope.

As a result of relating fixations to words displayed in the screen, it is assumed that one fixation is made per word (Noro, 1983). This hypothesis was verified by estimating the area of one word projected on the retina.

A detailed examination of the EOG waveform reveals that it is impossible to identify saccades fewer than the number of words presented on the VDU. This fact means that the eyes do not fixate on the individual letters that comprise one word.

3.2. **Search Efficiency**

3.2.1. *Parallel Processing of Information on Plural Words*

The mean of times taken by all subjects from fixating on the target word to pressing the response key was 1236·4 ms with a standard deviation (s.d.) of 616·2 ms. It is reported that simple reaction time is 200 ms and choice reaction time is 550 ms to 650 ms (Wargo, 1967). The time from discovery of the target word to depression of the response key in this experiment is twice as long as the choice reaction time. To analyse this point, the relationship between the time from reading the target word to making a response and the saccadic movements was investigated. Figure 1 gives saccadic movements, fixations and reaction times.

This result shows that the subject reads the words following the target word after reading the target word and before pressing the response key. It is thus presumed that the subject processes information on the other words in parallel with information on the target word after discovering the target word and before pressing the response key.

FIGURE 1. *Graphic representation of saccadic movements within time taken to make one response (from mean values of all subjects).* ▨ *is target word, and* ▭ *is word other than target word. Numbers indicate fixation time, and each circle corresponds to an area of central fovea.*

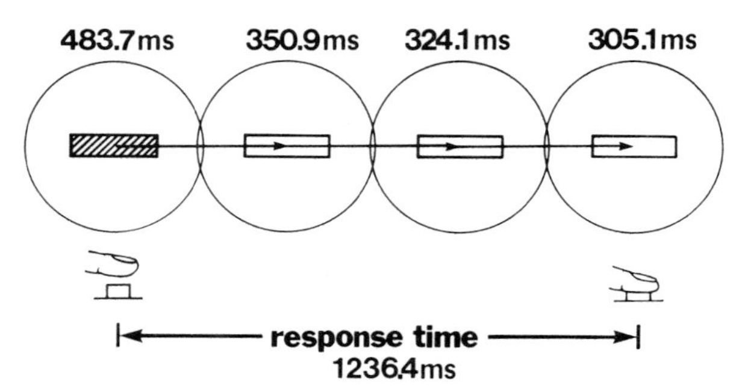

3.2.2. *Relationship Between Reaction Time and Saccade Regularity*

Figure 2 gives the results of the response process of Subject B who gave the best performance.

FIGURE 2. *Subject B's process from discovering target word to making response, and reaction time.*
* First word after target
** Second word after target

	response process				percent	response time (ms)
	target	word 1*	word 2**			
A1	○ →	△			11.7	980±102.3
A2	○ →	○ →	△		37.5	982.9±161.4
A3	○ →	○ →	○ →	△	8.6	989.1±214.9
B1	○ ⇄	○ △			5.5	1191.4±137.4
B2	○ ⇄	○ →	△		7.0	1095.6±181.9

○: fixation △: press down the response key

It is found from Figure 2 that the response process can be divided into two types. In one type, the subject processes information on the words following the target word in parallel and sequentially after discovering the target word. In the other type, the subject takes another look at the target word. This process is subdivided into two types according to the time when the subject presses the response key. It can be seen from Figure 2 that the reaction time under this process is 100 ms longer than under the sequential process. Also under this type, information on the words following the target words is processed in parallel.

The results of the subjects other than Subject B were investigated to see if the same were true for them and if they used different information processing methods. Their response processes for the respective targets were classified into the sequential and reread types, and the distribution ratios of the two types were determined as illustrated in Figure 3. These results indicate that each subject used mainly two different information processing methods, or the sequential and reread types, in different proportions. Subjects A, B and C used the sequential process more than the reread process, whereas Subjects D, E and F used the reread process more than the sequential process. The reaction time under the reread process was longer than under the sequential process.

FIGURE 3. *Distribution ratios of two types of information processing from discovery of target word to depression of response key.*

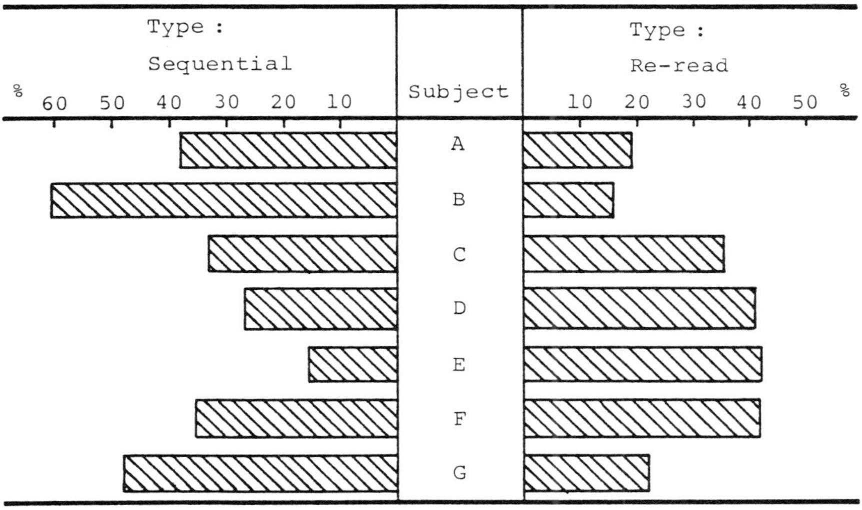

3.2.3. *Search Efficiency from Viewpoints of Fixation Time and Reaction Time*

Table 1 gives the fixation times of the fastest Subject B and the slowest Subject C for the words other than the target words.

TABLE 1. *Fixation time for words other than target word.*

Subject	Mean (ms)	s.d. (ms)	*n* [a]
Fastest Subject B	327·1	173·6	147
Slowest Subject C	341·8	131·7	99

[a] *n* = number of data items.

The fixation time of Subject C is slightly longer, but there is no significant difference between the two subjects ($p < 0{\cdot}05$).

Similarly, there is no difference in fixation time required for the target word ($p < 0{\cdot}05$). There is no difference either in fixation time between the subjects ($p < 0{\cdot}05$).

The reaction times of Subjects B and C are shown in Table 2. A highly significant difference is observed between the two subjects ($p < 0{\cdot}01$).

Analysis of the data of Tables 1 and 2 indicates that Subjects B and C are the same in the 'looking' behaviour but are different in the process from reading the target word to pressing the response key.

TABLE 2. *Reaction times of Subjects B and C.*

Subject	Mean (ms)	s.d. (ms)	*n*
B	1100·2	375·5	125
C	2321·6	1027·5	76

4. *Discussion*

4.1. **Evaluation of Subject's Performance by Reaction Time and Saccadic Movement Regularity**

Table 3 shows each subject's sequential response process distribution ratio and the number of frames processed per hour.

TABLE 3. *Sequential response process distribution ratio and number of frames processed per hour.*

Subject	A	B	C	D	E	F	G
Sequential process distribution ratio	38·1	60·2	33·3	27·0	16·0	35·5	59·7
Number of frames processed per hour	60	87	73	51	52	66	70

These data indicate that the order of the subjects according to the sequential response process distribution ratio agrees with the order of the subjects according to the number of frames processed per hour; the coefficient of correlation between these two orders is very high 0·8027 ($p < 0{\cdot}05$).

It is thus therefore found that the sequential response process ensures a higher search efficiency.

4.2. **Evaluation of Display Layout**

Display layout is considered to affect the regularity of eye movement. To confirm this effect, search patterns were investigated by adding a 12-line by 3-column display as another experimental condition to the 12-line by 5-column display used in the first experiment. Four subjects were used, and the experimental method was the same as the first one, except for the additional display layout.

The ratio of the sequential process is larger with the 12-line by 3-column display ($p < 0{\cdot}1$). It is, therefore, evident that the ratio of a response process changes when a different display layout is used.

5. *Conclusion*

The VDU screen search behaviour of VDU operators has been analysed from their saccadic movements and reaction times. It has been found as a result that VDU operators mainly use two information processing methods: sequential type and reread type. The VDU operators work by using these two methods. The operators who process information by the sequential method more frequently than the reread method performed better than those processing vice versa.

As far as saccadic movements are concerned, the operators who make regular saccadic movements performed better than those who do not. This finding points to the importance of training VDU operators on how to move their eyes.

References

Newell, F.W., 1982, *Ophthalmology*, 5th edition (St. Louis: C.V. Mosby).
Noro, K., 1983, A descriptive model of visual search. *Human Factors*, **25**(1), 93–101.
Wargo, M.J., 1967, Human operator response speed, frequency, and flexibility: review and analysis. *Human Factors*, **9**(3), 221–238.

An Appropriate Luminance of VDT Characters

M. TAKAHASHI, H. IIDA, A. NISHIOKA AND S. KUBOTA
Department of Work Physiology and Psychology, Institute for Science of Labour,
1544 Sugao, Miyamae-ku, Kawasaki, 213, Japan

Abstract

The purpose of this experimental study was to specify the VDT character luminance appropriate to prolonged VDT work. Eight subjects performed a visual search task for an experimental session two hours or more using CRT display with five conditions of the character luminance: 2, 5, 11, 27, and 64 cd/m^2.

The results showed that in the luminance conditions of 27 cd/m^2, search speed was faster and error rate was lower than in the other conditions. The results for subjective rating of fatigue showed the increase of visual fatigue as the experimental session progressed. The range between upper and lower limit of luminance acceptable to the subjects narrowed and the optimum level of luminance decreased to 22 $\pm$ 6 cd/m^2, at the late phase of the session.

These findings suggest that the luminance of VDT character should be designed to be easily and also adequately adjustable during VDT work to avoid unnecessary visual fatigue.

1. *Introduction*

Although every recommendation for the design of the VDT workplace states that glare reflected on the VDT screen is to be avoided, most of the recommendations give no value for the character luminance. Cakir *et al.* (1979) and Stewart (1980) recommend that the character luminance should be between 80–160 cd/m^2 and 45–100 cd/m^2 respectively. According to our observations the character luminance for the VDT workplace is rather lower than these recommended levels, and, in addition, VDT operators have a tendency to lower the luminance in the course of their work.

The aim of this experimental study was to ascertain the VDT character luminance appropriate to prolonged VDT work under control of visual environment without glare. To this end, the effects of character luminance on performance of a visual search task, and subjective visual fatigue was investigated in the laboratory experiment. The optimum level of character luminance was also examined by the method of self-adjustment during the experimental session. The character luminance adopted as the experimental variables were at the following five levels; 2, 5, 11, 27, and 64 cd/m^2.

2. *Method*

2.1. **Experimental Apparatus and Task**

The search materials and the targets were presented on the 14-in colour graphic display in which the character luminance is adjustable within the range 0–100 cd/m^2. The alphanumeric characters presented were capital letters only 3·1 × 4·0 mm in size (7 × 9 dots matrix) and green in colour. One page of the search materials contained 105 items arranged in 7 columns. Each item consisted of 5 characters chosen at random from the 26 letters of the alphabet. The target character of each row was presented under the fourth column of the search material. The subjects were told to indicate how many target characters were contained in each row by using the keyboard.

2.2. **Subjective Visual Fatigue**

The following eight items were used to assess subjective visual fatigue:

1. Difficulty of focusing.
2. Having difficulty keeping the eyes open.
3. Hardness of eye movement.
4. Watering of eyes (Epiphora).
5. Dryness of eyes.
6. Glare.
7. Soreness of eyes.
8. Sleepiness.

The assessment of 'glare' was made for the search material with a luminance of 64 cd/m^2. Subjects replied with the number corresponding to their current subjective estimate on a rating scale from 1, 'not at all', to 7, 'extremely intense' for each item. The intermediate points were only indicated by a numerical rating.

2.3. **Self-Adjustment of Character Luminance**

Subjects were asked to adjust the luminance of the search material by turning a know on the display terminal. The adjustment was made based on the following three levels of character luminance needed to perform the visual search task.

1. Acceptable lower limit.
2. Optimum level.
3. Acceptable upper limit.

2.4. **Visual Working Environment**

An experimental workplace was made in a laboratory. The workplace was designed to prevent the image of the wall being reflected on the display screen. An adjustable table and chair were adopted in order to reduce the subjects' postural load. Before the experimental session, the height of work surface and chair were adjusted according to the subject's preference. The artificial light was set to 400 lux in horizontal illuminance at the lower edge of the display screen. The average of the background luminance on the VDT screen was approximately 1 cd/m^2.

2.5. **Subjects**

Six males and two females were the subjects. Their ages ranged from 26 to 49 years (mean = 31). Two subjects wore corrective lenses. All subjects were familiar with visual display terminal operations. Prior to the experimental session, each subject had training experience of visual search task for two hours or more.

2.6. **Procedure**

In the first session and every block of the search task, the subjects were asked to make the assessment of subjective visual fatigue on a 7-point rating scale and by the self-adjustment of the character luminance. One block of the search task consisted of 6 pages with 5 different levels of the character luminance. The character luminance of page 6 was at a moderate level of 11 cd/m^2 to avoid excess character luminance. The subjects continued in the same way with subsequent blocks for two hours or more.

3. *Results*

The quantity of the search task performed was different for each subject, from 6 to 10 blocks. The duration of the tasks ranged from 120 to 155 minutes.

3.1. **Performance on the Search Task**

Subjects' performance was quantified by search time per row (the time duration between the presentation of target and return-key press) and the rate of error (the percentage of rows with incorrect key-input). Figure 1 shows the mean search time for each two blocks for the level of luminance at the early, middle, and late stages of the experimental session.

The result of two way ANOVA (luminance (5) × phase (3) × Subject (8)) indicated that only the main effect of luminance was significant ($F(4, 28) = 36{\cdot}51, p < 0{\cdot}01$). Although there

FIGURE 1. *Mean search time under 5 different luminance levels for the early, middle and late stages.*

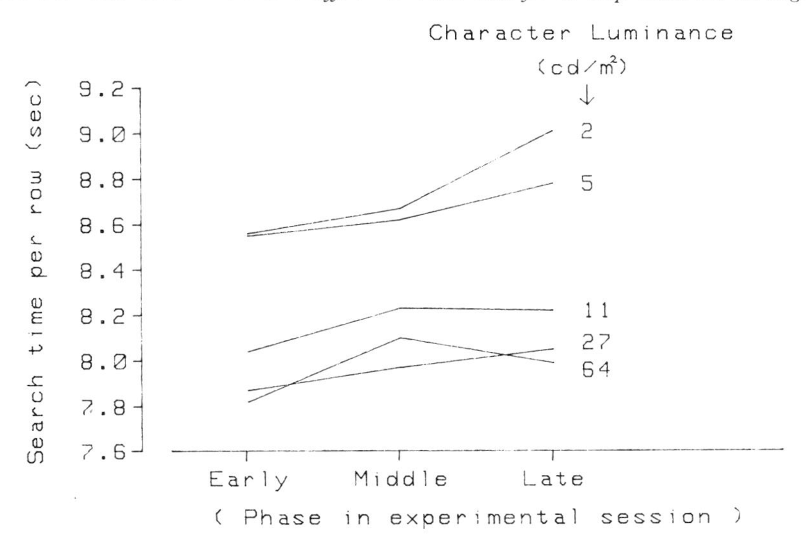

is no significant difference between the search time of 27 and 64 cd/m^2 levels, both were significantly faster than the other levels ($p < 0{\cdot}05$ by t-test). Moreover, at the level of 2 cd/m^2, the search time increased significantly at the late stage of the session (middle v. late: $t = 2{\cdot}64$, $p < 0{\cdot}05$). A similar effect was also manifested at the 5 cd/m^2 level, while there was no significance for the higher luminance levels.

The results of the mean error rate for the four subjects recorded are indicated in Figure 2.

FIGURE 2. *Mean error rate (%) for 4 subjects under 5 different luminance levels.*

Error Rate (%)
11
10
9
8
7
6
5
4
9.1%
8.2%
5.6%
5.5%
8.8%
2 5 11 27 64
Character Luminance (cd/m^2)

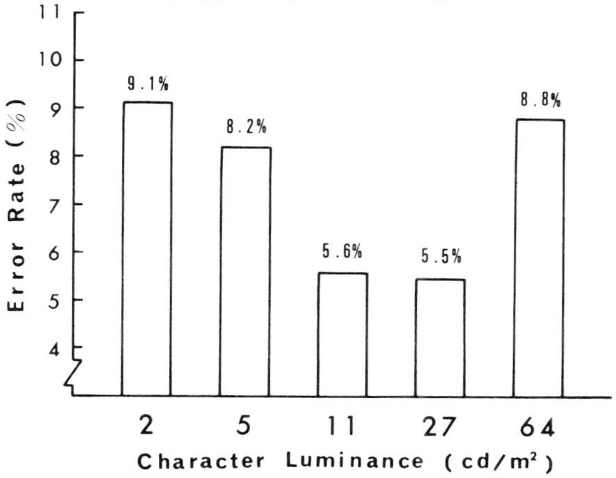

The higher error rate was obtained under the lower luminance levels except 64 cd/m^2. This higher error rate was recorded under the 64 cd/m^2 level, which was inconsistent with the results of search time. This error rate suggests a speed/accuracy trade-off.

3.2. Subjective Visual Fatigue

The mean rating scores over eight items of visual fatigue symptoms increased significantly with the duration on task. Each value of the three stages was 1·90, 2·74, and 3·16 respectively (early v. middle: $t = 3{\cdot}52$, $p < 0{\cdot}01$; middle v. late: $t = 3{\cdot}14$, $p < 0{\cdot}05$). This result clearly suggests that the subjects had become tired gradually.

3.3. Self-Adjusted Character Luminance

Figure 3 shows the mean and standard deviation for character luminance which the subjects adjusted according to three criteria. The upper level of luminance acceptable to the subject decreased gradually but, on the other hand, the lower limit of luminance acceptable to the subject increased.

FIGURE 3. *Mean and standard deviation of character luminances based on three criteria.*

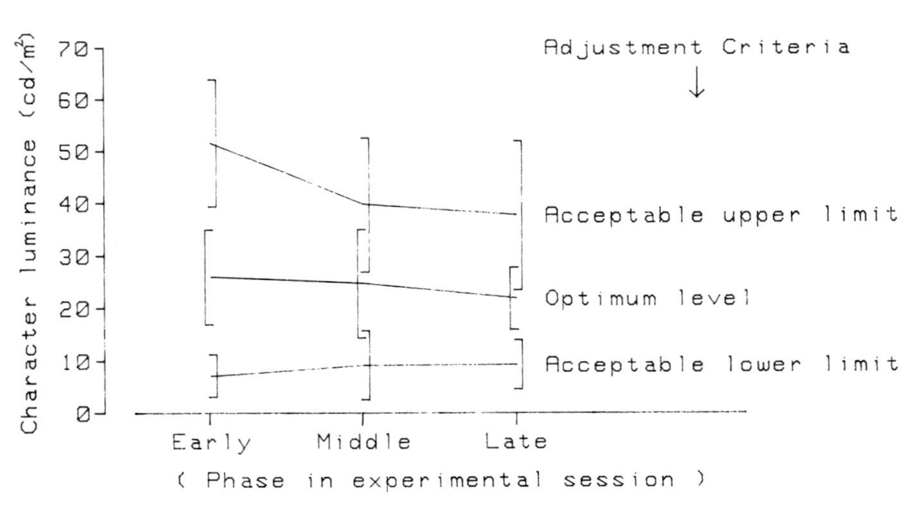

4. *Discussion*

The performance of the visual search task was best under the character luminance level of 27 cd/m^2.

The acceptable range of the self-adjusted luminance was made narrow and the optimum level decreased to 22 ± 6 cd/m^2 at the late stage of the experimental session, when the subject was tired; this luminance level is rather lower than the existing recommended levels. These results were recorded in a visual environment without bright reflections on the VDT screen; this is different from an actual working environment where there are bright reflections. Nevertheless, we think that the luminance of VDT character should be designed to be easily and also adequately adjustable during VDT work to avoid unnecessary visual fatigue.

References

Cakir, A., Hart, D.J. and Stewart, T.F.M., 1979, *The VDT Manual* (Darmstadt: IFRA).
Stewart, T., 1980, Problems caused by the continuous use of visual display units. *Lighting Research & Technology*, **12**, 26–35.

Reading from Microfiche, from a VDT, and from the Printed Page: Subjective Fatigue and Performance A Preliminary Report

W.H. CUSHMAN
Eastman Kodak Company, Rochester, New York 14650, USA

Abstract

Subjects read continuous text for 80 min using microfiche and VDTs (video display terminals) with negative and positive appearing images and printed paper copy. Measurements of reading speed and visual fatigue were obtained periodically, and a test of reading comprehension was given at the end of each session. Visual fatigue was significantly greater ($P < 0{\cdot}05$) when subjects read from negative microfiche (light characters, dark background) projected on a metal screen or from the screen of a VDT with positive appearing images (dark characters, light background). Reading speeds tended to be slower for conditions that were more fatiguing; however, reading comprehension scores were similar for all conditions.

1. *Methods*

In the first experiment reading continuous text from microfiche with positive and negative images is compared with reading from printed paper copy under conditions that were very similar.

The second experiment examines the positive/negative issue concerning the reading of continuous text from video displays.

1.2. **Subjects**

The 52 subjects (30 males and 22 females) were employees of Eastman Kodak Company. Most had some previous experience using microfilm readers and VDTs. All age groups from the early twenties to late fifties were represented.

1.2. Reading Materials

Subjects read articles of general interest from printed pages (the original source documents), microfiche (1:20 reduction ratio) produced by filming the originals, and the screen of a VDT. The articles were printed in 10-point Times Roman Medium type (a serif font) with approximately 55 characters per line and 2 columns per page. Each was accompanied by several validated multiple-choice questions for measuring reading comprehension. The articles were divided into 5 sets (designated as sets A–E, hereafter) with approximately 35 000 words in each of the sets. The Flesch 'reading ease' scores (Flesch 1949) for sets A–E were 60, 59, 60, 57, and 58, respectively.

1.3. The Microfiche Reader and VDTs

The microfiche reader was a modified front-projection model with two interchangeable screens (approximately horizontal screen orientation) and a lens suitable for enlarging the images of the original source documents (i.e. the printed pages) back to their normal size. One screen was a metal screen with a relatively diffuse surface; the other was a high-reflectance (HR) screen (titanium dioxide coating) with a matt surface. Images projected on the former screen (the metal screen) appeared to scintillate, but those projected on the latter (the HR screen) did not (see Cushman, 1978 for a discussion of scintillation).

Although the microfiche reader used in this study was a front-projection model, readers utilizing rear projection are currently more popular. However, since type of projection (front v. rear) has no effect on either reading speed or reading comprehension (Judish, 1968), the results reported below apply to rear-projection readers as well.

The two VDTs differed in only one important respect: the CRT display for one unit had a white (P4) phosphor, while the CRT display for the other had a green (P31) phosphor. Both had 30 cm (diagonally measured) screens, a 24 line × 80 character format, 7 × 9 dot matrix characters (4 mm in height) with descenders, a 60 Hz (non-interlaced scan) frame repetition rate, a smooth scrolling rate of 6 lines/s, a brightness control (up to 140 cd/m^2), and a means for generating both positive and negative images. The stroke width to character height ratio was near the 1:8 optimum for positive text images (dark characters, light background) but was significantly greater than the 1:10 optimum for negative text images (light characters, dark background).

1.4. Procedure: Experiment 1

Sixteen (16) subjects participated in five 80-min reading sessions, once for each of five conditions: paper copy and microfiche with positive and negative appearing images projected on the metal and HR screens. A counterbalancing scheme was used to determine the ordering of the five experimental conditions and five sets of reading materials (A–E) for each subject. The order of the articles within the sets was also varied from subject to subject. A minimum of 2 days rest was given between sessions.

When reading from the printed paper copy, the subject placed the individual pages directly on the screen of the microfiche reader in order to illuminate them with the same lamp used to project the film images. As necessary, the voltage to the projection lamp was raised or lowered in small increments so that the luminance of the light areas of the pages (i.e. the background for positive images and the characters for negative images) was 125 cd/m^2 with contrast modulation, defined as (L max − L min)/(L max + L min), between 0·76 and 0·88, regardless of the type of document (paper or film) and screen. An additional reason for placing the paper documents on the screen was to ensure that the subject had the same working posture for all conditions. The laboratory had indirect lighting; general illumination was approximately 270 lux.

Subjective visual fatigue was assessed with the Visual Fatigue Graphic Rating Scale (VFGRS, hereafter) described in an earlier paper (Cushman, 1978). The VFGRS consists of a 15 cm vertically-oriented line with eight verbal cues describing various gradations of eye discomfort from no discomfort to extreme discomfort. The verbal cues were place along the line from top to bottom in order of increasing severity. The subject indicated extent of visual fatigue by drawing a short horizontal line intersecting the vertical line at the appropriate point. For scoring, the vertical line was divided into 12 equal intervals and assigned values of 1–12. The VFGRS was administered at the beginning of each session and after 15 min, 30 min, 45 min, 60 min, and 80 min.

Reading speeds were calculated in the usual way—i.e. by dividing the number of words read in a fixed time period by the duration of that period—and expressed in words per minute. Reading comprehension scores were obtained by calculating the percentage of correct answers for selected reading comprehension questions administered at the end of each 80-min reading session.

1.5. **Procedure: Experiment 2**

In the second experiment 36 subjects read for 80 min from one of the two VDT screens (18 used the VDT with the P4 phosphor and 18 used the VDT with the P31 phosphor). Only one of the five sets of articles was used (set A), but the order of presentation of the articles was varied from subject to subject. The images were positive for half the subjects (dark characters on a light background, sometimes referred to as negative contrast) and negative for the others (light characters on a dark background, sometimes referred to as positive contrast). Each page consisted of 20 lines of text. To advance to the next page the subject pressed the 'return' key, and the text scrolled upward at the rate of 6 lines/s until the new page was completely visible. At the beginning of the session the subject adjusted the workplace, chair, and display luminance to reflect his or her individual preferences. The ambient illumination in the laboratory was approximately 750 lux. All other aspects of the procedure were the same as in the first experiment.

FIGURE 1. *Visual fatigue (increase in VFGRS score) as a function of reading time for microfiche and paper copy. MF−, microfiche with negative appearing images (light characters on a dark background) projected on the metal screen; MF+, microfiche with positive appearing images (dark characters on a light background) projected on the metal screen; PC, paper copy. Each point represents the mean of 16 observations. The curves for positive and negative microfiche projected on the HR screen are not shown here but were very similar to the PC curve.*

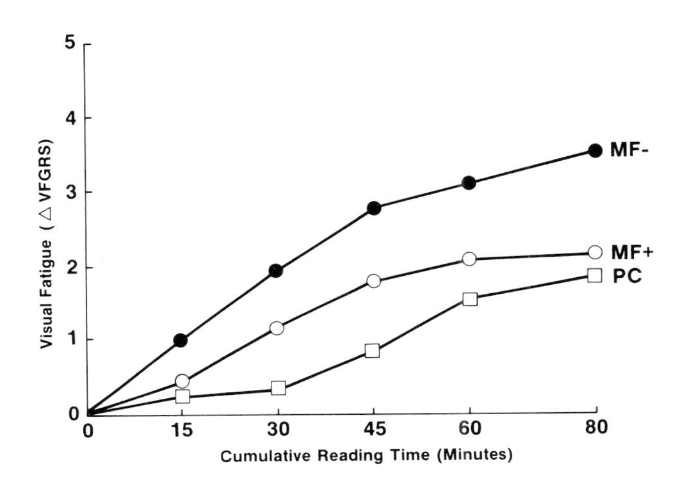

2. *Results*

Figure 1 shows that visual fatigue increased as subjects read from the printed paper copy (PC) and microfiche (MF) projected on the metal screen. The curves for positive and negative microfiche projected on the HR screen (not shown in Figure 1) were very similar to the PC curve. The vertical axis represents the increase in VFGRS score from the baseline value obtained before the reading session began. Each point represents the mean of 16 observations. After 80 min of reading the increase in visual fatigue was significantly greater for negative microfiche (light characters, MF− curve) than for either the paper copy (PC curve) or positive microfiche (dark characters, MF+ curve) (t-tests for paired observations, $p < 0{\cdot}05$). However, differences between the positive microfiche and paper copy were not significant.

Figure 2 shows that reading for 80 min from a VDT with a positive image (VDT+ curve, dark characters on a light background or negative contrast) was more fatiguing than reading the same material for 80 min from paper copy (t-test, $p < 0{\cdot}05$). However, the corresponding difference between the means for the positive and negative VDT modes (VDT+ and VDT−) and the difference between the means for paper copy and the negative VDT mode (VDT−, light characters on a dark background or positive contrast) were not significant ($p < 0{\cdot}08$ and $p < 0{\cdot}20$, respectively). (Because phosphor chromaticity did not appear to have affected visual fatigue, based on the number of observations reported here, the data for both VDTs were combined to increase the sample size.)

FIGURE 2. *Visual fatigue (increase in VFGRS score) as a function of reading time for the VDTs and paper copy. VDT+, VDT with positive image (dark characters on a light background, sometimes referred to as negative contrast); VDT−, VDT with negative image (light characters on a dark background, sometimes referred to as positive contrast); PC, paper copy. The PC curve is the same as in Figure 1. Each point on the VDT curves represents the mean of 18 observations. Data for the display with the P4 phosphor and the display with the P31 phosphor have been combined.*

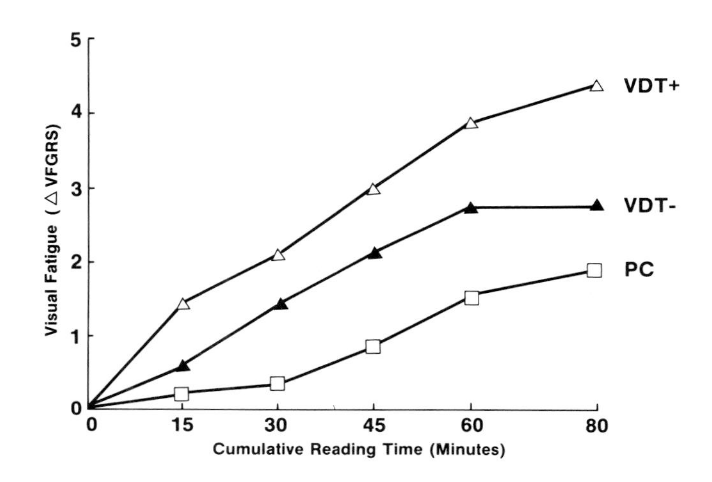

Reading speed and reading comprehension data are given in Table 1. The mean reading speed for microfiche with negative images (light characters) was significantly slower than the mean reading speed for paper copy (*t*-test for paired observations, $p < 0{\cdot}05$). However, the difference between mean reading speeds for microfiche with positive images (dark characters) and paper copy was not significant. Similarly, the difference between positive and negative modes of presentation for VDTs was also not significant.

Reading comprehension scores for all conditions were very similar. Differences between microfiche and paper copy and between the VDTs with positive and negative images were not significant.

TABLE 1. *Reading speed and comprehension.*

Condition	Mean reading speed (in words per minute)	Mean reading comprehension score (%)
Paper copy	218	76
Microfiche (positive image)	210	77
Microfiche (negative image)	199	79
VDT (positive image, negative contrast)	212	82
VDT (negative image, positive contrast)	220	81

3. *Discussion*

Although reading microfiche with either positive or negative images projected on the HR screen was no more fatiguing than reading the original paper copy, reading negative microfiche (light characters, dark background) projected on the metal screen was more fatiguing than reading either the paper copy or positive microfiche (dark characters, light background) projected on that same screen. The light areas of images projected on the metal screen appeared to scintillate, and this may account for the greater amount of visual fatigue that was associated with the use of that screen (Cushman, 1978).

In contrast to microfilm, reading continuous text from a VDT with dark characters on a light background (positive image, negative contrast) was more fatiguing than reading the same copy printed on paper, but reading from a VDT with light characters on a dark background (negative image, positive contrast) was not. This finding is contrary to the results of several experiments involving non-reading tasks in which performance was better (with less visual fatigue) when VDTs had positive images (Bauer and Cavonius, 1980; Radl, 1980). The displays used in these studies, however, had frame repetition rates that were considerably greater than 60 Hz and, hence, less flicker. Considering that flicker is a suspected cause of the visual fatigue often reported by VDT users and that frame repetition rates approaching 100 Hz may be necessary for positive image (negative contrast) displays to completely eliminate flicker (Bauer *et al.*, 1983), the results shown in Figure 2 are not surprising.

There was one other important difference between the Bauer and Cavonius (1980) and Radl (1980) studies and the experiments reported here. In those studies the subjects had to refer frequently to paper copy in addition to the VDT. The transient adaptational changes associated with such tasks are thought to be more fatiguing when the VDT has a dark background (negative images, positive contrast). Thus, the tasks in the Bauer and Cavonius (1980) and Radl (1980) studies may have been more difficult for the conditions involving negative images (positive contrasts).

4. *Conclusions*

The results reported here suggest that reading from microfilm is no more fatiguing than reading from printed pages if the user has a well designed reader with a non-scintillating screen and uses it in a relatively glare-free environment. Similarly, VDTs with a 60 Hz frame repetition rate and P4 or P31 phosphors are suitable for reading continuous text if operated in the negative image mode (light characters on a dark background, positive contrast). However, some users will find that operating these display in the positive image mode (dark characters on a light background, negative contrast) is more fatiguing than reading from printed pages. Apparently in the latter mode (dark characters, light background), the frame repetition rate must be considerably higher than 60 Hz in order to make the display flicker-free (Bauer *et al.*, 1983). Therefore, had the experiments described here been conducted with displays having higher frame repetition rates (or longer persistence phosphors), the results might have been considerably different.

Acknowledgements

I would like to thank Rise Segur, Shiela Accongio, Ken Corl, John Stevens, Debbie Mulcahy, and Karen Bell for technical sssistance. I also wish to thank Harry David for granting permission to conduct the experiments.

References

Bauer, D., Bonacker, M. and Cavonius, C.R., 1983, Frame repetition rate for flicker-free viewing of bright VDU screens. *Displays*, **4**, 31–33.

Bauer, D. and Cavonius, C.R., 1980, Improving the legibility of visual display units through contrast reversal, in *Ergonomic Aspects of Visual Display Terminals*, edited by E. Grandjean and E. Vigliani (London: Taylor & Francis) pp.137–142.

Cushman, W.H., 1978, Visual search using microfiche: Effects of microfiche reader screen scintillation on visual fatigue, general fatigue, and performance. *Proceedings of the Human Factors Society's 22nd Annual Meeting*, pp.299–302.

Flesch, R., 1949, *The Art of Readable Writing* (New York: Harper & Row).

Judish, J.M., 1968, The effect of positive-negative microforms and front-rear projection on reading speed and comprehension. *Journal of Micrographics* (NMA), **2**, 58–61.

Radl, G.W., 1980, Experimental investigations for optimal presentation-mode and colours of symbols on the CRT-screen, in *Ergonomic Aspects of Visual Display Terminals*, edited by E. Grandjean and E. Vigliani (London: Taylor & Francis), pp.127–135.

Doing the Same Work with Paper and Cathode Ray Tube Displays (CRT)

J.D. GOULD AND N. GRISCHKOWSKY
IBM Research Center, Box 218, Yorktown Heights, New York 10598, USA

Abstract

Much public concern has centred on possible fatiguing effects of using CRT terminals in one's work. Twenty-four participants proofread from a CRT on one day and from hard copy on another day. No change in participants' performance, feelings, or vision throughout the day could be attributed to using CRT terminals. However, participants did proofread hard copy about 20–30% faster than from a CRT, however.

1. *Introduction*

There is much international public concern about people using cathode-ray tube terminals (CRT) in their work. Newspapers and magazines contain many articles each month, the National Academy of Science sponsored a conference on the subject (Brown *et al.* 1982), and several studies have been conducted, including field opinion surveys and laboratory experiments (see Matula 1980, for a bibliography of 147 papers from 1977–81, Grandjean and Vigliani 1980, for 45 more papers).

The results of the studies are mixed. Some report that people have many complaints about how they *feel* as a result of using CRTs in their work (Dainoff *et al.* 1981, Smith *et al.* 1981), whereas others do not find any unusual number of complaints from users of CRT terminals (Sauter *et al.* 1983, Starr *et al.* 1982). Some studies report temporary effects on people's *vision* after using CRT terminals (Östberg 1980, Haider *et al.* 1980), whereas others have found no such temporary effects (Dainoff *et al.* 1981). Apart from specific effects of CRTs, one general hypothesis is that adult human vision may be temporarily affected by prolonged intensive close work of almost any sort. Experimental investigations do not always find this, however (see Brown *et al.* 1982).

There are two weaknesses in the studies which have attempted to assess the effects of CRTs on people's work. First, 'control groups' have been absent, inappropriate, or questionable. An exception is the recent survey results by Starr *et al.* (1982). No study has used subjects as their own control. Exceptions are a small experiment measuring eye movement durations by Mourant *et al.* (1981) and an unpublished Dutch study conducted by de Groot and cited by Starr *et al.* (1982). Second, studies have not simultaneously assessed people's *performance*, i.e. the quality and efficiency of their work, people's *feelings* and psychophysical measures of people's *vision*. Experiments have measured only one or another of these, and no study has measured all three and sought to understand the interrelations of these measures for people working with and without CRTs.

Thus, studies to date have not carefully distinguished the effects of CRT displays from the effects of the work itself on people. The purpose of our experiment was to investigate the role that the CRT display, apart from the work itself, plays in any changes in people's performance, feelings, or vision. The same participants did the same work on a CRT terminal and on hard copy. They looked for misspelled words in excerpts from magazines and newspapers (but did not correct the spelling or English). Proofreading was selected as the task because it is visually intensive and occurs in many real-world applications.

2. *Methods*

2.1. **Participants**

Participants were 24 clerk-typists hired from temporary employment firms. There were two age groups with 12 participants each, one with a mean age of 23 (range = 20–27) and the other with a mean age of 48 (range = 40–61). Thirteen had 20/20 or better corrected vision, 7 had 20/22–20/25, and 4 had 20/29–20/33. Fifteen had experience using CRT terminals.

2.2. **Design**

Participants worked for two days in six 45-min work periods each day. On one day participants proofread hard copy articles and circled misspellings. On the other day participants proofread similar material on a CRT terminal (IBM 3277) and pointed at misspellings with a lightpen connected to the display. Half the participants proofread hard copy on the first day and the other half proofread on a CRT on the first day. There were 12 sets of materials, and each participant proofread a different set in each work period. The order in which participants proofread them was balanced across with work periods and display mode with 6 × 6 latin squares. Each set contained more pages than the fastest participant completed in a work period. The mispellings were randomly inserted, on average one in every 150 words. The material was taken from articles on which Kucera and Francis (1967) based their word frequency counts.

2.3. Displays

Three different CRT displays (all IBM 3277s) were used so that the results would not be limited to one display tube. Hard copy materials were of good quality, being created with an APS5 computer-controlled photocomposer and reproduced with IBM copiers. The layout was the same for both the CRT and hard copy: each page contained 23 lines, with about 9 words on a line. Hard copy had dark characters on a light background, 10-point monospace Letter Gothic font, and 4 mm high characters. Participants could pick up the hard copy or read it on the table. The CRT had greenish self-luminous characters on a dark background, 3277 type font, and 3 mm high characters. Participants could sit at any distance from the CRT or the hard copy. For their comfort, participants were encouraged to adjust the lighting level of the room (25, 38, or 60 mL measured on the white hard copy paper), adjust the luminance and contrast on the CRT, and adjust their seating (on any of four optional chairs).

2.4. Measurements

Three classes of measurements were made. Participants' *performance* during each work period was measured in terms of proofreading times and errors.

Participants' *feelings* of comfort were measured with a 16-item rating questionnaire which they filled out at the beginning of each day and after each work period. Questionnaires of this

FIGURE 1. *Proofreading speed and misses for 24 participants on the same material shown on hard copy (HC) and on CRT displays. Hours are an approximation, consisting of a 45 minute work period followed by testing and an optional break. One page per min corresponds to about 207 words per min.*

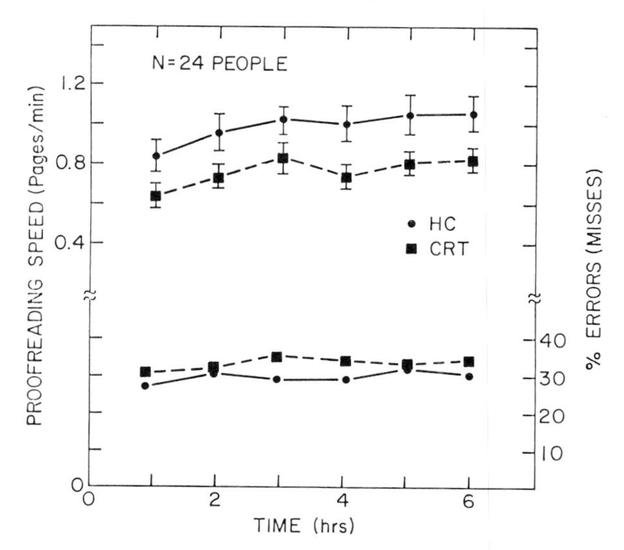

sort are sometimes referred to as 'health questionnaires' in the literature. The items reflected the most frequent comments people give as a result of using CRTs in their work (Dainoff *et al.* 1981).

Participants' *vision* was measured at the beginning of the day and following each work period. Binocular visual measurements were made under ambient lighting (60 mL), with participants wearing their corrective lenses. Contrast sensitivity and flicker sensitivity were measured using the Optronix Vision Tester (available from Optronix Corp, 2003 Orrington Ave., Evanston, Ill. 60201). Contrast sensitivity was measured for sine wave gratings of 0·5, 1, 3, 6, 11·4 and 22·8 cycles per degree, each presented in a 90 mL mean luminance field subtending 2·43 degrees at participants' 3 metre viewing distance. Here participants adjusted the contrast of each sine wave grating, using the von Bekesy threshold tracking method to determine their own sensitivity thresholds. Sensitivity to flicker was measured

FIGURE 2. *Results from 4 of the 16 items participants rated at the beginning of the day, after each work period, and after lunch. Analyses of variance showed that display mode had no significant effect on any of the feelings questions. The small differences observable between the hard copy and CRT curves for each question were not systematic across participants. Only 2 of 24 participants were more positive toward one display mode than toward the other on all four items shown. On none of the items was there a significant proportion of participants who gave more positive feelings on hard copy than on CRT.*

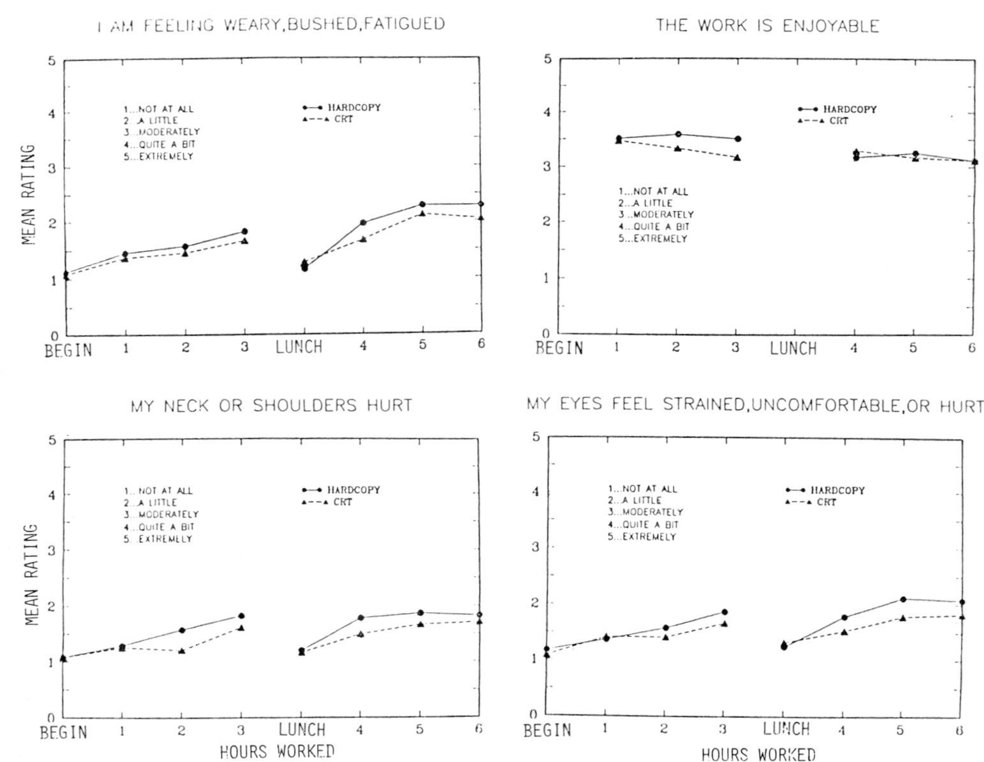

by presenting participants with a 90 mL mean luminance, 2·43° field 'flickering' at either 3, 8, 15, or 30 times per second. Here participants adjusted the contrast of each field, using the von Bekesy threshold tracking method to establish their own sensitivity thresholds. Visual acuity and phoria were measured with a Bausch & Lomb Vision Tester ('Orthorater') (available as Catalogue no. 71–22–41 from Bausch & Lomb, Rochester, NY 14602). These particular visual tests were used because they were relatively easy and quick to make on untrained participants, and because of their relevance to previous studies. The measurements had to be made simply and quickly so that they would not be perceived by participants as the main task, and so as not to induce any undue fatigue (which might dwarf any differential fatigue from the proofreading task itself). Participants worked in a room by themselves, and were knowingly videotaped while working. In fact, recordings were generally made only at the beginning of the first work period and near the end of each work period to reduce the amount of videotape needed.

Measurements of feelings and vision required about 15 min. Participants could take an optional break of a few minutes before beginning the next work period.

3. *Results*

Figure 1 shows that participants' proofreading *performance* was not differentially affected across the work day by using a CRT (i.e., the display mode × work period interaction was not significant ($F(5, 110) < 1{\cdot}0$). However, participants proofread significantly faster on hard copy than on CRT (0·98 *vs* 0·77 pages per minute, or about 203 *vs* 159 words per minute; $F(1, 22) = 13{\cdot}1$; $p < 0{\cdot}01$). Twenty-two of the 24 participants read faster on hard copy. There were slightly more errors on CRT (33%) than on hard copy (30%), $F(1, 22) = 19{\cdot}0$; $p < 0{\cdot}001$. False positives were rare. Participants' proofreading speed improved over work periods, $F(5, 110) = 8{\cdot}23$; $p < 0{\cdot}001$.

Participants *feelings* were generally positive, and were unaffected by whether they were proofreading hard copy or CRT. Figure 2 shows the results of four questions. None of the F-tests comparing feelings on hard copy and feelings on CRT was significant, and none of the display mode × work period interactions was significant either ($p > 0{\cdot}10$ in all cases). (The other items on the questionnaire showed even smaller differences between CRT and hard copy, and across work periods, than those shown in Figure 2.) The appropriate underlying scale of rating scales (rank order or interval) is arguable and therefore the appropriate statistical tests of significance are debatable. Inspection of the medians and histograms, which do not assume an interval scale, fully support these 'no difference' results.

Participants might have given unduly positive ratings about using CRTs because the experiment was done at IBM. To check this, we had five judges rate, from the videotapes of participants, how fatigued each participant looked near the end of each work period. The results of these ratings.indicated that participants looked equally fatigued while using either hard copy or CRT (F1, 22); $p < 1{\cdot}0$). However, these ratings by judges indicated more fatigue than did the self-ratings.

FIGURE 3. *Mean sensitivity (1/contrast) to sine wave gratings in cycles per degree (C/DEG) at the beginning of the day, at the end of the day after proofreading hard copy, and at the end of the day after proofreading on the CRT. Beginning of the day data were taken from Day 2, since it appeared that participants were still learning to use the vision tester and improving their sensitivity during their initial measurements on Day 1.*

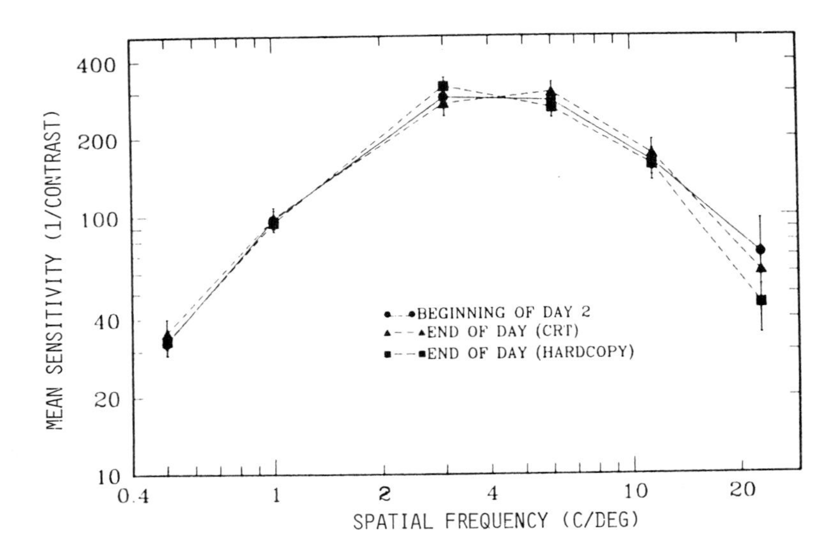

Participants' *vision* was not affected by display mode either. On average, near visual acuity stayed at 20/21 throughout the day and far visual acuity dropped from 20/21 to 20/22 ($F(1, 22) = 4{\cdot}75$; $p < 0{\cdot}05$). These acuity measures are generally thought to reflect people's ability to distinguish fine detail in normal ambient light. Contrast sensitivity to various sine wave gratings provides a threshold measure of sensitivity to gross as well as to fine detail. As shown in Figure 3, there was no significant change in contrast sensitivity from the beginning of the day to the end of the day when participants proofread hard copy (mean sensitivity = 154 and 151; $F(1, 22) < 1{\cdot}0$); or when participants proofread on a CRT (mean sensitivity = 154 and 156; $F(1, 22) < 1{\cdot}0$).* More importantly there was no display mode × work period interaction ($F(1, 22) < 1{\cdot}0$). Sensitivity to flicker, or to contrast modulation, did not change from the beginning to end of the day when participants proofread hard copy ($F(1, 22) = 1{\cdot}69$; $p > 0{\cdot}10$) or proofread on CRT ($F(1, 22) < 1{\cdot}0$) (Figure 4). There was no display mode × work period interaction ($F(1, 22) = 1{\cdot}31$; $p > 0{\cdot}10$).

Lateral phoria, which is the vergence angle of the two eyes when the stimulus to fusion is removed, changed throughout the day (far, $F(6, 132) = 3{\cdot}67$; $p < 0{\cdot}01$; near, $F(6, 132)$

* We observed a practice effect with participants providing more sensitive thresholds during the first 2–3 times they used this device. Consequently, we compared their results after using hard copy all day, and after using CRT all day, with those obtained at the beginning of Day 2, and still found no difference. Incidentally, this learning effect may be relevant to previous studies.

FIGURE 4. *Mean sensitivity (1/contrast) to a 2·43° flickering field at the beginning of the day, at the end of the day after proofreading hard copy, and at the end of the day after proofreading on the CRT. Beginning of the day data were taken from Day 2.*

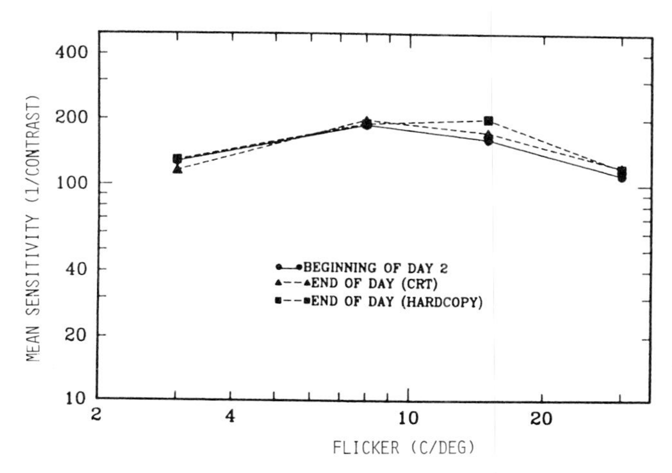

= 3·52; $p < 0{\cdot}01$). These changes were similar for hard copy and for CRT (i.e. no display mode × work period interactions). At an optical distance of 20 feet (lateral phoria far), 12 participants' phoria scores went down from the beginning to end of the day, i.e. their far non-fused vergence point moved inward, and 7 participants' phoria scores went up. At an optical distance of 14 inches (lateral phoria near), 15 participants' phoria scores went down, i.e., the near non-fused vergence point moved inward, and 4 participants' scores went up. Vertical phorias were unaffected by display or work period.

There were no significant differences between the two age groups on the measures studied.

Intercorrelation analyses showed that several measures within a class of measures (i.e. within performance or feelings or vision) correlated significantly. However, there were few correlations between measures in different classes. Results of factor analysis were consistent with this pattern of results, with ten factors having eigenvalues greater than 1·0. These results suggest that performance, feelings, and visual measures may have assessed three relatively independent classes of behavioural processes.

We measured body movements, since we felt that people might move around more when reading hard copy than when reading from a CRT. If true, this might relate to any differences in feelings of muscular fatigue. Analyses of the videotapes made at the end of each work period showed, however, that participants moved around similarly during each work period when reading hard copy and when reading from a CRT.

4. *Summary and Discussion*

In summary, the key finding is that any changes in participants' performance, feelings, and vision which occurred throughout the day were due to the work itself, and were unaffected by whether participants used a CRT in their work.

For interpreting 'no difference' results it is important to remember that one cannot prove the null hypothesis that no difference exists. One can, however, consider the validity and sensitivity of measurements made, and the consistency or error variance of the results. In the case of performance, speed and accuracy are generally accepted behavioural measures, and the units of measurement used are appropriately sensitive. The key performance result was the lack of a display mode × work period interaction. How confident can we be about this? Most participants proofread more rapidly on both hard copy and on CRT as the day went on. The difference in these rates of improvement, though small as can be seen by the relatively parallel curves of Figure 1, was about 1·5 words/min/work period more for hard copy than for CRT. With 95% confidence, the true difference between these rates of improvement is between 2·5 words/min/work period in favour of CRT and 6 words/min/work period in favour of hard copy.*

The feelings checklist contained items that are the major complaints and comments of people using CRTs in their work (Dainoff *et al.*, 1981). This suggests the checklist was valid. The five-point scale used was probably appropriately sensitive to changes that may have occurred, although a more differentiated scale might have been more sensitive. The visual tests were chosen in part because of their relevance to previous studies. We determined that the contrast sensitivity measuring device was appropriately sensitive by comparing the visual thesholds of four people when they wore lenses which slightly defocused their vision (1·25 diopter lenses) with their normal thresholds. We found about one-half log unit (i.e. a factor of three) difference in high frequency sensitivity.

It is not yet clear why participants read hard copy 20–30% faster, but the result is consistent with other work (Muter *et al.*, 1982). Body movements can be ruled out. The differences in the time to circle a misspelling on paper, versus to point at it on the screen, can account for only one-tenth of the 30% difference. The results of small follow-up experiments eliminated the possibility that it was due to reading the CRT vertically versus reading hard copy horizontally, and they seem to indicate that the difference is not greatly affected by years of experience at using CRTs. Viewing distance, and therefore visual angle, type font, and character sharpness (which affects perceived contrast) are other candidates.

4.1. Limitations

This study has limitations. It was only 2 days long. The work was not the participants' own everyday work. It was a proofreading task, and generalization of the results to other

* The 95% confidence interval was obtained by calculating the difference between the slope of each participant's hard copy curve (mean slope = 0·16) and the slope of that participant's CRT curve (mean slope = 0·13), using the means of work periods 1 and 2 and the means of work periods 5 and 6. (Twelve participants showed more improvement on CRT and 12 showed more improvement on hard copy.) The mean of these slope differences of each participant was 0·03 pages/min (S.D. = 0·197), or an improvement of about 1·5 words/min/work period more on hard copy than on CRT. The true slope difference (TSD) lies within a 95% confidence interval of about 2·5 words/min/work period more for CRT (0·013 pages/min/work period) and 6 words/min/work period more for hard copy (0·028 pages/min/work period). This is calculated by TSD = 0·03 + or − t(alpha = 0·05) × (0·197/square root of 24), where t(alpha = 0·05) = 2·069.

tasks must be thoughtful. The work was done in a friendly environment, with good ergonomics (lighting, seating, lack of glare). Good quality display and hard copy were used.

4.2. **Conclusion**

Given these limitations, we found no evidence that the CRT *itself* contributes to feelings of fatigue or affects visual functions, although it did lead to slower proofreading. Thus it appears that good quality hard copy or good quality CRT displays by *themselves* do not differentially produce 'fatiguing' effects. Two recent field studies, reported since the beginning of our experiment, support our main conclusions. One found no difference in the self-reported feelings of telephone operators who used CRTs in their work and those who looked up telephone numbers in a book (Starr *et al.*, 1982). In the other, the results of statistical analyses of questionnaire responses of several hundred government office workers led Sauter *et al.* (1983) to conclude that the CRT itself was not responsible for health complaints.

Although we were able to study a task done similarly on a CRT and on hard copy, we are aware that in practice the organization of work is usually affected by whether or not it is done on a computer. We believe that people can be placed in work environments and asked to do jobs which can lead to deleterious behavioural effects even if a good quality display is provided. Perhaps this is why most complaints of fatigue and stress about computer terminals originate where workers have little or no perceived control over their work lives (Johannsson and Aronsson, 1980). Few come from home computer owners. Ergonomics, thoughtfully designed and evaluated user interfaces, and humane management practices are required not only to arrive at exemplary workplaces, but also to achieve personally fulfilling work design and organization.*

Acknowledgements

We thankfully acknowledge assistance or suggestions from Jennifer Stolarz, Lizette Alfaro, Web Howard, Carol Thompson, Marc Donner, Clayton Lewis, Bernice Rogowitz, Bruce Rupp, Bob Mack, Jeff Kelley, and Mary Beth Rosson.

References

Brown, B.S., Dismukes, K. and Rinalducci, E.J., 1982, Video display terminals and vision of workers. Summary and overview of a symposium. Committee on Vision of the NRC. *Behaviour and Information Technology*, 1, 121–140.

Dainoff, M.J., Happ, A. and Crane, P., 1981, Visual fatigue and occupational stress in VDT operators. *Human Factors*, **23**, 421–438.

---

* These results were first reported at the Human Factors Society Meeting, Seattle, Washington, October 1982.

De Groot, J.P., 1981, Eye strain in video terminal users. Paper presented at the European Conference of Postal and Telecommunications Administrations, Symposium on Ergonomics in PTT-Administrations, The Hague, September (cited in Starr *et al.*, op. cit.).

Grandjean, E. and Vigliani, E. (eds), 1980, *Ergonomic Aspects of Visual Display Terminals* (London: Taylor & Francis), 300 pp.

Haider, M., Kundi, M. and Weissenbock, M., 1980, Worker strain related to VDUs with differently coloured characters, in *Ergonomic Aspects of Visual Display Terminals*, edited by E. Grandjean and E. Vigliani (London: Taylor & Francis) 53–64.

Johansson, G. and Aronsson, G. Stress reactions in computerized administrative work. Psychology Dept. Rept. Suppl. 5, Univ. of Stockholm, 1980.

Kucera, H. and Francis, W.W., 1967, *Computational Analysis of Present-Day American English* (Providence, Rhode Island: Brown University Press).

Matula, R.A., 1981, Effects of visual display units on the eyes: a bibliography (1972–1980). *Human Factors*, **23**, 581–586.

Mourant, R., Lakshmanan, R. and Chantadisai, R., 1981, Visual fatigue and cathode ray tube terminals. *Human Factors*, **23**, 529–540.

Muter, P., Latremouille, S.A., Treuniet, W.C. and Beam, P., 1982, Extended readings of continuous text on television screens. *Human Factors*, **24**, 501–508.

Östberg, O., 1980, Accommodation and visual fatigue in display work, in *Ergonomic Aspects of Visual Display Terminals*, edited by E. Grandjean and E. Vigliani (London: Taylor & Francis) 41–52.

Sauter, S.L., Gottlieb, M.S., Jones, K.C., Dodson, V. and Rohrer, K.M., 1983, Job and health implications of VDT use: initial results of the Wisconsin NIOSH study, *Communications of ACM*, **26**, 284–294.

Smith, W.J., Cohen, B.G. and Stammerjohn, J.W., 1981, An investigation of health complaints and job stress in video display operations. *Human Factors*, **23**, 387–400.

Starr, S.J., Thompson, C.R. and Shute, S.J., 1982, Effects of video display terminals on telephone operators. *Human Factors*, **24**, 699–713.

Pupillary Responses when Viewing Designated Locations in a VDT Workstation

H.T. Zwahlen
Department of Industrial and Systems Engineering, Ohio University, Athens, Ohio 45701, USA

Abstract

Four young and two old subjects were seated in a VDT workstation and their pupil diameter and eye fixations were measured and recorded using a Gulf & Western, Applied Science Laboratory 1998 computer controlled eye monitor system. The subjects were instructed to fixate their eyes at a designated point on the screen, the keyboard, the hard copy document and on the wall behind the VDT. Each subject made a sequence of 96 eye fixations in 2 separate trials. The results of this study indicate that the pupil diameter can vary considerably when fixating for about 5 s at a designated location. The average pupil diameters for the screen are usually slightly larger than for the other VDT locations. The pupil diameter changes during the transition from one location to another location are not excessive. Further, the overall pupil diameter averages for a given VDT location, between two trials conducted on separate days, appear to be closely the same. The two older subjects exhibited both considerably smaller pupil diameters and more blinks when compared with the four young subjects.

1. *Introduction*

A recent study (Rupp and Clauer, 1982) reported that pupillary responses when shifting the point of fixation between a source document of about 85 cd/m^2 and a VDT display screen with a background of 2·5 or 15 cd/m^2 were not excessive. A previous study (Cakir *et al.*, 1978) claimed that when changing the point of fixation between a document and a negative image display (light characters on a dark background), or vice versa, the initial pupillary response would be excessive and that the pupil would either almost completely open or close before reaching a more appropriate setting. A recent study (Zwahlen and Escontrella, 1982 and Zwahlen, 1983) where VDT operator performance, eye scanning behaviour and pupil diameter changes were measured, indicated no excessive pupil diameter changes between the transition from one VDT location to another. The objective of this research study is to use an actual VDT workstation and to determine under controlled conditions the magnitude of the pupil diameter as a function of time and VDT location, the reliability of the pupil diameter values from one trial to another trial, and the difference in the pupil diameter values between young and old subjects.

2. *Methods*

2.1. Subjects

Four young subjects (19, 19, 21, 21-years-old) and two old subjects (61 and 66-years-old) participated in the experiment during 2 different days. All subjects had normal uncorrected vision, were males, and were paid.

2.2. Apparatus

A Gulf & Western, Applied Science Laboratory 1998 computer controlled eye monitor system was used to collect and analyse the eye scanning and pupil diameter data (60 Hz sampling rate). A Digital VT 100 VDT was used. A Polaroid CP 70 filter was fitted over the VT 100 screen. Ergonomically designed furniture was used along with special lighting (Armstrong Tascon Fixture/Lens) and colour coordinated walls. The screen displayed a white character (X) in the centre on a dark background. The letter J on the keyboard served as a fixation point. A typed black letter X in the centre of a white sheet on the document holder and on the wall behind the VT 100 served as a fixation point. The average eye to screen distance was 655 mm, eye to keyboard distance 586 mm, eye to document distance 710 mm and eye to wall distance 1854 mm. The centre of the screen was 1170 mm above floor and the average subject eye height above floor was 1194 mm. The solid angle between the fixation points on the screen and document was 35°, screen and keyboard was 45°, screen and wall was 40°, keyboard and document 20°, keyboard and wall was 55°, document and wall was 32°. Luminance measurements were made for a number of selected locations in the VDT workstation. The dark screen area had 2·4 cd/m^2, the X character on the screen had 24 cd/m^2, the VT 100 frame had 32 cd/m^2 on the right side of the screen and 59 cd/m^2 below, the keyboard had 10·4 cd/m^2, the wrist rest had 108 cd/m^2, the white sheet on the slanted document holder slightly to the right of the keyboard had 141 cd/m^2, the wall to the left of the VDT had 104 cd/m^2, the wall behind the VDT had 98 cd/m^2, and the wall to the right of the VDT had 55 cd/m^2.

2.3. Procedure

Each subject was first made familiar with the equipment, the instructions and the task. Subsequently each subject was required to go through a predetermined sequence of 96 eye fixations (24 each on screen, keyboard, document and wall, 4 blocks, each 6 fix. per location, randomized) on 2 different days.

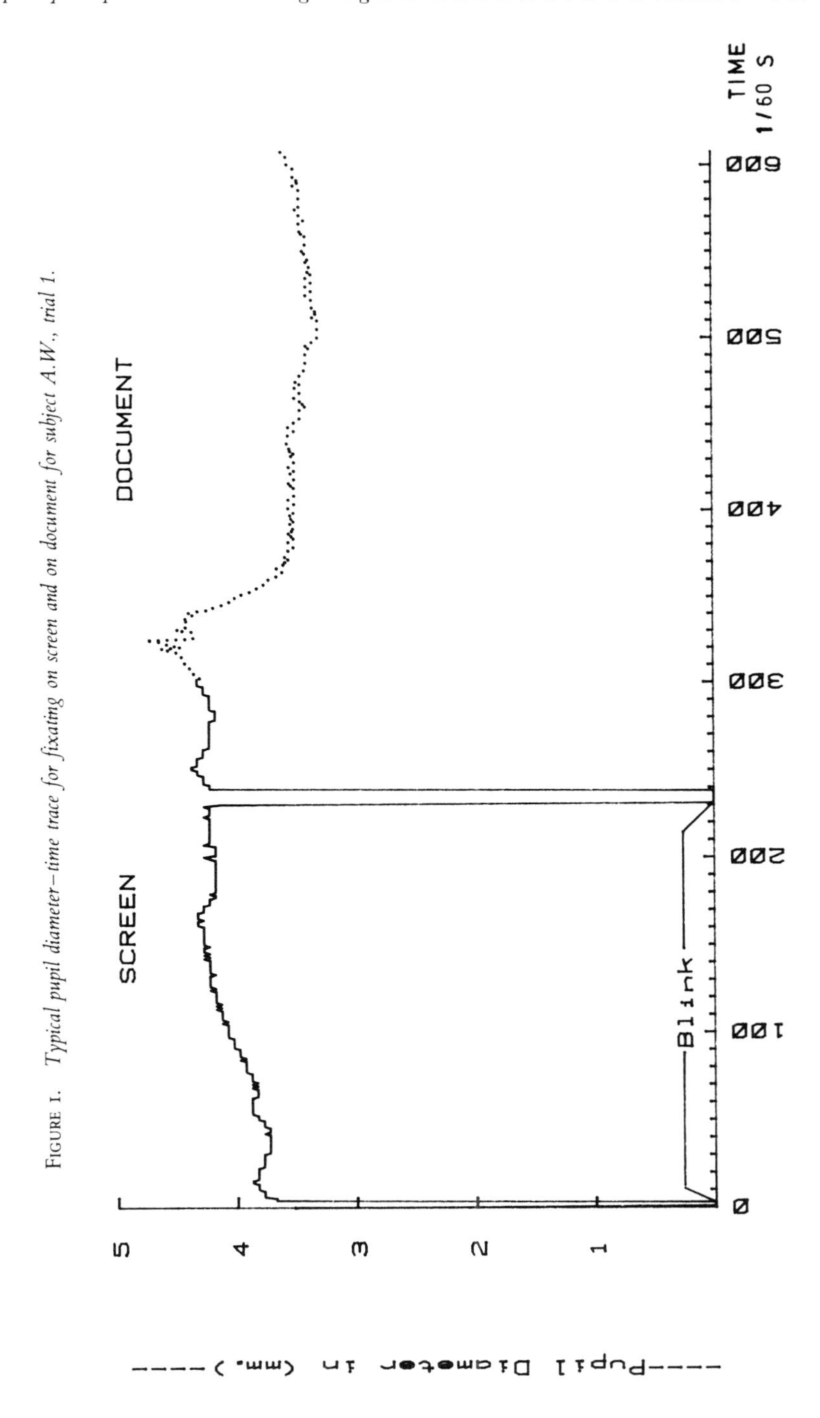

FIGURE 1. *Typical pupil diameter–time trace for fixating on screen and on document for subject A.W., trial 1.*

TABLE I. *Results for pupil diameters and eye blinks. F.LOC = fixation location, S = screen, K = keyboard, D = document, O = outside or environment.*

Subject/age/sex/ trial number	F. L O C	Total number of segments (*n*)	Average dur. of each segment (s)	Average mean pupil dia. (mm)	Average pupil s.d. (mm)	No of segments with at least one eye blink
A.H.	S	24	4·824	4·157	0·437	14
19 yrs	K	24	4·337	3·672	0·340	13
male	D	24	4·910	3·538	0·312	07
trial 1	O	24	4·760	3·347	0·347	18
A.H.	S	24	4·849	4·202	0·386	15
19 yrs	K	24	4·381	3·670	0·322	12
male	D	24	4·914	3·556	0·326	08
trial 2	O	24	4·785	3·396	0·355	16
A.W.	S	24	4·817	3·878	0·453	16
21 yrs	K	20	4·482	3·397	0·359	13
male	D	21	4·708	3·531	0·388	18
trial 1	O	23	4·743	3·464	0·335	18
E.H.	S	21	4·682	3·871	0·324	16
21 yrs	K	20	4·288	3·397	0·323	16
male	D	21	4·648	3·509	0·282	15
trial 1	O	22	4·684	3·680	0·324	19
E.H.	S	22	4·702	3·774	0·306	14
21 yrs	K	21	4·311	3·468	0·299	18
male	D	21	4·690	3·488	0·296	18
trial 2	O	21	4·703	3·596	0·307	17
H.Z.	S	24	4·413	3·526	0·357	15
19 yrs	K	21	4·025	3·409	0·336	15
male	D	22	4·612	3·348	0·314	10
trial 1	O	22	4·334	3·314	0·306	19
H.Z.	S	23	4·491	3·627	0·359	17
19 yrs	K	20	4·112	3·321	0·326	15
male	D	20	4·321	3·350	0·306	17
trial 2	O	21	4·635	3·426	0·326	09
G.C.	S	23	4·175	2·715	0·247	22
61 yrs	K	21	3·452	2·560	0·249	21
male	D	22	3·741	2·614	0·269	22
trial 1	O	20	4·017	2·541	0·241	20
G.C.	S	23	4·209	2·697	0·257	21
61 yrs	K	21	3·502	2·559	0·267	21
male	D	22	3·781	2·618	0·273	22
trial 2	O	20	4·170	2·559	0·257	20
T.S.	S	24	4·135	2·686	0·225	23
66 yrs	K	20	3·309	2·592	0·246	20
male	D	22	3·505	2·651	0·244	22
trial 1	O	20	3·754	2·569	0·267	20
T.S.	S	24	4·223	2·679	0·228	23
66 yrs	K	20	3·498	2·621	0·257	20
male	D	22	3·677	2·674	0·251	22
trial 2	O	21	3·922	2·611	0·285	18

FIGURE 2. *Typical cumulative distributions for mean pupil diameters for fixations on screen (SCR), keyboard (KEY), document (DOC), and wall (OUT) for a young subject (A.H.) and an old subject (G.C.) for trial 1 and trial 2.*

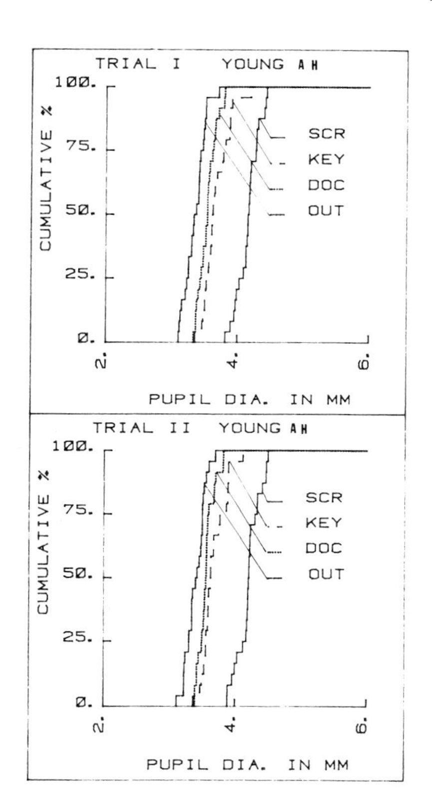

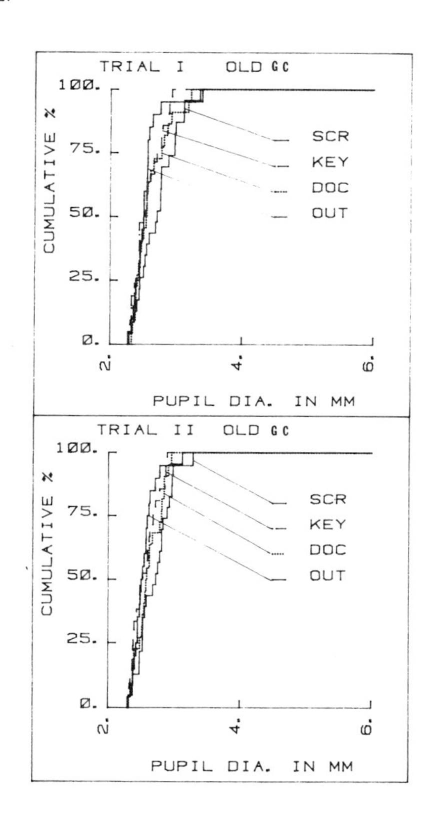

3. *Results*

Figure 1 shows a typical pupil diameter–time trace. Table 1 shows summary results for pupil diameters and eye blinks. Figure 2 shows the cumulative distributions for the average pupil diameters per fixation for the screen, the keyboard, the document, and for the wall (outside) for a young and an old subject (A.H. and G.C.) for trial 1 and for trial 2.

4. *Discussion*

The results of this study support the findings reported by Rupp and Clauer (1982). This study was conducted in an actual VDT workstation and does provide detailed quantitative pupil diameter values when fixation on the screen, the keyboard, the document, and the wall behind the VDT. The pupil responses during transitions from one VDT location to another are not excessive and there is no evidence that the pupil does overcompensate (dilates initially too much, or contracts initially too much) during the transition from a white sheet to a rather dark screen, or vice versa, as was claimed by Cakir *et al.* (1978). The results of this study also indicate that the pupil diameter is sensitive to the luminance level in the central visual field since the average mean pupil diameters are always slightly larger for the screen than for the other VDT locations (especially for young subjects). The average coefficient of variation for the mean pupil diameter within a fixation (which may include some adjustment activity from the preceding fixation) is about 10% suggesting that the pupil diameter does fluctuate considerably without a change in the luminance level of the visual field. The differences between the average minimum pupil diameters and the average maximum pupil diameters for a given VDT location are about 1 mm (higher for younger subjects) suggesting that the pupil area can easily change between 50% and 120% within a single fixation of about 5 s duration. Whether or not the pupil diameter fluctuations observed in this study could produced noticeable fatigue of the ciliary muscles (especially when working on a VDT for a whole day) and/or could be responsible for other VDT operator complaints such as an increased sensitivity to glare (Cakir *et al.*, 1978) cannot be answered. The two older subjects exhibited both considerably smaller pupil diameters and more eye blinks when compared with the four young subjects. Looking at Figure 2 one can observe that the cumulative distributions for the mean pupil diameters for the four VDT locations for the old subject are not only located at smaller values but also relatively close together when compared with the young subject. There appears to exist a relatively high consistency between the trial 1 and trial 2 cumulative distributions for both the young and the old subject.

Acknowledgement

The author wishes to acknowledge the contributions of Sudhakar L. Rangarajulu, graduate student, who was involved in the data collection and data analysis.

References

Cakir, A., Reuter, H.J., von Schmude, L. and Armbruster, A., 1978, Untersuchungen zur Anpassung von Bildschirmarbeitsplaetzen an die physische und psychische Funktionsweise des Menschen, Forschungsbericht April 1978, Der Bundesminister fuer Arbeit und Sozialordnung, Bonn, FRG.

Rupp, B.A. and Clauer, C.K., 1982, Pupillary responses as a source of stress during display viewing, paper presented together with 'Cognitive complexity related to image polarity in the etiology of visual fatigue', by Taylor, McVey and Emmons, at the IX International Ergophthalmological Symposium in San Francisco, October, 1982.

Zwahlen, H.T., 1983, Measurement of VDT operator performance, eye scanning behavior and pupil diameter changes, in *Proceedings of the 27th Annual Meeting of the Human Factors Society*, October 1983, Norfolk Virginia, Vol. 2, pp.723–727.

Zwahlen, H.T. and Escontrella, L.M., 1982, *Measurement of VDT Operator Performance, Eye Scanning Behavior and Pupil Diameter Changes*. Ohio University Research Laboratory Report, prepared for NIOSH, December, 222 p.

The Effects of Visual Ergonomics and Visual Performance upon Ocular Symptoms During VDT Work

R. BELLUCCI
Department of Ophthalmology, University of Verona, Italy
AND F. MAULI
INPS Medical Service and USL 25 Work Health Organization, Verona, Italy

Abstract

Work with VDTs is said to be annoying for the eyes of the operators, often causing ocular symptoms. Our aim was to check the relative importance of bad visual performance and bad visual ergonomics in the origin of these symptoms.

To this end, we asked 164 operators to reply to a specially designed questionnaire and compared their answers with the visual ergonomics of their VDT workplaces and with the results of ophthalmic examinations. In a group of these subjects we also considered the ocular symptoms reported two years before under worse visual ergonomic conditions. Psychological, postural and microclimatic aspects were not neglected.

The results seem to show that ocular symptoms are more related to bad VDT visual ergonomics than to bad visual performance of the operator, and confirm that factors other than visual ones are involved. The possible causes of these findings are briefly discussed.

1. *Introduction*

In most cases the introduction of VDTs into offices has led to numerous complaints from the office staff, and the conviction has gained ground that VDTs should only be used by subjects who have no visual defects of any kind. It is a well known fact that poor or bad visual functions may substantially increase visual discomfort when attempting a task which makes high demands at the eye.

Visual fatigue manifests itself subjectively particularly in the form of complaints about ocular disturbances. Fatigue, however, is not the only cause. The eye is an organ that serves for the release of problems and tensions of other origins. Eye symptoms are thus of multifactorial origin, and it is only by attacking the causes that these can be reduced to a minimum. We have still to define, in the VDT work environment, whether it is more important to study the visual ergonomics of the equipment more thoroughly, or work organization, or whether we should insist on careful selection of the operators.

In this regard, ophthalmologists generally propose increasingly high minimum standards of visual performance, such that subjects with corrected visual acuity only slightly less than 20/20 are normally judged unfit. In view of the high social, economic and organizational cost of such a policy, we were prompted to verify its validity, on the basis of our experience.

2. *Methods*

We therefore carried out a study on 164 VDT operators, working for various different companies, studied by our team over the past two years. A comparison was made between two work environments, with different ergonomic characteristics, occupied in successive periods by the same operators. The subjective symptoms were recorded during an interview, conducted on the basis of a questionnaire partly made up of multiple-choice questions. We used as controls groups of non-VDT office staff, working for the same companies and well matched regarding age and duties.

3. *Results*

Table 1 gives a summary of the results. As can be seen greater frequency, intensity and duration and earlier onset of eye troubles were observed in the case of VDT operators compared to the controls. In no case, in our research, did the symptoms of a group of VDT operators prove less than those of the corresponding non-VDT staff. These data confirm the high degree of ocular discomfort in VDT work.

3.1. **Ocular Symptoms and Age**

We divided the subjects into two groups, the over and under forties. In our series, the comparison with ocular symptoms proves inconclusive as the statistical analysis shows no influence of age. It should, however, be noted that this finding was certainly conditioned to a substantial extent by the absence of operators over 50. It is in fact such operators who present greater problems in VDT work.

3.2. **Ocular Symptoms and Visual Performance**

The operators had not been examined prior to taking up VDT work. All subjects were examined thoroughly at the university eye clinic, using routine clinical tests. Additional

TABLE I. *Ocular symptoms in the subjects studied.*

Ocular complaints	VDT (164) (%)	Controls (146) (%)
Burning, irritated eyes	66	11
Eye-strain	58	4
Blurred vision	45	5
Visual impairments	3	–
Red eyes	2	1
Photophobia	2	–
Onset		
Immediate	4	–
After 1 hour	27	9
After 2 hours	41	18
After 3 hours	15	36
After 4 hours or more	12	36
Duration		
Less than one work shift	7	25
Till the end of the shift	82	75
Longer than one work shift	11	–

tests were performed to ascertain fusional and accommodative amplitudes for the operating distance, as measured with the spectacle correction habitually used.

On the basis of the examination the subjects were divided into two categories:

1. Those with good visual performance, suitable for any type of visual task with standard predictable visual fatigue.
2. Those with poor or bad performance, suggesting an early onset of fatigue and of asthenopeic disturbances when faced with a difficult visual task.

This subdivision took no account of the quantitative visual demands of the VDT work, which in our cases were not high and were coped with by the operators with only a few exceptions. We adopted a stricter ophthalmological criterion based on clinical and ambulatory practice.

On comparing the ocular disturbances in the two groups, no significant differences were detected. Our findings are partly different from those of other published studies, and would appear to indicate that VDT induced visual discomfort is fairly independent on the visual performance of the operators, as analysed in our study. The importance of severe aptitude tests and selection is thus diminished. Good visual function would not appear to be enough 'in itself' to ensure minimum ocular discomfort, which must therefore be largely related to other factors.

3.3. **Ocular Symptoms and Visual Task**

When considering the duties of our operators, we find that a number of them usually perform data entry tasks, where the screen is rarely viewed or, at most, only for brief periods.

Others, however, have to enter data in a mask, which when completed flashes off and on at the moment of entry. In this case the screen is subject to greater scrutiny, with more frequent shifting of the gaze from screen to document and vice versa. Yet other operators have conversation tasks involving frequent fixation of the screen albeit for brief periods. Then there are others who correct a previously photocomposed text recalled on to the screen, which is viewed continuously. Most operators perform two or more types of task alternately.

TABLE 2. *Ocular symptoms of the operators compared with their task* ($p < 0{\cdot}01$).

	Type of task				
Ocular symptoms	Data entry	D.e. + mask	Conversation	Control	Total
None	40	39	48	27	154
Moderate	18	24	40	27	109
Severe	1	15	12	5	33
	59	78	100	59	296

Table 2 compares ocular symptoms and visual task and shows that the most tiring task is the checking of a text, while the least fatiguing is data entry. The differences are statistically significant and suggest that ocular symptoms are strictly related to fatigue in general and partly depend on protracted viewing of the screen. Here we are faced with that part of visual discomfort defined as 'necessary', which can only be reduced by improving upon existing technology.

3.4. **Ocular Symptoms and Visual Ergonomics**

The various different work environments we visited were classified on the basis of an assessment of the various aspects of visual ergonomics. We took into account the shape, dimensions and position of the rooms and windows, the VDT lay-out, the lighting, the presence of reflections or glare, the colours of visible surfaces, etc. which is usual in such studies.

This enabled us to divide the workplaces into two categories:

1. Those with good visual ergonomic conditions (2 work environments).
2. Those with poor or bad visual ergonomic conditions (3 work environments).

The importance of these elements in the genesis of discomfort and disturbances is generally held to be very substantial, and this is also the case in our studies. Table 3 shows that it is precisely the comparison between ocular symptoms and visual ergonomics which yields the highest significance test value. Further confirmation is provided by the reduction in symptoms observed in a group of operators on transferring to a new workplace with better visual ergonomics, other conditions being equal (Table 4). The subjective improvement appears, indeed, to be very substantial, inasmuch as the incidence of ocular symptoms is down from 80% to 30%.

TABLE 3. *Ocular symptoms of the operators compared with the visual ergonomics of their VDT workplaces* ($p < 0{\cdot}001$).

Ocular symptoms	Visual Ergonomics: Good	Visual Ergonomics: Poor or bad	Total
None	66	21	87
Moderate	31	25	56
Severe	5	16	21
	102	62	164

TABLE 4. *Effect of better visual ergonomics on ocular symptoms* ($p < 0{\cdot}01$).

Ocular symptoms	Visual Ergonomics: Old (poor)	Visual Ergonomics: New (good)	Total
None	5	20	25
Moderate	15	5	20
Severe	10	5	15
	30	30	60

3.5. Ocular Symptoms and General Ergonomics

As the eye is an organ for disturbances of various aetiology, we must also lastly consider elements of general ergonomics, i.e. work organization, relations with management, air conditioning, posture, etc. As before, we assessed the results as good or poor to bad. The fact that the statistical analysis of the data in Table 5 shows a high degree of significance confirms the influence of non-visual factors on the ocular comfort of VDT operators, as reported by various authors.

To achieve maximum levels of comfort, including visual comfort, we need therefore measures dealing with all the various aspects of the ergonomics of VDT activity.

4. *Discussion*

In the operators we studied, the ocular symptoms proved to be related primarily to ergonomic factors, especially of the visual type, and secondarily to the type of task. There was no correlation with the visual performance of the operators.

TABLE 5. *Ocular symptoms of the operators compared with the general ergonomics of their VDT workplaces* ($p < 0{\cdot}01$).

	General Ergonomics		
Ocular symptoms	Good	Poor or bad	Total
None	40	47	87
Moderate	18	38	56
Severe	2	19	21
	60	104	164

Our study confirms the close relationship between general and ocular fatigue, and in addition underlines the substantial influence of non-visual factors on ocular symptoms. The difficulties are considerably accentuated by bad ergonomics and particularly by bad visual ergonomics, which would appear to be the most important element in the genesis of the disturbances experienced. This does not mean that aptitude tests are useless or that subjects with poor visual performance will never have problems with VDT work. It would simply not appear to be advisable to insist on perfection of ocular functions in assessing the suitability of people for this type of work.

We believe tht it would be more worthwhile to devote further study and research to the continuous improvement of ergonomic conditions, especially visual, as well as to making technological advances in the field of tasks and VDT equipment.

References

Anfossi, D.G., Maina, S., Grignolo, F.M., Romano, L. and Sonnino, A., 1982, Aspetti oftalmologici ed ergonomici negli addetti ai video terminali. *Giornale italiano di oftalmologia occupazionale*, Vol. I.

Bonomi, L. and Bellucci, R., 1982, Problemi ergooftalmologici connessi al lavoro con video terminali. *Bollettino di Oculistica*, **61**, 9–19.

Cakir, A., Hart, D.J. and Stewart, T.F.M., 1980, *Visual Display Terminals* (Chichester: John Wiley).

Dubois-Poulsen, A., 1969, La fatigue visuelle. *Ophthalmologica*, **158**, 157–180.

Grandjean, E. and Vigliani, E., (eds), 1980, *Ergonomic Aspects of Visual Display Terminals* (London: Taylor & Francis).

Verriest, G. and Hermans, G., 1975, Les aptitudes visuelles professionnelles. *Bulletin de la Société Belgique d'Ophtalmologie*, supplement n. 1.

Changes in Saccadic Eye Movement Parameters Following Prolonged VDU Viewing

E.D. MEGAW AND T. SEN
Department of Engineering Production, University of Birmingham,
Birmingham B15 2TT, England

Abstract

An attempt was made to identify objective measures of visual fatigue by quantifying a variety of parameters reflecting the behaviour of the saccadic eye movement control system while subjects performed a number comparison task. The fatigue was induced by two methods. The first required subjects to continuously perform over one hour a Neisser search task displayed on a VDU. The other required subjects to spend over 60% of their working day at a VDU. Although several significant changes in the estimates of the eye movement parameters were observed following the fatiguing task, they were comparatively small changes and were not always in the expected direction. Generally, there were large individual differences in the responses of the subjects to the fatiguing tasks.

1. *Introduction*

Many recent studies, reviewed by Dainoff (1982), have reported that operators complain of symptoms of visual fatigue following prolonged VDU viewing. Although a satisfactory definition of visual fatigue remains as elusive as ever, these reported complaints have precipitated a revival of interest into the development of objective measures of fatigue. Many of the reported symptoms, such as double vision and aching behind the eyes, suggest that visual fatigue can be investigated in the context of a temporary reduction in function of the ocular-motor control systems. Because of the problems arising from the close interdependence of the accommodation, vergence and pupil control systems, some studies have chosen to investigate changes in saccadic eye movement parameters. As early as 1917, Dodge demonstrated a general reduction in the velocity of saccades as a result of fatigue. More recently, Bahill and Stark (1975) reported both a reduction in the peak velocity of saccades and an increase in the frequency of overlapping saccades with the onset of fatigue. These were identified by changes in the main sequence diagrams which plot either peak velocity or duration against amplitude of movement. On the other hand these results were not confirmed by Schmidt *et al.* (1979). Bahill and Stark also reported an increase in the frequency of glissades with fatigue. A further parameter that might reflect fatigue is dynamic overshoot (Bahill *et al.* 1975). Finally, if fatigue affects the accuracy of the step component leading to overshoot or undershoot of the main saccade, an increase in the frequency of execution of corrective saccades would be expected.

A major criticism of all the studies on fatigue of the saccadic control system is that the fatigue was induced by having subjects perform a continuous series of large amplitude horizontal saccades with an emphasis on accurate target acquisition. This contrasts to practical tasks where saccades greater than 15° are rarely executed and where there is a large cognitive element. The two experiments reported in this study attempted to induce visual fatigue by having operators perform normal VDU tasks although changes in saccadic parameters were evaluated from the subjects peforming blocks of number comparison trials which only required the execution of horizontal saccades. It was hypothesized that prolonged VDU viewing would lead to a reduction in saccadic peak velocity, a lengthening of saccadic duration and increases in the frequencies of overlapping saccades, glissades, dynamic overshoot and corrective saccades.

2. *Method*

Eye movements were recorded while subjects performed blocks of number comparison trials displayed on a VDU with a P31 phosphor and a screen luminance of 0·24 cd/m^2. On each trial a pair of 3-digit number strings was displayed, one either side of a fixation point. The separation of the middle digits of the two strings varied randomly between 9° and 16°. On half the trials the strings differed by one digit and on the other half they were the same. Subjects had to decide as quickly as possible whether or not they were the same. Typically this required the execution of 2 or 3 main saccades. To prevent subjects developing a strategy of always executing their first saccade to either the left or right string, the onset of the two strings was separated by 100 ms and the presentation of the first string was randomly assigned to the left or right position. Head movements were discouraged by providing a chin support and a head rest, giving a viewing distance of 30 cm.

2.1. **Eye Movement Recording and Analysis**

These have been detailed elsewhere (Megaw and Sen, 1983). Horizontal eye movements were recorded separately for each eye using standard infrared photoelectric techniques. Pairs of emitters and receivers were mounted on a pair of trial frames. The outputs were amplified and sampled at 1000 Hz by a 12-bit A–D converter and then stored on an Altos 8000–2 microcomputer. A calibration task was presented before each block of 10 trials and required subjects to perform a self-generated target pursuit tracking task.

The stored data were filtered and analysed off-line. To register a saccade, the velocity had to remain above 20° s for longer than 10 ms. The same routine was used to identify corrective saccades and dynamic overshoot. To register a glissade, the velocity had to remain above 2°/s for longer than 30 ms and had to start within 25 ms from the end of the preceding saccade. To obtain estimates of peak velocity and duration of the main saccades from any block of trials or combination of blocks, two sets of linear regressions were performed. The first was between log peak velocity and log amplitude values and the other was between duration and amplitude. From these, estimates of the peak velocities and durations of 5°, 10° and 15° saccades were obtained along with 95% confidence limits.

2.2. **Experiment 1**

Five subjects with uncorrected normal near vision performed a Neisser search task continuously for one hour. The task was generated by the micro and displayed on the VDU. During this time the subjects searched between 84 and 130 pages of letter strings. They were free to move their heads. Eye movements were recorded in the manner outlined above during two blocks of number comparison trials given immediately before and after the Neisser search task.

2.2.1. *Results*

Figure 1 illustrates the relationship between peak velocity and amplitude for the main saccades before and after completing the one hr search task. Although the results are in the

FIGURE 1. *Estimates of peak velocity as a function of saccadic amplitude before and after completing the one hr search task. Results are shown for the left eye and have been averaged over the 5 subjects.*

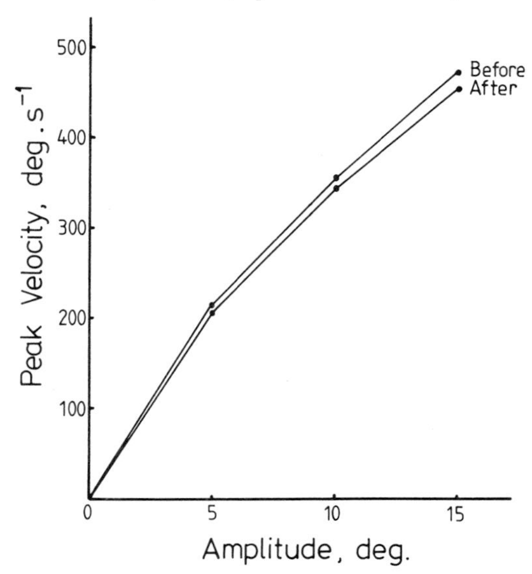

predicted direction, the differences are not significant. On the other hand, for 3 subjects the variability in the peak velocity for 15° saccades were significantly greater after the search task ($p < 0.05$). The pattern of results on the probabilities of dynamic overshoot, of glissadic overshoot and undershoot and of corrective saccades was very idiosyncratic. One subject showed a significant increase in probabilities for all the measures while another subject showed a decrease in all of them. Two subjects showed no significant effects of the search task. A close look at individual records failed to identify one example of an overlapping saccade of the type described by Bahill and Stark (1975). All the subjects reported an increase in symptoms of fatigue following the search task, mostly related to general fatigue such as overtiredness, boredom and difficulties in maintaining attention.

2.3. **Experiment 2**

Two experienced secretaries with normal uncorrect near vision were tested on two blocks of number comparison trials at four different times of the working day, 09.00, 11.30, 14.00 and 16.30. They were instructed to spend at least 60% of their time working at a VDU.

2.3.1. *Results*

Figure 2 shows how the peak velocities varied over the day for the 3 amplitudes of movement. The results for subject AI are in the predicted direction with the velocities for the larger movements decreasing during the morning, recovering following the lunch break and decreasing again during the afternoon. For the other subject the estimates remain comparatively constant through the day although there is a slight decrease over the afternoon. For subject AI, contrary to predictions, the probabilities of dynamic overshoot and glissadic overshoot and undershoot were higher at 09.00 and 14.00. For the other subject there was a general decline in the probabilities over the day. The levels of complaints of symptoms of fatigue were much lower than those reported in the first experiment.

3. *Discussion*

The effects of the fatiguing task on saccadic peak velocity were small. In the first experiment, the variability for large amplitude saccades was significantly increased following the fatiguing task for some subjects. One of the two subjects from the second experiment showed a pattern of results over the working day which would be expected if excessive VDU viewing did fatigue the saccadic control system. However, the effects on peak velocity were relatively small compared to the reported effects of drugs such as alcohol and diazepam. Similarly, the effects on the probabilities of dynamic overshoot, glissades and corrective saccades following the fatiguing tasks were not uniform, some subjects showing an increase in all or some of these probabilities while others showed decreases or no changes.

FIGURE 2. *Estimates of peak velocity as a function of time of day for two experienced secretaries. Results are from the left eye.*

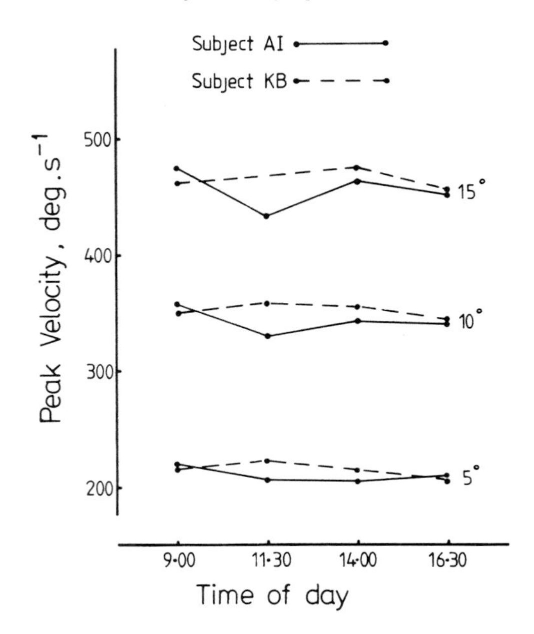

Had there been evidence of large uniform changes in any of the parameters, there would have been grounds for including control groups of subjects. In this study, random fluctuations could be assessed by comparing estimates from the two consecutively presented blocks of number comparison trials. Generally, these fluctuations were smaller than the changes reported in the results section. However, there still remains the possibility that the observed changes reflected factors such as arousal and boredom rather than visual fatigue *per se*.

Finally, it should be said that, unlike in the case of accommodation control, there is no *a priori* reason why the saccadic eye movement system should be fatigued by prolonged VDU viewing, unless possibly indirectly by fatigue to the vergence system. It could be argued that despite the complaints of the subjects, the task demands were insufficient to fatigue the saccadic system. For example, Haider *et al.* (1980) found that temporary myopisation occurred only after 3 hours of continuous VDU viewing.

References

Bahill, A.T., Clark, M.R. and Stark, L., 1975, Dynamic overshoot in saccadic eye movements is caused by neurological control signal reversals. *Experimental Neurology*, **48**, 107–122.

Bahill, A.T. and Stark, L., 1975, Overlapping saccades and glissades are produced by fatigue in the saccadic eye movement system. *Experimental Neurology*, **48**, 95–106.

Dainoff, M., 1982, Occupational stress factors in visual display terminal (VDT) operation. *Behaviour and Information Technology*, **1**, 141–176.

Dodge, R., 1917, The laws of relative fatigue. *Psychological Review*, **24**, 89–113.

Haider, M., Kundi, M. and Weissenbock, M., 1980, Worker strain related to VDUs with differently coloured character. In *Ergonomic Aspects of Visual Display Terminals*, edited by E. Grandjean and E. Vigliani (London: Taylor & Francis) pp.53–64.

Megaw, E.D. and Sen, T., 1983, Visual fatigue and saccadic eye movement parameters. *Proceedings of the 27th Annual Meeting of the Human Factors Society, Norfolk*, 728–732.

Schmidt, D., Abel, L.A., Dell'Osso, L.F. and Daroff, R.B., 1979, Saccadic velocity characteristics: intrinsic variability and fatigue. *Aviation, Space and Environmental Medicine*, **50**, 393–395.

The Effects of Various Refresh Rates in Positive and Negative Displays (Introductory paper)

S. Gyr, K. Nishiyama, R. Gierer, T. Läubli and E. Grandjean
Department of Hygiene and Ergonomics, ETH-Center, CH–8092 Zurich, Switzerland

1. *Introduction*

For most VDTs the refresh rates are 50 Hz in Europe and 60 Hz in USA and Japan.

At present little is known about the effect of the frequency of oscillating luminance on visual functions and subjective feeling.

The present study deals with the effect of various refresh rates on visual functions and subjective feeling; it is a summary of two studies described elsewhere in detail (Nishiyama *et al.*, 1984; Läubli *et al.*, 1984).

2. *Methods*

An apparatus generating oscillating luminances of various frequencies on a screen was developed. The apparatus is described elsewhere by Nishiyama *et al.* (1984). The apparatus is based on the principle of a light ray projected on a screen of 170 × 210 mm^2. Chopper discs generate oscillating luminances between 0 and 180 Hz. With a modified slide projector a text with dark characters on a bright screen or with bright characters on a dark screen can be produced. The oscillating luminances have a decay time of about 4 ms to drop to 10% of the peak luminances.

Thirty subjects were engaged in the bright screen and 29 in the dark screen (bright characters) experiments. The subjects had to read aloud the text from the screen. The results of the reading task lasting one hour are presented here.

Before and after the reading task the following visual functions were measured: critical fusion frequency (CFF), near point distance, visual acuity (near and far) and lateral heterophoria (near and far). Furthermore, the subjective feelings were recorded in questionnaires; in the bright screen experiments a semantic seven step scale was used and in the dark screen experiments a non-semantic stepless scale.

Five frequencies of oscillating luminance and a printed text, which are indicated in Figure 1, were used in the experiments.

3. *Results*

Figure 1 shows the result of the CFF for the bright and dark screen experiments. The ordinate indicates the mean shift of the CFF after a reading time of one hour and the abscissa indicates the six experimental conditions. This figure reveals—for the bright as well as for the dark screen:

1. A significant decrease for each experimental condition. For the bright and the dark screen the decrease is clearly at its highest level for the 30 Hz condition.
2. A significant difference between the decrease for the 0 Hz condition and the ones for the 30 and 60 Hz conditions; the decrease is higher for the 30 and 60 Hz conditions than for the 0 Hz condition.

There is no significant difference between the bright screen and the dark screen experiments.

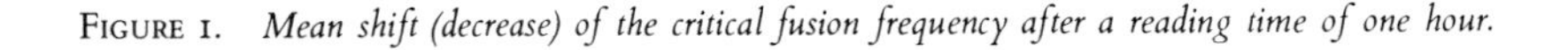

FIGURE I. *Mean shift (decrease) of the critical fusion frequency after a reading time of one hour.*

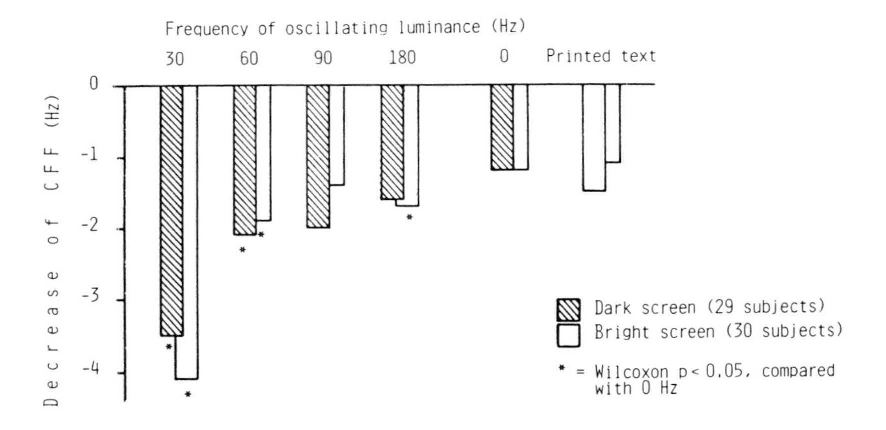

Figure 2 shows for the symptom 'I have tired eyes' the mean shift for each experimental condition. This figure reveals a significant increase for the bright and dark screen for each experimental condition and a significant difference between the increase for the 0 Hz condition and the one for the 30 Hz condition; the decrease is higher for the 30 Hz condition than for the 0 Hz condition.

Figure 3 shows the results of the various items related to possible visual complaints. This figure reveals a similar result for the bright and dark screen: the highest increase is in general observed for the 30 Hz condition and the lowest one for the printed text.

For the ten items the following index was calculated: number of increased symptoms (items) divided by the total number of items.

Figure 4 shows a significant higher index value for the dark screen for the 30 and 60 Hz conditions than for the 0 Hz condition.

FIGURE 2. *Mean shift (increase) of the symptom 'I have tired eyes' after a reading time of one hour.*

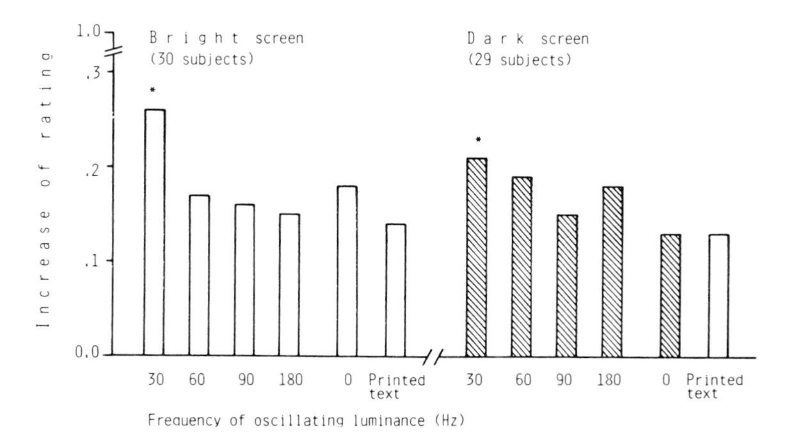

FIGURE 3. *Mean shift (increase) of ten symptoms related to subjective feeling after a reading time of one hour.*

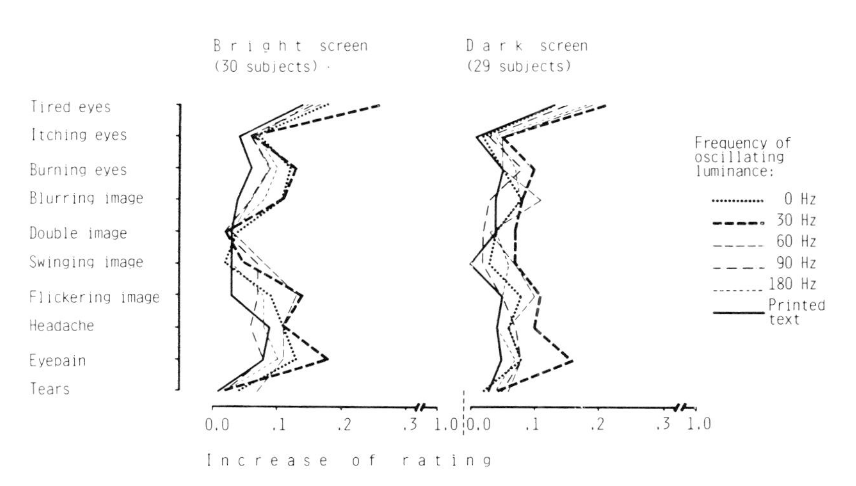

The significant changes at the 30 and 60 Hz conditions compared with the 0 Hz condition are recorded in Table 1. The results do not reveal a clear picture except that they show a higher incidence of eye troubles with the dark than with the bright screen but this difference cannot be considered as relevant.

FIGURE 4. *Mean index of ten symptoms related to subjective feeling after a reading time of one hour.*

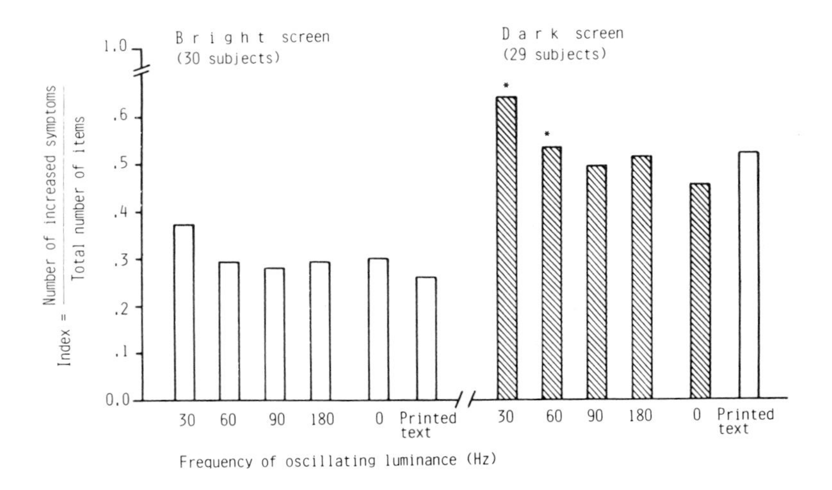

TABLE I. *Variables showing a significant (Wilcoxon $p < 0{\cdot}05$) higher decrease (dec) respectively increase (inc) for the 30 and 60 Hz conditions than for the 0 Hz condition after a reading time of one hour.*

Screen	30 Hz Bright[a]	30 Hz Dark[a]	60 Hz Bright	60 Hz Dark
Critical fusion frequency	dec	dec	dec	dec
Near point distance			inc	
Visual acuity near	dec			
Tired eyes	inc	inc		
Itching eyes				inc
Headache		inc		
Eyepain		inc		
Index (subjective feeling)		inc		inc

[a] Bright and dark are related to the background

4. *Discussion*

The results of the present study reveal a higher decrease of the CFF figures and a slightly higher increase of some symptoms related to subjective feeling for the 30 and, to a smaller extent, for the 60 Hz conditions than for the 0 Hz condition. It can be deduced that oscillating luminances with 30 Hz and, to a smaller extent those with 60 Hz are associated with a certain amount of eye-strain. The results, however, give no clear picture of the nature of this strain. In this relation a result of 39 subjects engaged in the study with the dark screen might be of interest: the CFF of the bright characters was determined when subjects focused on the screen centre and when looking at an edge of the screen. The mean CFF was 48 respectively 49 Hz. For eight subjects the CFF was within 50 and 60 Hz and one subject had a threshold above 60 Hz. Bauer *et al.* (1983) determined the CFF on 31 subjects exposed to a bright screen. The measured CFF lay between about 55 and 85 Hz. Another recent publication of Isensee and Bennett (1983) reveals that subjects viewing a bright screen with a 60 Hz oscillation notice a higher sensitivity to flicker than when viewing a dark screen (bright characters).

Taking the results of the present study as well as those mentioned in the literature it is concluded that VDTs using phosphors with usual decay times should be provided with refresh rates of more than 60 Hz. VDTs with dark characters on a bright screen should be equipped with a refresh rate of at least 100 Hz.

Acknowledgement

The research reported in this paper has been supported by the Swiss National Science Foundation, Grant No. 3.810–0.81/3.810–1.81.

References

Bauer, D., Bonacker, M. and Cavonius, C.R., 1983, Frame repetition rate for flicker-free viewing of bright VDT screens. *Display*, January, 31–33.

Isensee, S.H. and Bennett, C.A., 1983, The perception of flicker and glare on computer CRT displays. *Human Factors*, 25(2), 177–184.

Läubli, T., Gyr, S., Nishyama, K., Gierer, R., Bräuninger, U. and Grandjean, E., 1984, Effects of various refresh rates in a VDT reading task (bright characters on dark screen), in the press.

Nishiyama, K., Bräuninger, U., de Boer, Hildeke, Gierer, R. and Grandjean, E., 1984, Physiological effects of oscillating luminances in the reversed display of subjects, in the press.

Causes of Flicker at VDUs with Bright Background and Ways of Eliminating Interference

D. Bauer
Institut für Arbeitsphysiologie an der Universität Dortmund, Ardeystrasse 67, 4600 Dortmund 1, FR Germany

Abstract

Flicker perception may occur when temporally modulated signals stimulate the retina. Three sources of flicker are particularly troublesome when VDUs with bright screens are used:

1. Local oscillation of lines or characters.
2. Pure temporal variations of local luminance distributions.
3. Eye-movement induced modulation of retinal illuminance.

All three classes of flicker may be reduced or eliminated by proper selection of equipment parameters. The most important of these are low jitter, an adequate frame repetition rate and reducing interactions between the screen and its surround.

1. *Introduction*

VDUs with bright backgrounds have some advantages over those with dark backgrounds. In 1980 we showed (Bauer *et al.* 1980) that for difficult visual tasks dark characters on a light background are better, whether subjectively or objectively tested. In 1983 we measured (Bauer *et al.* 1983) this by a presentation of dark characters on a 100 cd/m^2 bright background disturbing reflections can be suppressed to a degree which is adequate for office applications.

These and other advantages may be lost, however, if

1. An image repetition rate that is too low, causes (large field) flicker.
2. Other faults in the technical construction—as, for example, jitter—cause apparent movement or line flicker.
3. Wrong design of the border of the screen in connection with a too low image repetition frequency causes border flicker effects. These 3 points will be considered below.

2. *Method*

2.1. **Image Repetition Frequency**

Bauer *et al.* (1983) measured the critical flicker frequencies (CFFs) of 30 observers for raster screens without interlacing and with fast phosphor (P104). The distribution of CFF-values for a mean luminance of 80 cd/m^2 is shown in Figure 1. The distribution may be approximated by a Gaussian with a mean of 73 Hz.

FIGURE 1. *Distribution of CFF-values for 30 observers. 1 cross = 1 observer-CFF, dashed line: gaussian approximation.*

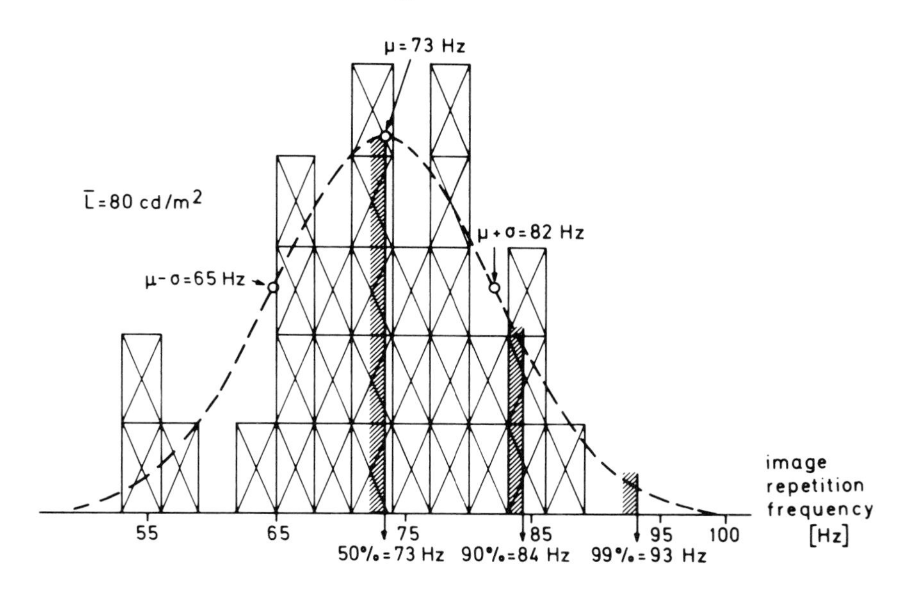

To define the image repetition frequency in a VDT, the manufacturer must decide how great a percentage of the observers are to be without large field flicker. In Figure 1, the image repetition frequencies ω for 50, 90, and 99% flicker-free observers are marked by vertical hatched borders. They are at 73, 84, and 93 Hz, respectively.

2.2. **Jitter (Temporal Image Instability)**

There may be two possible sources for local instabilities of image points:

1. The first possibility is that noise in the electronic line and image deflection circuits causes irregular displacements of the single image point.

2. The second possibility for producing jitter are a.c.-fields of external sources, which may be superimposed onto the deflection field.

By improper mechanical and electronic design, both kinds of jitter may introduce visual interferences: the observer reports irregular movements or an additional stroke unsharpness.

In order to be able to present worst-case jitter limits, we performed some measurements, using a high quality monitor which was adapted to fit the experimental requirement.

Test experiments showed that, within a wide range of difference frequencies between image repetition and external field, i.e. 5 to 20 Hz, even minor a.c.-amplitudes cause apparent movement. The main experiment is sketched in Figure 2.

FIGURE 2. *Measurements of jitter interference threshold. Observer distance d = 30 cm, image repetition without interlacing: 90 Hz.*

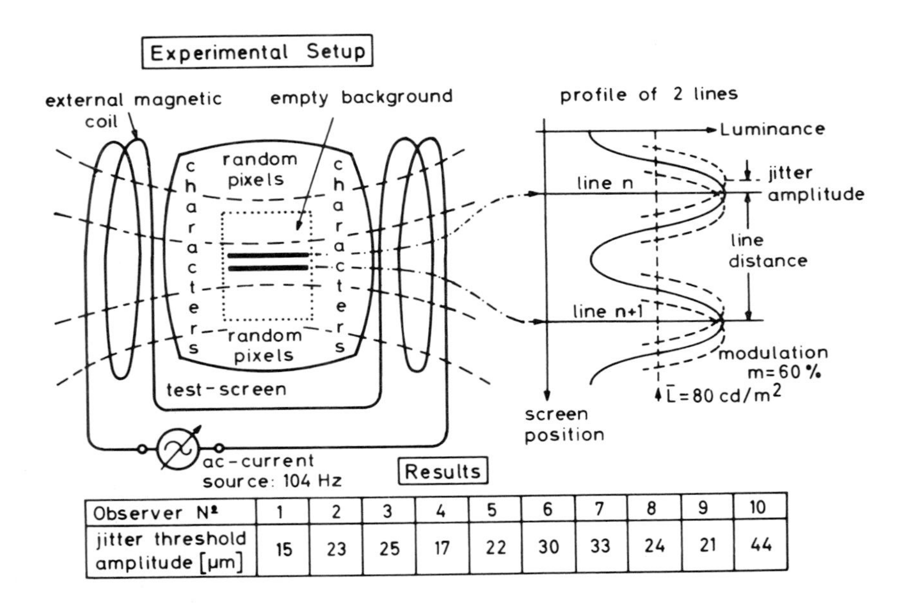

Observer Nº	1	2	3	4	5	6	7	8	9	10
jitter threshold amplitude [μm]	15	23	25	17	22	30	33	24	21	44

In the middle of the test screen—as an example—two of the horizontal lines are drawn which could be set into a jitter-like vibration by applying a magnetic a.c.-field induced by external magnetic coils.

The test screen—with P104 phosphor—had a line density of 9 lines/18 minarc with a spatial modulation of 60% (Figure 2, right). Without external field, the jitter was unmeasurably low.

At the beginning of the experiment, the observers adjusted the coil current so that they clearly saw the jitter as a movement in the background raster. They then decreased the amplitude of the magnetic field until movements just disappeared.

In Figure 2 (bottom), the displacements that yielded just not noticeable movement (method of adjustment) are tabulated for 10 observers. The mean threshold jitter is 25·4 μm which is 17·5 secarc of visual angle. Movements with 15 μm jitter amplitudes were just below threshold for the most sensitive observers. We may conclude that if movement interferences are to be suppressed for VDU-workers using 80 cd/m^2 bright screens, the physical jitter (at 10 Hz) must be smaller than about 15 secarc. Test experiments show that approximately the same limits hold for other frequencies near 10 Hz. Visual sensitivity drops for very low difference frequencies — 1 Hz and for higher frequencies $\geq$ 30 Hz; at higher frequencies, perception of movement changes into perception of unsharpness—which will not be considered here.

2.3. Border 'Flicker' Induced by Eye Movements

Even if jitter induced movements and large field flicker are suppressed other flicker-like interferences may occur with the bright background VDU. They are sketched in Figures 3a and 3b.

FIGURE 3A. *Steady fixation causes irregular shifting or jumping of the border.* B. *Saccadic jumps of the eye cause bright and/or dark flashes on or near the border.* C. *Suppression of border effects by a border-contrast L_{su} : L_{sc} 2:1 or/and a dark separation line with $d \geq$ 5 mm.*

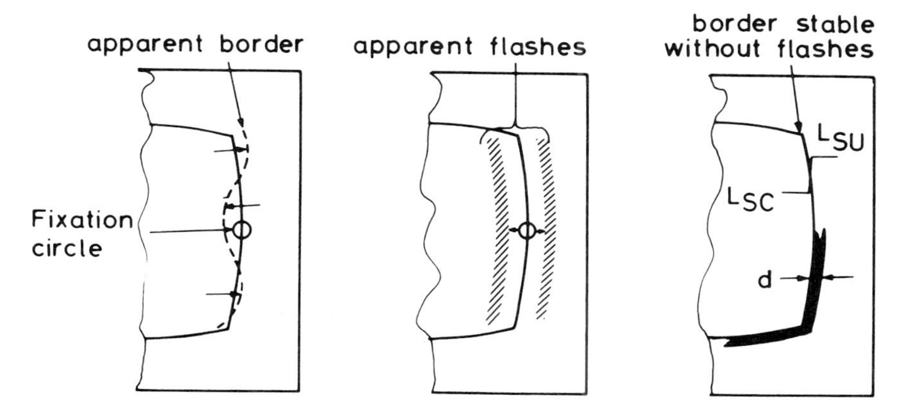

The observers report if

1. The border is fixated without strong effort (cf. Figure 3a): 'The border between screen and surround moves; the whole (vertical) border line or parts of it may jump to and fro.' Often the line is seen as an irregularly oscillating cord.
2. Saccadic eye movements are involved either voluntarily or by applying slight hammer taps to the head (cf. Figure 3b): 'The border region is flashing.' Bright and/or dark flashes are reported.

All interferences are strongest if the border contrast is around 1 (i.e. no contrast). They decrease if the contrast increases. They also decrease if the width of a black line, which separates screen and surround, increases.

Therefore, we performed an experiment in which we asked our observers to adjust the contrast between screen and surround so as just to suppress the border effects described. These adjustments were performed with different widths of the border separation line and with different image repetition frequencies.

FIGURE 4. *Suppression of eye movement induced border movement (solid lines) or border flashes (hatched lines).*

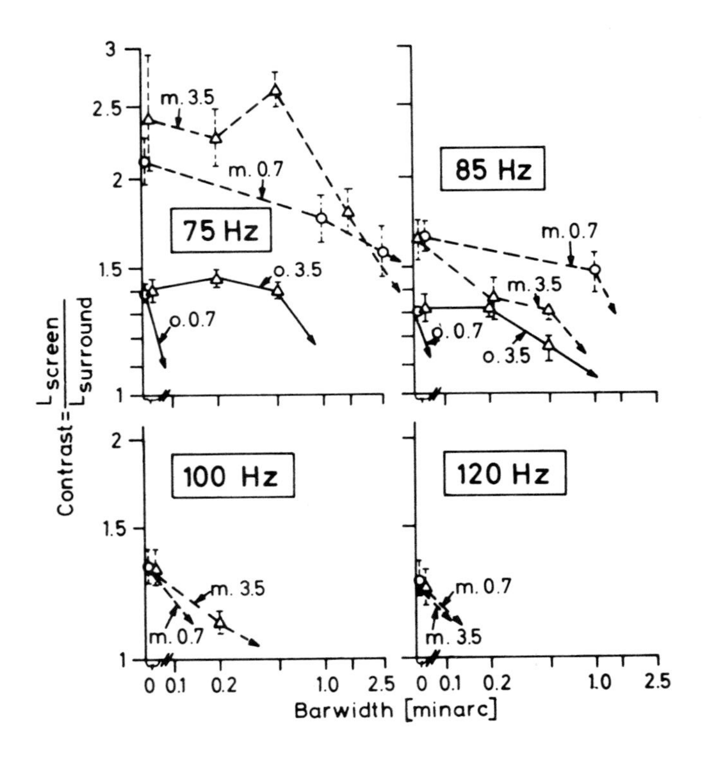

In Figure 4, the data of 1 individual are shown. On the ordinate is plotted the contrast that was needed to suppress the border effects, as a function of the width of the border separation line. The measurements were carried out for image repetition frequencies of 75, 85, 100 and 120 Hz. Solid lines connect all those measuring values which result by relaxed border fixation (condition a, above). Dashed lines give those values that are the result of suppressing flashes which were induced by saccadic eye movements (condition b, above). It can be seen that in order to suppress flashes, higher contrast or wider separation lines were necessary. If we compare the values for different image repetition frequencies, we see that

FIGURE 5. *Suppression of eye movement induced border movement (solid lines) or border flashes (dashed and dashed-dotted lines), numbers are observer distances in m.*

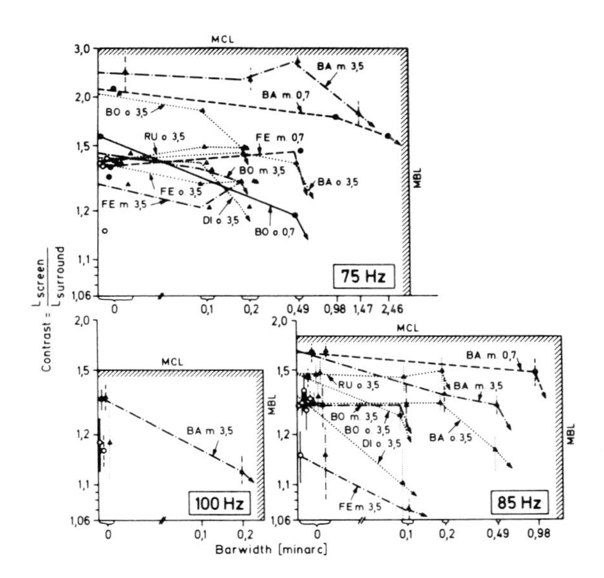

the interference area shrinks with increasing frequency. At 120 Hz, an effect could only be seen for zero bar width. It is interesting to note that at a specific bar width, the border effects disappear rather abruptly.

In Figure 5, data similar to that in Figure 4 are superimposed for 5 observers. All data points are concentrated in an area which may be limited by a 'minimum contrast' line MCL and a 'minimum bar width' line MBL. Interferences are suppressed at 85 Hz, if a contrast higher than 1·8:1 or a bar width larger than about 1·5 minarc visual angle are provided. Since it may be difficult in practice to always maintain the contrast condition, it is recommended to use a minimum separation bar width to suppress boundary interference effects.

3. *Conclusion*

For the elimination of temporal visual interferences, which may be induced by the temporal nature of the electron–raster of VDU screens with 'fast' phosphors and bright background, 3 conditions should be met:

1. It has to be flicker free—which is achievable for 90% (99%) of all observers, if an image repetition rate $\geq$ 84 Hz (95 Hz) is chosen for a raster without interlacing and 'fast' phosphor.

2. The line jitter has to be low enough to suppress movement interferences (or micro-line flicker)—which in our (typical) case is achievable, if jitter is less than 15 secarc visual angle.
3. The optical transition screen–surround must have a minimum contrast and/or a black separation bar with a minimum width—at $\geq$ 85 Hz image repetition a width larger than about 5 mm is recommended.

References

Bauer, D. and Cavonius, C.R., 1980, Improving the legibility of visual display units through contrast reversal, in *Ergonomic Aspects of Visual Display Terminals*, edited by E. Grandjean and E. Vigliani (London: Taylor & Francis), 137–142.

Bauer, D., Bonacker, M. and Cavonius, C.R., 1980, Influence of VDU screen brightness on the visibility of reflected images. *Displays*, April, 242–244.

Bauer, D., Bonacker, M. and Cavonius, C.R., 1983, Frame repetition rate for flicker-free viewing of bright VDU screens. *Displays*, January, 31–33.

Information Display on Monochrome and Colour Screens

P. HAUBNER AND C. BENZ
Applied Ergonomics, Siemens Research Centre, Germany

Abstract

If related pieces of information are combined to form blocks in screen masks, it is possible to make these blocks stand out against each other by means of spacing, colour or by the redundant use of both types of coding. Experiments were carried out to investigate the influence of the coding on the performance and subjective appraisal of the users. Except in the case of redundant coding, colour had no significant influence on performance, but is subjectively thought of as an effective means of configuration.

1. *Introduction*

In work systems with visual display units (VDU), the information is frequently presented to the user by means of masks. Masks are basically forms on the screen consisting of fixed and variable fields. Locating and reading off information from such masks can be improved by means of suitable structuring. Here, basic principles of cognitive psychology, e.g. the so-called Gestalt laws, may provide useful configuration aids (Zwerina, Benz and Haubner 1983).

In earlier investigations with monochrome representation of light characters on a dark background, screen masks were structured by combining related information contents to form blocks which were visually separated from each other by distinct spacing. In this way, it was possible to increase both the work output as well as subjective comfort (Benz and Haubner 1983 a).

The question now arises—particularly in the case of information blocks which are relatively close to one another—to what extent can the use of colour assist visual information processing?

Furthermore, it is necessary to check the hypothesis that the behaviour of the user is improved by redundant coding of the information blocks, i.e. by spacing and colour.

TABLE 1. *Experimental design.*

C / O	m		p	
	ws	ns	ws	ns
1				
2				
3				
k		X_{ijkl}		

C: Coding
m: monochrome
p: polychrome
ws: wide spacing
ns: narrow spacing
O: Observer
X_{ijkl}: Performance of the observer k in the trial l in the coding situation ij

TABLE 2. *A few data on the experimental set-up.*

Horizontal illumination on the working level		approx. 500 lx
Vertical illumination on the screen		approx. 210 lx
Luminance of the character background		approx. 14 cd · m^{-2}
Character contrast (white)		approx. 8 : 1
(yellow)		approx. 6 : 1
(green)		approx. 3 : 1
(orange)		approx. 3 : 1
Visual angle of the characters		approx. 30′
Chromaticity coordinates of the characters		
White:	x = 0.33	y = 0.37
Yellow:	x = 0.42	y = 0.51
Green:	x = 0.30	y = 0.59
Orange:	x = 0.51	y = 0.41

To answer these questions, an experimental study was carried out with a performance test and with subjective appraisals of various types of representation.

2. *Experimental Investigation*

The object investigated was a user program from the motor vehicle trade which was also easy to learn for laymen. It had already been used in a previous study of information coding (Benz and Haubner 1983 b). This user system consisted of a series of screen masks with technical and commercial data relating to motor vehicles.

The data were visually combined to form information blocks; however, the different masks contained the same amount of information. The coding of the block formation was varied by means of spacing, colour and a combination of spacing and colour.

The monochrome representation with a minimum just practical space between the information blocks was simultaneously used for reference and checking purposes. The experimental design is shown in Table 1.

In the test, the subject's task was to find a relevant piece of information from each mask, e.g. the engine output. Search time and errors were recorded by computer, and from this the effective performance was calculated as the number of correctly recognised search items per unit of time.

In addition to performance, the masks were subjectively evaluated by means of psychometric scaling. In particular, an attempt was made to record the experienced strain by means of comparison with everyday situations (Bartenwerfer 1969; Benz and Haubner 1983 a). The test was controlled and evaluated by a process computer. Table 2 contains a few data relating to the experimental set-up.

3. *Results and Discussion*

The evaluation of the data from the performance test produced the identification performances shown in Figure 1.

FIGURE 1. *Identification performance in relation to the coding of the information blocks.*

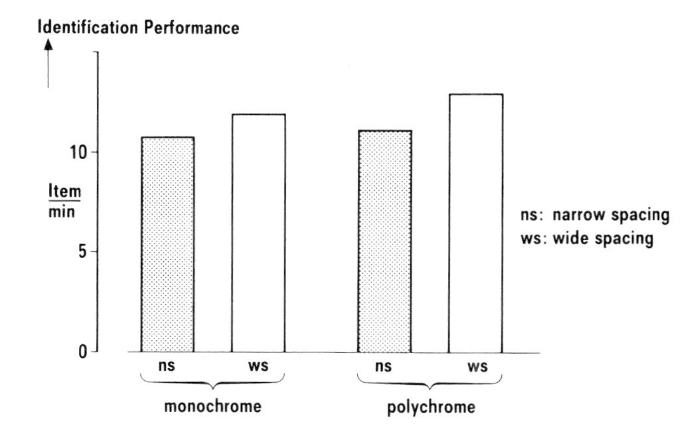

For the relative difference in performance compared with the reference situation, the coding parameter 'spacing' is considerably more important than colour. A clear improvement in performance can, however, be achieved by using colour as an additional coding dimension for the redundant coding of the blocks (Table 3).

From a statistical point of view, however, only the spacing and redundant coding have

TABLE 3. *Alteration in performance, related to the reference situation.*

Coding	Relative Performance Increment
Colour (C)	3.2%
Spacing (S)	9.1%
Redundant (C+S)	19.0%

a significant influence on the identification performance, whereas the same was not true of colour. Moreover, there are significant differences between the observers, which can for the most part be attributed to the users' different methods of working (quick v. slow reactions, cf. Benz and Haubner 1983 b).

FIGURE 2. *Subjective evaluation of the differently coded screen masks.*

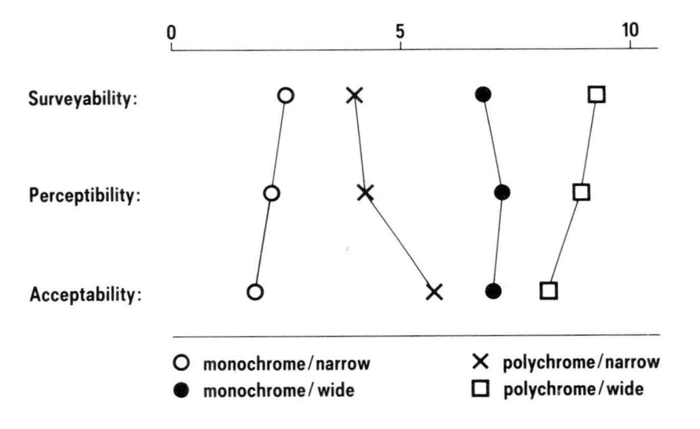

The results of the psychometric scaling are shown in Figures 2 and 3. The items of a questionnaire used for subjective appraisal can essentially be described by three aspects (Figure 2):

1. Surveyability.
2. Perceptibility.
3. Acceptability.

Their scaling, as well as the scaling of the experienced strain (Figure 3), exhibit the same order of rank of block coding as the performance test. However, the subjective reactions permit greater differentiation between the individual representations than the identification performance. All four situations differ from one another significantly. In keeping with the

FIGURE 3. *Psychometric scaling of the experienced strain.*

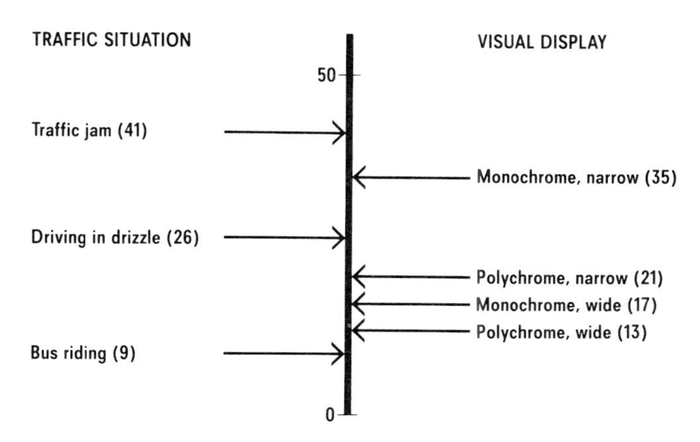

performance test, the effect of the coding by spacing is rated above the coding effectiveness of colour. In addition, the redundant coding of the blocks is thought to be the most effective variant. Subjectively, however, colour is assumed to have a significant influence, in contrast to the performance evaluation. This applies both to the perceptual and cognitive aspects 'surveyability' and 'perceptibility' and, to an even greater extent, to the more affective aspect 'acceptability'. Here, the observers attach almost as much importance to colour as they do to spacing.

It should be stated that in the statistical decisions a level of significance of max. 0·05 was permitted.

4. *Final Remarks*

Previous investigations into representation by means of masks had shown that both the performance and the subjective judgements of the users can be considerably improved by dividing up the masks into information blocks following rules of cognitive psychology.

Additionally, it can be shown from the present test series that performance is not improved if the blocks are delimited by means of colour only. A distinct space between the blocks is much more suitable for this purpose. Performance can, however, be improved even further by means of redundant coding (colour + spacing). Furthermore, colour is subjectively felt to promote performance and increases user acceptance. Finally, it should be pointed out that the coding of masks is only one aspect which should be seen in the general context of the design of VDU workplaces as an entire system (Benz, Grob and Haubner 1983).

References

Bartenwerfer, H.G., 1969, Einige praktische Konsequenzen aus der Aktivierungstheorie. *Zeitschrift für experimentelle angewandte Psychologie*, **16**, 195–200.

Benz, C., Grob, R. and Haubner, P., 1983, *Designing VDU Workplaces* (Cologne: TÜV Rheinland).
Benz, C. and Haubner, P., 1983 a, Gestaltung von Bildschirmmasken. *Office Management*, **31**, 36–39.
Benz, C. and Haubner, P., 1983 b, Codierungswirksamkeit bei Informationsdarstellungen in Bildschirmmasken. In *German Chapter of the ACM Berichte 14: Software-Ergonomie* (Stuttgart: B.G. Teubner).
Zwerina, H., Benz, C. and Haubner, P., 1983, *Kommunikations-Ergonomie: Benutzerfreundliche Anwenderprogramme in Maskentechnik* (Munich: Siemens AG, Bereich Datentechnik).

A Method for Measurement of Misconvergence on a Colour VDU

A. Castaldo
Olivetti, I–10015, Ivrea, Italy

Abstract

This paper deals with a method of measurement of the misconvergence on a colour VDU; this is the deflection of system mistake in the convergence of three electron beams on their own phosphor deposits. This method consists of the measurement of the reciprocal distance of the luminous barycentre of each of three luminous spots, generated by the electron beams, by a microphotometer. This method is easy to use in the laboratory. The purpose is to give to the ergonomist, for the evaluation of the misconvergence, an objective method coherent with the complexive evaluation of the optical response of a colour VDU. This method shows that the actual definition of convergence is not the better to use in ergonomic problems.

1. *Introduction*

The colour video display unit (VDU) has been used, until now, only in particular applications, such as research of town-planning, architecture, meteorology, stress analysis, etc. Now it is beginning to be used in EDP applications. The use of the colour VDU is helpful for formatting or editing and in the development or debugging of a document. Also, the colour VDU is helpful in redundancy of information, i.e. by colouring different sets of data with a different colour. Of course, the output characteristics of colour VDU as luminance, contrast, sharpness, absence of flicker must be at the same level as in a monochromatic VDU; moreover the colours must be clearly different among them.

2. *Colour CRT*

The CRT (cathode ray tube) of a colour VDU is composed of:

1. A complex layer of phosphors on the front side (screen).

2. A shadow mask.
3. An electron gun assembly with three beams (three guns).

The layout of the screen is a regular array of phosphor deposits (dots); the standard colours of phosphors are red, blue, green; optionally red, green, white; or red, green, light blue; they are called primary colours. The dots are set in an elementary triad (RGB), that is repeated on the whole screen. Each dot is separated from the next dot (Figure 1); the

FIGURE 1. *Array of the phosphors on the screen; the dashed circles show the array of the holes on the mask (delta system). Most of the colour CRT on the actual market are in-line system.*

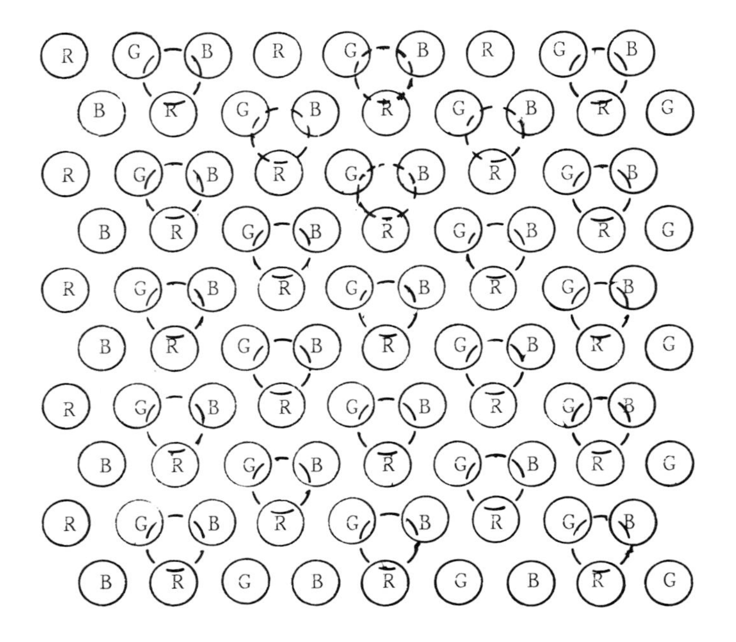

distance between the centres of two close dots is about 0·140 mm (for a shadow mask of 0·31 mm). As a consequence, the luminance distribution of any character is not continuous, but it is matrix-type. Due to the human visual acuity limits, the dots distance and the distance from screen to operator, the luminance appears as continuous. Each single hole on the shadow mask corresponds to one triad RGB. The purpose of the shadow mask is to allow each beam to land only on its own phosphor. The electron gun is a complex assembly of electrodes that generates, controls and focuses the three electron beams. Each beam excites only one phosphor.

In order to certify the visual output quality of a colour VDU it is necessary to carry out the same tests as in the monochromatic VDU, typically white, and the convergence. Further convergence problems will be studied in an attempt to explain the meaning of good or bad convergence; finally a laboratory test method for convergence will be explained.

FIGURE 2A. *Correct convergence of a white spot.*

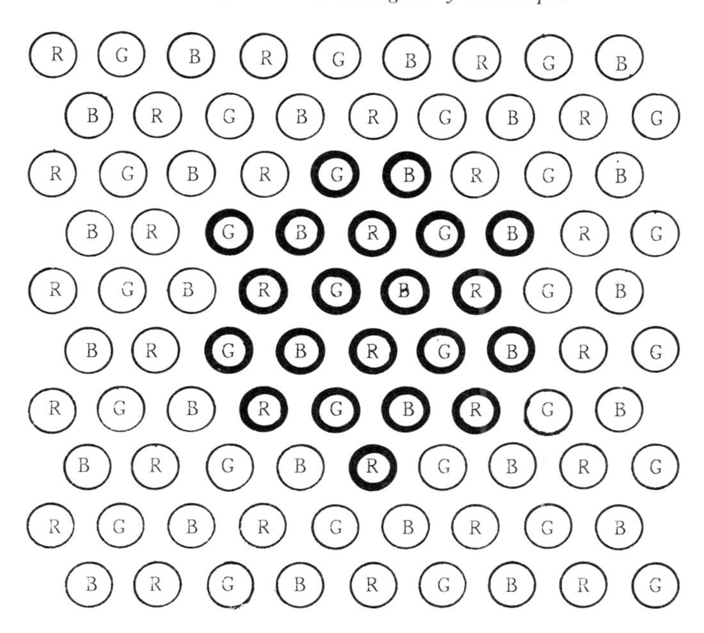

FIGURE 2B. *Misconvergence of a white spot. The centre of luminance of each monochromatic spot is the central dot of each spot.*

3. *Convergence Problem*

In order to have composed colour, i.e. white, the three electron beams shall be completely overlapped, so that the eye recognizes only one colour. Since in colour VDU the phosphor dots are not overlapped but close, instead of overlapping there is a convergence of the three points. As the dimensions of each beam are larger than a single hole, it covers several holes and excites several dots of the same phosphor making a spot. In white spot, there are three single monochromatic spots, RGB, overlapping. In order to have convergence the energetic centres of the three electron beams must go through the same shadow mask hole. If this does not happen, there is no convergence of one or more beams. The lack of convergence is called misconvergence (Figures 2a, b). In case of misconvergence there is a splitting of each spot of a character, it is hard to recognize the character colour and, if there are two near characters, they are partly overlapped. In this situation it is not possible to use a colour VDU for a different editing or formatting of information, or to use a colour code, and moreover it increases the operator's visual fatigue. It is then necessary to recognize the maximum allowed misconvergence without which the characteristics of colour VDU are inferior in quality to the characteristics of the monochromatic VDU. This is an ergonomic problem, but, till now, there is not an objective measurement of the misconvergence precise enough.

FIGURE 3. *White spot with comet-effect on the blue spot. The dots of the comet-effect have a luminance smaller than ten per cent of the luminance of the central dot blue.*

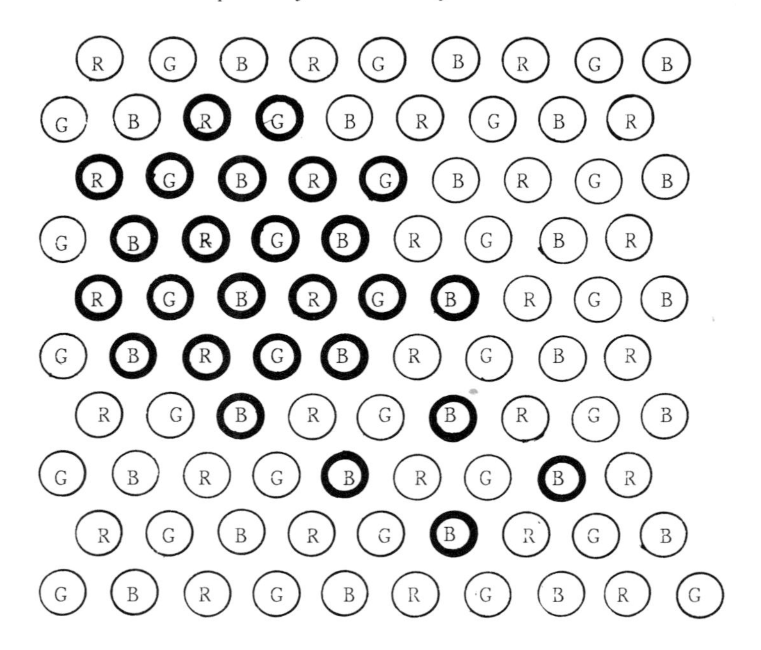

4. *Proposed Measurement Method*

The measurement method, which is explained now, has been derived from the previous definition of convergence, assuming that the energy of the electron beams have a Gaussian distribution and the luminance of excited spots also have a Gaussian distribution. The maximum luminance of a single excited deposit of each phosphor is measured by a microphotometer covering an area with a diameter of 0·060 mm. This luminance is transferred on a map of the phosphor array and then the luminous barycentre of each spot of the primary colour is calculated by the rules of mechanics. This calculation is necessary because if there is misconvergence, the energetic centre of a monochromatic beam does not fit with the more luminous deposit of phosphor. The maximum distance between two of the three barycentres is the misconvergence of the electron beams at that point. The area covered with the microphotometer is sufficient to avoid meaningful luminous influences by the close excited dots. This method has been completely automated; both measurements and calculations are carried out with a computer.

A number of measurements have been made, each tested VDU had an in-line electron gun. Then a statistical evaluation of misconvergence is carried out, utilizing several workers of the laboratory and EDP operators in order to give an ergonomic meaning to the effectuated measurements; this evaluation is still in progress.

However, from the first results, it seems that a misconvergence of 0·3 mm is acceptable for white spot. But while such a misconvergence is acceptable for certain colours, it is not the same for the other colours; i.e. the white-misconvergence of 0·3 mm is not acceptable for a magenta spot (red and blue). In fact, the lack of green, which overlaps the red dots and the blue dots in the white spot, reduces the feeling of melting of colour by human eye.

However, the reciprocal array of the elementary triad (RGB) and the in-line gun, where the electron beam for green is in central position, does not favour the magenta.

To fix an acceptable misconvergence for each composed colour is not only helpful for the ergonomist, but also therefore there are some hard technological problems for reducing the misconvergence of the actual commercial VDU, it shall be helpful for a better use of colour in software programs, avoiding problems due to the misconvergence.

5. *Conclusions*

I think this method is of value in fixing acceptable misconvergence, but it revealed that the actual definition of convergence, which is accepted and used by manufacturers and users in their business, is not the best one for ergonomic purposes. In this definition the dimensions of the primary spot are not fixed, but it assumes the diameter of the three spots is the same for each one. It is not always true, because the three phosphors have a different quantic efficiency; for achieving the requested white, it is necessary that the luminance of the three primary spots is more exactly in reciprocal ratios. Both these phenomena are related to the energy of the beams, and then to the physical dimensions of the same beams and consequently to the dimensions of each monochromatic (primary) spot.

See the following example: two transparent disks of different colours and diameters are overlapped; these overlapped disks have the same axis. An observer sees a composed disk with the part of the same diameter as the smaller disk with a colour which is the addition of the two primary colours, and the external circular ring, which is the remainder of the larger disk with its same colour. This situation is not rare on a colour VDU, and the circular ring of the largest spot is visible even if there is convergence; misconvergence makes matters worse. It is not rare that the ratio between the diameters of two primary spots is 1:2.

Another problem is the comet effect on the only one monochromatic spot. This effect, though it has low luminance and it is not able to shift the luminous barycentre of the monochromatic spot and consequently the convergence, is clearly visible and causes much confusion (Figure 3). The same situation is with monochromatic spots that are not circular, but stretched in a direction, and their largest axes have not the same direction. Then I think the previous definition of convergence, based on overlapping of the spatial luminance of monochromatic primary colours, is not the best to use in ergonomic problems. But it would be probably more meaningful, for ergonomic purposes, to give a definition of convergence from the spatial homogeneity of composed colours in tristimulus terms.

Lighting Characteristics of VDTs from an Ergonomic Point of View (Introductory paper)

U. BRÄUNINGER, E. GRANDJEAN, G. VAN DER HEIDEN, K. NISHIYAMA AND R. GIERER
Swiss Federal Institute of Technology, Department of Ergonomics, CH–8092 Zurich, Switzerland

1. *Measuring Equipment*

Equipment and measuring methods are described in detail by Bräuninger (1983) and by Bräuninger and Grandjean (1983). Only the principles are mentioned here.

A microscope picked up the luminance of a small dot of 0·1 mm of a character and led it to a photomultiplier. In some cases the luminance was also recorded with a camera from a display surface of 5 × 7 cm. The signals from the photomultiplier were transferred to several instruments to measure the lighting characteristics. All measurements were carried out under standardized lighting conditions and mostly with an adjusted luminance level, the so called 'preferred' luminance. Most of the preferred figures were between 20 and 50 cd/m^2 and were assessed by the experimenter. (Operators adjusted luminances in practice between 9 and 77 cd/m^2 with a median value of 33 cd/m^2.)

2. *Results*

2.1. **Sharpness of Characters**

To measure the sharpness of characters, the microscope was moved across capital letters 'U' with a thrust of 5 mm/min. The sharpness was determined by the slope of luminance in the border area of characters according to the formula shown in Figure 1.

The evaluation was based on the observation, that a border zone r less than 0·3 mm is not perceived, while larger figures of r reveal increasing blurred edges of characters.

Results and evaluation of 33 VDT models are reported in Table 1.

FIGURE 1. *Sharpness of characters is expressed in mm according to the above mentioned formula.*

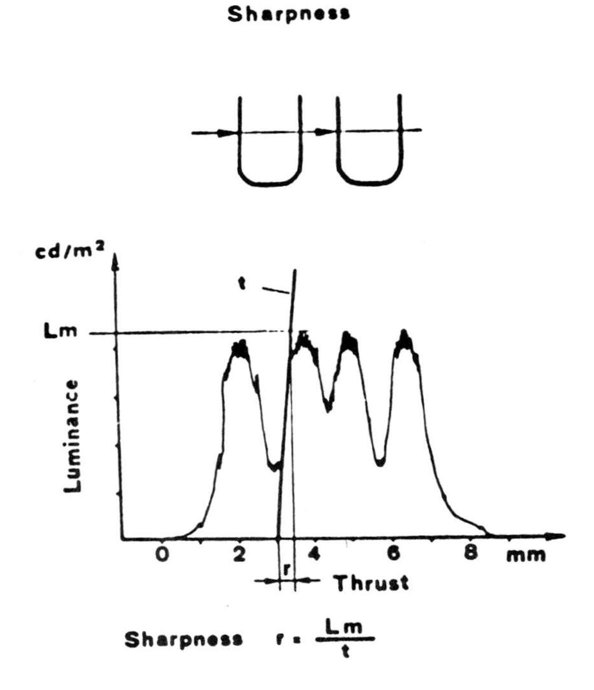

TABLE 1. *Sharpness of characters of 33 VDT models expressed as the border zone* r *(mm). The figures are mean values of nine measurements taken from nine locations on the screen and at preferred luminances.*

Number of models	Border zone r (mm)	Evaluation of sharpness
5	0·1–0·29	Good
9	0·3–0·39	Unsatisfactory
12	0·4–0·5	Poor sharpness
7	> 0·5	Unacceptable

2.2. Character Contrasts

The character contrasts are determined by measuring the luminance in the space between two 'U's and expressed as percentage of the preferred luminance measured inside the legs of 'U'. This space luminance is called rest luminance.

The evaluation of character contrasts is based on the following consideration: *A ratio of 10:1 can be considered as suitable for a contrast between the blank screen background and the luminance of character legs.*

However, if the luminance of the space between letters is taken into consideration, it seems that the above mentioned standard is too severe. Based on the observations on 33 different VDT models, it can be concluded *that the rest luminance in the space between two characters should not exceed 15%, which corresponds to a contrast ratio of about 1:6.*

Results and evaluation of 33 models are shown in Table 2.

TABLE 2. *Character contrasts of 33 VDTs expressed as rest luminance in the space between two 'U's measured with preferred character luminances.*

Number of models	Rest luminance (%)	Approximate contrast ratios	Evaluation
8	< 17	More than 6:1	Good
10	17–25	6:1 to 4:1	Acceptable
6	26–33	4:1 to 3:1	Insufficient
9	> 33	< 3:1	Very poor contrast

Snyder and Maddox (1980) studied the reciprocal relationship between sharpness and contrasts of characters. They concluded that a poor sharpness requires an increase of character contrasts if legibility is to be kept at a good level. It is concluded that a sharpness r of less than 0·3 mm combined with a character contrast of 1:6 or more should guarantee a reasonable legibility.

2.3. **Stability of Characters**

If the electron beam is well regulated, the characters show a good stability. If the regulation is poor, the character become unstable, appearing as drift, jitter or disturbances of linearity. Drift is a change in the position of a symbol and can cause a merge of characters, illustrated in Figure 2.

Jitter is an abrupt and repetitive change in the position of a symbol while disturbances of linearity refer to troubles of straightness of lines.

In order to measure the stability, the luminance of a dot in the middle of a leg of a letter is continuously recorded.

The evaluation is based on the requirement that the movements of characters should not be perceived; this is achieved when the deviations of the luminance of a dot do not exceed 20% of the mean luminance. Results and evaluation are reported in Table 3.

FIGURE 2. *Three cuttings of faces of three VDTs.* Upper cutting: *'legs' of letters are sharp and have good contrasts with their near vicinity. The letters have a good relationship between height and width, and the space between letters and lines is large.* Middle cutting: *The relationship between height and width is not appropriate. The space between letters and between lines is too small.* Lower cutting: *The space between letters is not constant and the 'legs' are often merged. This phenomenon is, to some extent, due to the instability of characters. The whole face is poor.*

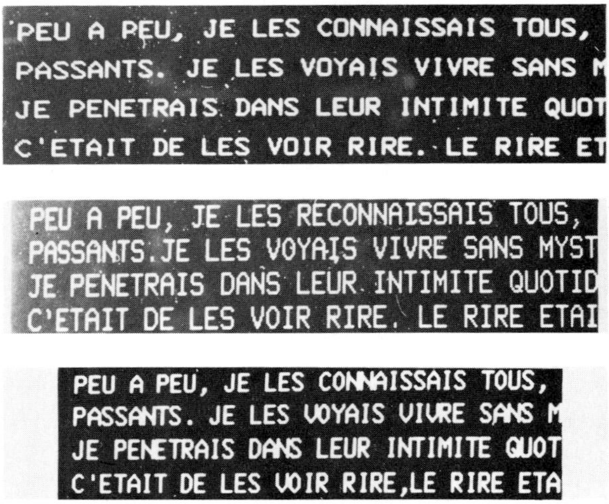

TABLE 3. *The stability of characters, expressed as luminance deviations in % of the mean luminance of a leg. The figures are mean values of nine measurements taken from nine locations on the screen and at preferred luminances.*

Number of models	Deviation (%)	Evaluation of stability
8	< 5	Good
15	5–20	Acceptable
3	21–40	Insufficient
8	> 40	Unacceptable, letters merge

2.4. **Brightness Contrasts of Surfaces at the VDT Workstation**

A brightness contrast between screen background and source document should not exceed 1:10. This has become a generally accepted standard for display design. All other surfaces in the visual field, like the frame around the screen, the VDT box or the keyboard should show luminances lying approximately in the middle of those of the screen and those of the source documents. Table 4 shows the measured brightness contrasts between screen and source document (50 cd/m^2).

TABLE 4. *Luminance contrast ratios between screen background and source document (50 cd/m²).*

Number of models	Brightness contrasts	Evaluation
19	< 1:10	Good
5	1:10 to 1:15	Unfavourable
9	1:16 to 1:25	Unacceptable

2.5. **Oscillation Degree of Character Luminances**

The majority of VDTs today have a refreshing rate of 50 Hz. With a phosphor of fast extinction (decay time: 5 ms to zero cd/m²) it was observed that subjects perceive the oscillations of characters in a range of 45 to 55 Hz (see Gyr *et al.* this volume, pp.359–363).

The oscillation degree of a light source can be determined by the quotient of the amplitude of the oscillation over the mean luminance. A more adequate procedure is based on the following formula:

$$a = \frac{1}{Lm} \sqrt{\sum_{n=1}^{20} . A_n^2 \text{ eff}}$$

whereby a = oscillation degree
Lm = mean luminance
A_n eff = amplitude of the groundwave and of 20 first harmonics of a Fourier transformation.

In the present study the oscillation degree a was measured for display surfaces of 5 × 7 cm with standardized lighting conditions in the room.

The evaluation is based mainly on the requirement that visible flickering characters must be avoided by all means. It is reasonable to call for a flicker-free screen with a degree of oscillation corresponding to that of phase shifted fluorescent light with figures of a below 0·2. The results of 33 models (of 20 different makes) and the evaluation are reported in Table 5.

TABLE 5. *Oscillation degrees* a, *according to the above mentioned formula.*

Number of models	Oscillations degree a	Evaluation
11	0·02–0·19	Low oscillation, no flicker visible
10	0·2–0·39	High oscillation like unshifted fluorescent lamps
12	0·4–1·0	Unacceptable, flicker strongly perceived

The higher the refreshing rate and the slower the phosphor extinction, the higher is the threshold of perceived flicker. It is recommended that preference should be given to VDTs with phosphors showing a decay time of at least 20 ms or with refreshing rates of 80 Hz and more.

2.6. Reflections on Screen Surfaces

Bright reflections reduce character contrasts and disturb legibility. Furthermore reflections can be a source of distraction (Elias and Cail 1983).

Many manufacturers have developed antiglare systems for the screen. To measure their efficiency a standardized procedure was adopted: a light source of 100 W is arranged in a given distance and angle to the screen and the reflected luminances are measured. The results of measurements and observations are reported in Table 6.

TABLE 6. *Reflected luminances of a light source of 100 W on screen surfaces and observed drawbacks.*

Applied system	Number of models	Reflected luminance (cd/m^2)	Drawbacks
No protection	5	525–2450	
Lambda/4 coatings	1	26	Easily soiled
Micromesh	7	32–72	Dark background, poor sharpness
Etching or roughening	18	46–360	Some reduced sharpness
Polarization filters	2	160–1480	Easily soiled double images
Coloured glass	2	205–470	Poor sharpness

All anti-reflective technologies have serious drawbacks. If efficiency is weighed against drawbacks, the Lambda/4 coatings and the etching–roughening procedures should be preferred. It is obvious that the correct arrangements of lights and windows in relation to the screen as well as a certain adjustability of the screen angle are still the most important preventive measures.

2.7. Characters and Face

Characters and face are measured with the microscope moving at a given thrust across letters and lines. In Table 7 the mean figures of 26 models are compared with generally accepted recommendations. It is shown that many models present too narrow letters and too small spaces between lines. Some of these insufficient characteristics are illustrated in the middle cutting of Figure 2.

TABLE 7. *Dimensions of characters and face of 26 VDT models.*

Dimensions	Recommended figures	Mean of 26 models	Range
Height of capital letters (mm)	3–4·2	3·4	2·5–4·4
Width (% of height)	75	55	31–81
Space between letters (mm)	0·7–1·0	1·0	0·7–1·7
Space between lines (mm)	4–6	3·3	2·1–5

3. *Conclusions*

According to present knowledge, a VDT with good legibility and visual comfort should show the following lighting characteristics:

1. The degree of oscillation of character luminances should be comparable to figures shown by phase shifted fluorescent tubes; *a* should be lower than 0·2. Refresh rates of 80 to 100 Hz with phosphor decay times of about 10 ms for the 10% luminance level are recommended.
2. The blurred border zone of characters should not exceed 0·3 mm.
3. The luminance in the space between letters is not to exceed 15% of the brightness of character legs (contrast ratio $> 1:6$).
4. Movements of characters should not be perceived. (The deviations of luminance of a dot is not to exceed ± 20%.)
5. Reflected glare on the screen should be reduced by 5 to 10 times, but the applied techniques are not to decrease sharpness nor darken the screen background too much.
6. Luminance contrasts of surfaces at VDT workstations should be low; the contrast ratio between screen background and source documents are not to exceed 1:10. All other surfaces in the visual field should disclose luminances lying approximately in the middle between those of screen and source documents.
7. In order to guarantee good legibility, the dimensions of characters as well as the space between letters and between lines should correspond to recommended figures.

References

Elias, R. and Cail, F., 1983, Contraintes et astreintes devant les terminaux à écran cathodique. *Travail Humain*, **46**, 81–92.

Bräuninger, U., 1983, Lichttechnische Eigenschaften der Bildschirmgeräte aus ergonomischer Sicht. Thesis of the Swiss Federal Institute of Technology edited by the Department of Hygiene and Ergonomics, ETH, 8092 Zürich.

Bräuninger, U. and Grandjean, E., 1983, Lighting characteristics of VDTs from an ergonomic point of view. *Proceedings CHI–83 Human Factors in Computing Systems* (Boston, Dec. 13–15 1983) ACM, New York, pp.274–276.

Läubli, T., Hünting, W. and Grandjean, E., 1981, Postural and visual loads at VDT workplaces, Part II. Lighting Conditions and visual impairments. *Ergonomics*, **34**, 933–948.

Shurtleff, D.A., 1980, *How to Make Displays Legible* P.O. Box 134, La Mirada, California 90638: Human Interface Design.

Snyder, H.L. and Maddox, M.E., 1980, On the image quality of dot-matrix displays. *Proceedings of the SID*, **21**, 3–7.

Some Experiences in the Field of Design of VDU Work Stations

A.M. Paci and L. Gabbrielli
Olivetti, Ivrea, Italy

Abstract

The experiences acquired in the 1970s with the introduction of visual display terminals into the office, have brought to light a series of problems of both a psychological and physiological nature caused by the inadequacies of the equipment and the environment. Many studies and research have pointed out the problems and the possible solutions have been indicated. On this basis we designed a new line of VDU work stations. We thought it useful to bring to the attention and appraisal of ergonomists the results achieved; this will make possible an evaluation of the degree of adequacy of an industrial product line and the verification of what remains to be done.

1. *Introduction*

At the end of the 1970s, Olivetti decided to design a new generation of VDU work stations. At that time it was already evident that there were a number of problems involved with the introduction of new methods of working into the office. Therefore, we decided to wait until sufficient research had been carried out on this subject. Our aim was that operators should be able to use the new product line with very few problems in different fields of application and environmental conditions.

2. *Work stations*

The first step was to collect all existing documentation and analyse it with particular attention to the most complete and significant studies, such as those conducted by Stewart *et al.* (1974), Östberg (1975, 1976), Cakir *et al.* (1978, 1979), Grandjean (1979, 1980).

Furthermore we carried out some surveys with users, particularly with Northern

Europeans, who at that time were experiencing to a greater degree the new problems of VDU work stations.

This led us to believe that, if we wanted to give the operator the most suitable configuration, both in personal requirements and in the variable nature of the work, the design would have to have a closer and more complete relationship with ergonomics than in the past, and in particular would have to follow some fundamental points (Figure 1).

FIGURE 1. *The correct VDU workplace.*

1. The work station had to consist of: keyboard, VDU, document holder, table, chair, footrest, conceived as independent but co-ordinated modules.
2. The VDU had to be tiltable and rotating.
3. The table had to be preferably adjustable or equipped with a foot-rest.
4. The chair had to be adjustable in height and have a high, adjustable backrest.
5. The colour aspects and surface quality of the modules had to be carefully considered from the physiological point of view also.

The subsequent, detailed analysis of the collected documentation showed general agreement for some features and disagreement or lack of consistent indications for others.

In the first case, the job of organizing the specifications was simple. In the other cases, where the information was incomplete or not in agreement, it was necessary to take some autonomous decisions, both on the basis of research carried out in our laboratories, and from outside consultants.

We shall now comment on such decisions regarding the keyboard and the VDU in particular.

3. *Keyboard*

The points that required particular attention were the determination of the keyboard height, the keyboard profile and the key force–stroke diagram.

3.1. **Keyboard Height**

Regarding this point, whereas the European school of thought states that for visual and postural reasons (correct position for forearm), the height should not be over 50 mm (Cakir *et al.* 1979) or even 30 mm (DIN 1980), the American school of thought (Rupp 1981) states that a keyboard of 30 mm is certainly acceptable but that the benefits given to the operator are minimal.

In our opinion, we were soon convinced that the keyboard should be as low as possible because, even if the visual and postural advantages are minimal, the operator can have, for a given working level (home row level) a larger space for the legs and can therefore adopt a more relaxed and comfortable position during the breaks or waiting time.

On this basis we designed a keyboard with a height (at the level of the home row) of 30 mm.

3.2. **Keyboard Profile**

The information about keyboard profile was rather scarce and incomplete: it was said in fact (Cakir *et al.* 1979) that the keyboard could be sloped, stepped or dished (Figure 2), pointing out a possible superiority of the dished profile, since it would improve the performance and accuracy of the skilled operator.

In order to check such a hypothesis a high-speed cinematographic test was performed with three operators working on two alpha-numerical keyboards, the first with a dished profile and the second with a stepped profile.

Figure 3 shows the average values of the inclination of the finger tip movement at the moment of impact on the keys of different rows. It should be noted that the average movement of the finger tip varies as the operator passes from the nearest row to the one furthest

FIGURE 2. *Keyboard profile.*
A: sloped keyboard, B: stepped keyboard, C: dished keyboard.

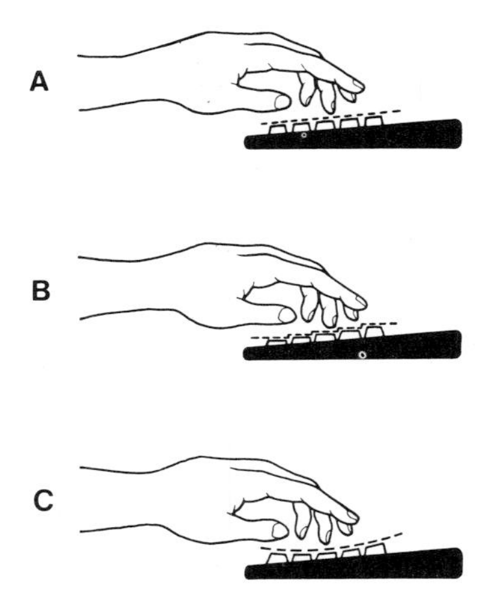

away. In particular it can be seen that with the dished keyboard, the angle between the average movement of finger tip and the normal to the surface of the keys is fluctuating between 8° and 11°, whereas with the stepped keyboard, it fluctuates between 2° and 13°.

Such results, together with the improved performance of the operators and their expressed preferences for a dished profile, persuaded us to design a dished keyboard in the alpha-numerical area; the stepped profile, typical for calculators was considered suitable for the numerical area, whereas the sloped profile was considered the best solution for the function keys which require better readability and more space for the labelling.

3.3. **Key Force–Stroke Diagram**

An area of uncertainty was revealed regarding the features of the key force and stroke, both because of the values indicated (Cakir *et al.* 1979) referred to a rather wide range (25–150 gr. for the force and 0·8–8 mm for the stroke) and because their correlation with the operator's performance was not indicated.

Also in this case, it was necessary to programme and implement specific tests with a representative sample of users through which all the involved parameters were considered.

The obtained results allowed us to define a force–stroke diagram that represents the best compromise between the reduction of muscular effort and limited error tax at high typing speed.

Such a diagram, represented in Figure 4, is for a stroke of 3 mm, a force of 70 gr and the presence of tactile feedback thought to be useful, especially for the unskilled operator (Scagliola 1981).

FIGURE 3. *Inclination of the fingertip movement at the moment of impact for three operators (D, C, F) and its average value. A: stepped keyboard, B: dished keyboard.*

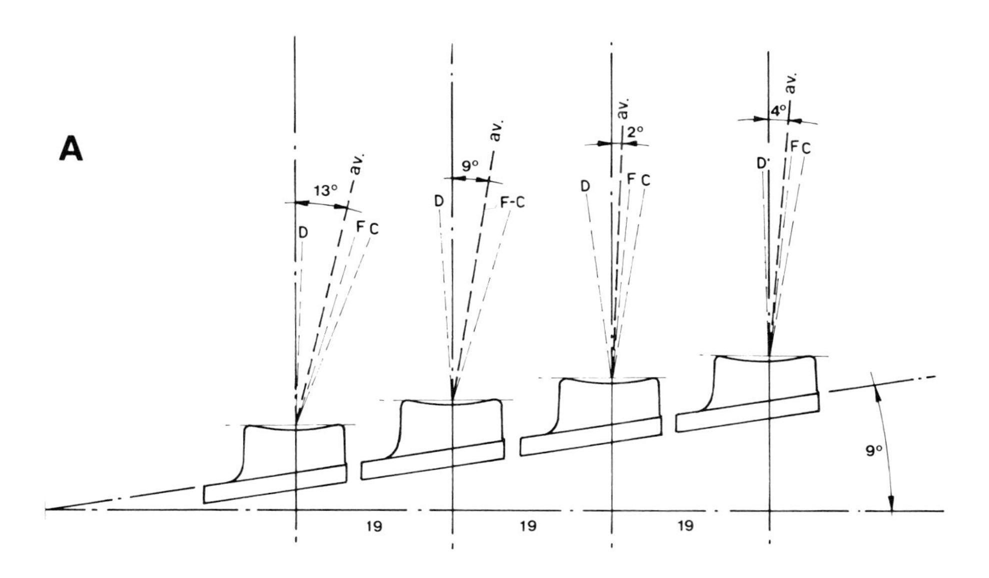

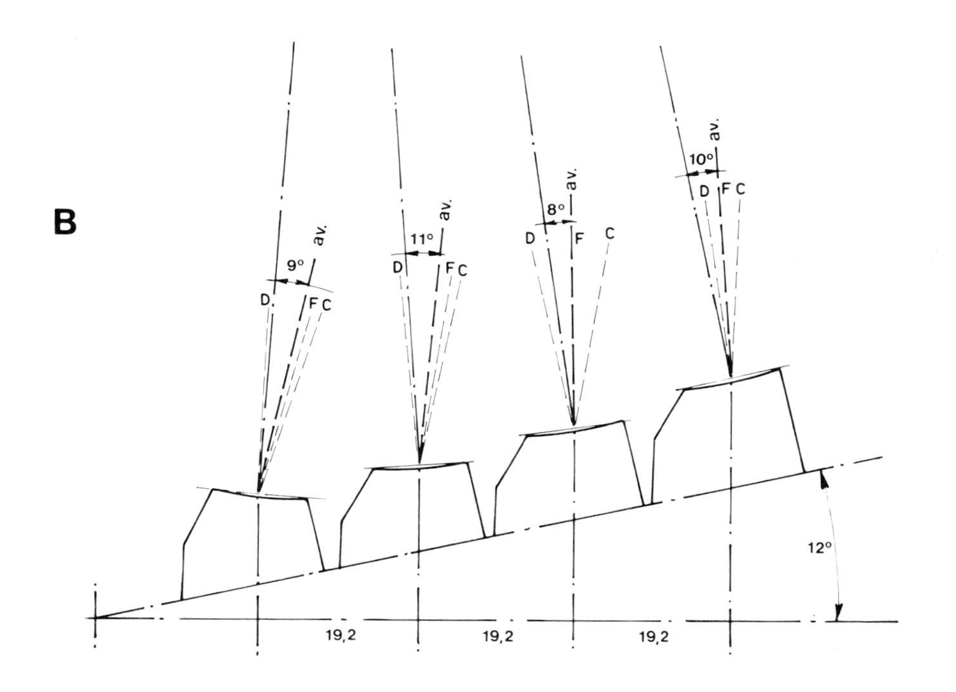

FIGURE 4. *Key force–stroke diagram.*

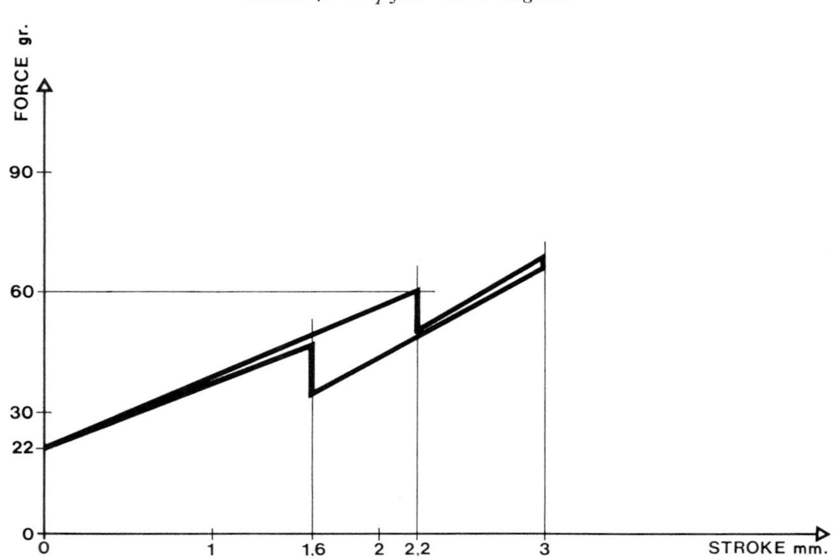

3.4. Discussion

The decisions made concerning the design of the keyboard were judged positively by the users. On the other hand, it is honest to say that the task of defining the keyboard features was developed in an environment that benefited from experience in such a field over the years and therefore was facilitated by this collected experience. For example, the experimental methodology of evaluating the degree of comfort offered to the operator by a keyboad had already been set up (Fubini 1974) for other keyboards and was used then for the study of the new features.

More difficulties were met in facing the problems related to the VDU, since the experience we could use in this field was much less.

4. *VDU*

The points which concerned us most were the determination of the values of the background luminance and contrast and the choice of the type of phosphor and polarity.

4.1. Background Luminance

For the background luminance, values greater than 10 cd/m^2 (DIN 66234 1980) were suggested on the basis of the hypothesis that the sensitivity of the eye is a function of the background luminance level or of the average screen luminance level.

In fact, rather high values of background luminance would reduce the difficulties of the eye adaptation in passing frequently and sequentially from one surface of high luminance (in this case the document) to a surface of low luminance (the screen).

Such considerations led us to design a VDU that excluded the use of the black filter previously used. On the other hand, since the information coming from the users did not always agree with our choice, we decided to change the design such as to allow an easy setting of extra filters that could be added by the customer when appropriate.

Recently, on the other hand, a different hypothesis has been formulated (Rupp 1981) according to which, looking at a VDU with light characters on a dark background for a certain period of time, the adaptation level of the eye is determined by the level of maximum screen luminance and not by the average screen luminance or the background luminance. Probably, both hypotheses are valid depending on the period of screen observation and therefore on the particular type of work or the particularly extreme situations.

4.2. **Polarity and Type of Phosphor**

Also for the polarity, if we agree with the hypothesis that the adaptation level of the eye is determined by the background luminance level or by the average screen luminance level, the positive polarity (dark characters on a light background) has better results. Positive and negative polarity would have equally acceptable results in the case of the second hypothesis, but with positive polarity a greater perceptibility of screen flicker results, being the flicker perception proportional to the screen luminance and to the width of the visual field (Cakir *et al.* 1979).

On the basis of these considerations and taking into account that flicker can be partly reduced by operating on the refresh rate and using a high persistence phosphor, we designed a VDU with negative polarity and the availability of positive polarity only for fields.

The refresh rate was raised to values above 60 Hz, while we preferred to use a low persistence phosphor in order to avoid, during the scrolling, the image smearing effect which we found to be rather tiring.

Professor Grandjean, after the evaluation of our products in his laboratory suggested that we should attempt a better compromise between the smear effect and the high oscillation degree value of the low persistence phosphors.

It was for this reason that the low persistence phosphor was used only on some VDUs while a greater persistence phosphor was introduced on others.

4.3. **Character–Background Contrast**

A point we wanted to go into further was the determination of a maximum level of character-background contrast available by adjusting the character luminance. The indications coming from literature gave as a minimum a ratio of 3:1, as an optimum a ratio of 10:1, and as a maximum a ratio of 15:1, with background luminance greater than 10 cd/m^2 (DIN 66234 1980).

To check what would be the contrast values used by operators working at a VDU for several consecutive hours, we took a survey of 54 operators who used different VDUs for different jobs and in different lighting conditions.

Together with some physiological and psychological parameters, we took the background luminance and the spot luminance of the different VDUs and we calculated then the contrast values ($C = (L_O - L_B)/L_B$ where L_O = spot luminance and L_B = background luminance).

Since the indications coming from literature refer to VDUs with a background luminance greater than 10 cd/m^2, we consider now only the values relative to the VDUs which, measured at 300 lux, minimum lighting value advised (Cakir *et al.* 1979), presented background luminance of at least 10 cd/m^2 (60% of the examined modules).

FIGURE 5. *Contrast value of character–background (C) versus background luminance (L_B) taken from VDUs with L_B greater than 10 at 300 lux, used in different lighting conditions..*

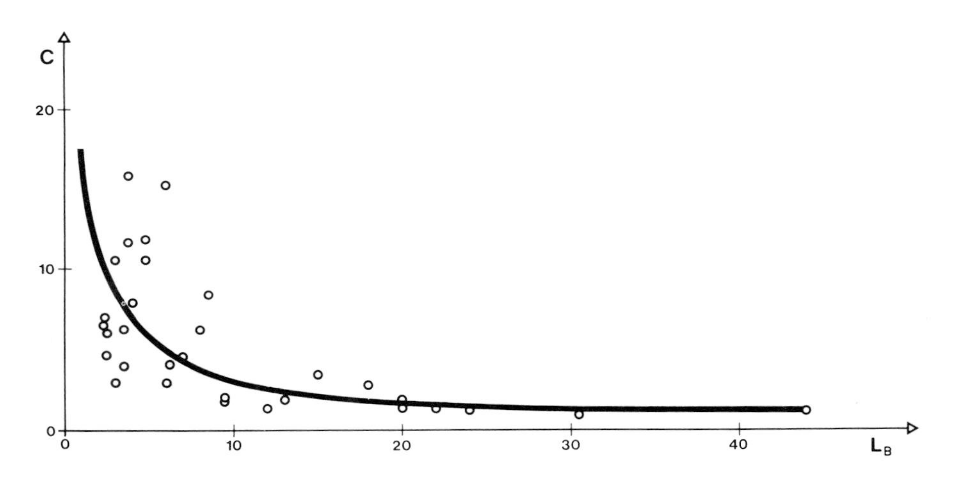

The diagram drawn in Figure 5, whose trend can be fairly well fitted with an hyperbole derived from the least square method, shows the contrast versus the background luminance values found in the real lighting conditions.

We can see that for lighting greater than 300 lux and therefore L_B greater than 10 cd/m^2 (11 VDUs) the required contrast tends to stabilize around an average value of two.

For lighting less than 300 lux and therefore L_B less than 10 cd/m^2, an unadvisable but rather frequent situation (22 VDUs), we can see from Figure 5 that the operator tended to increase the contrast value up to 16 even if he could get on his VDU much higher levels.

On the basis of such results, we designed a VDU with background luminance of 13 cd/m^2 and maximum contrast of 12 at 300 lux.

With this VDU we allowed the operator to receive a contrast greater than 3 up to lighting values of 900 lux.

Furthermore, since during the survey we also found some VDUs working in lighting conditions greater than 900 lux, which, in order to achieve a sufficient contrst, were provided with a dark filter, we decided to give to the user the possibility of setting a Micro-mesh filter suitable for particularly high lighting conditions.

References

Cakir, A., Hart, D.J. and Stewart, T.F.M., 1979, *The VDT Manual* (Darmstadt: Inca-Fiej Research Association).

Cakir, A., Reuter, H., v. Schmude, L. and Armbruster, A., 1978, *Untersuchungen zur Anpassung von Bildschirmarbeitsplätzen an die physische und psychische Funktionsweise des Menschen* (Bonn: Federal Ministry for Work and Social Order).

DIN 66234, 1980, *Display Workstations*, Part 6.

Fubini, E., 1974, Studio dinamico della digitazione su alcuni tipi di tastiere. In *Atti del congresso nazionale di Rimini 2 Ottobre 1974*.

Grandjean, E., 1979, *Arbeitsmedizinische und ergonomishe Probleme der Arbeit an Bildschirmgeräten* (Zürich: ETH-Zentrum).

Grandjean, E., 1980, Ergonomics of VDUs: review of present knowledge. *Ergonomic Aspects of Visual Display Terminals*, edited by E. Grandjean and E. Vigliani (London: Taylor & Francis).

Läubli, Th., Hünting, W. and Grandjean, E., 1980, Visual impairments in VDU operators related to environmental conditions. *Ergonomic Aspects of Visual Display Terminals*, edited by E. Grandjean and E. Vigliani (London: Taylor & Francis).

Östberg, O., 1975, Health problems for operators working with CRT displays. *International Journal of Occupational Health and Safety*, Nov./Dec.

Östberg, O., 1976, Office computerisation in Sweden: worker participation, workplace design considerations, and the deduction of visual strain. Paper read at the NATO Advanced Study Institute on Man-Computer Interaction, held at Mati, near Athens.

Rupp, B.A., 1981, Visual display standard: a review of issues. *Proceedings of the SID*, 22(1).

Scagliola, B., 1981, *Ergonomics at Olivetti*, edited by Olivetti (Milan).

Stewart, T.F.M., Östberg, O. and MacKay, G.J., 1974, *Computer Terminal Ergonomics*. Rapport ONR 170/72–5.

Screen Design

W.O. GALITZ
Galitz, Inc., 4N961 Stonebridge Lane, St. Charles, Illinois 60174, USA

Abstract

Screen design often fails to reflect human perceptual and information processing capabilities. Many screens lack visual clarity and are difficult to use. Poorly designed screens are also contributing to the visual problems associated with VDT use. The considerations in screen design are presented and significant guidelines for various kinds of screens summarized.

1. *Introduction*

Design objectives for modern information systems are ease-of-learning and ease-of-use. For the visual display terminal-based system, the design of the display screen, and how a person interacts with a computer through it, significantly contributes to achieving these objectives.

Unfortunately much screen design occurs with little to guide it. Technical considerations have received the most attention and the human factors involved are not well understood or neglected entirely. It often tends to be unsystematic, inconsistent and fails to adequately reflect human perceptual and information processing capabilities. As a result, many screens in today's systems are difficult to use and lack visual clarity. They are packed with information, have a cluttered look, and the various components are poorly differentiated.

The frequent result is office systems far less productive than need be. For large systems, a few extra seconds processing each system screen can translate into many people-years of lost productivity. An instance has been reported where improving a screen's readability yielded a 20% increase in productivity (Dunsmore 1982). At worst, discretionary users of systems (managers and professionals) may reject a system entirely if screens give the impression that understanding them will take more time than they have available.

Screen design may also be contributing to the visual fatigue now reported by some system users. Eye movement studies of data entry operators have shown instances where visual movements between screen and source documents exceed several thousands for one work day. A significant difference in the brightness level between display screen and source document can fatigue the muscle of the eye if performed excessively. This has led to attempts to brighten the display screen, or lower the illumination to try to achieve the proper balance. But what about the design of the screen? Several thousand eye movements a day may reflect poor screen design rather than an unsatisfactory environment or terminal. The symptoms of a problem rather than the cause may, in some cases, be being addressed.

What can be done then, to improve the screen design process? While screen design is not yet a precise science, the body of knowledge derived from experimental studies is growing. A wealth of information derived from printed material research (e.g., books and newspapers) is available to provide interim guidance until more research questions are answered. This material simply awaits conscientious application to the screen design process.

2. *Most Desirable Features of Screens*

What are people looking for in the design of screens? One organization asked a group of users. Their response is summarized below:

1. An orderly, clean, clutter-free appearance.
2. An obvious indication of what is being shown and what should be done with it.
3. Expected information where it should be.
4. A clear indication of what relates to what (headings, captions, data, instructions, options, etc.).
5. Plain simple English.
6. A simple way of finding what is in the system and how to get it out.
7. A clear indication of when an action could make a permanent change in the data or system operation.

The desired direction is toward simplicity, clarity, understandability, the qualities lacking in many of today's screens.

3. *Screen Design Considerations*

Screen design must consider a number of factors. A well-designed screen format:

1. Reflects the needs and idiosyncracies of its users.
2. Is developed within the physical constraints imposed by the terminal.
3. Fully utilizes the capabilities of its software.
4. If used for data entry, is developed within the constraints imposed by related source materials, such as worksheets, forms or manuals.
5. Is consistent within itself, with related screen formats, and with other screens within the application and the organization.
6. Achieves the business objectives of the system for which it is designed.

Human considerations—the needs and requirements of people—include clarity, meaningfulness and ease of use. A vital human consideration in screen design is design consistency. Design consistency enables a screen user to learn concepts and apply these concepts to a family of screens, and a family of systems. Learning capacities may then be devoted to how to use the system to enhance one's job, and are not consumed in understanding meaningless differences.

Hardware and software considerations reflect the physical constraints and characteristics of the terminal on which the screen appears and the characteristics of the controlling program. They provide a framework within which the screen design must occur and specify the design tools available to the designer (highlighting, inverse video, lower case characters, etc.). Application considerations reflect the objectives of the system for which the screen is being designed. They are the screens data or information building blocks.

4. *Screen Design Guidelines*

Eyeball fixation studies indicate that, initially, the human eye usually moves to the upper-left centre of a display, then proceeds in a clockwise direction. During and following this movement, people are influenced by the display's symmetrical balance and the weight of its titles, graphics, and text. The human information processing system also tries to impose a structure when confronted with uncertainty. It seeks order and meaning, quickly discerning whether a screen has a meaningful or cluttered form.

Screens should provide cohesive groupings of screen elements so that people perceive identifiable pieces. People prefer viewing groups of data. Using contrasting elements serves as an excellent way to call attention to differing display elements, since contrast forms a major source of human understanding. The screen's structure should also provide a symmetrical balance to help viewers establish a meaningful form. Reserving specific areas of the screen for various screen components also aids in establishing meaningful form. An excellent way to determine if a screen is structured properly is to ascertain if all display elements (title, captions, data, error messages, etc.) can be identified *without* reading the words which comprise the element. Display features or characteristics and/or location should be a sufficient cue.

Screens should display only relevant information, because the more information, the greater the competition among screen components for a person's attention. Screens that flood a person with too much information prolong visual search times and make meaningful patterns more difficult to perceive.

No one has yet determined what quantity constitutes 'too much' information. An ultimate answer must reflect, among other things, the application requirements and the screen structure. People tend to have subjective preferences for the amount of information presented on a display, and subjective ratings decline as the amount of information displayed deviates either way from the preferred amount (Vitz 1966). Danchak (1976) reports that a well-designed page of printed material has a density loading of only 40% and that qualitatively judged 'good' screens possessed a loading of about 15%. These values are not constructed as guidelines or absolutes, but as illustrations. The ultimate determination of guideline values will depend upon various complicated factors, many of which still remain poorly understood.

5. *Kinds of Screens*

5.1. **Data Entry Screens**

Data entry screens are designed to collect information quickly and accurately. Also called data collection screens, they usually contain a large number of captioned fields into which users key in data. Data entry screens are sometimes referred to as fixed form or form fill-in screens. All data entry screens, however, are not alike. Whether or not data keying is performed from an especially designed source document or form will cause fundamental differences in screen design.

With a dedicated source document. When keying from a dedicated source document, the document becomes the visual focus of the user's attention. The document itself contains keying aids, and the design of the screen will be interwoven with the design of the document. The two will be in the image of one another.

Without a dedicated source document. When no dedicated document exists, the user's primary visual focus is the screen itself. Screen clarity will assume a much more important role in the design process.

5.2. **Inquiry Screens**

Inquiry screens are those developed for displaying the contents of computer files. Data on these screens is permanent, and the screens are structured for ease in visual scanning Most often this visual search is through the data fields themselves, which should be visually emphasized. A columnar orientation will minimize eye movements and call attention to differences in data field sizes and structures, important cues in information location.

5.3. **Menu Screens**

The purpose of a menu screen is to permit a person to select one or more alternatives from a variety presented. It combines in a unique way the characteristics of both data collection and inquiry screens. Menu screens are particularly effective because they also utilize the more powerful human capability of recognition rather than recall. The primary design objective for menu screens is ease of visual scanning. The secondary objective is ease of alternative or choice selection.

6. *Conclusion*

Screen design, then, is a detailed process with a wide variety of considerations whose importance in the system design and acceptance process cannot be overestimated. Proper consideration of the human element in screen design will significantly close the gap between people and computers, resulting in enhanced human efficiency and effectiveness, and increased comfort. For more information on screen design considerations and guidelines see Galitz (1981).

References

Danchak, M.M., 1976, CRT displays for power plants, *Instrumentation Technology*, 23(10), pp.29–36.

Dunsmore, H.E., 1982, Using formal grammars to predict the most useful characteristics of interactive systems, *Office Automation Conference Digest*, San Francisco, 5–7 April, pp.53–56.

Galitz, W.O., 1981, *Handbook of Screen Format Design* (Wellesley, Massachusetts: Q.E.D. Information Sciences).

Vitz, P.C., 1966, Preference for different amount of visual complexity, *Behavioral Science*, II, pp.105–114.

A Comparison of Anti-Glare Contrast-Enhancement Filters for Positive and Negative Image Displays Under Adverse Lighting Conditions

B.W. McVey, C.K. Clauer and S.E. Taylor
Human Factors Center, IBM Corporation, San Jose, CA 95193, USA

Abstract

A slightly higher (1·4%) rate of random character recognition occurred with dark character than with light character presentation with high ambient illuminance and adverse glare conditions. There were no performance differences between four types of contrast-enhancing filters. However, typists agreed very significantly on an identical preference ranking of these filters regardless of image polarity.

1. *Introduction*

A study by Radl (1980) reports both performance and preference advantages for dark character versus light character VDT presentation. Unfortunately this report fails to define the ambient illuminance or display luminance adjustments. A report by Bauer and Cavonius (1980) also found improved speed and accuracy with a dark versus light character presentation. However, either the display image polarity or the reduced ambient illuminance for the light character display can explain the differences which were obtained. A later study by Cavonius (1982) with a more appropriate design for polarity comparison resulted in no significant performance differences. A study of VDT performance under high ambient glare conditions (Habinek *et al.* 1982) resulted in no polarity differences of significance and only minor differences between different anti-glare methods.

A report by Isensee and Bennett (1983) indicates no polarity difference for reflected glare discomfort. Finally, a thorough review paper by Campbell and Durden (1983) suggests that there are many overlooked factors contributing to the humorous accusation that VDT means 'Visual Discomfort Terminal'.

The purpose of this study was to evaluate contrast-enhancing filter treatments under high ambient glare conditions with both VDT image polarities. While no filter differences emerged significant on performance, they did produce a consistent preference order across polarity. A further analysis of the performance data by polarity resulted in an unexpected speed difference.

2. *Method*

2.1. **Subjects**

Subjects were 24 female typists engaged for one day each from a temporary employment agency. The agency designated that their text typing speeds are greater than 40 words per minute.

2.2. **Apparatus**

Four different contrast-enhancing filters were interchanged over the faceplate of an IBM 3101 Display Terminal. The description of the filters and their photometric transmission (T) of the display luminance follows:

AR—antireflection optical coating on grey glass (34% T)
3101—Product filter, cylindrically concave grey plastic (37% T)
Mesh—Black nylon fine mesh filter (41% T)
Matte—Green plastic with matte first surface (37% T)

The filters will be identified by the brief designations underlined above.

The experimental room was arranged to have fluorescent lights in a horizontal line on the wall behind the typist to intentionally reflect over the displayed image. The overhead lighting was also increased so that the ambient illuminance measured on the horizontal terminal table was 1076 lux. The display was inclined 15° back from the vertical and so received an incident 861 lux including the glare sources which reflected over the fifth or sixth line of the display.

2.3. **Procedure**

An IBM System/7 Computer was employed to present random upper-case alphabetic characters in four-letter groups. The typists' task was to key the displayed characters, and these keystrokes were displayed on the bottom line of the display. The RETURN key would clear the feedback line as well as the uppermost of the 20 lines of potential input to be keyed. A timer in the computer stopped each session at the RETURN keystroke next occurring after 10 minutes. Sessions alternated between dark character and light character presentations. The typists were forced to readjust the display luminance to begin each session. The adjustment left at the end of each session provided a voltage which was recorded and translated to display luminance. Five sessions were performed sequentially with each filter. The filter treatment sequence for a given typist was one of four balanced Latin square orders.

Preference data were obtained after collecting the speed and accuracy performance data from two sessions each for each filter and image polarity (the first session with each filter was ignored). First, each typist was asked whether she preferred the light or dark characters. Then two displays were placed side-by-side for a series of six paired comparisons of filters for light characters. The typists were forced to adjust the display luminance for both displays before making each preference decision. This sequence implied a rank ordering which was next checked by demonstration in side-by-side order. This procedure was then repeated for dark character polarity.

3. *Results*

3.1. **Performance**

Computation and analysis were performed for a variety of performance measures of speed and accuracy. The principal speed measure is called Free Keying Rate, which consists of the count of all keystrokes not related to operator detected error corrections divided by the time, not including such error correction keystroke times. The principal error measures are operator detected errors and operator undetected errors as a percentage of effective keystrokes.

Significant differences ($p < 0{\cdot}05$) did not occur among filters for either polarity on any performance measures. However, Free Keying Rate was significant between polarities and across filters [$F(1, 20) = 13{\cdot}3$, $p < 0{\cdot}005$]. The following tabulation shows the measured values, their difference, and the 95% confidence intervals and limits for their difference (keystrokes/second).

2·553	Dark character polarity
2·517	Light character polarity
0·036	Difference
± 0·021	95% confidence interval
0·015	Least difference 95% confidence limit
0·057	Greatest difference 95% confidence limit

Expressed as a percentage of the lower value, the difference measured 1·4%, and we have a 95% confidence that the difference is no less than 0·6% and no greater than 2·3%.

3.2. **Preference**

The polarity preference conclusion is that 17 typists favoured light characters while 7 favoured dark characters. This ratio is not regarded as statistically significant beyond chance.

The overall ranking of the four filters was exactly the same for both polarities. The preferred order was AR, 3101, Mesh and Matte. Kendall's Coefficient of Concordance (W) was computed for both polarity rankings, and the results showed very substantial agreement. For light characters $W = 0{\cdot}27$ [$\chi^2(3) = 19{\cdot}44, p < 0{\cdot}001$], and for dark characters $W = 0{\cdot}32$ [$\chi^2(3) = 23{\cdot}04, p < 0{\cdot}001$]. Friedman rank sum intertreatment comparisons with an experiment error rate of 0·05 allow the following significant difference statements: AR is preferred over both Mesh and Matte, and 3101 is preferred to Matte.

4. *Discussion*

The most surprising result of this study was the lack of a significant performance difference between contrast-enhancing filters. The strong subjective preference simply lacks a performance correlate. That might be explained by the typist's freedom to move the head and accordingly shift the position of the glare image for one or both eyes. However, the study by Habinek *et al.* (1982) constrained the head and yet only found slight differences under severe reflection conditions. That study also showed very significant preference ordering. We are forced to recognize that filter preferences develop into strong emotional attitudes without much empirical evidence of superiority!

The lack of a performance correlate with preference that is seen in the filter variable is also found again in the polarity variable. Despite the trivial performance difference favouring dark characters, three typists showed overall Free Keying Rate advantages with light characters. One of these subjects was 60-years-old. It should be pointed out that there is a considerable advantage to light character polarity for retinal image contrast because of optical scattering of light within the eyes, particularly for older people. This suggests an age discrimination bias favouring light character presentation.

There are basically two kinds of tasks performed with visual displays: (1) character recognition, and (2) word recognition. The identification of random letters with foveal acuity is a prototype for the first kind and was used in this study. The second kind is exemplified in reading and visual search tasks where the parafoveal acuity ($\pm 5°$) is more important than the foveal ($2°$). The relationship between performance and display image quality is probably more apparent for those tasks that are more dependent on parafoveal inputs, tasks such as word or phrase recognition and visual search. Further studies should emphasize the parafoveal tasks and should take age as an independent variable. It should also be recognized that the effective stroke width of the character font design should be greater with dark character polarity. All of this suggests that there is much work to do before polarity conclusions can be empirically supported and generalized to benefit users. Certainly the trivial recognition speed advantage which we found under severely adverse conditions has no practical utility.

Acknowledgement

The authors wish to thank our colleague, Dr Michael J. Darnell, for providing the complete statistical analysis of the performance data.

References

Bauer, D. and Cavonius, C.R., 1980, Improving legibility of visual display units through contrast reversal. In *Ergonomic Aspects of Visual Display Terminals*, edited by E. Grandjean and E. Vigliani (London: Taylor & Francis) pp.137–142.

Campbell, F.W. and Durden, K.., 1983, The visual display terminal issue: a considerations of its physiological, psychological and clinical background. *Ophthalmology and Physiological Optics*, **3**, 175–192.

Cavonius, C., 1982, Changes in contrast and sensitivity from VDT usage. *IX International Ergophthalmological Symposium* (San Francisco).

Habinek, J.K., Jacobson, P.M., Miller, W. and Suther, T.W., 1982, A comparison of VDT anti-reflection treatments. *Proceedings of the Human Factors Society—26th Annual Meeting*, 285–289.

Isensee, S.H. and Bennett, C.A., 1983, The perception of flicker and glare on computer CRT displays. *Human Factors*, **25**, 177–184.

Radl, G.W., 1980, Experimental investigations for optimal presentation-mode and colours of symbols on the CRT-screen. In *Ergonomic Aspects of Visual Display Terminals*, edited by E. Grandjean and E. Vigliani (London: Taylor & Francis) pp.127–135.

Measurements of Character Contrast and Luminance Distribution on Data Screen Workstations

L. Agesen
Brüel & Kjær,
Naerum, Denmark

Abstract

The character contrast and distribution of luminance at a work place which includes a data screen are two important visual parameters which can be helpful for designing and setting up lighting in offices and computer rooms. A method for measuring the character contrast on a data screen is proposed. The method is based upon representing the character luminance of homogenous dot-fields on the screen or of fields composed of characters. Furthermore, measurement of mean luminances for the different areas of the work field is mentioned. Finally, a measurement report suggestion is illustrated.

1. *Character Contrast*

1.1. **Screen**

Many parameters influence the readability of a screen. For instance, the dimensions of the characters and the sharpness of their contours, and the colour of the characters and background. However, one of the most important parameters undoubtedly is the luminance contrast between the letters on the screen and the background. The contrast of a visual paper task is often defined as

$$C = (L_P - L_C)/L_P$$

where L_P is the luminance of the paper and L_C is the character luminance. However, for VDUs another definition of contrast is normally used:

$$C = (L_C/L_B)$$

where L_B is the background luminance. Thus the simple *luminance ratio* between character and background luminance is used.

If a reflection is superimposed upon the luminance generated by the screen itself, the resultant contrast will be:

$$C = \frac{L^*_C + R}{L^*_B + R}$$

where L^*_C and L^*_B are the luminances generated by the screen and R is the luminance of the reflection. It is clear from this expression that the larger R is, the poorer the contrast will be.

Measuring this luminance contrast is not a straightforward task. The screen character is built up of small 'dots' or line elements, and the obvious approach might be to measure the luminance of just one dot or line element.

However, measurements with very fine luminance meters show that a picture element does not have a very uniform luminance with a sharp cut-off at the borders. On the contrary, the dot luminance has a maximum in the middle from which it rolls off rather slowly. Moreover: The shape of this curve is not constant for different luminance settings of the screen.

Brüel & Kjær therefore proposes a measuring technique based on the generation of uniform dot fields on the screen. The fields should be composed of dots with the same luminance as those forming the characters. On the assumption that the field luminance is representative of the character luminance, the measuring problem is reduced to the measurement of luminance from areas of a comfortable size.

Moreover, contrast mapping of the screen in e.g. 9 measuring fields is proposed (see Figure 1). Of these, the 5 upper fields are most important, since the contrast of the lower

FIGURE 1. *A data screen with 9 measuring fields and the operator reflected in the screen.*

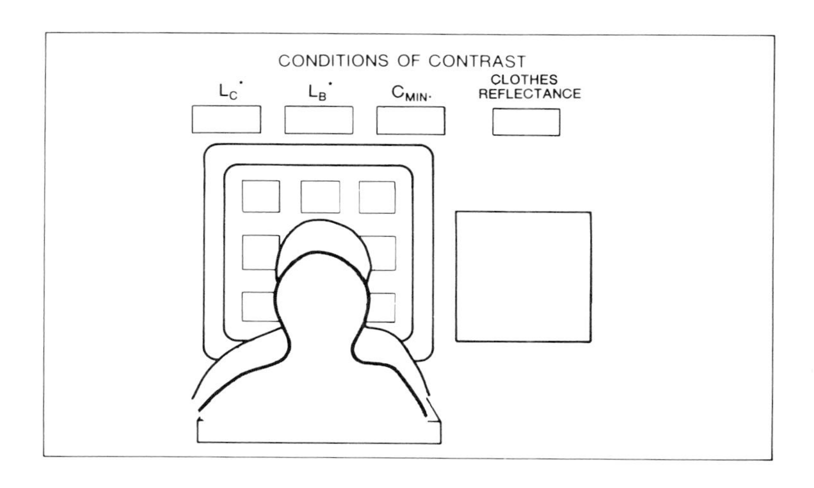

fields is highly influenced by the screening effect and clothing of the operator. Consequently, the reflectance of the clothing of the measuring technician should be stated in the measurement report. The contrast of the 9 fixed fields should be supplemented by the minimum contrast for the whole screen. The minimum contrast will be found where a particularly pronounced reflection is situated.

L^*_C and L^*_B, the self-generated luminances should be measured either by removing the ambient illumination, or by placing the luminance detector directly on the screen. These values are important for later comparative measurements.

It should be remembered that the contrast seen by the observer's eye is composed of the screen generated luminances and the luminance of objects reflected in the screen surface.

Thus the measurement technique must be referred to the operator's eye position. The luminance detector should be placed in the typical eye position of the operator. From there the luminance detector should be aimed at the different measurement fields (Figure 2).

FIGURE 2. *The luminance detector should be situated at the operator's eye position. From there, it can be aimed at any area of the total work field.*

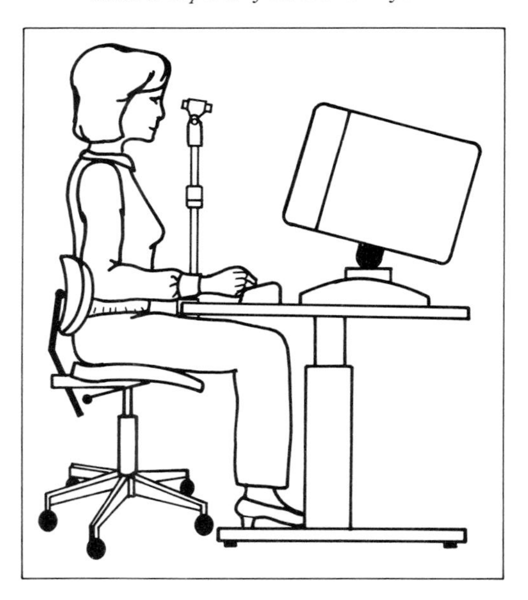

To measure the desired luminances, $(L^*_C + R)$ and $(L^*_B + R)$, it is necessary to ensure that R is the same for both measurements. Thus the luminance detector should be aimed at the same position on the screen for both measurements. The two desired luminances are obtained by measuring with and without the dot-field (Figure 3).

To find the minimum contrast for the screen, it is necessary first to find the reflection producing the maximum luminance. Having found the field of maximum luminance (R_{max}),

FIGURE 3. *Reflection of a luminaire. To be measured with and without dot-field.*

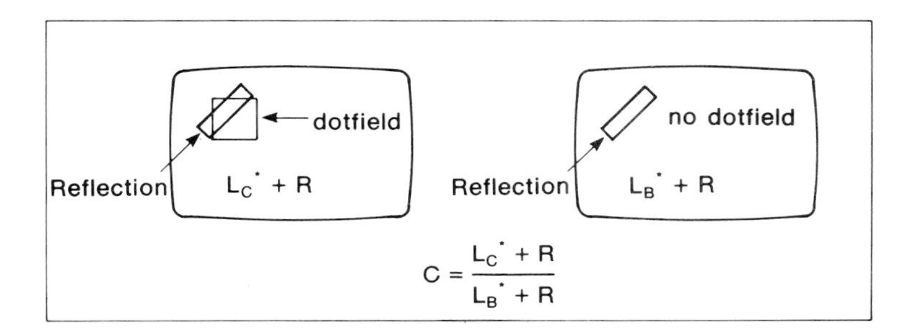

the remaining task is to generate a uniform dot-field 'behind' this reflection, and then measure R_{max} with and without the dot-field.

The measurement results should be stated in a measurement report. So far, very little standardization exists. In a DIN standard, West Germany has specified that C should ideally be between 6 and 10, but contrast from 3 to 15 'might' be tolerated. Thus the measured screen here (Figure 4), just about fulfills 'Class 2', not 'Class 1'. The situation of C_{min} on the screen is also shown.

FIGURE 4. *Example of character contrast mapping of a data screen.* L^*_C *and* L^*_B *are the character and background luminances generated by the VDU itself.*

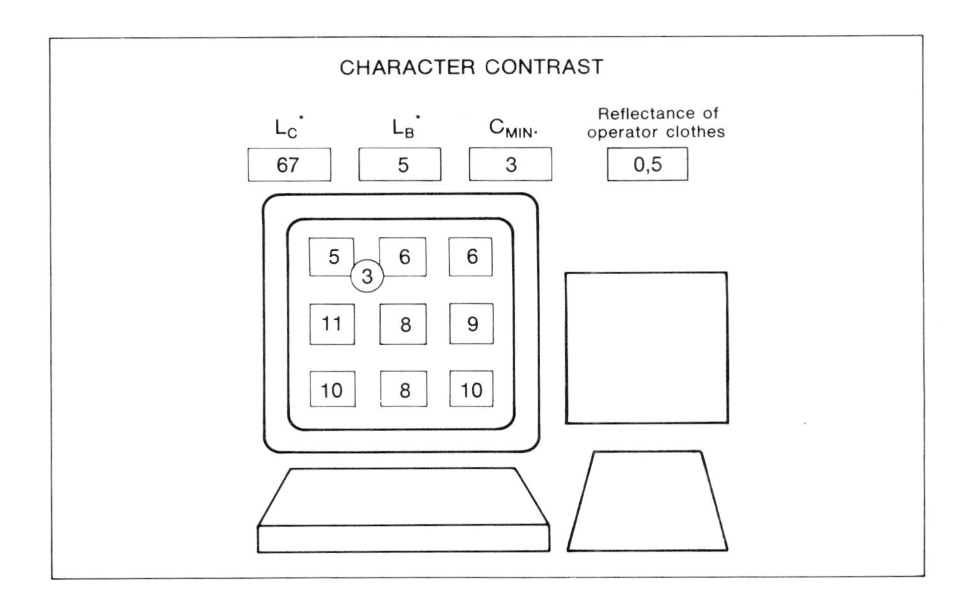

1.2. **Keyboard**

It is difficult to assess the keyboard character contrast. A light source might reflect itself in a corner of the keys. To measure these luminances, quite fine optics are necessary, and even then, this will not tell you the extent to which the measured luminance means a reduction in contrast.

2. *Luminance Distribution*

The other important aspect of the visual environment of a data screen work station is the luminance distribution of the different surface components of the visual environment.

The main components are the screen, the manuscript, the keyboard, and the surroundings. Since the surroundings might be changing, more than one measurement field might be chosen.

The distribution of the luminances of the different fields is of interest, because these luminances control the adaptation state of the eye.

In order to assess such 'adaptation luminances', a luminance average of a certain field size should be measured. A reasonable field size is obtained by using e.g. a ±10° or ±15° opening angle from the observer's eye position. The resulting field size is suggested in the

FIGURE 5. *Approximate field size for measuring average luminances using wide angle adaptor. Examples of average luminances with calculated ratios between those referred to the screen.*

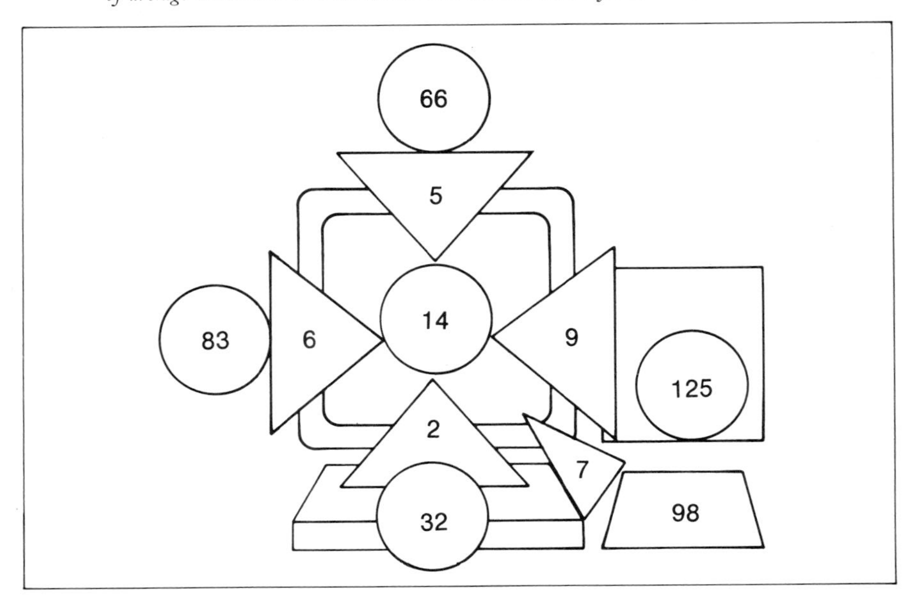

measuring report example (Figure 5). The average screen luminance should be measured with a 'typical' record on the screen.

Having measured the different average luminances and written them into the report sheet, it only remains to calculate the ratios between these average luminances. The most critical one here—and this is typical of screens with positive contrast—is the ratio of 9 for manuscript/screen. This figure, in fact, is lower than the average figure found in actual situations, but yet too high to be recommended. A ratio of 3 or perhaps 5 should be the maximum allowable.

A high ratio is also seen between screen and desk top document, but shifts between these two fields might be more scarce.

3. *Conclusion*

A measurement report composed of Figures 4 and 5, plus a description of the ambient conditions for the measurements would be useful. The suggested measuring methods for character contrast and luminance distribution are open to discussion. The need for such methods is generally recognized.

References

DIN 66234, 1981, *Bildschirmarbeitsplätze*. Part 2, Entwurf.

Kokoschka, 1980, Visual criteria for lighting video display positions. *International Lighting Review*, 4.

Läubli, T., Hünting, W. and Grandjean, E., 1981, Postural and visual loads at VDT work places. *Ergonomics*, 24(12).

Snyder, H.L., 1983, Visual ergonomics and VDT standards. *Digital Design*, Feb.

Travaux devant ecran cathodique, 1981, Recommendation R 198, INRS Publication, ND 1359–105–81.

Visual-Photometric Problems of VDUs in Relation to Environmental Luminance (Introductory paper)

L.R. Ronchi and F. Passani
Istituto Nazionale di Ottica, Arcetri, Florence, Italy

1. *Introduction*

The photometric relations within the display and between the display and the environment are usually approached in terms of contrast ratio, surround contrast and discomfort glare, the overall adaptational state of the retina being more or less momentarily affected by maximum luminance ratios within the operator's visual field.

The foveal–extrafoveal vision dichotomy, from the ergo-ophthalmological point of view, is being considered by an appropriate 'working group' on occupational functional visual field (Verriest *et al.* 1982). Amongst others, the 'colour' is a relevant factor. Some authors have considered its emotional impact on employees, and subjective preferences have been assessed. Others (Bouma 1980, Engel 1980) have noticed that since the colour itself may be well visible far from fixation, it contributes to a wide conspicuity area, which is a significant tool for assessing the influence of colour in displays, as well as for research on information selection from VDUs in general. Specific works have been recently devoted to the usefulness of colour as an aid for visual search (Boynton 1978, Bouma 1980).

The colour on a display is generated by the phosphor, the emission of which is confined to a spectral band. Now, official photometry is a tool *ad usum* of foveal vision, at photopic levels, for broad band stimuli. Hence, two stimuli, with different bandwidths, matched in luminance, do not necessarily match in brightness (Kaiser 1971). The VDU screen seems to be narrow-band enough to imply a brightness–luminance discrepancy. This is shown by some preliminary observations (Ronchi and Cicchella 1980), and can be predicted by the estimates of the J-factor (Boynton 1978), as is shown, for instance, in Table 1.

The present report aims at describing some experimental data, recorded in a laboratory situation, where the observer was presented, in foveal vision, with a white test-spot, while a monochromatic test spot was presented at a given eccentricity. Since extrafoveal brightness perception is biased by rod intrusion, we expect both a wavelength dependence and a dependence on retinal adaptational state (as dictated by environmental luminance) as a function of the dominant wavelength of peripheral stimulus. Of course, the fact that extrafoveal

TABLE I. *Dominant wavelength, chromaticity coordinates and J-factor for a number of phosphors.*

Phosphor	λ_{dom} (nm)	x	y	J-factor
P_1	533	0·218	0·712	1·208
P_3	579	0·507	0·485	1·106
P_4	481	0·270	0·300	1·040
P_{11}	478	0·139	0·148	1·390
P_{15}	507	0·221	0·438	1·209
P_{20}	566	0·427	0·549	1·006
P_{22B}	440	0·153	0·048	1·490
P_{22R}	Red/purple	0·442	0·202	1·800
P_{24}	528	0·263	0·476	1·167
P_{25}	576	0·492	0·505	1·240

colour perception also differs from the foveal one, together with sensitivity and brightness perception (Ernst, 1968) is expected to play a role.

2. *Apparatus and Method*

The observer was seated in a booth where internal walls were kept at a constant luminance, from 30 cd/m^2 to complete darkness, according to a pre-arranged schedule. The head was kept in position by means of a chin rest. The left eye was occluded, while the right (viewing) eye looked at a fixation point. After a warning signal, two circular test spots, subtending at the eye 1° each, were presented. The foveal spot was white (A-ill.), that presented in the temporal horizontal meridian, at 10°, was monochromatic (an interference filter was used). The exposure time was 3 sec. The observer was requested to tell, for each filter setting varying white luminance, whether the white spot appeared brighter or less bright than the monochromatic one. A threshold, at the brightness match, was estimated (constant stimulus method and Probit Analysis). Being w and c the white and monochromatic luminances, at the brightness match, the deviation from unity of c/w ratio is assumed as a measure of the brightness luminance discrepancy.

Five normal young adults, highly skilled and cooperative, took part in repeated sessions.

3. *Experimental Findings and Discussion*

Figure 1 shows the wavelength dependence of the ratio of coloured (peripheral) to white (foveal) test fields, the eye being dark adapted, and the environment being dark. Although individual differences in the absorption of macular pigment cannot be ruled out, the spread of data is bound to the intrinsic variability of simultaneous heterochromatic brightness

FIGURE 1. *Lower plot. Abscissae: dominant wavelength of the extrafoveal stimulus. Ordinates: ratio of luminance (c) of peripheral coloured stimulus to luminance (w) of the white foveal stimulus, at the brightness match. Data from different observers are put together. Environmental luminance: darkness. Dark-adapted eye. Upper plot: Optical density for macular pigment and eye lens (standard correction data).*

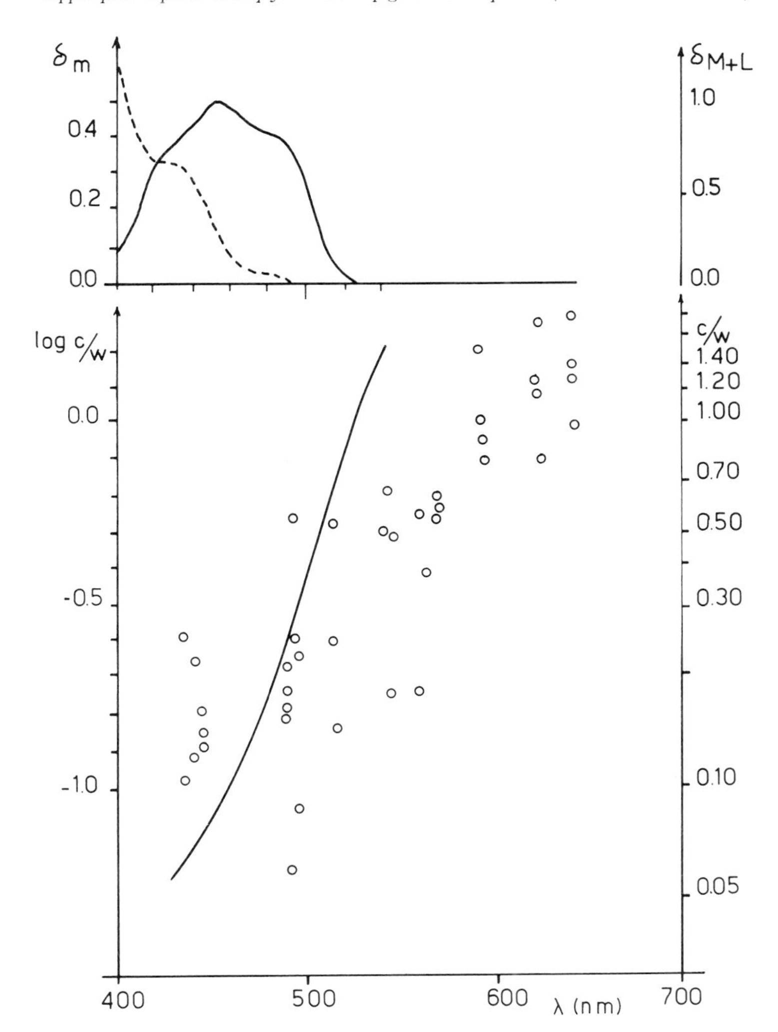

matches. For the sake of reference, the wavelength dependence of ratio V_λ / V'_λ (thin full line, in the figure), is also drawn. These data indicate the effect of rod intrusion on brightness perception, according to the expectations based on classical retinal duplicity theory. As a further support to this explanation, we may adduce (Figure 2) the time dependencies of c/w ratio, during the recovery in the darkness after the offset of an adapting field (10 000

FIGURE 2. *Abscissae: time (in minutes) after the offset of a 10 000 cd/m² white adapting field. Ordinates: log ratio of colour-to-white luminances, at the brightness match. Data from one observer. Label denotes the dominant wavelength of the extrafoveal test field.*

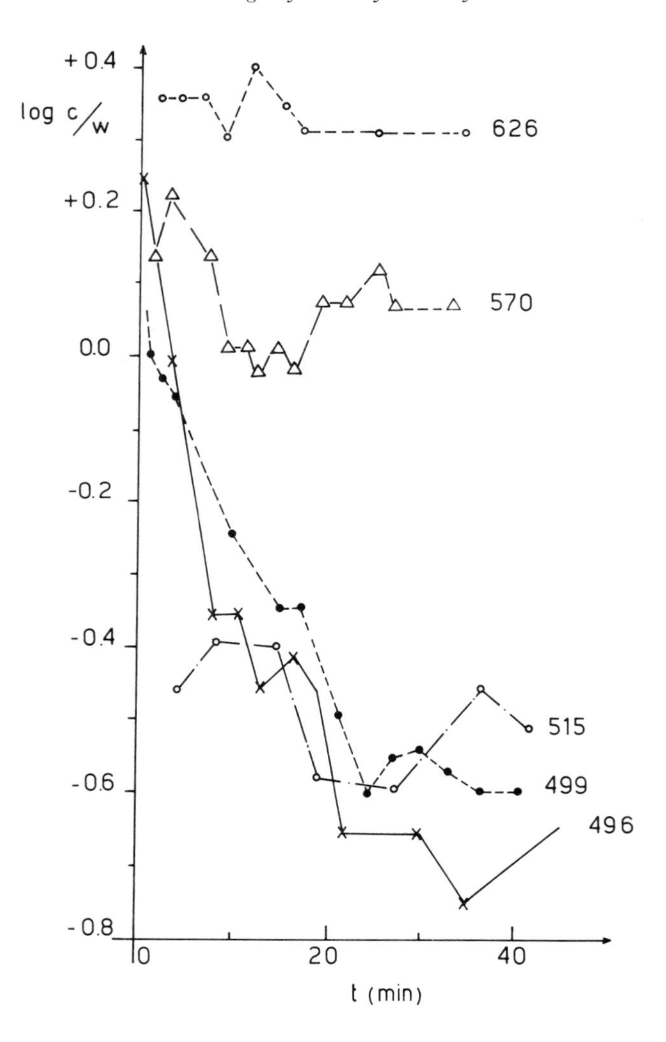

cd/m², A-ill.). When the dominant wavelength of the (peripheral) field is long, *c/w* ratio is practically independent of adaptational state. This seems to indicate that the time changes (if any) of the contribution to brightness on the part of foveal cones runs parallel to that of the cones stimulated by the peripheral test spot. This is no longer the case at shorter wavelengths, because rods are becoming increasingly active in the darkness, so that *c*, at the brightness match, decreases.

FIGURE 3. *Abscissae: dominant wavelength of the extrafoveal test field. Ordinates: c/w ratio at the brightness match. Data from one observer. Bars: Standard Deviation for repeat variability. Crosses and full point: environmental luminance 3 and 30 cd/m², respectively.*

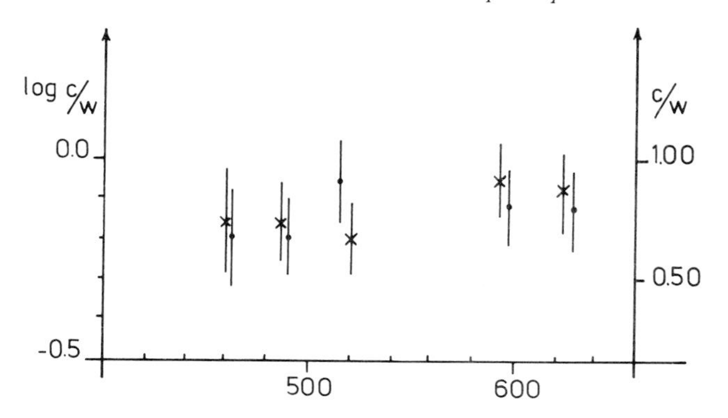

Let us consider now the case where the environment is kept at a given luminance, instead of being dark. As is shown, for instance, in Figure 3, the wavelength dependence of *c/w* is now minimized. Apparently (Table 2), rod contribution can be neglected as long as environmental level remains photopic.

TABLE 2. *Dependence of coloured field luminance (at 10° eccentricity) to white field luminance (foveal vision) on environmental adapting (white) luminance,* L.

L (cd/m²)	c/w
30	0·71
3	0·63
1·5	0·58
0·19	0·38
Darkness	0·25

In conclusion, the foveal J-factor prediction can be extended to extrafoveal vision, at least within our experimental conditions, provided rod contribution is excluded.

References

Bouma, H., 1980, Visual reading processes and the quality of text displays. In *Ergonomic Aspects of Visual Display Terminals*, edited by E. Grandjean and E. Vigliani (London: Taylor & Francis) pp.100–114.

Boynton, R.M., 1978, *Human Color Vision* (New York: Holt, Rinehart & Winston).

Engel, F.L., 1980, Information selection from visual display units. In *Ergonomic Aspects of Visual Display Terminals*, edited by E. Grandjean and E. Vigliani (London: Taylor & Francis), pp.121–126.

Ernst, W., 1968, The dependence of critical flicker frequency on rod threshold and on the state of adaptation of the eye. *Vision Res*, **8**, 889–895.

Kaiser, P., 1971, Luminance and brightness. *Applied Optics*, **10**, 2768–2770.

Ronchi, L. and Cicchella, G., 1980, Dioptric problems in connection with luminance-brightness relationships on VDUs. In *Ergonomic Aspects of Visual Display Terminals*, edited by E. Grandjean and E. Vigliani (London: Taylor & Francis), pp.65–70.

Verriest, G., Barca, L., Dubois-Poulsen, A., Houtmans, M.J., Johnson, C.A. and Ronchi, L., 1982 The functional visual field. I. Theoretical aspects. In *Proceedings of the IPS Symposium* (Sacramento, Ca), 20–23 Oct.

Two New Visual Tests to Define the Visual Requirements of VDU Operators

J.J. Meyer, A. Bousquet, P. Rey and J. Pittard
Laboratoire d'Ergonomie, University of Geneva, Switzerland

Abstract

This paper refers to laboratory data obtained with two experimental devices which are aimed to simulate the visual conditions which are encountered in visual display units. Operators with slight visual defects were tested. It is demonstrated that to satisfy the visual requirements of older or handicapped operators without disturbing more sensitive subjects, the opportunity should be given to settle a high level of luminance on the TV screen without producing flicker. The recommended values are then: adjustable luminances from 1 to 100 cd/m^2 either for white or coloured letters (without signal distortion) and 90 c/s for refreshing rate. Furthermore, it is suggested that similar testing procedure should be applied when investigating the visual adaptation of VDU operators.

1. *Introduction*

Our objective, in the present paper, is to suggest that the combination of two testing procedures which are aimed to investigate luminance sensitivity, either in terms of spatial or in terms of temporal resolution power, may help to understand the visual troubles of VDU operators and design better material and work places. The presented data were obtained with two prototypes and an experimental set-up which had been developed in our laboratory to simulate the working conditions at VDUs and have already been referred to in previous papers (Rey and Meyer 1980, Meyer *et al.* 1982 and Meyer 1982).

2. *Methods*

Three experiments were performed by 7 operators who had been selected for slight visual defects (Table 1). In the first experiment, the C45 Rey-Tagnon simulator was used.

TABLE I. *The visual characteristics of the 7 operators.*

N: emmetropic subject
Hy: hyperopic subject
Ast: astigmatic subject
Pr: presbyopic subject
1d, 2d, 3d: 1, 2, 3 dioptria
Dp: Deuteranope (green-blind)
PN: Protan (red-blind)

Subjects		D1	S1	P1	S2	P2	S3	S4
Age		21	28	26	24	45	62	56
Visual acuity	Far	12	12	12	12	10	7	8
	Near	12	12	12	12	10	8	9
Near point		10	19	12	10	32	22	28
Red–green	Far	R +	R = G	R = G	R +	R +	G +	G +
test	Near	R = G	G +	R = G	G +	G +	G +	R = G
Diopt. anom.		N	HY	N	MY	HY AST	PR	PR
Correction		—	—	—	1d	3d	2d	1d
Colour vis.		Dp	—	Pn	—	Pn	—	—

Subjects were requested to adjust the luminance of a Landolt ring of decreasing size located at 50 cm from the eye, the background luminance being settled at 0·1 and the panorama luminance at 10 cd/m^2. With such a procedure, visual acuity can then be plotted as a function of the test object luminance (= luminance–visibility function).

In the second experiment, the operator measured his flicker sensitivity with a Meyer-Richez device which produces automatically a DeLange curve (= luminance–flicker function). Fusion thresholds were established for areas of increasing size (1°, 3°, 6° and 12°).

In the third experiment, the operator was required to adjust the luminance of letters on a conventional TV screen (display A) at threshold or for comfortable reading and also to increase the luminance of the background up to flicker (display B). This procedure was repeated for two frequencies of the refreshing rate: 50 and 60 Hz.

In all three experiments, measurements were performed with white, red and green light (Kodak wratten filters 25 and 58 respectively).

3. *Results*

In Figure 1, the luminance–visibility function is presented for two cases: a green-blind subject of 20 years (D1) and a red-blind subject of 45 years (P2), facing white or green or red Landolt rings in the C45 simulator. Figure 1 illustrates at first the ageing effect: that is a shift of the curves from left to right (see also Figure 2). One sees also that the green-blind

FIGURE 1. *Examples of luminance–visibility curves as obtained with the C45 simulator in three colour conditions.*

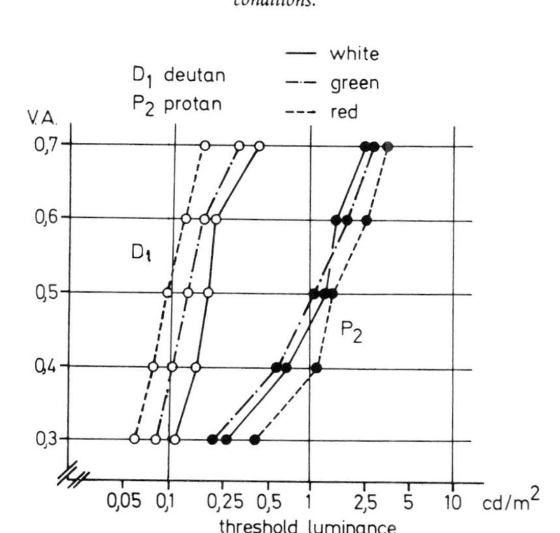

operator's sensitivity (D1) is favoured by red light, while the red-blind operator's sensitivity (P2) is favoured (if not in the same amount) by green light. But for the subject P2 (see table 1) who is also characterized by hyperopia and presbyopia, the colour effect may also be due to dioptric factors. Indeed, Figure 2 shows that inadequate dioptric correction for myopia (subject S2) or for near sight at 50 cm (subject S3) may produce an appreciable sensitivity improvement for red light, while the reverse could be predicted when testing them with a mass-screener Visiotest (red–green test) (Table 1). Such a result may question the validity of testing devices which are designed for near vision at 30 cm, for the eye examination of VDU operators.

Light sensitivity at VDU in the same subjects, when tested with a conventional display, was shown to fit on the whole their C45 performance (Figure 2). Considering, on the one hand the threshold light level range characterising operators aged from 18 to 65 (up to 16 cd/m^2; Meyer *et al.* 1982) and, on the other hand, the supplement light requirement for comfort (5 to 10 times), one can deduce from these data that to satisfy a majority of operators, the upper luminance limit of a display screen (without distortion) should reach 100 cd/m^2.

Figure 3 displays for our two subjects (green-blind and red-blind) the range of luminances at which, for different frequencies of the Meyer-Richez apparatus, flicker is not perceived (we call them flicker-free luminances). The age effect is again a loss in sensitivity (see also Figure 5). In other words older operators will be less sensitive to VDU fluctuations than are younger operators but they will need more light to read letters properly. The colour effect previously described is here too noticeable.

Both age effect and colour effect are accentuated when the size of the stimulating area

Figure 2. *Light requirement as measured for 7 operators on both display screen and simulator with red and green light. The C45 simulator's thresholds are expressed in mean luminance calculated on the luminance–visibility curves. Horizontal lines mark the limits of the brightness interval without distortion (without filters this interval would have been situated five times higher). Subjects symbols D1 S1 . . . correspond to those of Table 1. To be noted: for the three older people the simulator luminance thresholds overlap the screen adjustment values, indicating that for them, the screen brightness interval is too small.*

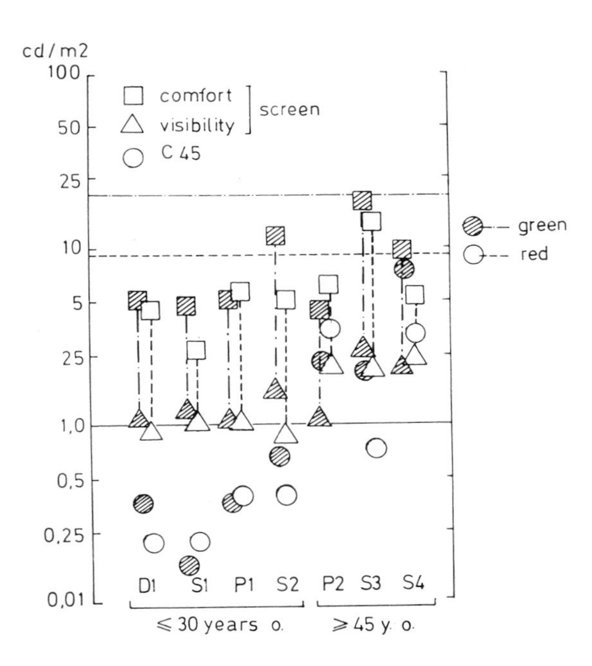

is increased (Figure 4). It is the reason why acceptable luminance levels of VDUs should be settled with reference to the visual angle under which the oscillating area is perceived.

According to the DeLange model, it should be possible, from the knowledge of flicker fusion frequencies in a given population to deduce the luminance for flicker-free perception at a given sweeping rate. It may also be predicted what would be the appropriate sweeping rate for flicker-free perception even in very sensitive operators. To verify this hypothesis, we plotted the observed flicker-free luminances, obtained with a conventional TV screen (characterized by a 50 Hz sweeping rate and a short light emission decay), as a function of flicker fusion frequencies at 100 cd/m^2 (deduced from flicker-test data). Several conclusions can be deduced from Figure 5:

1. The screen may be perceived as flickering even in people with a flicker fusion frequency (CFF) lower than the sweeping rate (CFF as low as 45 Hz). This phenomenon has to do with the flashing aspect of the screen enlightenment. When expressed in its

FIGURE 3. *Examples of luminance–flicker perception curves as obtained with the Meyer-Richez flicker test in three colour conditions and a 6 degrees stimulating area.*

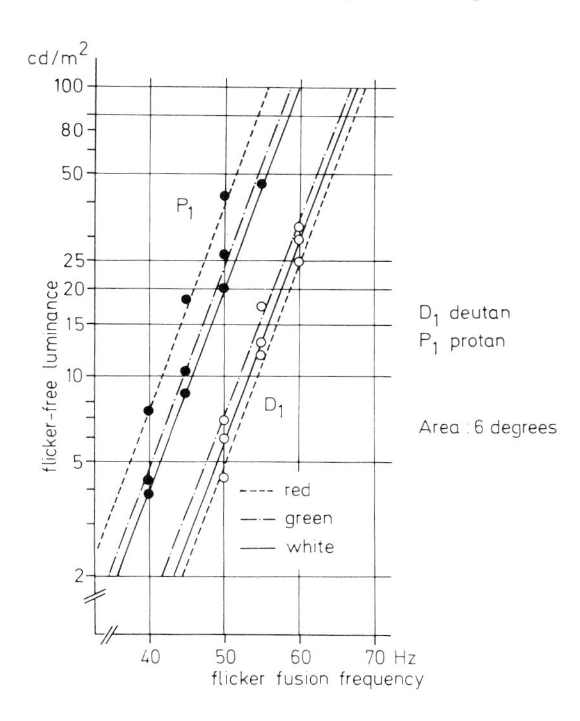

sinusoidal DeLange equivalent, the corresponding modulation overpasses 100%, it is supermodulated (cf. Kelly 1954).

2. When increasing the sweeping frequency from 50 to 60 Hz, the increment of flicker-free luminance was approximately the same for all subjects in all conditions, as expected from Figure 3 (five times).
3. For all individual CFF higher than 45 Hz, i.e. 5 Hz below 50 Hz (the sweeping rate in our case), the corresponding available flicker-free luminance decreases in inverse proportion to the flicker sensitivity: for CFF situated between 45 and 60 Hz, the slope of the flicker-luminance relation matches expectancy; for CFF higher than 60 Hz the slope becomes steeper, indicating that flicker sensitive people react to the screen oscillations more than expected. When increasing the sweeping frequency, the flicker-luminance is shifted upward in a parallel direction, with a gain of five times per 10 Hz sweeping rate increase.

To explain all these results, our hypothesis is close to the one of Bolusset and Devauchelle (1980), that means that flicker perception on the VDU screen is determined not only by the periodic refreshment of the phosphorus, but also by the combined effects of the complex sweeping process and eye movements.

FIGURE 4. *Flicker-free brightness which could be settled by two subjects tested with a conventional VDU screen for four areas and three colours. To be noted: the sensitivity difference, which is mainly due to age, increases with the perceived area.*

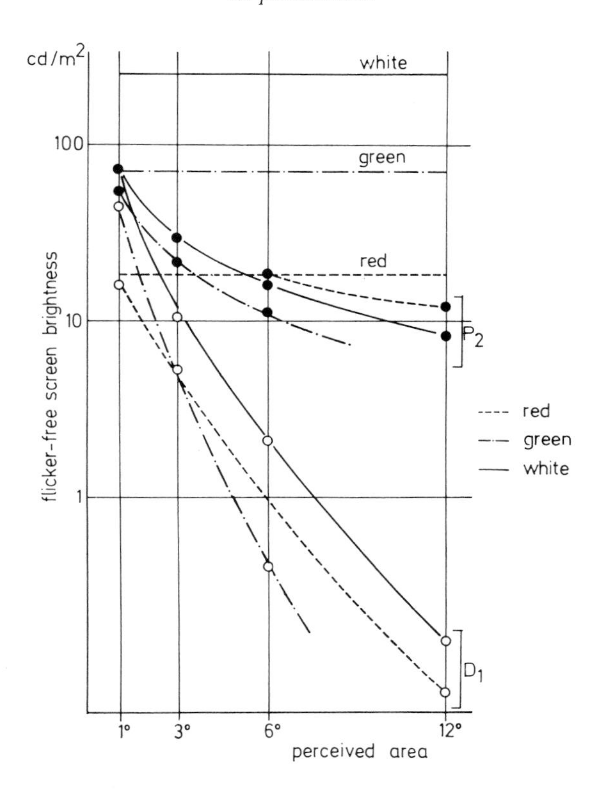

From a practical point of view, this relation between flicker fusion frequency and screen flicker-free luminance allows us to predict what should be the screen sweeping rate warranting a flicker-free perception at 100 cd/m^2 for the most sensitive people: for example an operator whose CFF may be as high as 70 Hz (for 100 cd/m^2 and 6 degree area, cf. Kelly 1954) would require a refreshment rate increase up to 90 Hz, to shift his flicker-free luminance from 0·3 to 100 cd/m^2. These values are comparable with those observed by Bauer and Cavonius (1980).

4. *Conclusion*

With our prototypes, the visual performance at the work place is expressed by parameters which are broadly used in visual investigation, either by physiologists or by eye doctors.

FIGURE 5. *Flicker-free luminances in relation to flicker fusion frequency for 7 subjects tested for a 6 degrees area and two colours on both simulator and display screen (contrast reversal situation). Expected values (parallel lines) and observed values for a 10 Hz increase of the screen refreshment rate.*

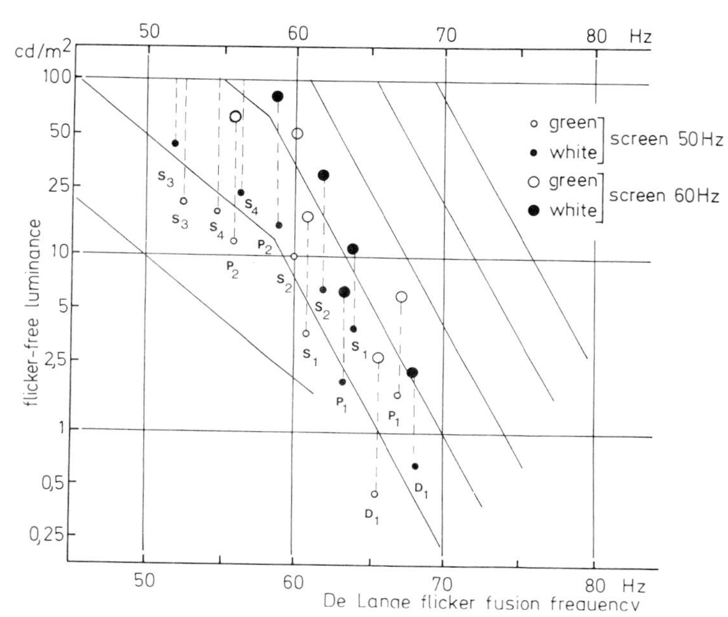

Such a procedure presents two advantages: on one hand, results are independent from the specificity of materials and may be applied to any type of VDU; secondly, our testing methods are appropriate for the eye doctors to evaluate and predict visual capabilities and difficulties of VDU operators, at the time of recruitment.

Since with our methodology, the visual performance is investigated for various physical conditions which are components of the VDUs' images and screen, physiological data may be directly translated in concrete data for the designers.

References

Bauer, D. and Cavonius, C.R., 1980, Improving the legibility of VDUs through contrast reversal. In *Ergonomic Aspects of Visual Display Terminals*, edited by E. Grandjean and E. Vigliani (London: Taylor & Francis), pp.137–142.

Bolusset, C. and Devauchelle, P., 1980, Lecture sur écran de visualisation. Influences dues aux caractéristiques du terminal. *L'Echo des Recherches*, 16. (Issy Les Moulineaus, France: CNET–DICET).

Kelly, D.H., 1954, Sine wave and Flicker Fusion. *Proceedings ISCERG* (The Hague: W. Junk) pp.16–35.

Meyer, J.J., 1982, An automatic intermittent light simulator to record flicker perceptive thresholds in patients with retinal diseases. 2nd International Symposium on Visual Optics, Tucson, Arizona 1982. To be published.

Meyer, J.J., Rey, P., Bousquet, A. and Pittard, J., 1982, Contribution à l'étude de la charge visuelle aux postes de travail aved écran de visualisation. *Klin. Mbl. Augenheilk.*, 180, pp.373–376.

Rey, P. and Meyer, J.J., 1980, Visual impairments and their objective correlates. In *Ergonomic Aspects of Visual Display Terminals*, edited by E. Grandjean and E. Vigliani (London: Taylor & Francis), pp.121–126.

Considerations on Ocular Motility and Refractive Errors in VDU Operators

F.M. GRIGNOLO, F. VITALE BROVARONE, D.G. ANFOSSI AND G. VALLI
Istituto di Clinica Oculistica dell'Università, Turin, Italy

Abstract

The authors have examined a population of operators who use VDUs for their whole working time. These subjects constitute a uniform population sample as they all work in the same place, devised and realised ergonomically, using the same type of VDU for the same number of hours. Subjective and objective ocular symptoms were analysed and related to the refractive and orthoptic conditions.

1. *Introduction*

It is by now common knowledge that a good number of VDU operators report a series of more or less intense complaints, which clinically speaking are not always clear. Finding the answers and determining the cause of these complaints has always been difficult.

Analysis of this ergonomic aspect is further complicated by the fact that individual operators present different working conditions (office environment, type of job, working hours, type of VDUs, etc.) and this great number of variables has meant great complexity in any intended evaluation.

We were offered the opportunity to analyse a large and extremely uniform group of subjects, from which we felt significant results could be obtained in understanding the ophthalmic behaviour and attitude of VDU operators.

2. *Material and Methods*

The group consisted of 143 VDU operators working in the same environment, designed according to the latest ergonomic standards, for the same number of hours and with identical terminals.

TABLE 1. *Refraction: the score of each eye was added to the one obtained from the other eye.*[a]

Hypermetropia	Astigmatism	Myopia	Score
< 1 D	< 1 D	< 2 D	0
1–2 D	1–2 D	2–3 D	3
2–3 D	2–3 D	3–6 D	5
> 3 D	–	7–12 D	8
> 5 D	> 3 D	> 12 D	10

[a] The score was doubled when anisometropia higher than 2·5 D was found.

Each subject underwent a refraction test and, if present, visual defects (hypermetropia, myopia, astigmatism and anisometropia) and their correction were recorded and evaluated according to the degree of the defect (Table 1). This was followed by a thorough orthoptic examination to determine the possible presence of phoriae or tropiae, the subjective and objective angle of deviation using the major amblioscope, the convergence fusional power, the possible suppression using the Worth four dots test, the stereopsis using the Titmus stereotest, the convergence power and the punctum proximum convergentiae (PPC). Scores were assigned to each of the above parameters following the degree of the defect, as can be seen from Tables 2, 3, 4, 5, 6, 7.

Following the examination of each subject the refraction scores were summed to those

TABLE 2. *Cover test: the far vision score was added to the near vision one.*

Orthoptic alteration	Score
Ortophoria	0
Exophoria (1–10$^{\Delta}$)/esophoria (4– 6$^{\Delta}$)	1
Exophoria (11–24$^{\Delta}$)/esophoria (8–15$^{\Delta}$)	2
Exophoria-tropia/esophoria (> 15$^{\Delta}$) microtropia (1–8$^{\Delta}$)	3
Exotropia/esotropia	4

TABLE 3. *(Major amblioscope) objective angle of the deviation.*

Degrees	Score
0–2°	0
3–4°	1
5–6°	3
6°	5

TABLE 4. *Fusional power (major amblioscope).*

Degrees	Score
15°	0
12°–15°	1
9°–11°	2
5°–8°	3
2°–4°	4
Absence	5

TABLE 5. *Vision suppression (worth four dots test): the far vision score was added to the near vision one.*

Results	Score
Four lights	0
suppression	3
Diplopia	5

TABLE 6. *Stereopsis (Titmus test).*

Four rings test number	Score
7–9	0
5–6	2
3–4	4
1–2	6
Absence	10

obtained from eventual orthoptic defects. This gave us a precise numerical indicator of the efficiency of the subject's visual capacity.

At the end of the examination we then handed the subjects a questionnaire in which they were asked to specify which of the following ocular complaints they experienced at the end of a day's work (visual fatigue, itchy eyes, eyes watering, conjunctival hyperaemia, blinking, headache, double vision, unfocused vision, ocular pain) and with which intensity (from 0 to 4) they were felt. They were also asked to specify after how many hours of work the complaints appeared and whether they had ever been experienced previous to the subject's working with a VDU. Lastly, the questionnaire asked how long the subject had been working with a VDU and when he or she had last undergone a complete ophthalmic examination.

TABLE 7. *Convergence power and punctum proximum convergentiae: the scores of the two parameters were evaluated separately and added to each other.*

Distance	Score
Up to 10 cm	0
From 11 to 15 cm	2
From 16 to 20 cm	4
From 21 to 30 cm	6
Higher than 30 cm	10

3. *Results*

Following the above mentioned parameters we then divided the subjects into two groups (Table 8): in the first group, which we called group A, we classified operators with a good tolerance to their job (i.e. those who had reported less than three subjective ocular complaints of a given degree); in the second group, called group B, we included subjects reporting considerable disorders and difficulties in the performance of their tasks (i.e. those having reported a greater number of, or more severe, complaints). In our series, 121 subjects (84·62%) were classified in group A, and 22 (15·38%) in group B.

TABLE 8. *Results.*

A: Operators with a *good* tolerance = 121 (84·62%)

1. Good visual efficiency = 102 (71·33%)
   mean score = 7·13 (from 0 to 14)
2. Poor visual efficiency = 19 (13·29%)
   mean score = 6·05 (from 16 to 44)

B: Operators with *intolerance* = 22 (15·38%)

2. Poor visual efficiency = 12 (8·39%)
   mean score = 25·9 (from 16 to 48)
1. Good visual efficiency = 10 (6·99%)
   mean score = 6·3 (from 0 to 10)

We then analysed the numerical value assigned to each subject's visual efficiency for each group and a clear correlation between these two parameters was found: the demarcation being an overall score of 15.

In group A, comprising a population reporting few subjective defects, 102 operators (71·33%) presented minimal pathological signs and therefore a good score for visual efficiency ($\bar{M}$ = 7·13), whereas only 19 (13·29%) presented an evident pathological picture ($\bar{M}$ = 26·05).

In group B instead, comprising subjects reporting considerable subjective complaints, 12 (8·39%) presented a very poor efficiency score ($\bar{M}$ = 25·9), while 10 (6·99%) presented practically no pathological state and are, therefore, the group of false positives.

4. *Discussion and Conclusion*

From an analysis of this data one can clearly demonstrate a close correlation between the visual efficiency score and the reported complaints. Obviously, many ocular alterations either of refractive or especially of orthoptic nature may, if taken individually, seem insignificant for the onset of ophthalmic disorders, but their interaction, when they are present together, may result in visual fatigue and disorders such that VDU work becomes impossible. On the basis of this correlation, we feel that this study can also be used in job screening to assess the fitness of subjects applying for VDU associated positions.

References

Anfossi, D.G., 1983, Affaticamento visivo e disturbi oculari in soggetti addetti ai video terminali (V.D.T.). Considerazioni sui risultati di uno screening condotto nella città di Torino. *Bollettino di Oculistica*, 62, 11–12.

Anfossi, D.G., Grignolo, F.M., Maina, S., Romanco, C. and Sonnino A., 1983, Aspetti oftalmologici ed ergonomici negli addetti ai video terminali. *Giornale Italiano di Oftalmologia Occupazionale*, 1, 1.

Anfossi, D.G. and Vitale Brovarone, F., 1981, Disturbi oculari e loro prevenzione in soggetti addetti ai video terminali (V.D.T.). Primi risultati di uno screening condotto nella città di Torino. *Proceedings of the 11 Ergophthalmology Congress*, Bari, 1981.

Pesce, F., Vitale Brovarone, F. and Anfossi, D.G., 1984, Risultati di un'indagine condotta sull'utilità di filtri anteposti ai videoterminali nella riduzione dei disturbi oculari e dell'affaticamento visivo. *Bollettino di Oculistica* 63, 11–12.

Vitale Brovarone, F., Anfossi, D.G. and Pesce, F., 1983, Studio preliminare sulla prevenzione dell'affaticamento visivo mediante applicazione di filtri ai videoterminali. *Giornale Italiano di Oftalmologia Occupazionale*, 1, 1.

Effect of Methylcobalamin (Vitamin B_{12}) on Asthenopia Induced by VDT Work

S. Kurimoto and T. Iwasaki
Department of Ophthalmology, University of Occupational and Environmental Health, Tokyo, Japan
K. Noro and S. Yamamoto
Department of Human Factors Engineering, University of Occupational and Environmental Health, Tokyo, Japan

Abstract

The effect of methylcobalamin on the accommodative function of 6 subjects with asthenopia induced by experimental VDT work was studied by measuring accommodation time, near point distance and small fluctuation of accommodation. In the first experiment, the subjects were administered a placebo 7 days before the experiment; in the second, they received methylcobalamin (3000 μg/day). Accommodative function was found to deteriorate after VDT work in the placebo group, but not in the active group.

1. *Introduction*

The number of people who complain of subjective symptoms such as eye pain, deteriorating vision, lacrimation, conjunctival hyperaemia, headache, nausea, dizziness, etc., so-called symptoms of ocular fatigue, has increased in recent years.

The visual display terminal (VDT) which is an intelligence terminal is being introduced into many general job functions. As VDT use expands, numerous reports are appearing and complaints of symptoms of asthenopia by the users, whose symptoms had been regarded as accommodative depression by us (Kurimoto *et al.* 1982, Iwasaki *et al.* 1982, Kurimoto *et al.* 1983a and Kurimoto *et al.* 1983b).

In this paper, methylcobalamin was administered as a means of countering the depression of the visually taxing VDT work.

2. *Subjects and Method*

The subjects were selected 6 males (mean age 32·0), who had no ophthalmological anomalies. All cases had corrected visual acuity of greater than 1·0.

The task performed by the subjects was a name search job. First of all, one name written in katakana (Tr. note: a Japanese phonetic syllabary) was presented on the VDT screen (TRS–80, TANDY RADIOSHACK), called the target. Next, a list of 80 names, including the individual target, was presented on the screen. The subjects were to find the target in the list and push a button. This task was continued for 2 hours with 1–2 minute breaks every 15 minutes. The room used for working was a twilight room (about 8 lux).

The accommodative function was estimated by measurement of the near point distance, the contraction time and the relaxation time by using an accommodo polyrecorder (HS–9D, KOWA), and was estimated by measurement of the small fluctuation of accommodation by using an infrared optometer (OPTRON 2 AS–157, KOWA).

The near point distance is the nearest point that the black Landolt ring on a white background can be seen clearly. It was measured in centimetres. The contraction time means the time taken to focus the eye from the subject's far point to the Landolt ring located at the subject's near point. The relaxation time is reversed. They were measured in seconds. The average of 10 measurements at each period was recorded. The small fluctuation of accommodation, one of the residual oscillation in the body movement system, was measured objectively by this optometer. In measurement of the small fluctuation, a far visual target was placed at optical infinity and watched vaguely in an unaccommodated state as a visual line fixing target. Recording was conducted for about 30 seconds. The peak of the obtained wave form vertex interval was measured on the digital computer, which was converted into a frequency (Hz), and, except for the frequency component of over 3·0 Hz, classified them into 5 stages (0–0·5, 0·5–1·0, 1·0–1·5, 1·5–2·0, 2·0–3·0 Hz). The variation of each component was calculated as a percentage.

The fluctuation in accommodation which has the dominant frequency component of about 1·5 Hz seems obviously due to the normal activity of the ciliary muscle. However, generally, the low frequency component of 0·5–1·0 Hz increase in the frequency spectra of the fluctuation of persons with disordered accommodation and asthenopia. Therefore, the variation of the low frequency component (0·5–1·0 Hz) was selected from other components in this study.

In the first experiment, the subjects were administered a placebo (lactose) 7 days before the experiment but in the second experiment they were administered methylcobalamin (3000 μg/day) (METHYLCOBAL® , EISAI). A double blind method was used (Figure 1).

3. *Results*

Figure 2 shows the variations in the near point distance at 5 points in time. In the placebo group, the near point distance tended gradually to prolong and finally reached a level of maximum. In the active group on the other hand, it tended to shorten. The near point

FIGURE 1. *Experimental procedure; a double blind method.*

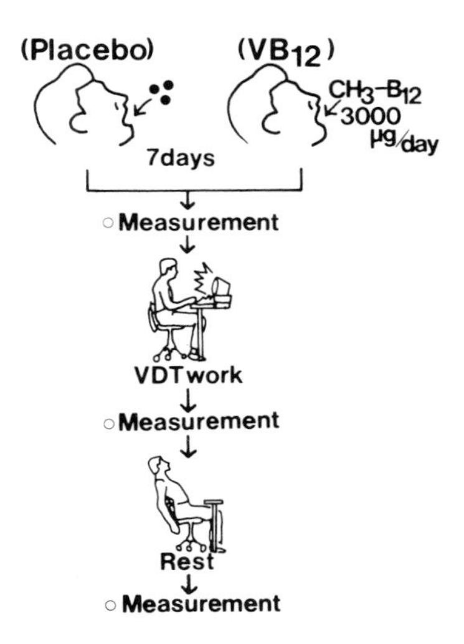

FIGURE 2. *Variations of near point distance at 5 points in time.*
———— : *methylcobalamin* (n = 6)
– – – – – : *placebo* (n = 6)

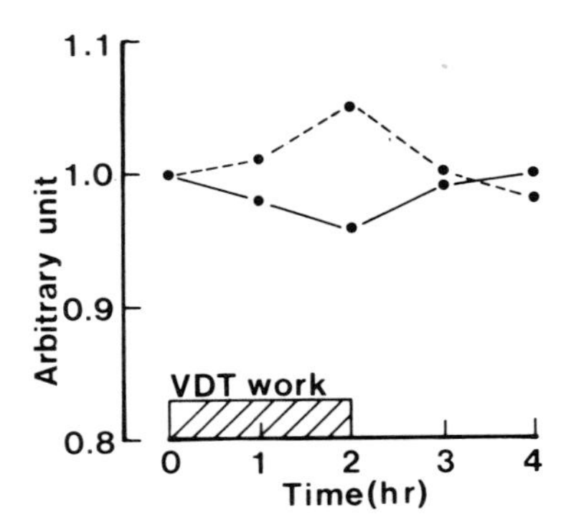

distance in both groups which had been changed after loading, returned to the preloading values after 2 hours of rest.

Figure 3 shows the accommodation time of the respective averages of the contraction time and the relaxation time in the placebo group and the active group. The contraction time was unchanged before and after the task load in both groups (Figure 3b), but as can be seen in Figure 3a, the relaxation time tended to delay in the placebo group after task load, the change of which was the mean value of about 0·6 sec. In the active group, there is not an aggravation in accommodation time.

FIGURE 3. *Changes in accommodation time at 4 points in time. (a) relaxation time, (b) contraction time.*
———— : *methylcobalamin* (n = 6)
– – – – – : *placebo* (n = 6)

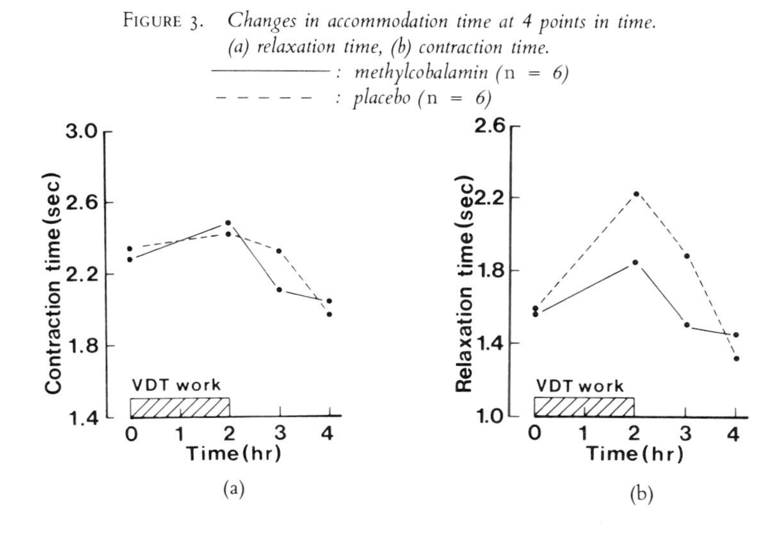

Frequency analysis was performed based on the recordings obtained from small fluctuations of accommodation changes. In the movements of the ciliary muscle during resting accommodation, seeing a target at optical infinity, more than half is occupied by frequencies up to 1·0–2·0 Hz.

FIGURE 4. *Small fluctuation of accommodation movement chart obtained when subject (34-year-old male, right eye) viewed the target placed at his optical infinity. Before VDT work (upper), after VDT work (middle), and after 2-hour rest pause after working (lower). (a) administered placebo, (b) administered methylcobalamin.*

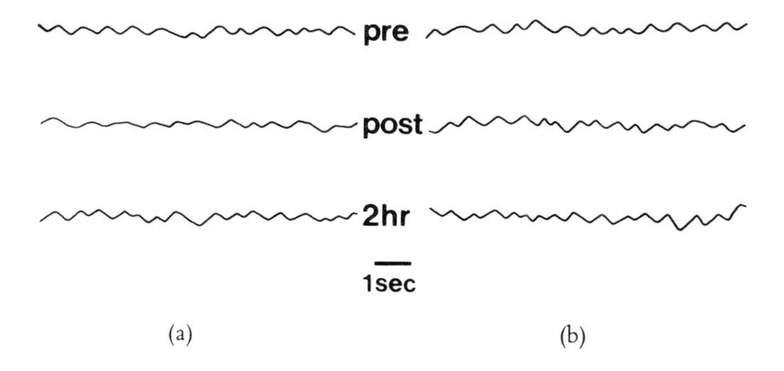

Figure 4 is a recording of the small fluctuation of accommodation at the 3 points in time of the placebo and active administered subject, who is a 34-year-old male. In Figure 4a, when the subject was administered the placebo, there was a decrease in the high frequency component and an increase in the low frequency component after the task load. On the other hand, when the subject was administered methylcobalamin, the recording did not show such a tendency. The results with the other subjects were similar to his. Then, the variation of the low frequency component (0·5–1·0 Hz) at 4 points in time were selected from other components in order to compare the active group with the placebo group in Figure 5.

FIGURE 5. *Changes in the low frequency component (0·5–1·0 Hz region) of small fluctuation at 4 points in time.*
———— : *methylcobalamin* (n = 6)
- - - - - : *placebo* (n = 6)

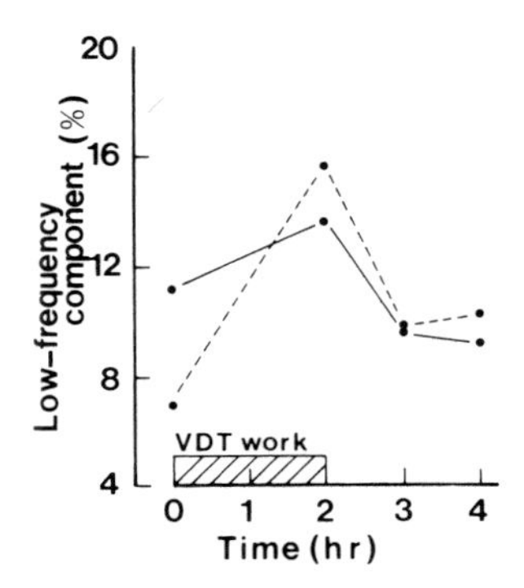

The results of small fluctuations were also in accord with the results of the near point and the relaxation time. The low frequency component of 0·5–1·0 Hz was not found to increase after VDT work in the group receiving methylcobalamin. However, a statistical significant difference was recognised in the group receiving placebo with an increase of about 9% of low frequency component ($p < 0{\cdot}05$, by t-test).

4. *Discussion*

Accommodation function can be objectively measured and analysed by the infrared optometer and subjectively by the accomodo polyrecorder. The effect of VDT tasks on accommodation could be confirmed using them.

A prolongation in the near point distance and the relaxation time and an increase in the low frequency component were seen after loading in the group receiving placebo, but not in the active group. In this experiment it was ascertained, because this programme was carried out by using a method of double blind, that Vitamin B_{12} (methylcobalamin) produced a good effect on asthenopia induced by VDT work. This is the first paper to examine the effect of methylcobalamin on asthenopia induced by experimental VDT work.

Although it was not easy to measure asthenopia objectively, using the infrared optometer and the accomodo polyrecorder we were able to take an objective view of it induced by VDT work as accommodation depression.

The methylcobalamin is involved in the methyl transfer reaction converting homocystein to methionine (Nakasawa 1970a, b), and facilitates peripheral neural regeneration (Fischer *et al.* 1958, Yamazu *et al.* 1976). If small doses of it were applied, active potential of sensory and motor nerve was potentiated (Takeshige *et al.* 1971). It was also reported that it was rapidly metabolized in the living system, and was utilized for the synthesis of RNA, DNA and lipids (Walerych *et al.* 1966, Misra *et al.* 1967).

Based on these reports, it is quite conceivable that the disordered accommodation caused by the physiology activity can be improved by methylcobalamin. However, no biochemical and physiological examinations on the individual ciliary muscles have been made.

It is anticipated that the use of VDT will continue to expand in the future and that the accompanying disorders will become more numerous. The visual factors related to asthenopia are image quality, illumination image, glare, luminous wave length and visual distance. Therefore, the suitable countermeasure must be worked out from the ergonomical standpoint. Although one way is that methylcobalamin is administered to counter the depression of the visual taxing VDT work, it is more important to maintain VDT operators' health by a better environment.

References

Fischer, J. *et al.*, 1958, A histoautoradiographic study of the effect of section of the facial nerve on the uptake of methionine-^{35}S by the cells of the facial nerve nucleus. *Nature*, 174, 341–342.

Iwasaki, T. *et al.*, 1982, The influence of VDT work on accommodation. *Folia Ophthalmologica Japonica*, 33, 90–95.

Kurimoto, S. *et al.*, 1982, Accommodation and eye strain in cathode ray tube (CRT) work. *Japanese Journal of Clinical Ophthalmology*, 36, 1155–1160.

Kurimoto, S. *et al.*, 1983, Asthenopia in visual display terminal (VDT) and paper work. *Japanese Journal of Clinical Ophthalmology*, 37.

Kurimoto, S. *et al.*, 1983a, Accommodation in VDT operators and clerical workers. *Japanese Journal of Traumatology and Occupational Medicine*, 31, 31–38.

Kurimoto, S. *et al.*, 1983b, Influence of VDT work on eye accommodation. *Journal of UOEH*, 5, 101–110.

Misra, U.K. *et al.*, 1967, Incorporation of ^{14}C activity of ^{14}C-methylcobalamin into lipids, RNA and DNA of regenerating rat liver. *Indian Journal of Biochemistry*, 4, 132–133.

Nakasawa, K. *et al.*, 1970a, Vitamin B_{12}-dependent methionine biosynthesis in brain cells, cultivated in vitro. *Vitamin*, 41, 333–337.

Nakasawa, K. *et al.*, 1970b, The active of methyltransferase induced by Vitamin B_{12} in cell line established from rat brain. *Vitamin*, 42, 193–197.

Takeshige, C. *et al.*, 1971, Effects of Vitamin B_{12} and of aldosterone on the conduction of sensory and motor nerve impulse. *Vitamin*, 44, 272–282.

Walerych, W.S. *et al.*, 1966, The methylation of transfer RNA by methyl cobamide. *Biochemical and Biophysical Research Communication*, 23, 368–374.

Yamazu, M. *et al.*, 1976, Pharmacological studies on degeneration and regeneration of peripheral nerves (II). *Folia Pharmacologica Japan*, 72, 269–278.

Visual Fitness for VDU Operators

B. Boles-Carenini*, G.F. Rubino†, F.M. Grignolo* and G. Maina†
Istituto di Clinica Oculistica dell'Università, Turin, Italy*
Istituto di Medicina del Lavoro dell'Università, Turin, Italy†

Abstract

The rapid diffusion of visual display units (VDU), the use of which can eventually induce visual stress, has led the authors to set some parameters (presence of emmetropia, astigmatism, anisometropia, monophthalmia, visual field and chromatic sense defects, binocular vision and ocular motility deficiency, glaucoma and other ocular pathology) to advise terminal operators what limits should be set for using VDUs.

VDUs are rapidly expanding into all occupational areas and we know through our own experience that their prolonged use may lead to visual fatigue. We, therefore, examined the fundamental visual parameters to set down, on a theoretical basis and as a possible working standard for future research, threshold values above which VDU activity is permissible and below which it should be limited or even discouraged.

The visual parameters we considered were the following:

visual acuity (natural and/or corrected, monocular and/or binocular);
astigmatic or spherical refraction defects (myopia, hypermetropia), when present;
anisometropia;
visual field;
chromatic and light sense;
binocular vision;
ocular motility; and
chronic and/or acute ocular pathologies, when present.

VDU activity is *unadvisable in subjects whose visual acuity with each eye in single vision is below 8/10*, either in natural vision or with the best tolerated correction, if ametropic (see Table 1). In anatomically or functionally *monocular* subjects and in *ambliopic* monocular subjects, visual acuity in the only eye or in the best one must be 9/10 or better, in natural vision or with the best tolerated correction; activity on a VDU for the latter must not, however, claim more than 50% of working hours.

TABLE I. *Visual acuity standards.*

VDU working hours	
100%	50%
≥ 8/10 in each eye in single vision	≥ 9/10 in the best eye < 8/10 in the worst eye

All *emmetropic* subjects who meet the above mentioned standards, with appropriate working distance correction if necessary (*presbyopia*), present no contra-indication to VDU activity during working hours.

Myopic subjects with refraction defects up to −6·00 D, suitably corrected, require no limitation as long as visual acuity with best tolerated correction meets the minimum values stated above. In myopias higher than −6·00 D, in which the above threshold requirements are met, VDU activity should anyway be limited to 50% of working hours.

Hypermetropic subjects with refraction defects up to +1·50 D, with correction, need no limitation as long as no latent ocular motility defects are present. If hypermetropia is higher than +1·50 D, VDU activity should be limited to 50% of working hours, even with correction.

Hypermetropic or myopic *astigmatism* up to 2·00 D requires no VDU limitation. Mixed astigmatism or astigmatism higher than 2·00 D requires the 50% limitation.

A refractive difference between the two eyes up to ± 2·00 D is considered to be normal as long as the subject uses appropriate correction and meets minimum required visual acuity; for subjects presenting anisometropia higher than ± 2·00 D and up to ± 3·00 D, the 50% limitation is applied as long as visual acuity is sufficient. Greater degrees of anisometropia are classified as for monocular or ambliopic subjects: if adequate correction can be obtained with corneal lenses, the extent of VDU activity can be increased on the basis of case by case evaluations (generally speaking, correction of a refraction defect with corneal lenses is considered as being equivalent to spectacle correction; clearly, in this specific case, tolerance must be good for at least 8–9 hours).

As to *visual field*, *chromatic sense*, *light sense*, *ocular motility* and *binocular vision*, despite their representing very different anatomical structures and functional systems, we consider that individual defects require the 50% limitation to VDU activity, and that combined defects must discourage it completely. These alterations must, however, be evaluated individually and, according to their severity, may be considered as requiring total limitation.

Amongst *ocular diseases*, aphakia and glaucoma of any type are considered as being incompatible with VDU activity (in mass screening, subjects with one IOP value higher than 20 mmHg—even obtained with a non contact tonometer—are considered to be suspects and shall require a full examination in order to exclude the possibility of glaucoma). All other ocular diseases will also be considered as being totally incompatible during acute phases; once the patient has recovered, he will require a thorough re-evaluation.

To conclude, we wish to submit a set of guidelines, as balanced as possible with our state of knowledge, so as to open up this area of activity to as homogeneous a group as possible and isolate the many variables that might contribute towards the onset of intolerance phenomena. This will also enable us to identify the specific influence of VDU activity on each ocular parameter.

References

Anfossi, D.G., 1983, Affaticamento visivo e disturbi oculari in soggetti addetti ai video terminali (V.D.T.): Considerazioni sui risultati di uno screening condotto nella città di Torino. *Bollettino di Oculistica*, 62, 11–12.

Anfossi, D.G., Grignolo, F.M., Maina, S., Romano, C. and Sonnino, A., 1983, Aspetti oftalmogici ed ergonomici negli addetti ai video terminali. *Giornale Italiano di Oftalmologia Occupazionale*, 1, 1.

Anfossi, D.G. and Vitale Brovarone, F., 1981, Disturbi oculari e loro prevenzione in soggetti addetti ai video terminali (V.D.T.): Primi risultati di uno screening condotto nella città di Torino. *Proceedings of the 11 Ergophthalmology Congress*, Bari, 1981.

Pesce, F., Vitale Brovarone, F. and Anfossi, D.G., 1984, Risultati di un'indagine condotta sull'utilità di filtri anteposti ai videoterminali nella riduzione dei disturbi oculari e dell'affaticamento visivo. *Bollettino di Oculistica*, 63, 11–12.

Vitale Brovarone, F., Anfossi, D.G. and Pesce, F., 1983, Studio preliminare sulla prevenzione dell' affaticamento visivo mediante applicazione di filtri ai videoterminali. *Giornale Italiano di Oftalmologia Occupazionale*, 1, 1.

Postural Problems at Office Machine Work Stations (Introductory paper)

E. GRANDJEAN
Swiss Federal Institute of Technology, Department of Ergonomics, CH-8092 Zurich, Switzerland

1. *Man–Machine Systems in Offices*

The introduction of machines, such as accounting machines and more recently VDTs, leads to an integration of employees in man–machine systems. The operator and the machine have a reciprocal relationship with each other. Such a system is a closed cycle. One of the consequences is a restriction of the space of physical activities with more constrained postures, associated with long-lasting static contractions of muscles. Static effort reduces blood irrigation of the muscles and causes localised fatigue. Tiredness, pains and even cramp are the symptoms.

2. *Medical Aspects of Postural Efforts*

Postural efforts do not only decrease performance and productivity, in the long run they also affect well-being and health. In fact, if postural efforts are repeated daily over a long period, more or less permanent aches will appear in the limbs concerned, and may involve not only muscles but also other tissues. Thus, long-lasting postural efforts can lead to deterioration of joints, ligaments and tendons. Several field studies as well as general experience show that postural efforts are associated with an increased risk of:

1. Inflammation of the joints.
2. Inflammation of the tendon-sheaths.
3. Inflammation of the attachment-points of tendons.
4. Symptoms of chronic degeneration of the joints in the form of chronic arthroses.
5. Painful induration of the muscles.
5. Disc troubles.

Persistent pains in the overloaded tissues appear particularly among older operatives.

3. *A Field Study on Accounting Machine Operators*

A summary of some results of the field study on 119 accounting machine operators of Maeda *et al.* (1980) and of Hünting *et al.* (1981) will be reported here.

The main task of female accounting operators was to read figures from coupons and to type them with the right hand on a keyboard. The main visual axis was directed to the coupons, causing some forward bending of the neck. The keying speed varied between 8000 and 12000 strokes/hour. These operators complained frequently about tiredness, pains or even cramps in the neck–shoulder–arm–hand area (see Table 1). The impairments were significantly more frequent on the right than on the left side.

In order to check possible effects of body postures on the incidence of impairments, 57 subjects were divided in three groups of different body posture characteristics. The comparison of these data showed relationships between three ranges of body angles and impairments, which are shown in Figures 1 and 2.

Not all of the reported relations are statistically significant. Nevertheless, the general tendency is obvious and the following statements from the two figures can be deduced:

1. The incidence of stiffness and pains in the neck is increasing with the degree of forward bending of the head (Figure 1).
2. The incidence of tiredness, pains and cramps in the right hand is increasing with the degree of ulnar deviation of the same hand (Figure 2).

The analyses of the work station showed some weak points: the keyboards were fixed and could not be adjusted to the natural arm–hand position; forearms and wrists of the right hand could not rest neither on the table nor on a special support equipment. The source documents had a poor legibility; in order to facilitate the reading task, the operators

FIGURE 1. *Incidence of impairments in the neck related to the neck–head angles of 57 accounting machine operators.*

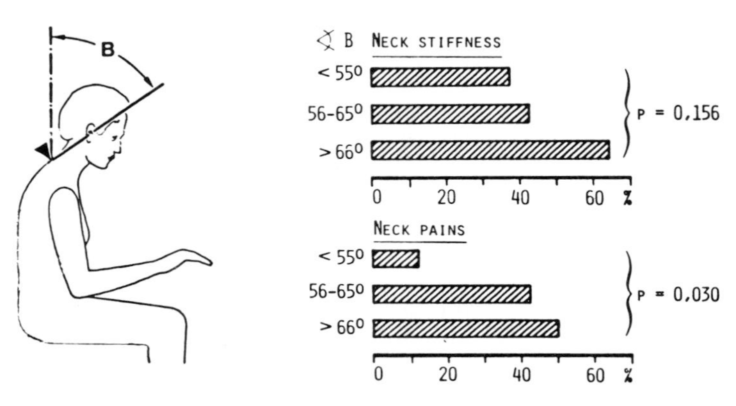

FIGURE 2. *Incidence of impairments in the right hand related to the ranges of the lateral deviation of the right hand (angle H) of 57 accounting machine operators.*

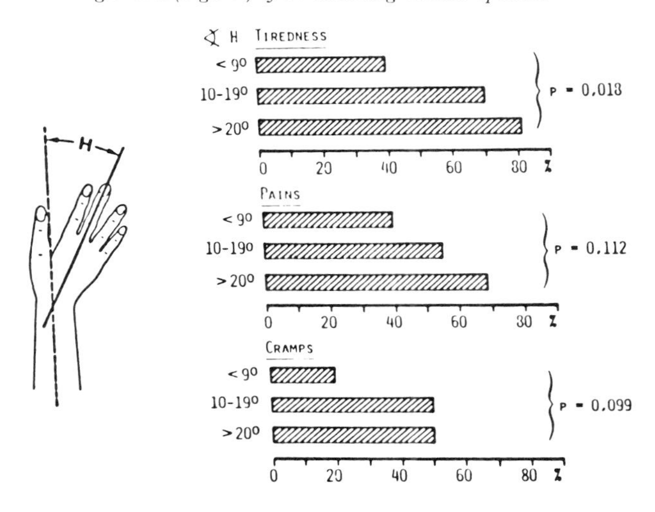

100 % = SUBJECTS IN EACH OF THE THREE "ANGLE-GROUPS"

were rotating trunk and head to the left and they were forced to keep the head in a forward position. Position and quality of the source documents caused postural efforts which were in any case already badly determined by the fixed positions of the keyboards.

4. *Postural Efforts at VDT Work Stations*

In another field study, 159 VDT operators and 133 other office employees were examined. In Table 1 the results of reported 'almost daily pains' are shown for all examined groups, including those of the above mentioned accounting machine operator.

TABLE I. *Incidence of 'almost daily' pain in five office jobs.*

Groups	*n*	Incidence of 'almost daily' pain			
		neck %	shoulder %	r. arm %	r. hand %
Data entry terminals	53	11	15	15	6
Accounting machine operators	119	3	4	8	8
Conversational terminals	109	4	5	7	11
Typists	78	5	5	4	5
Traditional office work	55	1	1	1	0

FIGURE 3. *Medical findings in a VDT field study: painful pressure points at tendons, joints and muscles in the shoulder area.*
r = right, l = left.

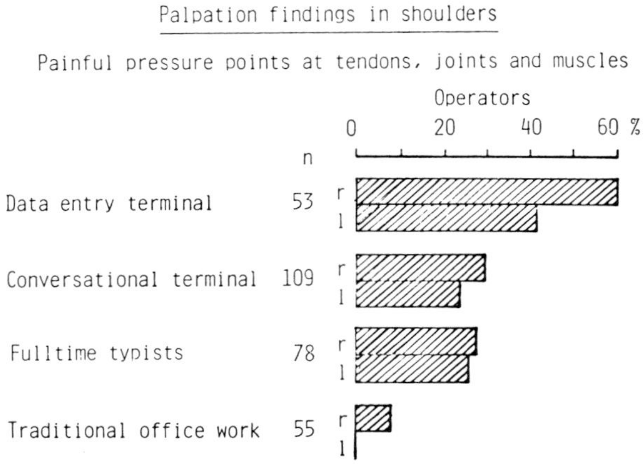

Palpation findings in shoulders

Painful pressure points at tendons, joints and muscles

Operators

0 20 40 60 %

n

Data entry terminal 53 r l

Conversational terminal 109 r l

Fulltime typists 78 r l

Traditional office work 55 r l

The results reveal that serious impairments were observed in each group, but it is obvious that the highest figures are seen in the group operating data entry terminals and the lowest figures in the group occupied with traditional office work. These 'almost daily pains' must be considered as relevant injuries, since about 20% of the VDT operators visited a doctor because of the reported troubles.

All operators of the VDT field study were also examined by a medical doctor. One result of the medical findings is reported in Figure 3 which shows the incidence of painful pressure points on tendons, joints and muscles in the shoulder area.

The clinical symptoms in the shoulder area are frequent at data entry terminals and rare in the control group of traditional office work. The analysis of the relationship between the work station design or body postures on one side and physical impairments on the other side gave the following results:

The incidence of subjective symptoms of localised fatigue and/or of clinical findings in the neck–shoulder–arm–hand area is increased

1. If the desk supporting the keyboard (and the source documents) is low.
2. If forearms and wrists cannot rest on an adequate support.
3. If the height of the keyboard above the desk level is high.
4. If the operators show a marked head inclination.
5. If the operators show a marked lateral deviation of the hands while operating keyboards.

5. *Studies with Adjustable VDT Work Stations*

These results led to the assumption that a VDT work station with adjustable dimensions would reduce constrained postures and grant relaxed body postures with less physical

FIGURE 4. *The adjustable VDT work station with the ranges of possible settings.*
A = keyboard height above floor (62– 88 cm)
B = screen height above floor (90–128 cm)
C = screen distance from table edge (40–115 cm)
D = screen inclination (75–110°)
source document holder inclination (0– 90°)

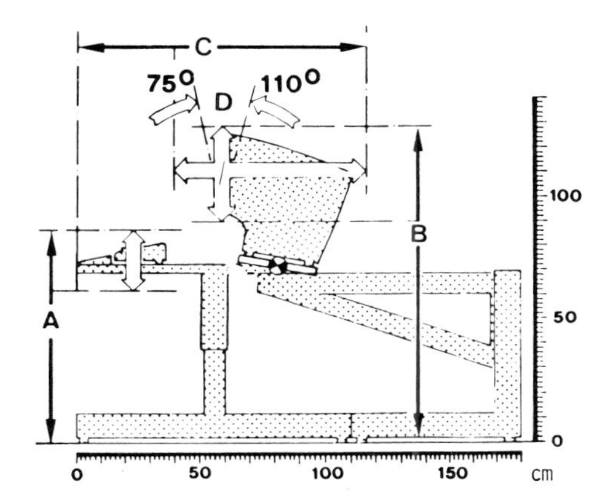

discomfort. A study was carried out by Grandjean *et al.* (1983) to assess the preferences of 68 VDT operators with regard to their body posture and the settings of an adjustable work station as shown in Figure 4.

The adjustable work stations were tested by operators while they performed their usual daily job. Subjects came from four different companies; 45 were engaged in conversational VDT jobs and 17 in data entry operations. The preferred work station settings are reported in Table 2.

TABLE 2. *Preferred work station settings (236 observations on 59 operators).*

Settings	Mean	Range
Keyboard[a] height above floor (cm)	79	71–87
Screen height above floor (cm)	103	92–116
Screen distance to table edge (cm)	64	50–79
Visual distance to screen (cm)	76	61–93
Screen inclination[b] (°)	94	88–103
Seat height above floor (cm)	48	43–57
Home key to table edge (cm)	20	9–43

[a] Key top level of the home row.
[b] Angle in relation to a horizontal plane.

The operators reported less physical discomfort when working with the preferred work station settings and a majority of them rated their body postures as relaxed.

However, the most striking result concerned the measured body postures, which were characterised by a marked trunk inclination, i.e. a pronounced backward leaning. Furthermore the upper arms were kept higher and the elbow angles slightly opened. The mean figures of preferred angles are shown in Figure 5.

FIGURE 5. *The 'mean body posture' under practical conditions at preferred settings of the VDT work station. Results of 236 observations on 59 VDT operators.*

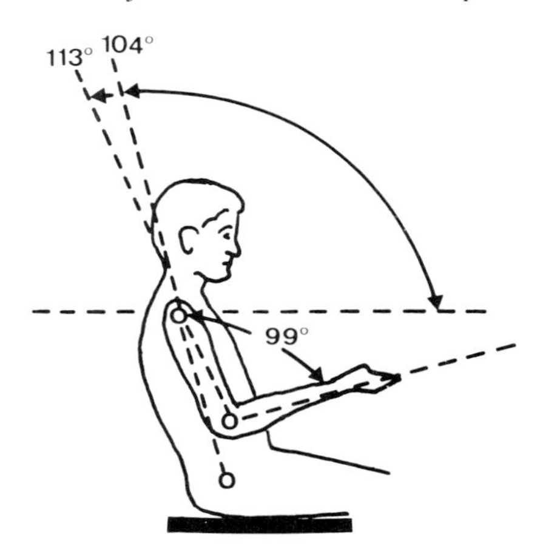

This study confirmed a general impression when observing the sitting posture of VDT operators: the majority of them lean backwards and often extend the legs forwards. They seem to put up with forward bending of the head, with no support of the lower part of the spine and with lifting their arms.

In fact, the VDT operators disclose body postures very similar to those of car drivers.

The preferred body posture does not at all correspond to the commonly published and recommended postures. Figure 6 illustrates the great gap between 'wishful' recommendations and actual postures.

An important question suggests itself: is the upright posture healthy and therefore recommendable or is the relaxed position with the backward-leaning trunk to be preferred? Interesting experiments of the Swedish surgeons Nachemson and Elfström (1970) and Andersson and Ortengreen (1974) offer an answer to this question. These authors measured the pressure inside the intravertebral discs as well as the electrical activity of the back muscles in relation to different sitting postures. When the backrest angle of the seat was increased from 90° to 110°, subjects exhibited an important decrease of the intervertebral disc pressure and of the electromyographic activity of the back. Since increased pressure inside intervertebral discs means that they are being stressed and will wear out more quickly, it is concluded that a sitting posture with reduced disc pressure can be considered healthy and desirable. The results of the Swedish

FIGURE 6. *Recommended and actual postures at VDT work stations.*
Left: The upright trunk posture required in many brochures and standards is 'wishful thinking'.
Right: The actual body posture mostly observed at VDT work stations allows the operators a good relaxation of back and shoulders.

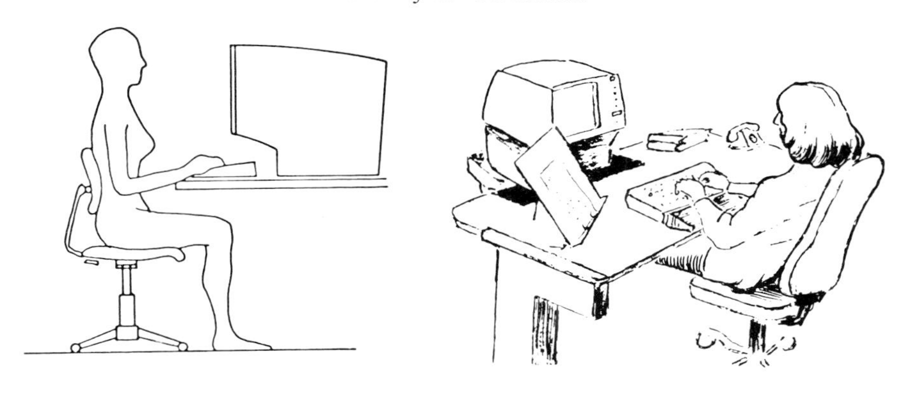

Wishful thinking   Preferred body posture

group indicate that resting the back against a sloping backrest, transfers a portion of the weight of the upper part of the body to the backrest and noticeably reduces the physical load on the intervertebral discs as well as the static strain of the back muscles. It is concluded that the VDT operators show good sense when they prefer a backward leaning trunk posture and ignore the recomended upright trunk position of brochures and standards!

Several recommendations related to VDT work stations are based on anthropometric data of upright trunk positions and on the assumption that the forearms must be kept on a horizontal plane. These rather theoretical assumptions lead to the opinion that the height of the home row of keyboard above floor should be between 72 and 75 cm. The study on preferred settings of adjustable work stations clearly disclosed that the VDT operators prefer higher keyboard levels.

If all results of this study are taken into consideration the preferred settings and body postures of VDT operators can be summarised as follows:

1. They like to lean backwards and to rest the trunk on a backrest.
2. They like to keep the upper arms in an upright inclination of approximately 14° and to open the elbow angles to about 100°.
3. They try to rest forearms and wrists on a support which can be the desk itself, provided that the keyboard has a flat design.
4. Keyboard levels (home row) between 75 and 85 cm above floor correspond to these preferred postures.
5. They like to have the screen about at the level of the head, though the neck is bent only moderately.
6. They like an adjustable screen inclination to avoid, to some exent, annoying reflections.

6. *The Keyboard*

The keyboard for typing letters was invented in 1868. It was a mechanical device which required a design of 4 parallel rows of keys. To operate these keys quickly, the typist must keep the hands in a parallel position to the rows. This requires an unnatural posture of wrists and hands, characterised by an inward rotation of forearms and wrists and a lateral deviation of hands. These constrained postures often cause physical discomfort and in some cases even inflammations of tendons in the forearms of the keyboard operators. Figure 7 illustrates such constrained postures of wrists and hands at a keyboard. With the development of electronics, the mechanical typewriter was replaced by the electrical one. The mechanical resistance of keys decreased and the operation of the keyboard was made easier, but the unnatural position of wrists and hands was maintained.

FIGURE 7. *Wrist and hand postures at keyboards, characterised by a pronounced inward rotation of forearms (pronation) and a lateral deviation of hands (ulnar deviation).*

At VDT work stations the typing activity is similar to the traditional operation of typewriters. But some slight differences occurred: first of all the number of keys increased with specially arranged numerical keys and several functional keys for operating the computer. Secondly, in conversational jobs, operators must frequently wait for the response of the computer. These response times last from one to several seconds. According to the Swedish study of Johansson and Aronsson (1980) response times of more than 5 seconds were experienced as annoying and stressing. During these unwanted pauses, operators like to rest forearms and wrists on suitable supports. This induced some VDT designers to develop flat

FIGURE 8. *An ergonomic design of a keyboard which allows a more natural position of wrists and hands: the inward rotation as well as the lateral deviation are reduced in the wrists.*

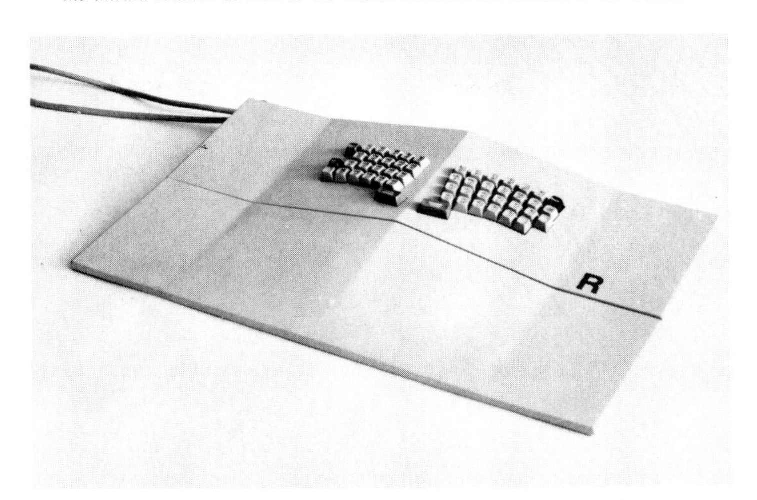

keyboards which allow operators to rest forearms and wrists on the desk. For this reason many ergonomists recommend today *a flat keyboard with a home row not higher than 3 cm above the desk and the possibility to move the keyboard on the desk as operators like.*

The next step should be an ergonomic design of the keyboards which will avoid a constrained posture of hands by decreasing the inward rotation of the lateral deviation of hands as well as by providing a large support for resting forearms and wrists. Such an ergonomic design of a keyboard is shown in Figure 8.

There is a general agreement among experts about other dimensions of keyboards which can be summarized as follows:

Keyboard height above desk (middle row)	30 mm
Keyboard height (front side)	less than 20 mm
Inclination	5–15°
Distance between key tops	17–19 mm
Resistance of keys	250–800 mN
Key displacement	3–6 mm

The operator should feel when the stroke has been accepted; this is called the tactile feedback. The best feedback characteristics are achieved when the point of acceptance and pressure is located about halfway of the key displacement.

7. *The Chair at a VDT Work Station*

For thousands of years designing chairs was mainly a question of exterior form. Even in the early part of this century, chairs tended to be more a means of heirarchical status symbol rather than useful chairs. Only in the last decades sitting posture and seats became a target of scientific research especially of ergonomics and of orthopaedics. These studies revealed that sitting is associated with a decrease of static muscular efforts in legs and hips, but on the other side with increased physical load on the intervertebral discs in the lumbar region of the spine. It is not surprising that the frequency of lumbar back complaints is high in all sedentary jobs. From these findings the most important recommendations for office and many other chairs could be deduced: the profile of the backrest must have a 'lumbar cushion' to support this area of the back.

A properly designed backrest gives a good lumbar support in the forward sitting position as well as in the backward position, as shown in Figure 9.

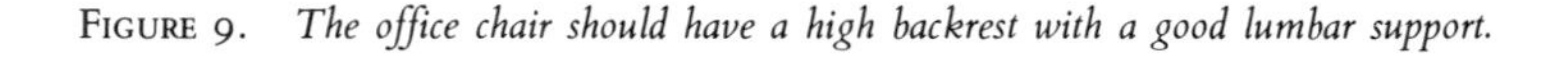

FIGURE 9. *The office chair should have a high backrest with a good lumbar support.*

The studies on VDT work stations revealed that operators sit like car drivers and they adjust the angles of the backrest in a range between 90 and 120°. From these observations it can be concluded that a *VDT chair needs a backrest reaching a height of 50 cm above seat level and a range of adjustable inclination from 90 to 120°. It should be possible to fix the inclination at any position desired.*

Traditional office chairs with a relatively small back support are not suitable for VDT jobs, since they do not allow relaxation of the full back.

It is furthermore obvious that a chair for office machine work stations must fulfill all requirements of a modern office chair: adjustable height, swivel, rounded front edge of the seat surface, castors or glides, 5-arm base and user friendly controls.

8. *Conclusions*

To reduce constrained postures and to secure a physical comfort of VDT operators, the designers and the manufacturers of office furniture should take into consideration the following recommendations.

The furniture should, in principle, be conceived as flexible as possible. A proper VDT work station should be adjustable in the following ranges:

Keyboard height (middle row to floor)	70–85 cm
Screen centre above floor	90–115 cm
Screen inclination to horizontal plane	88–105°
Keyboard (middle row) to table edge	10–26 cm
Screen distance to table edge	50–75 cm

A VDT work station without an adjustable keyboard height and without an adjustable height and distance of the screen is not suitable for a continuous job at VDTs.

The controls for adjusting the dimensions of a work station should be easy to handle, particularly at work stations for rotating shift work.

It is nearly impossible for an operator to adjust the work station dimensions by himself. Another person should be in charge of handling the controls, while the operator is working at the VDT work station.

Insufficient space for the legs causes unnatural body postures. The space at the level of the knees should be at least 60 cm from table edge, and at least 80 cm at the level of the feet.

The chair should have a 50 cm long backrest (above the seat surface) and an adjustable inclination. The backrest should show a lumbar support (10 to 20 cm above seat level) and a slightly concave form on the thoracic level. It should be possible to fix the inclination of the backrest at any position desired.

The design of the keyboard must be flat and movable on the desk. An ergonomic design of keyboards allowing a natural position of both hands is recommended.

References

Andersson, B.J.G. and Ortengreen, R., 1974, Lumbar disc pressure and myoelectric back muscle activity. *Scandinavian Journal of Rehabilitation Medicine*, 3, 115–121.

Grandjean, E., Hünting, W. and Piderman, M., 1983, VDT workstation design: preferred settings and their effects. *Human Factors*, 25(2).

Hünting, W., Läubli, Th. and Grandjean, E., 1981, Postural and visual loads at VDT workplaces. Part 1: Constrained postures. *Ergonomics*, 24, 917–931.

Johansson, G. and Aronsson, G., 1980, Stress reactions in computerized administrative work. Report from the Department of Psychology, University of Stockholm, Suppl. 50, November.

Maeda, K., Hünting, W. and Grandjean, E., 1980, Localized fatigue in accounting machine operators. *Journal of Occupational Medicine*, 22, 810–816.

Nachemson, A. and Elfström, G., 1970, Intravital dynamic pressure measurements in lumbar discs. *Scandinavian Journal of Rehabilitation Medicine*, Suppl. 1.

Posture Analysis and Evaluation at the Old and New Work Place of a Telephone Company

S. CANTONI, D. COLOMBINI
Service of Occupational Health of the USSL of Milan, Italy
E. OCCHIPINTI, A. GRIECO
University of Milan, Italy
C. FRIGO, A. PEDOTTI
Center of Bioengineering, Foundation 'Pro-Juventute Don Gnocchi', Milan, Italy

Abstract

Using an original method to analyse postures, an old and new (VDT) work station in a switchboard control room were examined. The spinal pathology of 300 workers was studied by means of an appropriate questionnaire. It was observed that the change from the old to the new work station produced an average decrease in lumbar load of 37 kg/h and a significant increase in postural fixity. The discussion of the results suggests how this disadvantage can be avoided.

1. *Introduction*

The increasing application of ergonomics is changing many work places, particularly those requiring sitting postures. In these cases, the use of correct ergonomic criteria has led to considerable improvements with a reduction in the arthro-muscular load, especially on the spine and shoulders.

Besides the need for quantifying the extent and quality of these improvements more precisely, our aim was also to verify if ergonomically designed work places increase postural fixity. A fixed posture, maintained throughout the whole working shift, can in itself be considered a risk factor, particularly for the lumbar spine, where correct intervertebral disc nutrition depends mainly on alternating hydrostatic pressures, above and below the 'critical' value of 80 kg for optimum time periods not exceeding a few hours (Kroemer 1977).

The posture in which the disc pressures constantly exceed this 'critical' value produce alterations of the disc nutrition mechanism, which lead to both short-term disorders, such as numbness or a sense of heaviness in the lumbar region, and long-term, discal degeneration processes. Furthermore, if prolonged, supported sitting postures do not strengthen the spine fixing muscles, which are a protective factor against degenerative disease of this region.

The present paper briefly describes an investigation carried out in the switchboard control room of the National Telephone Company (NTC) in Milan where a gradual transformation of work places from the traditional electromechanical switchboard to a VDT-operated switchboard is in progress. The new work stations have been designed and accomplished according to advanced ergonomic criteria with the advice of the Occupational Health Service of the USSL of Milan.

The purposes of the investigation were:

1. To quantify the damages to the locomotor apparatus in a population that had assumed unsuitable postures for several years.
2. To identify and quantify the differences in loads on joints between the old and new postures, particularly as regards the lumbar spine.
3. To verify whether the work stations designed according to ergonomic criteria increase postural fixity.

2. *Methods*

About 300 persons work in the NTC switchboard control room, of whom about 50% are males. At the present all the workers use the same traditional electromechanical switchboards that gradually will be replaced by VDTs. The switchboard (Figure 1) consists of a fixed work surface and a vertical panel in which the operator inserts the jacks; the chairs have fixed backs and a rotating seating surface resting on a 4 point base; the leg area is not sufficient.

The VDT work station was recently designed according to ergonomic criteria (Figure 2). All the parameters and particularly the height and horizontal and sagittal planes of the VDT are easily adjustable. The position of the keyboard, which can be adjusted backwards and forwards, is such that the forearms can be supported. The keyboard is encased to avoid extreme wrist extension movements during typing.

To satisfy the aims of the investigation an appropriate data sheet was developed in order to analyse the postures at old and new work stations with standardized criteria and record posture duration times (Figure 3).

The data sheet covers the possible postures of the trunk on horizontal, sagittal and frontal planes, the way of sitting on the seat, the position of the legs. The cervical region, which is a substantially independent variable and the upper limbs, which almost always rest on the work surface, are not considered. To each variable was assigned a code number so that each posture is identified by 6 figures followed by the respective duration expressed in seconds.

The whole postural sequence of a subject is determined by the temporal succession of his different postures. The alteration of even only one variable was considered a 'change'.

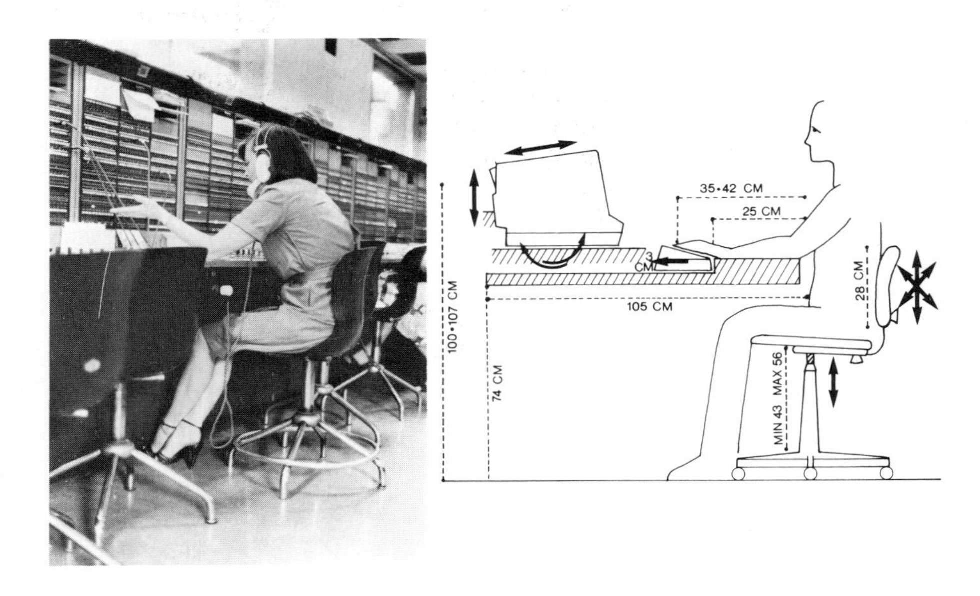

FIGURE 1. *The old work place.* *Figure 2. The new work place (VDT).*

The study was carried out on 4 subjects, 2 female (A: 41 years, 54 kg, 170 cm; B: 37 years, 52 kg, 159 cm) and 2 males (C: 37 years, 82 kg, 170 cm; D: 44 years, 73 kg, 165 cm) who were not suffering from any serious disorder of locomotor apparatus. The individuals were unaware of the aim of the study. The observations were carried out by 4 different experts and 4 observations were made on each subject: 2 during work at the old switchboard, from 10 to 11 a.m. (the peak period for telephone traffic) and from 1 to 2 p.m. (at the end of the shift). The other 2 observations were carried out in the same periods the following day at the VDT switchboard. Each subject was always observed by the same expert. Postures maintained for less than 5 seconds were not considered for analysis of postural sequences. To compare the number of postural changes in different subjects and in different work stations the number of observed changes was multiplied by 3600 seconds and divided by the real observation time.

To identify the prevalent postures we rated each posture, on the basis of its duration, in the sequence observed.

Lastly, a study of spinal disease in all the workers was carried out by means of a suitably designed questionnaire, since it was not possible to perform medical examinations. The results of the questionnaire were analysed according to interpretative criteria recently developed by the authors (Colombini *et al.* 1983).

The questionnaire was distributed to the 248 workers present at the time of the investigation and 201 answers (81%) were received. Age, sex and main anthropometric characteristics of the subjects examined are given in Table 1.

FIGURE 3. *Data sheet to register postural changes.*

Name	Surname		Age	Weight	Height		
Trunk				**Trunk**		**Legs**	**Time**
	[1] TRUNK IN FRONT OF THE		[1]	[1] UPRIGHT	[1] SUPPORTED	[1] AT 90°	Min. ....
	SWITCH BOARD		OCCUPYING		[2] UNSUPPORTED	[1] CROSSED	Sec. ....
[1]			ALL THE		[3] LAT. BENT R	[3] FEET UNDER	
UNTWISTED	TRUNK AT SIDE OF SWITCH	[2] R			[4] LAT. BENT L	THE SEAT	
	BOARD	[3] L	SEAT				
	LEGS AT SIDE AND TRUNK	[4] R	[2]	[5] KIPHOSIS	[6] LAT. BENT R	[4] EXTENDED	
[2]	IN FRONT OF THE SWITCH	[5] L	OCCUPYING		[7] LAT. BENT L	KNEES	
TWISTED	BOARD		A SMALL	[8] BENTED	[9] LAT. BENT R	[5] STRADDLED	
	LEGS IN FRONT AND	[6] R	PART OF		[10] LAT. BENT L		
	TRUNK AT SIDE OF THE	[7] L	THE SEAT		[10] LAT. BENT L		
	SWITCH BOARD			11 SLIDING POSITION			

R. = Right
L. = Left

TABLE I. *Main characteristics of the examined population.*

Age groups	Number		Age ($x \pm s$)		Working age		Height ($x \pm s$)		Weight ($x \pm s$)	
	M	F	M	F	M	F	M	F	M	F
26–35	13	9	32 ± 4	32 ± 3	9 ± 6	13 ± 5	171 ± 6	161 ± 6	61 ± 12	60 ± 13
36–45	64	62	40 ± 3	41 ± 3	15 ± 3	17 ± 4	172 ± 6	161 ± 6	73 ± 11	58 ± 9
> 45	18	35	52 ± 4	50 ± 3	16 ± 7	24 ± 5	170 ± 7	161 ± 6	78 ± 14	62 ± 9

3. *Results and Discussion*

3.1. Evaluation of Lumbar Spine Load

Not all the variables included in the data sheet are relevant, in our specific case, to the study of lumbar disc loads. Therefore we can limit observations to the following variables for the lumbar spine: supported, upright unsupported, kyphosis, bent. Although analysed, the variants 'bent sideways' and 'twisted', were not considered for the synthesis because of their limited influence on the lumbar load (Schultz *et al.* 1982): these variants were therefore incorporated with equivalent variants on the sagittal plane.

By means of the duration rating of the postures it was possible to calculate the duration of each postural variant for all 8 postural sequences examined at the old switchboard and for the 8 similar postural sequences at the VDT switchboard (Figure 4). The duration is expressed as a percentage of the total observation time (20 706 seconds at the old switchboard; 26 493 at the VDT).

We should like to emphasize that during work at the VDT we registered a considerable increase in the duration of the positions with the spine resting on the back-rest, while the duration of kyphosis and trunk bending, positions which were extensive during work at the old work place, dropped to nearly zero.

In Figure 4, for each postural variant, we have expressed the load on L_3 disc (kg) calculated with an original biomechanical study method (Molteni *et al.* 1983), with due account taken of the results of recent experiments aimed at better evaluating loads in postures with upper limbs supported (Occhipinti *et al.* 1983). Finally we calculated a synthetic index, the weighted load on L_3 per hour (kg/h), which was 97·5 at the old work place and 60·2 at the new work place. All values on L_3 are referred to a standard subject weighing 70 kg.

3.2. Evaluation of the Incidence of Postural Changes

The data pertinent to the number of postural changes per hour registered for each subject at the old and new switchboard are shown in Table 2.

FIGURE 4. *Average time of permanence in four different positions and load on lumbar spine (L_3) at the old and new work place.*

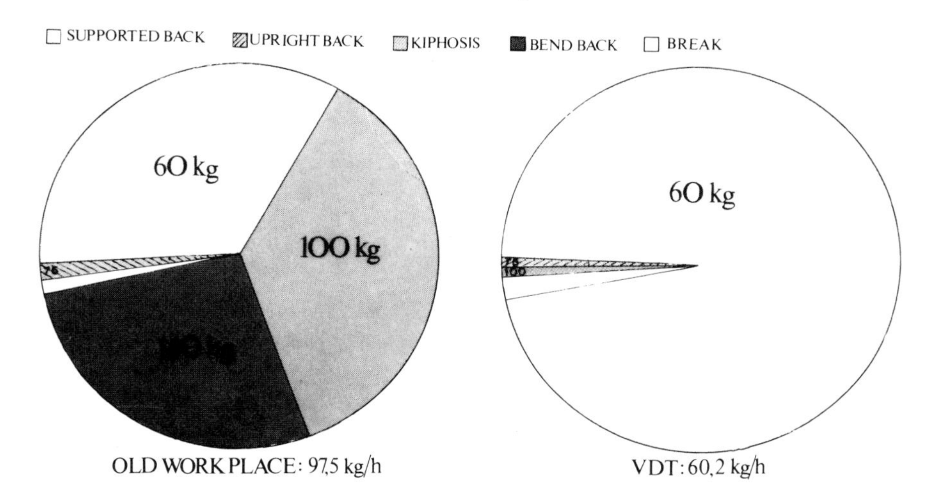

TABLE 2. *Number of postural changes per hour observed in 4 subjects at the old and new work place.*

	Subject			
Old work place (h. 10–11)	31	83	22	14
Old work place (h. 13–14)	58	65·5	12	10
New work place (h. 10–11)	13	28	8	7·5
New work place (h. 13–14)	21	23·5	8	6

The differences in behaviour at various times during the work shift were not significant, which makes it possible to compare the data concerning the old and new switchboards independently of the working hours.

Tests of statistical inference were therefore carried out to compare the mean values and the variances of the data at the old and new switchboards. The results show a highly significant difference ($p < 0{\cdot}001$) between the mean values of the number of postural changes at the old and new work places (Table 3). The difference between the variances were equally significant: in fact, whereas at the old switchboard a considerable dispersion of values around the mean was registered, at the VDT, behaviour was more homogeneous.

A significant difference was also observed between the behaviour of the females and males. But in view of the small number of observations further studies are needed.

Since coupled data were available in this case, we studied the correlation between the number of changes per hour at the old and new switchboards. The correlation coefficient was higher than 0·99. The equation of the regression line is

$$y = 2{\cdot}98 + 0{\cdot}31\ x.$$

TABLE 3. *Results of inference tests for differences between means and variances of number of changes per hour in the old and new work place.*

$\bar{X} = 36\ 93$	$SX = 26,25$	$\bar{Y} = 14,5$	$SX = 8,23$
DIFFERENCES BETWEEN MEANS		DIFFERENCES BETWEEN VARIANCES	
$\bar{D}_{x-y} = 22,43$	$S_{x-y} = 18.1$		$F = \frac{S_x^2}{S_y^2} = 10,1$
$t = 3,5$	$p < 0,01$		$p < 0,01$

Figure 5 reports the number of changes per hour at the old and new switchboards. It can be observed that the change from the old to the new work station produced a reduction in the number of changes (and therefore an increase of fixity) that is much more marked for the subjects who showed a high index at the old switchboard: that is to say, the mean decrease per cent in the number of changes per hour was 71% in subject B, 62% in subject A, 53% in subject C and 44% in subject D.

FIGURE 5. *Number of postural changes per hour observed in the subjects at the old and new work place.*

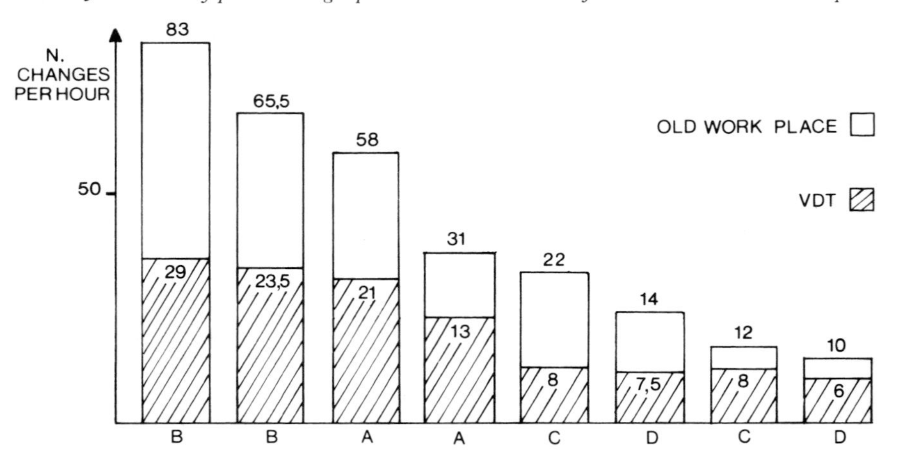

3.3. Study of Prevalence of Spinal Disorders in the Population Under Study

Figure 6 reports the prevalence of anamnestic cervical and lumbar osteoarthro-muscular pathology, divided by age and sex. Besides an age-related trend, attention is drawn to a constant higher prevalence of this pathology in females.

FIGURE 6. *Prevalence of anamnestic cervical and lumbo-sacral osteoarthro-muscular pathology in relation to age and sex.*

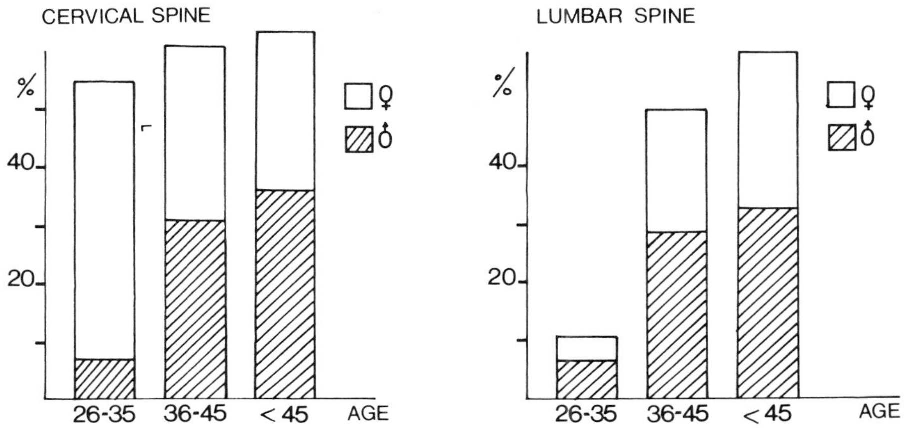

4. *Conclusions*

We can draw the following conclusions:

1. The prevalence of anamnestic spinal disorders was similar to that which we have observed in other groups of workers exposed to a similar risk. It is, however, fairly high thus confirming the diffusion of spinal disorders in relatively young people. Nevertheless, in this case, the results do not permit a precise interpretation of the relationship between the previous postural risk and the observed damage.
2. We utilize methods to quantify the loads on different segments of the body, especially on the spine, which also take account of the time and mode of exposure to postural risk. By means of these methods, in the observed situation, we recorded an average decrease in lumbar load of about 37 kg/hour changing from the old to the new work station.
3. We found that the new work station produced a significant increase in postural fixity. To avoid this disadvantage ergonomic improvements must be applied both to machines and work organization. In our case, the introduction of new work stations was accompanied by the introduction of a 15-minute break every 2 hours. For similar cases a break of between 5 and 15 minutes every 2 hours maximum is suggested as a suitable measure to prevent the disadvantage due to excessive postural fixity.

References

Colombini, D., Occhipinti, E., Menoni, O., Bonaiuti, D. and Grieco, A., 1983, Posture di lavoro incongrue e patologia dell'aparato locomotore (Constrained Work Postures and Damages of Locomotor Organs). *Med. Lav.*, 74, 198–210.

Kroemer, J., 1977, *Bandscheiben-Schaden*. Goldmann Medizin.'

Molteni, G., Grieco, A., Colombini, D., Occhipinti, E., Pedotti, A., Boccardi, S., Frigo, G. and Menoni, O., 1983, Analisi delle posture. CEE, 6 June.

Occhipinti, E., Colombini, D. and Frigo, C., 1983, Sitting posture: analysis of lumbar trunk loads with supported upper limbs (unpublished).

Schultz, A.B., Anderson, C.B.J., Haderspeck, K., Ortengren, R., Nordin, M. and Bjork, R., 1982, Analysis and measurements of lumbar trunk loads in tasks involving bends and twists. *Journal of Biomechanics* 15(9).

Design of a VDT Work Station for Customer Service

M. LAUNIS
Institute of Occupational Health, Department of Physiology, Laajaniityntie 1,
SF–01620 Vantaa 62, Finland

Abstract

Serving a customer has a significant effect on the employee's postural behaviour and thus also on the design of the work station. This concept was confirmed in a project carried out to accomplish the model work station for the selling of tickets. The optimum dimensions for the work station were defined in adjustment experiments, where 30 women employees simulated the most central work phases. The follow-up inquiry on the actual work done at the model work station revealed that, despite the optimum dimensions of the work station, the employee's attitude with regard to the customer affected the employee's posture, thus making the optimization of the working situation impossible.

1. *Introduction*

The design of a work station for customer service is always ergonomically problematic for several reasons. First of all, it is unrealistic to have individually adjustable massive constructions and, if there is some adjustability, it is seldom used because the employees at the work station change continuously. Secondly, the numerous equipment and activities often set conflicting requirements for the construction of the work place. In addition to these difficulties, the contact between the employee and the customer can severely restrict the adoption of an appropriate posture.

All these conditions predominated in the redesign project which was carried out in the State Railways of Finland. The aim of the project was to accomplish a model work station for the computer-aided selling of tickets.

The preliminary studies at the old work site included an analysis of the task and the work station and interviews with the employees and their supervisors. It was found that the employees must constantly be oriented to the customer so that they could 'read from the lips'. In the noisy conditions the working posture became extended forward and upward to the utmost. The equipment and activities were numerous and spread over a wide sector (about 180°). Rotating the trunk and twisting the back and neck was considered uncomfortable and strenuous, especially among elderly employees.

The postural problems were considered to be caused primarily by the unsuitable dimensions of the work station. To determine the optimum dimensions, adjustment experiments were arranged. The more general objective of the experiments was to clarify the relationship between the individually optimum dimensions of the work station and body size so that more relevant and more up-to-date recommendations for modern office work places could be derived.

2. *Methods*

2.1. Adjustment Experiments

In the experiments 30 women employees aged ($\bar{x}$ = 32, SD = 14) simulated the most central work phases in an electrically manoeuvrable mock-up and simultaneously defined their optimum dimensions and layout for the work station according to their sensations of comfort. For the vertical dimensions, such as the height of the seat and the height of the table, the individual tolerance limits were also studies using a modification of the method of fitting trials by Jones (1969). The criteria for judging the maximum and minimum values were verbalized as the level of comfort at which the subjects believed they could work comfortably throughout their entire shift. The judgements were repeated eight times to compensate for the short-term nature of the method.

The height of the seat was measured from a 20 mm polyurethane foam pad (when compressed), 270 mm from the front edge. The inclination of the seat was 2·5° backwards, and the clearance between the front edge of the seat and the popliteal space was 100 mm. In the experiments concerning table height the home row of the keyboard was set 32 mm above the table.

The body reference measures, e.g. lower leg length and elbow height, were recorded according to DIN standard 33402. The elbow height was also measured in a relaxed working posture, with the back leaning against a normally adjusted backrest.

2.2. Follow-Up Inquiry

Half-a-year after assembling the model work station, 17 employees were questioned for their subjective opinions about the fitness of the new work station. Each single dimension was rated on a five point scale: too small/slightly small but tolerable/suitable/slightly large but tolerable/too large. Also other ergonomical features were questioned.

3. *Results and Discussion*

3.1. Adjustment Experiments

The average results within subjects are shown in Table 1. The repeated judgements brought about no systematic changes in group averages during the experimental session.

TABLE I. *The means, standard deviations and ranges of subjective judgements and body reference measures (N = 30).*

	$\bar{x}$	SD	range
Optimum seat height	442 mm	39	360–518
Optimum table height for keying	226 mm	26	168–265
Popliteal height, shoes included	425 mm	29	369–495
Elbow height, measured in the standardized way	240 mm	23	183–495
Elbow height, measured in the relaxed posture	201 mm	21	162–244
Heel height of the shoe	49 mm	28	0–83
Stature	1632 mm	72	1500–1800
Weight	62·1 kg	8·5	45–80

There was great individual variation between the optimum sitting height and the popliteal height. On average, the subjects preferred sitting quite high, at a height of $\bar{x}$ = 17 mm (SD = 30, range = −52 . . . +76) higher than their popliteal height. This may be explained by the fact that employees are used to sitting with the feet almost dangling. The correlation between the popliteal height and the optimum height of the seat was r = 0·64. Correspondingly, the optimum table height for keying was, on average, $\bar{x}$ = 14 mm lower than the elbow height measured in the standardized way (SD = 27, range = −58 . . . +48). The correlation was quite weak, r = 0·36, whereas the elbow height in the relaxed posture was clearly more strongly correlated with the table height, r = 0·57.

In an attempt to explain the great individual variation, three groups of 10 subjects were formed according to the ratio weight/height^3. The subjects of light build preferred a sitting height about 20 mm higher than the other two groups. This is likely due to fewer soft tissues on the subjects' thighs. Correspondingly, the subjects of heavy build preferred a slightly higher table height from the seat than the other groups, which indicates that they have adopted a more upright sitting posture.

The average group results were adapted in the design of the model work station. For the table a height of 680 mm was chosen, and the height of the customer's desk was placed 260 mm above the table. In addition, a 70 mm high footrest easily removable by foot, was assembled to the work station. Figure 1 shows how well or poorly the table height (horizontal line) suited the subjects. The redesigned work station is shown in Figure 2.

Figure 1. *Individual optimum table heights and ranges for adaptability, a combination of the results concerning seat height and table height from the seat. The horizontal line is the final fixed table height.*

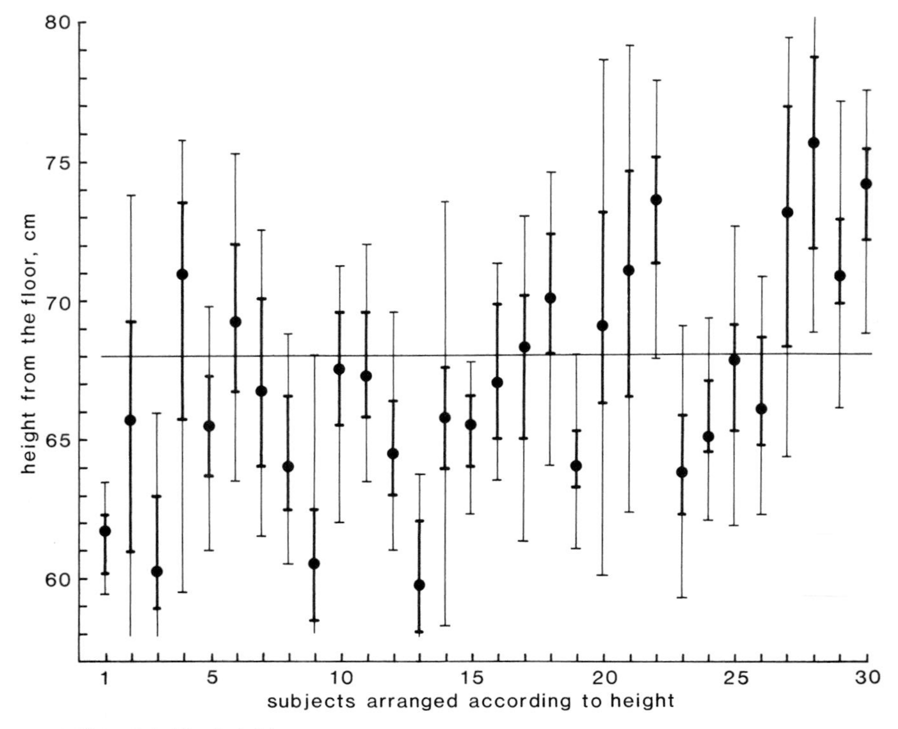

3.2. **Follow-Up Inquiry**

The new work station was generally considered improved, but the sitting height was on average considered too low, the distribution of answers being 5/3/7/1/1. This conflicted with the fact that 13 persons of total 17 voluntarily used the removable footrest. It was also observed that, especially during the rush hours, some employees adjusted the seat to its maximum height and leaned with their stomach against the front edge of the table. Obviously, the employees still attempted to get closer to the customer and considered the relaxed posture less important.

4. *Conclusions*

The adjustment experiments confirm the concept that individual preferences for the vertical dimensions of the work place vary greatly. In a right-angled upright working posture, the

FIGURE 2. *The redesigned work station. The compact layout, into the table inset equipment and the lowest possible height of the desk are accomplished to minimize postural problems.*

variation due to several physiological factors (e.g. the demands of the task, slump of the back and pressure on soft tissues against the seat) is greater than the anthropometrical variation, which is normally considered the base of ergonomical planning. Therefore there is considerable risk in proposing that optimum dimensions for unadjustable work sites be based only on anthropometrical data.

In an actual working situation poor audibility and, especially in some cases, also showing readiness to serve the customer, decisively affected the working posture. It can be assumed that the strongly backwards leaning postures commonly preferred among VDT operators according to Grandjean *et al.* (1983) cannot be generally adopted in customer service.

Therefore, to ensure relaxed posture and natural arm movements, it is essential that more attention be paid to the contact between employees and customers. At the very least this implies that employees and customers are in close contact, can communicate without obstructions, and are face to face. These conditions require that VDT devices are manufactured in minimum-sized cases and, especially, that the monitor is much smaller and more backwards tiltable than those now normally available.

References

DIN, Deutsches Institut für Normung, 1978, *Teil 1: Körpermasse des Menschen. Begriffe, Messverfahren.*

Grandjean, E., Hünting, W. and Pidermann, M., 1983, VDT workstation design: preferred settings and their effects. *Human Factors*, 25(2) 161–175.
Jones, J.C., 1969, Methods and results of seating research. *Ergonomics* 12(2), 171–181.

What Is the Correct Height of Furniture?

A.C. MANDAL
Finsen Institute, 2100 Copenhagen Ø, Denmark

Abstract

Man's average height has increased by 10 cm during the last century and the table height has in the same period been reduced by about 10 cm. This inevitably leads to more constrained postures and probably more backache. Eighty persons were asked which height of furniture they preferred and everybody wanted to sit much higher. This automatically leads to a more upright posture with reduced flexion and tension of the back. It is suggested that the consumers are consulted—not only the architects and the standardisation organisations.

1. *Introduction*

During the last century the average height of man has increased about 10 cm, but for some incomprehensible reason the height of the tables has *decreased* about 10 cm. As the visual distance has remained the same, namely about 30–40 cm during a major part of adult life, the natural consequence of this is that we have to sit more bent over. To sit like this for many hours every day produces a long term strain on the bones, tendons and muscles of the spine.

Backache is becoming a serious problem all over the industrialised world and nothing will strain the back for nearly as long a time as the fact that we spend a good deal of our time hunched over tables. In 1981 the European committee for standardisation, CEN, sent out for approval new standards for office furniture. The recommended height of the tables was 720 mm for all non-adjustable tables. No explanation was given why this height should be better than any other, nor were the names of the persons responsible for this idea given.

There is a world-wide unanimity of opinion with respect to 'correct' sitting posture, namely that the body should be very upright and with a flexion of the hip joint of 90° and a concavity in the small of the back (lordosis). However, no normal person is able to work in this posture. Sketches representing models for sitting postures from various countries are presented in Figure 1. These sketches constituted the basis of:

FIGURE 1. *On the left: A) Standardization of furniture (CEN). In the middle: B) Training of designers (Dreyfuss). On the right: C) Training of posture (Snorrason).*

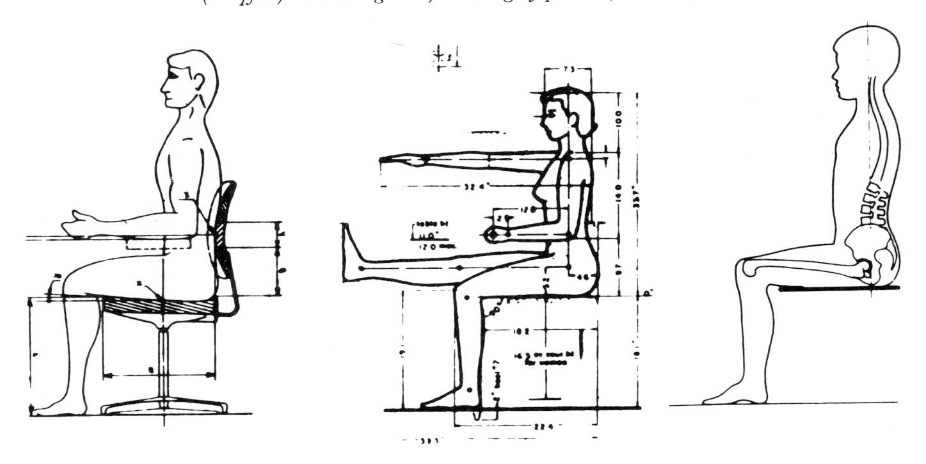

A. Standardisation of furniture (CEN, 1981).
B. Training of furniture designers (Dreyfuss, 1955).
C. Training of people to sit 'correctly' (Snorrason, 1968).

It is surprising that no explanation appears to have been given substantiating why one should sit in this particular manner, nor is information supplied regarding the method by which the sketches were constructed. Nevertheless they have been accepted quite uncritically by experts all over the world. If you base prevention of backache on such illusions, you are doomed to fail.

During the past 20 to 30 years there has in Denmark been great interest in preventing backache by training people in better sitting postures. Figure 1C has formed the basis of this training. It clearly recommends that one should sit with a 90° flexion of the hipjoint and with a concavity of the small of the back. The sketch has simply depicted a skeleton sitting on a chair. But the sitting problems of living persons have very little in common with the problems of a skeleton. No one is able to sit like this while working.

To evaluate this I have taken pictures of the Danish physiotherapists during their final examinations. The pictures were taken with an automatic camera with 24 minutes interval during a 4 hour examination. Not even the physiotherapists, who usually give the instructions in 'correct' sitting postures, are able to sit with a straight back (Figure 2). They all sit hunched over their tables (height 72 cm) to bring the eyes at 30–40 cm distance from the book and the axis of vision at a reasonable angle.

2. *Methods*

To evaluate the influence of table height on the flexion of the back, a person (height 171 cm) was asked to sit reading on a chair with 5° backward sloping seat of 43 cm height and at a table of

FIGURE 2. *Danish physiotherapists during a 4 hour examination. 3 Photographs at 24 minute intervals.*

72 cm with horizontal desk as advocated by CEN. The feet were supported by a transverse bar under the table 20 cm above ground level (to achieve the desired table height of 72 cm; Figure 3). To control the flexion of various parts of the body, well-defined anatomical points were marked with spots:

1. Capitulum fibulae.
2. Trochanter major, which is located very close to the hipjoint.
3. A point midway between spina iliaca sup anterior et posterior to mark the axis of the pelvis.
4. The acromion.
5. Finally the ear and the eye were used as marks.

For 20 minutes the woman was sitting and reading at table height 72 cm/chair 43 cm, and during this period 5 pictures were taken with 4 minutes interval by an automatic camera (Figure 3A). For the next 20 minutes she was sitting at table height 82 cm (achieved by placing the feet on a transverse bar 10 cm above ground level. Chair height was 53 cm. The desk and the seat were sloping about 15° towards one another. In this position 5 photographs were also taken with the same interval, Figure 3B.

Finally she was asked to place the feet on the floor to achieve a table height of 92 cm measured at nearest edge. Chair height was 63 cm, measured at axis of rotation (Figure 3C). In all 3 situations she was asked to sit in the position she found most comfortable. No instructions were given concerning body posture or eye distance. The experiment was repeated for 10 days.

During the first day she found the table height 82 cm to be the most comfortable but during the rest of the experiment she definitely preferred 92 cm.

3. *Results*

At the end of the experiment 50 photographs of each of the situations were available. The skin marks were connected by lines on the pictures. The resulting angles between these

FIGURE 3. *Flexion in lumbar region. Height of table/chair:* (*a*) 72/43 cm, (*b*) 82/53 cm, (*c*) 92/63 cm.

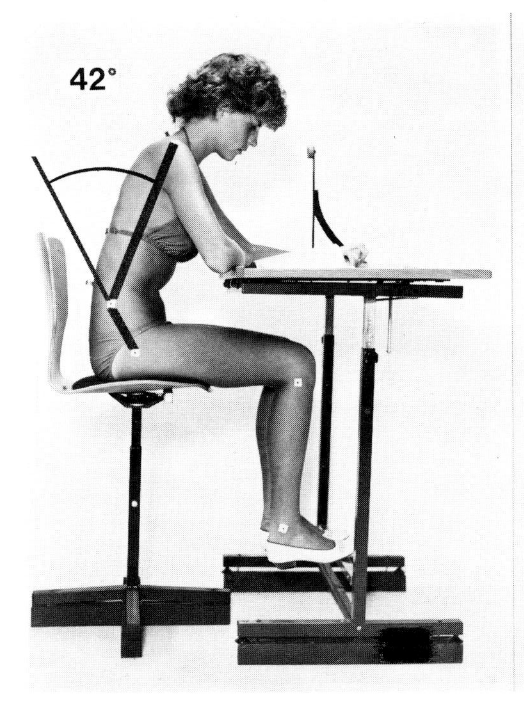

(*a*)

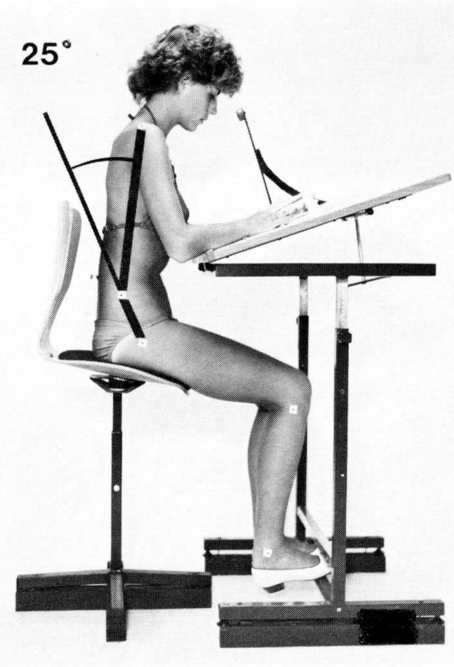

(*b*)

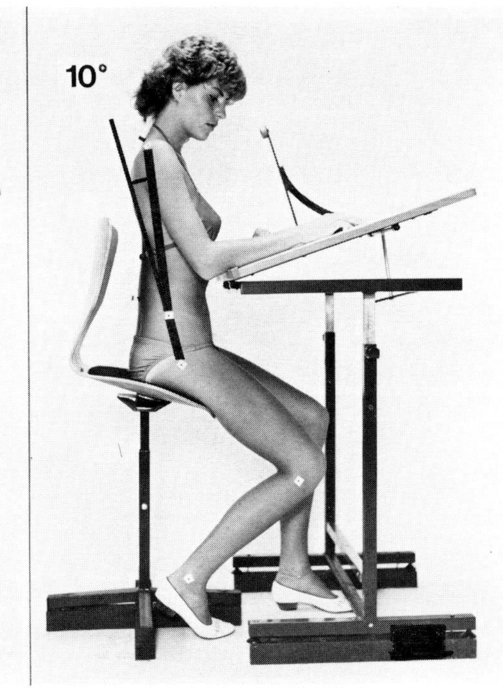

(*c*)

marks (hip angle, lumbar angle, neck angle and head angle) were measured. The flexion of the lumbar region was found to be an average:

A: 42° at table height 72 cm
B: 25° '' '' 82 cm
C: 10° '' '' 92 cm

The 17° reduction of flexion in the lumbar region when changing from 72 cm to 82 cm is highly significant ($p = 2 \times 10^{-9}$). So is the 15° reduction from 82 cm to 92 cm ($p = 10^{-8}$). Besides the flexion in the hipjoint was found to be:

A: 57° at table height 72 cm
B: 50° '' '' 82 cm
C: 42° '' '' 92 cm

FIGURE 4. *Preferred height of chair and table (marked ●). CEN recommendation marked by solid line.*

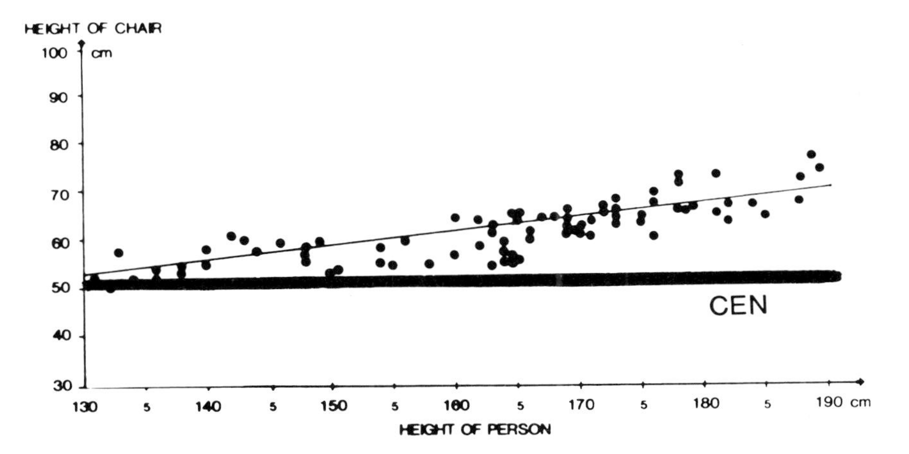

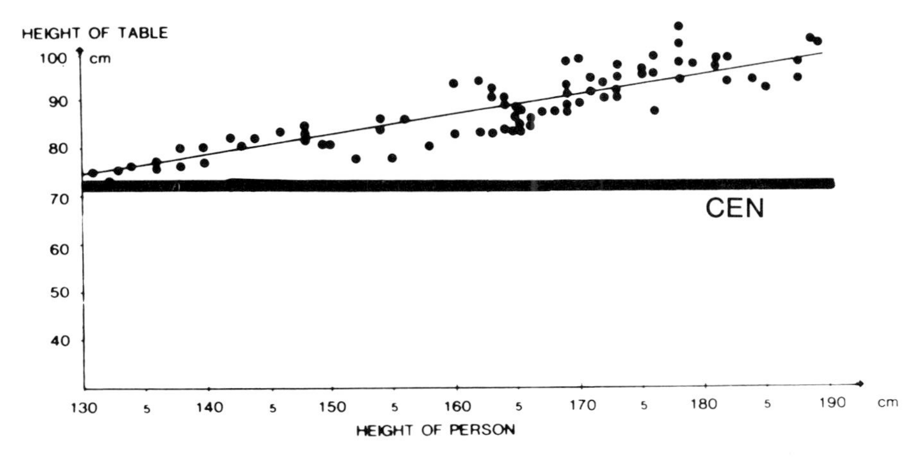

The 7° reduction of flexion in the hip joint from 72 cm table height to 82 cm is also highly significant ($p = 5 \times 10^{-6}$). So is the 8° reduction from 82 to 92 cm ($p = 7 \times 10^{-7}$). In all the reduction of flexion of the hipjoint is 15°. This means that the total flexion of the back (lumbar region and hipjoint) can be reduced by 47° by increasing the height of table (and chair) by 20 cm providing the seat and table top is tilted 12–15° towards one another.

In a similar experiment I have asked 80 persons (age 7 to 50 years) which height of chair and table they preferred to use if they had to sit reading or writing for longer periods.

At the beginning the persons were placed on a chair 43 cm high and 5° backward sloping seat at a table of 72 cm with a horizontal top. Then a brake under the front part of the seat was removed allowing the seat to tilt forward 10–15°. The table top was adjusted to a slope of about 12°. Both the chair and the table were based on hydraulic columns so that wile seated they could be pumped slowly to the height the persons themselves preferred. In Figure 4 the height of chairs which was preferred is marked with black spots; it also shows the table height and it is surprising that *everybody* prefers to sit much higher than the solid line advocated by CEN.

4. *Conclusions*

Office furniture has so far primarily been constructed to satisfy the architects and manufacturers. It is suggested that the consumers should be consulted in the future.

References

CEN, 1981, *Meubles de bureau*, pr EN 91 (Paris: La Defence).
Dreyfuss, H., 1955, *Designing for People* (New York: Simon & Schuster).
Snorrason, E., 1968, *Hvilestolsproblemer* (Copenhagen: Tidsskrift for Danske Sygehuse).

The Effects of Various Keyboard Heights on EMG and Physical Discomfort

A. WEBER, E. SANCIN AND E. GRANDJEAN
Swiss Federal Institute of Technology, Department of Ergonomics, CH-8092 Zurich, Switzerland

Abstract

Twenty trained typists performed 6 consecutive 10 min typing tasks on an adjustable VDT work station. The experiments were conducted with preferred as well as with imposed keyboard heights, each setting once with and once without forearm–wrist support. Body postures, EMG of the trapezius, pressure load exerted on the support, and subjective feelings of tension and pain were recorded.

The preferred keyboard levels are distinctly higher than those recommended in standards. Two-thirds of the subjects prefer a support to rest forearms and wrists. For comparable keyboard height conditions, EMG of the trapezius is lower when there is a support than when there is none. With the support, the pressure load increases with increasing keyboard level. The feelings of tension and pain in the neck, shoulders and back have a tendency to be lower with support than without. For a constant keyboard height with support, there is a significant negative correlation between EMG and exerted pressure load.

1. *Introduction and Methods*

Field studies with adjustable VDT work stations showed that VDT operators preferred keyboard heights markedly above usual ergonomic recommendations. Furthermore, the operators preferred body postures characterised by a backward leaning trunk and by slightly elevated arms and hands (Grandjean *et al.* 1983). The aim of the present study was to compare physiological effects of preferred and imposed keyboard heights under experimental conditions.

Twenty subjects had to imitate VDT work by operating the keyboard and watching alternatively a source document and the screen. In a first session, all preferred dimensions of the adjustable work station were determined for each subject (keyboard height, inclination, distance, height of screen, source document holder and seat). In the second session, the following 6 experiments—each one lasting 10 minutes—were carried out: work at the preferred keyboard height as well as 5 cm above and 5 cm below it; each condition with and without forearm–wrist support. All other dimensions of the work place were kept constant at the initially determined preferred levels—except the location of the chair which could be moved by the subject toward or away from the table. During work, EMG was recorded by surface electrodes from the right trapezius pars descendens. Furthermore, the pressure load of forearms, wrists and hands on the support and the keyboard was recorded. Different elements of the body posture were measured. At the end of each working period the subjects had to rate on a questionnaire (4-grade-scale) and on an ungraded scale (between 'very relaxed' and 'very tense') their subjective feelings of pain and tension in the neck–shoulder–back–arm area.

2. *Results*

2.1. Electromyography (EMG)

Since in these experiments EMG activity does not change over time, mean values over each 10 min period are considered for each subject. Mean EMG values of 20 subjects are represented in Figure 1.

When working with forearm–wrist support, the electrical activity of the trapezius is—for each keyboard height condition—lower compared to the experiment without support.

FIGURE 1. *Mean EMG activity during each experimental condition. The integrated EMG values/10 s are expressed as a proportion of tension during standard position with arms stretched laterally at 90°. Means by 10 min periods for 20 subjects.*

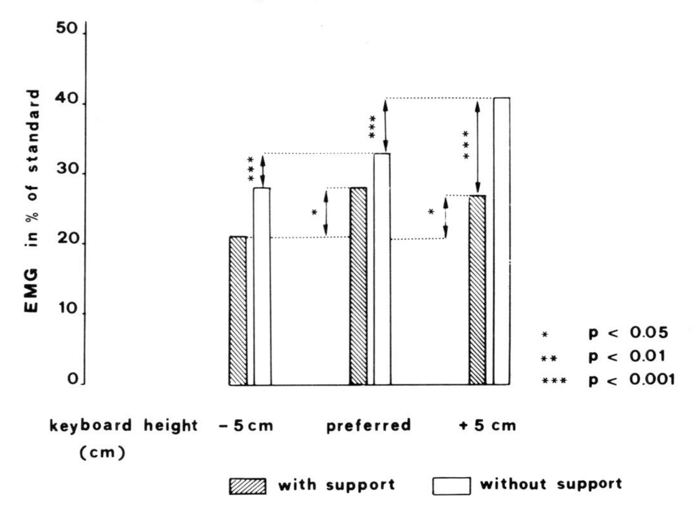

This difference is significant for the highest height. Furthermore, without support EMG activity increases significantly with increasing keyboard height. Moreover, it appears that EMG activity is not at its lowest at the preferred keyboard height. This is comprehensible since the locomotor system must be in a dynamic state of equilibrium. To prevent overload of ligaments and tendons, some muscle tension is required.

2.2. **Pressure Load**

Mean pressure load remains extremely constant over the 10 min periods. When working without forearm–wrist support, it equals nearly zero for all keyboard heights. When working with support, it ranges between 15 and 35 N and increases significantly with increasing keyboard height. This result is represented in Figure 2.

FIGURE 2. *Mean pressure load on forearm–writs support and keyboard during each experimental conditions with support. Means by 10 min periods for 20 subjects.*

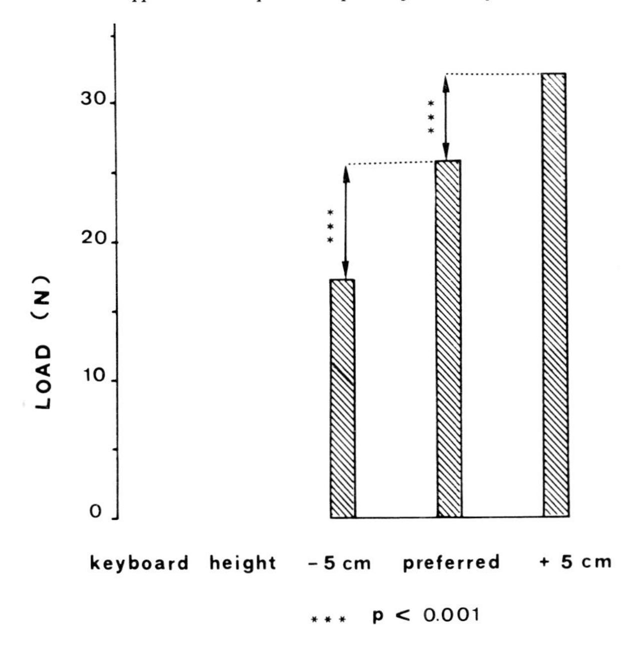

In each working condition with forearm–wrist support, there is a significant negative correlation between EMG and exerted pressure load (see Figure 3, all $p < 0{\cdot}001$). It means that the more arms and hands rest on the support, the lower is the electrical activity in the trapezius.

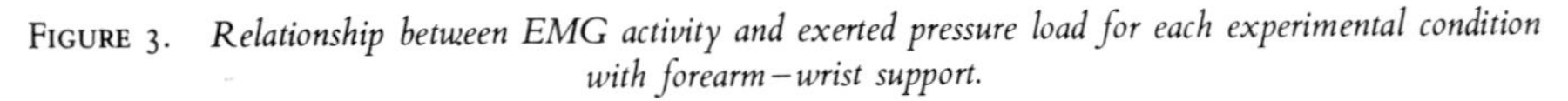

FIGURE 3. *Relationship between EMG activity and exerted pressure load for each experimental condition with forearm–wrist support.*

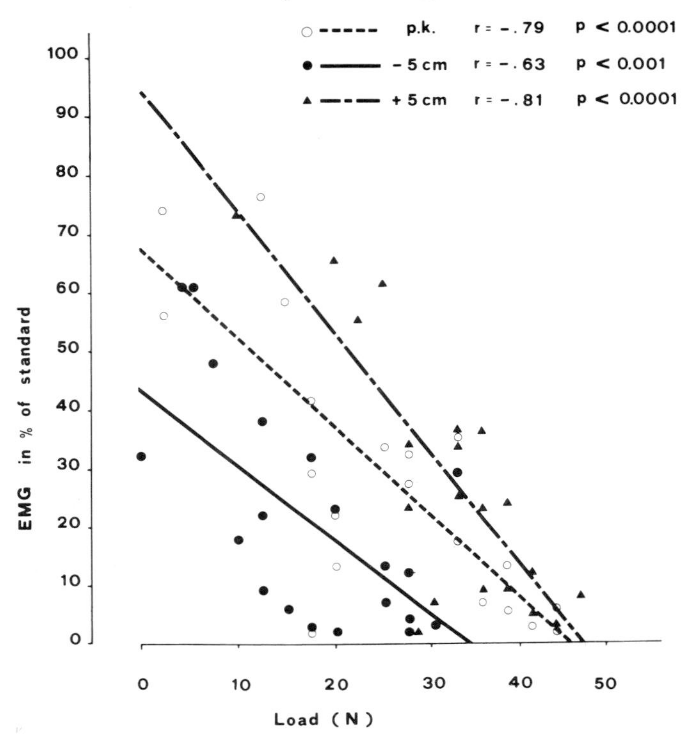

r = Pearson correlation coefficient. p.k. = preferred keyboard height, −5 cm = 5 cm below it, +5 cm = 5 cm above it. Means by 10 min working periods for each subject. Total 20 subjects.

2.3. Preferred Work Station Dimensions

The mean values of all preferred dimensions, the standard deviation as well as the range between minimal and maximal values are reported in Table 1.

The results show that the ranges between the minimal and maximal values are important. Two results are striking: the preferred keyboard heights with and without support are quite similar: the mean and the extreme values differ only by 1 cm. Furthermore, all persons like to incline the chair, the screen and the source document holder backwards. When comparing the preferred dimensions with those of the field study by Grandjean *et al.* (1983), it appears that all settings are very similar except those of the screen: in the present experiment the inclination backwards is larger and the height lower. We assume that in the field study the reflections from light sources are responsible for the different screen settings. Indeed, in the present experiment every kind of reflection was carefully avoided.

The analysis of correlations reveals no relationship between body length and preferred

TABLE I. *Preferred work station dimensions of 20 subjects.*
(all heights are reported to the floor).

Dimensions	Mean	S.D.	Range
Keyboard height[a]			
with support	78 cm	3·4	74–84 cm
without support	77 cm	2·8	73–83 cm
Screen height[b]	97 cm	6·4	85–108 cm
Screen distance[b] from table edge	71 cm	7·4	60–96 cm
Screen inclination[c]	101°	5·3	90–111°
Source document holder height[b]	96 cm	6·5	85–105 cm
Source document holder inclination[c]	111°	5·8	94–122°
Height of seat surface	47 cm	3·2	43–55 cm
Seat backrest inclination[c]	99°	4·2	92·106°

[a] Top level of middle row to floor.
[b] Centre of the screen or source document holder.
[c] > 90° means backward inclination of upper edge.

keyboard height or between eye level and preferred screen height. There is only a significant correlation between body length and preferred seat height ($r = 0{\cdot}56$, $p < 0{\cdot}005$). The results of the judgement at the end of all experiments show that 12 out of 20 persons prefer a keyboard with forearm–wrist support.

2.4. **Body Posture**

With the preferred settings, the body posture is characterised by a mean trunk inclination of 97° backwards (line acromion-crista iliaca vs. horizontal) and a mean head inclination of 52° forwards (line ear-C7 vs. vertical). The comparison of the 6 working conditions shows that trunk and head are kept nearly in the same position in every condition. However, obvious differences appear for the distance of the seat to the table and for the arm positions. With forearm–wrist support, subjects sit nearer to the table than without support (mean difference: 3 to 4 cm; $p < 0{\cdot}001$ for each keyboard height condition). The arm positions in the different conditions are shown in Table 2.

The following results can be deduced from Table 2:

1. *With support*, elbow angle, arm abduction and anteversion are significantly larger than without support ($p < 0{\cdot}001$).
2. *With support*, arm abduction and anteversion increase markedly with increasing keyboard height ($p < 0{\cdot}001$).
3. *Without support*, elbow angle decreases with increasing keyboard height ($p < 0{\cdot}001$).

It appears, therefore, that the subjects mainly adapt themselves to different keyboard heights with arm abduction and anteversion when they have a support, and with elbow angle when they have no support.

TABLE 2. *Mean values of postural elements for all 6 conditions. 20 subjects. p.k. = preferred keyboard height (all heights are reported to the floor, lateral arm abduction and arm anteversion to the vertical)*

Postural elements Conditions	Elbow angle (°)	Arm abduction (°)	Arm anteversion (°)	Epicondylus lat. height (cm)
−5 cm, with support	99	14	25	72
−5 cm, without support	80	12	10	70
p.k., with support	99	18	31	76
p.k., without support	75	13	11	71
+5 cm, with support	106	28	41	81
+5 cm, without support	69	15	15	72

2.5. Subjective Feelings

The feelings of tension in the neck–shoulder–back–arm area, reported on the ungraded scale, are shown in Figure 4.

FIGURE 4. *Mean subjective tension in neck–shoulder–back–arm area at the end of each experimental condition.*

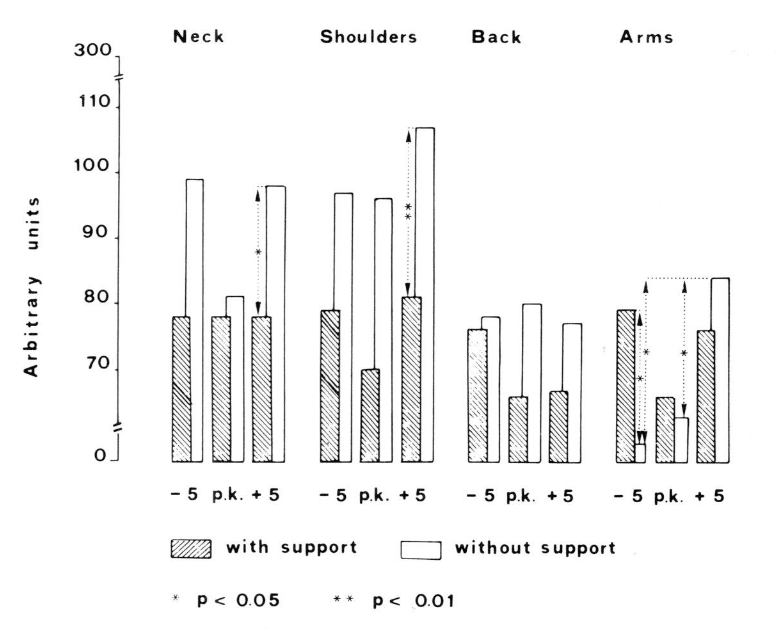

Ungraded scale from 0 = very relaxed to 300 = very tense. p.k. = preferred keyboard height, −5 cm = 5 cm below it, +5 cm = 5 cm above it. Means for 20 subjects.

When working with a forearm–wrist support, the mean subjective tension in neck, shoulder and back is—for each keyboard height condition—lower than without support. The mean tension in the shoulders and arms is at its lowest at the preferred height with forearm–wrist support, but the differences are not significant. A similar, but less pronounced result is obtained for pain in the neck–shoulder–back–arm region with the 4-grade scale questionnaire. There is no significant correlation between the subjective tension nor the subjective pain in the neck–shoulder region and the EMG of the trapezius.

Reference

Grandjean, E., Hünting, W. and Piderman, M., 1983, VDT workstation design: preferred settings. *Human Factors*, 25(2), 161–175.

VDT Work Place Design and Physical Fatigue: a Case Study in Singapore

ONG, C.N.
Department of Social Medicine and Public Health, National University of Singapore, Outram Hill, Singapore 0316

Abstract

This paper illustrates a case study on how ergonomics could help to reduce the physical fatigue and improve performance on workers using VDTs.

Thirty-six female data-entry operators in a computer centre of age below 40 were studied over a period of 18 months on their performances and subjective symptoms. Work place design and work organisation were also studied.

The results indicated that visual and musculo-skeletal complaints were more common among the VDT operators as compared with a group of 41 conventional office workers. Assessment on the work environment suggested that the high prevalence of subjective symptoms were related to the illumination level, improper working posture, badly designed table and chairs, and insufficient rest pauses.

A new work area was designed and the work–rest regime revised. It was observed that a better working environment led to significantly less visual and muscular complaints. In addition to the reduction in physical fatigues, the efficiency and performance of the operators also improved remarkably. The average keying speed increased from 9480 key stroke/hour to 13 002 key stroke/hour and the average monthly error declined from 1·54% to 0·11%.

1. *Introduction*

The development of visual display terminals (VDTs) and their incorporation into the work environment have been so rapid that some confusion and misunderstandings exist. This is especially so in many developing countries. In addition to the problem of how to fit the latest technology to assist production output, the right matching of man to machine has also become a topic of concern. In many cases, machines and tools designed to accommodate operators in the West were introduced to local Asian workers without anthropometric considerations. This usually led to physical fatigue, muscular strains and unhappiness among local workers who are usually smaller in size (Hoong and Ong 1980).

The first report of operator's discomfort in the use of VDTs came from Hultgren and Knave (1974). It was found that VDT operators, who remained for long periods in the same position and paid attention to a specific task were more prone to feelings of tiredness.

There is at present, however, no conclusive evidence that working with VDTs can cause damage to health, but there are indications that prolonged use of VDTs has frequently been associated with complaints of fatigue and other symptoms of stress at work (Binaschi 1980, Smith 1980). On the other hand, several studies have also suggested that the high prevalence of physical fatigue among VDT operators was related to poor work place design (Hünting *et al.* 1980, Hünting 1981). Ergonomics were often overlooked in offices where VDT work was carried out. In this study, attempts were made to change the physical working environment and conditions in a computer centre in the hope that work fatigue among the operators would be reduced.

2. *Methods*

2.1. **Subjects**

This study is part of an on-going project conducted in an airline computer centre in Singapore. The performance of 36 data-entry operators using VDTs was studied for a total period of 26 months. The subjective symptoms were also assessed before and after changes were carried out in the work place. The main task of these operators was to enter alpha-numeric information through the VDT to the main computer. The source materials included airline tickets, written documents, computer printouts, etc.

The performance of the VDT operator was determined by both the average keying speed (in key stroke/hour) and key-punching error (in percentage). These were assessed and recorded daily by the computer.

The subjective symptoms of a group of 41 conventional office workers engaged in traditional sedentary work were also studied.

2.2. **Questionnaire Survey**

Each operator was required to complete a questionnaire concerning subjective symptoms, working conditions, job stress and aspects of VDT equipment. The questionnaire on subjective symptoms was adapted from the Japanese Association of Industrial Health (Maeda 1977).

2.3. **Site Survey**

Ergonomic evaluations were conducted at the computer centre on seating height, work area and visual distance, etc. A Photo Research Model FC 200 light meter was used to measure the illumination level.

Table I. *Work station dimensions.*

Desk height	Work area	Seating height	Leg room
a. 69·85 cm	122 × 66	38–44 cm	13–19 cm
b. 68·6 cm	91 × 60		

3. *Findings*

3.1. **Work Station Design**

The work station dimensions of this computer unit are shown in Table 1. Chairs provided were adjustable with backrests. However, most operators did not make use of the backrests. Operators in this centre read figures from the source materials placed on the left and typed them on to a keyboard. The operators had a continuous sitting position with the neck leaning forward and inclined to the left for orienting the visual lines to the source materials. Sometimes they brought their heads near to the display screen momentarily to check if the information had been correctly entered. Only the right hand was used to operate the keyboard while the left hand handled the source materials. No work stations were equipped with copy-holders.

3.2. **Visual Distance and Illumination Level**

The visual distance of eye–source document and eye–VDT screen as well as distance between source documents and display screen are as shown in Figure 1. A remarkable visual distance difference of 6·6 cm was noted. The mean distance of eye–source document was 49·87 cm and the mean eye–VDT distance was 43·22 cm. This difference called for frequent visual adjustments between VDT and source materials. Furthermore, it was observed that the source materials used in this centre were usually of poor quality computer printout or sketchy handwritings and verifications were usually done under high contrast and low illumination level ranging from 45 lux to 260 lux (Table 2).

3.3. **Visual and Muscular Fatigue**

The prevalence of visual fatigue is shown in Table 3. Over 90% of the VDT operators claimed to have some visual problems. Only 38% of the conventional office workers complained about eye-strain.

TABLE 2. *Illumination level.*

	Illumination on source material	Illumination on keyboard
Mean	131·57 ± 45·3 lux	117·27 ± 50·56 lux
Range	65–260 lux	45–250 lux

FIGURE 1. *Sitting posture of VDT data-entry operators. A = Eye–source material visual distance. B = eye–screen visual distance.*

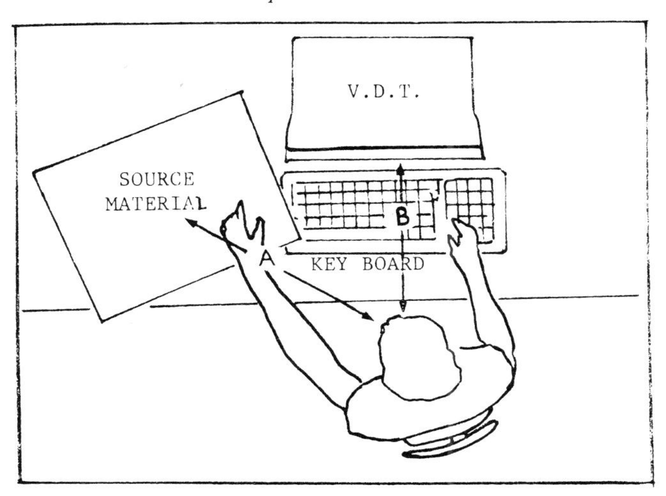

TABLE 3. *Prevalence of visual problem.*

Groups	*n*	Percentage of operators with visual problems (*n*)	Prevalence		
			Almost daily (%)	Occasionally (%)	None (%)
Conventional office work	41	43·9 (18)	9·8	34·1	56·1
VDT data entry	36	94·4 (34)	45·4	49	5·6
VDT conversational terminals	26	76·9 (20)	26·9	50	23·1

3.4. Symptoms in the Torso

The prevalence of symptoms in the torso region in conventional office workers and VDT operators is shown in Figure 2. In general, VDT operators had a higher prevalence than conventional office workers. In addition, it was noted that for the data-entry operators, the prevalence of stiffness on the left side was slightly higher than that of the right. However, it was particularly significant in the neck.

FIGURE 2. *Prevalence of symptoms in the torso of conventional office workers and VDT operators.*

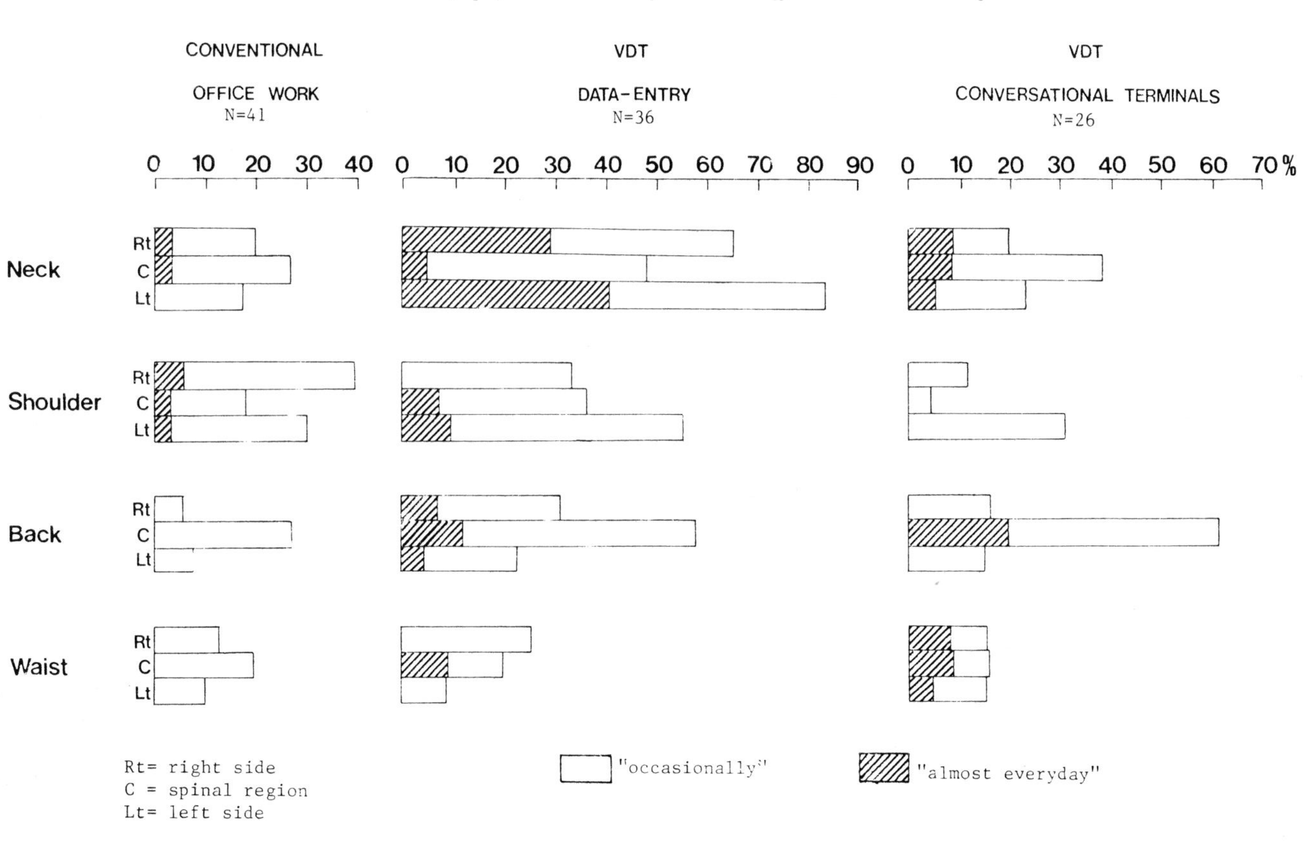

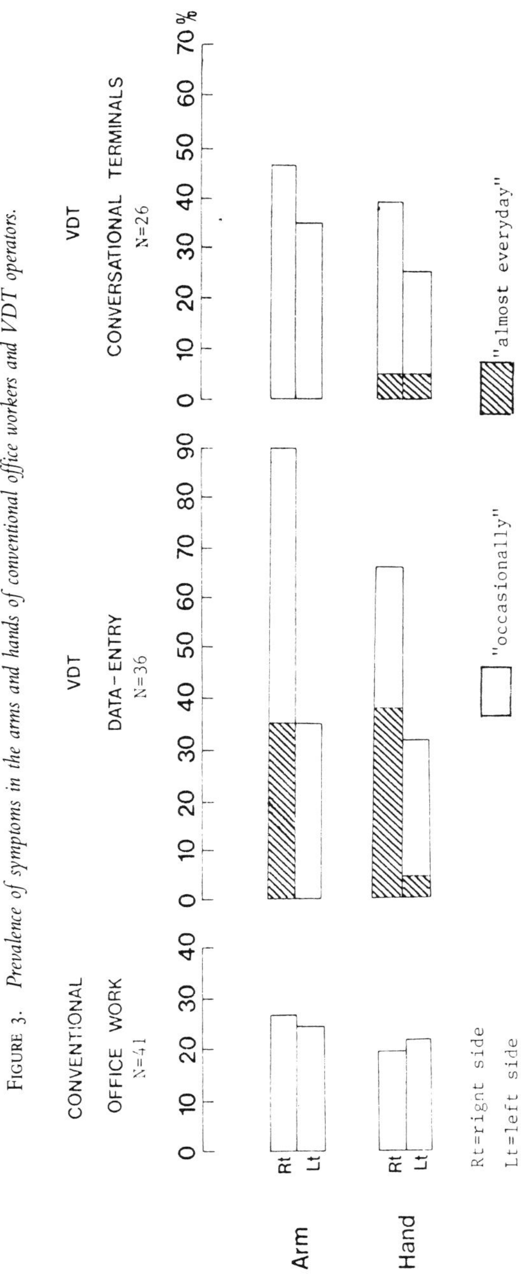

FIGURE 3. *Prevalence of symptoms in the arms and hands of conventional office workers and VDT operators.*

3.5. Symptoms in the Arms and Hands

The prevalence of symptoms in the arms and hands is shown in Figure 3. VDT data-entry operators have a much higher prevalence than conventional office workers. It is interesting to note that the frequency of symptoms in the right hands and arms was higher than that of the left hands and arms. This phenomenon indicated that the symptoms were related to the physical work load placed upon the right hand of operators when using the VDT keyboards.

3.6. Working Hours and Performance

The work–rest regime in this computer centre is shown in Table 4. The total breaktime was 45 minutes in a day of 8 hours operation with 2 breaks in the morning and one in the afternoon. If the performance of the operators as monitored by the average key/stroke/hour was to be considered, the hourly output had actually declined as the hour goes by. Highest keying speed was noted in the morning, gradually declining in the afternoon and dropping further during overtime work after 1530 hours (Table 5).

TABLE 4. *Working hours.*

8.30–10.00
break 20 min
10.20–11.50
break 10 min
12.00–13.00
lunch break 60 min
14.00–15.15
break 15 min
15.30–16.30
16.30–17.15

TABLE 5. *Performance of key punch operators in three sessions.*

Time	Mean key stroke/hour
0830–1300 hr	11 883
1400–1530 hr	11 513
1530–1915 hr	10 775

3.7. **Modification of Work Place**

It was obvious that ergonomics were overlooked in this workroom. In aiming to have a better work environment and work organization a new work area was proposed with the physical environment redesigned. Changed included illumination, colour scheme and dimensions of work area. Noise and thermal environment were also attended to. A new work–rest regime was proposed to have more recess time after lunch break. Ergonomics principles were brought into consideration whenever possible. To minimize visual fatigue due to frequent accommodation of the eyes a document holder was recommended. To maintain proper work posture, a foot rest was also recommended. Certain work habits of operators were also adjusted. Detailed recommendations for this new workroom were reported elsewhere (National Productivity Board 1981).

4. *Results and Discussion*

4.1. **Visual and Muscular Fatigue**

Figure 4 shows the prevalence of fatigue in the torso region of the same group of operators before August 1980 and after October 1981. There was an obvious reduction of musculo-skeletal complaints among the operators in the new work place. Only a small proportion (10%) of the operators claimed to have 'almost daily' musculo-skeletal fatigue. Similar observation was noted in the arms and hands (Figure 4). However, it was interesting to note that the right hands and arms were still having a higher prevalence of muscular fatigue than the left. This was probably related to the keyboard arrangement of having the numeric keypad on the right of the operators, which required only the use of right hand for entering the data.

The prevalence of visual fatigue had also reduced from 94% to 62%. This figure was still higher than conventional office workers which was usually around 40% (Hunting 1981).

4.2. **Performance and Error**

With regards to the performance of workers, Table 6 shows that the average keying speed had increased from 9480 keystroke/hour (in average during the period of 1979–80)

FIGURE 4. *Prevalence of symptoms in the torso before and after changes.*

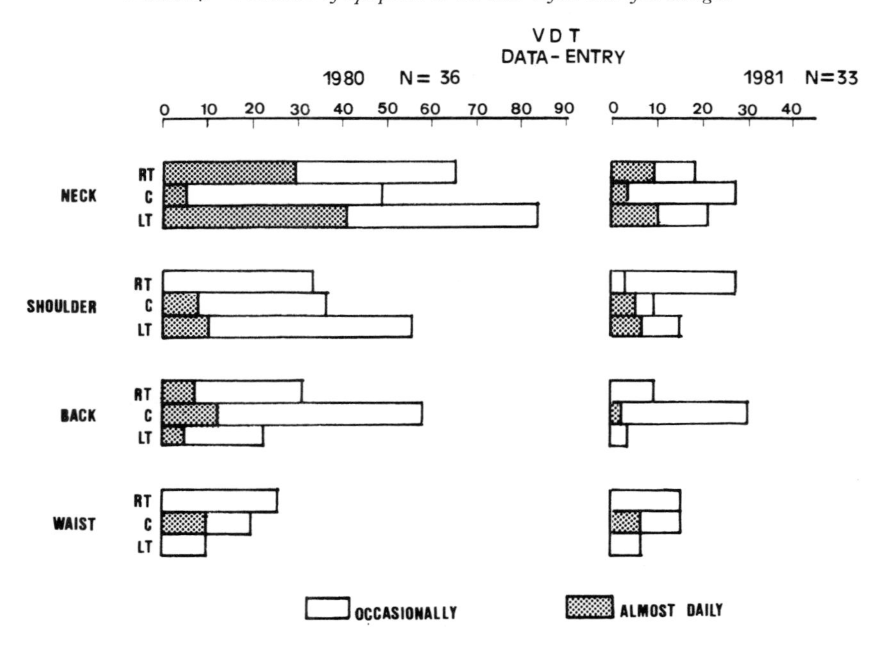

to 13 002 keystrokes/hour (in average for 12 months after the changes). A difference of about 2500 keystroke/hour. It is important to note also that besides the increase in keying speed, the error rate which was usually around 1% had declined to 0·11%.

FIGURE 5. *Prevalence of symptoms in the arms and hands before and after changes.*

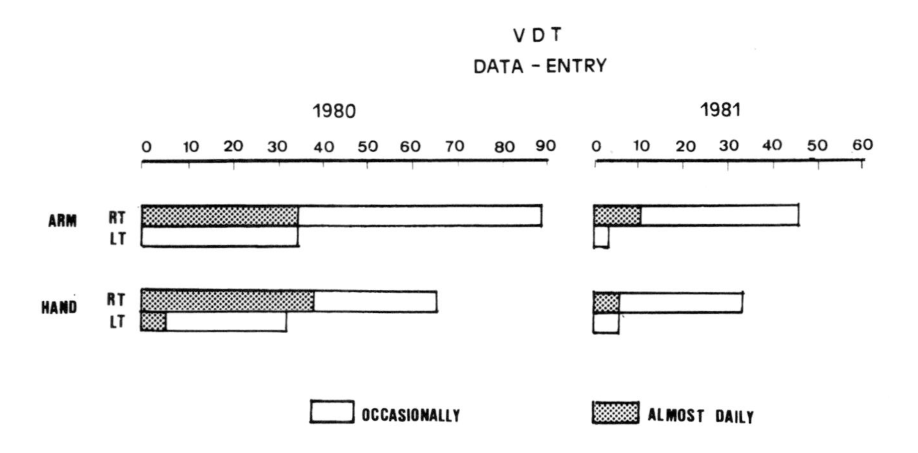

TABLE 6. *Performance and error.*

Speed and error	Original (1980–81)	New (1982)	Difference	
Key stroke/hour	9480 (7166–14 000)	13 002 (12 700–13 392)	+ 3·512	> 25%
Average error %	1·54 (0·19–4·73)(	0·11 (0·10–0·13)	– 1·43%	< 92%)

It was interesting to note that provision of more rest pauses did not cause any deteriorating effects on the productivity but speeded up the work. It may be interpreted that for this type of work which required vigilance, a well-designed work–rest regime would help to keep the operators in an alert and fresh stage. Short recess time would help to minimise the visual and muscular fatigue of the operators.

5. *Conclusion*

Introduction of new equipment or technology often leads to an increase of efficiency, but sometimes may also give rise to adjustment problems for operators (Ong *et al.* 1981). In this study, the results clearly indicated that neglecting ergonomics gave rise to adverse effects on health and performance. On the other hand, the results also showed that a better work organization and work environment led to higher performance of the operators and lesser physical fatigue.

Acknowledgement

The authors wish to thank Mr Philip Soh, Miss Alice Lim for the technical assistance and Miss Lim Poh Choo for typing the manuscripts.

References

Binaschi, S., 1980, Study of subjective symptomatology of fatigue in VDU operators. In *Ergonomic Aspects of Visual Display Terminals*, edited by E. Grandjean and E. Vigliani (London: Taylor & Francis).

Hoong, B.T. and Ong, C.N., 1980, Improving Our Working Environment: the Ergonomics Approach. National Productivity Board, Singapore.

Hultgren, G.V. and Knave, B., 1974, Discomfort glare and disturbance from light reflections in an office with CRT display terminals. *Applied Ergonomics*, 5, 2–8.

Hünting, W., 1981, Postural loads at VDT workstations. Health Hazards of VDUs, No. 2 Husat Research Group, UK.

Hünting, W., Läubli, T. and Grandjean, E., 1980, Constrained postures of VDT-operators. In *Ergonomic Aspects of Visual Display Terminals*, edited by E. Grandjean and E. Vigliani (London: Taylor & Francis).

Maeda, K., 1977, Occupational cervicobrachial disorder and its causative factors. *Journal of Human Ergology*, 6, 193–202.

National Productivity Board, 1981, *Improving the working environment in the Data Preparation Room*. A report to Singapore Airlines, Singapore.

Ong, C.N., Hoong, B.T. and Phoon, W.O., 1981, Visual and muscular fatigue in operators using visual display terminals. *Journal of Human Ergology*, 10, 48–53.

Smith, M., 1980, Job stress in video display operations. In *Ergonomic Aspects of Visual Display Terminals* (London: Taylor & Francis).

Data Entry Performance and Operator Preferences for Various Keyboard Heights

W.H. Cushman
Eastman Kodak Company, Rochester, NY 14650, USA

Abstract

Twenty experienced female operators of VDTs used for word processing entered text for 50 minutes from paper copy using a standard VDT keyboard. Subjects performed the task for 10 minutes for each of five keyboard heights: 70 cm, 74 cm, 78 cm, 82 cm, and 86 cm. Keying rate and error data plus subjective judgements pertaining to keyboard height and posture were obtained for all five conditions.

Keying rates for all five keyboard heights and for each of the five 10-minute intervals were very similar. Error rates were lowest for the 74 cm height. Subjects made fewer errors when they reported that the keyboard was 'at about the right height' and when the working posture was very comfortable. The preferred height for the keyboard home row was 5–10 cm above elbow height rather than at elbow height as recommended ergonomic texts.

1. *Introduction*

Until recently most ergonomists have recommended that typists and visual display terminal (VDT) users adjust their chairs and work places so that the home row of keys is approximately at elbow height when seated so that the forearms are parallel to the floor. In addition, an erect seated posture with upper arms vertically oriented and a 90° angle between forearms and upper arms has also been advised. According to Arndt (1983), the basis for these recommendations can be traced to several early non-keyboard studies which reported that performance, discomfort, and energy expenditure for manipulative tasks were minimized when the lower arms are parallel to the floor.

Using anthropometric data and assuming that the posture described above is the most favourable posture for keying, Cakir *et al.* (1979) have recommended that the height of the keyboard (vertical distance between the floor and home row of keys) should not exceed 75 cm and that the height of any desk used to support a keyboard should be 72 cm. These recommendations later became part of ZH 1(618 (TCA 180), one of several ergonomic standards for VDTs now in effect in the Federal Republic of Germany.

Recently, however, a growing number of field studies and laboratory studies have shown that keyboard heights are often greater than 75 cm and that many users prefer these higher heights when given adjustable work places and chairs (e.g., Brown and Schaum 1980, Stammerjohn *et al.* 1981, Grandjean *et al.* 1982, Grandjean *et al.* 1983). One study reported fewer posture-related complaints among VDT operators with keyboard heights exceeding 84 cm (Hünting *et al.* 1981).

Previous studies, however, have not examined the relationship between keyboard height and performance (i.e., keying rate and errors); neither have they attempted to determine whether user preferences and discomfort ratings are related to keying performance. The objective of the study described below is to determine these relationships.

2. *Methods*

2.1. **Subjects**

All 20 subjects were female employees of Eastman Kodak Company. Seventeen were full-time operators of VDTs used for word processing. Each had at least one year of experience as a VDT user. The remaining three subjects were frequent users of VDTs, but they also performed some non-VDT tasks during a typical work day. All 20 subjects had typing rates of at least 50 words per minute (wpm, hereafter). Most had rates above 60 wpm. Their average stature was 164·3 cm with a standard deviation of 8·0 cm.

2.2. **Materials**

The keyboard was a standard detachable VDT keyboard. The vertical distance from the base to the top of the home row of keys was 7 cm, and the slope of the keyboard was 14 degrees. The keyboard had standard-size keys (13 mm × 13 mm square keytops with rounded corners) and 19 mm centre-to-centre spacing. Key activation required a displacement of about 2·5 mm and a force of about 0·75 N. All values given above are well within the limits given in various ergonomic standards and guidelines (Helander, 1981, Alden *et al.* 1972).

Keyboard height was varied by raising and lowering the keyboard shelf of a bi-level adjustable VDT work place. The keyboard shelf could be set at any height between 73 and 79 cm. An adjustable chair with separate seat height and fore/aft/height backrest adjustments was also provided. The chair did not, however, have a backrest tilt adjustment feature. The height of the seat was continuously adjustable from 42 cm to 55 cm. An adjustable document holder was provided, but a palm rest was not.

2.3. **Procedure**

After completing the chair adjustments, the subject was shown how to adjust the workplace and document holder and asked to make the necessary adjustments to obtain the preferred workplace configuration (i.e. preferred keyboard and display heights and distances, preferred screen angle, and preferred document holder position and angle). For half the subjects the chair was initially at its lowest position and the keyboard shelf at its highest position; for the remaining subjects the chair was initially at its highest position and the keyboard shelf at its lowest position. At the end of the experiment the subject was again asked to set the keyboard at the preferred height.

During the text entry phase the subject entered text from paper copy for five 10-minute intervals with a 5-minute break after the third interval and a 5 – 10 minute warm-up period preceding the first. Subjects were asked to try for both a high keying rate and a low error rate. Error correction was optional; however, most subjects chose to correct some of their errors by immediately re-keying any word that had been initially keyed incorrectly. Thus, error data reported below are probably better estimates of undetected errors than total errors.

After the initial warm-up period and after each of the first four 10-minute trials, the experimenter raised or lowered the keyboard shelf to the designated height for the next 10-minute trial. The five keyboard heights that were selected to be used in this study were as follows: 70 cm, 74 cm, 78 cm, 82 cm, and 86 cm. Keyboard height refers to the vertical distance between the floor and the home row of the keyboard. The order of the conditions was varied from subject to subject. Sequences were obtained from four 5 × 5 Latin squares where the rows represented individual subjects and the columns represented order of treatments.

After each 10-minute trial the experimenter asked the subject to complete a brief questionnaire concerning the keyboard height and working posture. The rating scale for keyboard height had seven categories: from much too low (1) to much too high (7).

The rating scale for working posture had six categories: from very comfortable (1) to very uncomfortable (6).

3. *Results*

Figure 1 shows the relationship between keyboard height and keying performance. (See section 2.3 for a definition of keyboard height.) Keying rates were similar for all keyboard heights (Figure 1, upper curve). Differences were not significant. Apparently there is no single keyboard height that will maximize the mean keying rate for the group as a whole. Differences in mean errors for the five keyboard heights (lower curve) were also not significant ($p < 0{\cdot}05$, F-test, ANOV, 5 × 5 Latin square with 4 replications). However, the data suggest that a minimum may occur at about 74 cm.

Figure 2 (upper curve) shows that keying rates for each of the five 10-minute intervals were very similar. In other words, a typical subject was able to maintain her initial keying rate for 50 minutes. There were exceptions, of course. At one extreme, one subject's keying

FIGURE 1. *Keying performance as a function of keyboard height. Note that the height of the keyboard, when placed on a standard desk, is approximately 82 cm. The vertical line segment passing through each mean represents a range of two standard errors.*

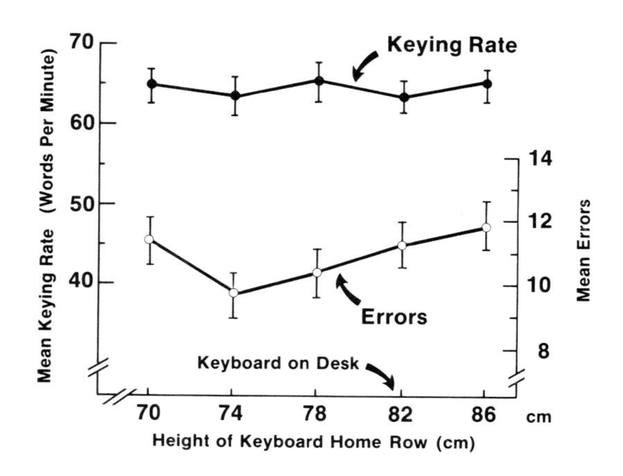

FIGURE 2. *Changes in keying performance as a function of time. The 5-minute break is represented by the discontinuity that occurs between 30 min. and 40 min. The vertical line segment passing through each mean represents a range of two standard errors.*

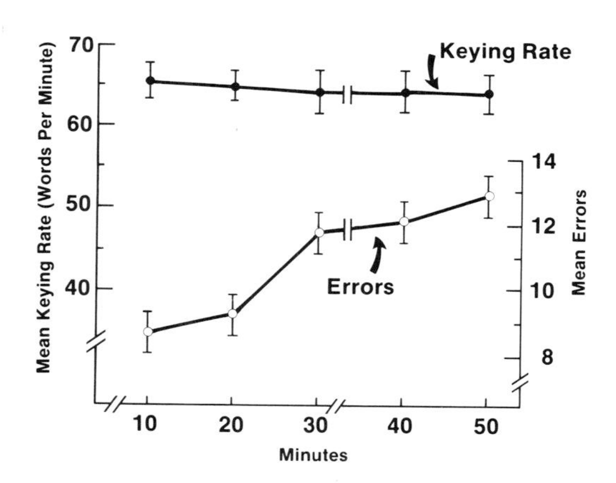

rate declined 38%. At the other, another subject's rate increased 20%. Note that keyboard height was changed at 10-minute intervals and that a 5-minute break was given after 30 minutes (indicated by the discontinuity between 30 minutes and 40 minutes). Keying errors, on the other hand, increased significantly (Figure 2, lower curve). Mean errors for 30 min., 40 min., and 50 min. were significantly greater than mean errors for 10 min. and 20 min. ($p < 0{\cdot}05$, Duncan Multiple Range Test). Although the break did not reduce the error rate for subsequent 10-minute trials, it appears to have prevented any further significant increase.

Figures 3 and 4 are scatterplot diagrams showing that the difference between the operator's seated elbow height and the height of the keyboard is of little value in predicting either keying rate or keying errors. The arrows indicate the average difference between seated elbow height and preferred keyboard height. The dashed lines indicate the outer boundaries of the envelopes. The often cited ergonomic recommendation that the keyboard home row should be approximately at elbow height so that the forearms are parallel to the floor appears to have little support from these data. High keying rates and low error rates were observed for conditions in which the keyboard was as much as 15–20 cm above the elbow.

FIGURE 3. *Scatterplot diagram with vertical distance between the elbow and keyboard home row as the independent variable and keying rate as the dependent variable. The arrow indicates the mean difference between the preferred keyboard height (home row) and seated elbow height for the 20 subjects.*

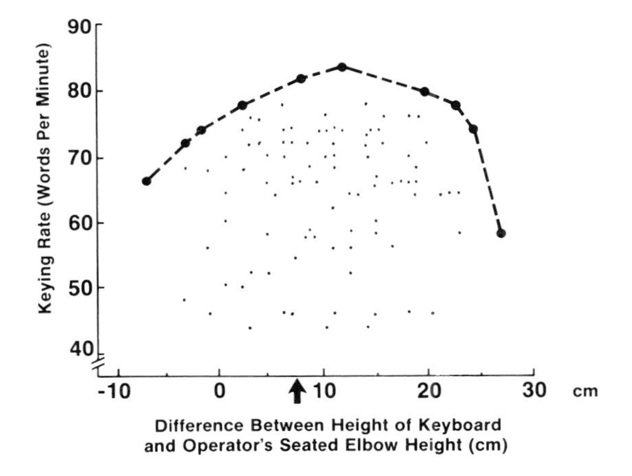

FIGURE 4. *Scatterplot diagram with vertical distance between the elbow and keyboard home row as the independent variable and keying errors as the dependent variable. The arrow indicates the mean difference between the preferred keyboard height (home row) and seated elbow height for the 20 subjects.*

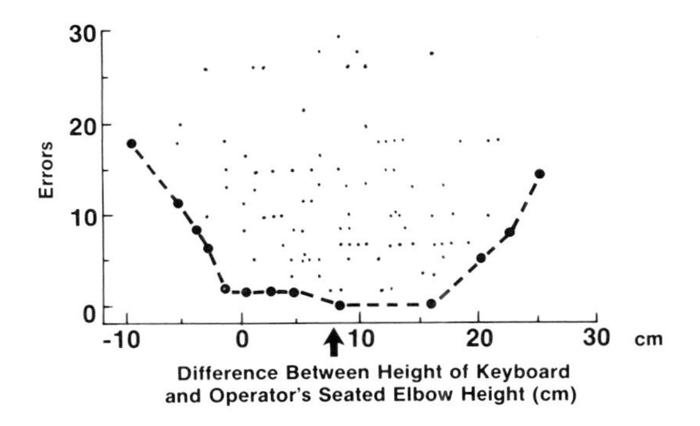

The mean preferred keyboard height (based on all measurements obtained before and after the text entry phase) was 75·7 cm with a standard deviation of 2·62 cm. The mean preferred chair seat height was 49·8 cm with a standard deviation of 2·24 cm. However, those subjects that began the chair adjustment by lowering the seat from its highest position had a significantly higher mean preferred seat height than the group that began by raising the chair from its lowest position ($t = 6{\cdot}56$, $p < 0{\cdot}01$). The means for the two groups were 51·8 cm and 48·0 cm, respectively, for a difference of 3·8 cm. This difference cannot be explained by differences in the stature of the two groups. The group that had the lower preferred seat height was significantly taller ($t = 2{\cdot}26$, $p < 0{\cdot}05$). The most likely explanation for this finding is that the subjects lowered or raised their chairs until they were just within the comfortable range of heights or 'comfort zone'. Thus, subjects that began with the chair in the high position set their chairs near the highest point within the comfort zone, while those that began with the chair in the low position set their chairs near the lowest point.

Data relating keyboard height, the subjective ratings, and errors are given in Figures 5–8. The upper curve in Figure 5 shows the relationship between keyboard height and mean keyboard height rating. Category 4 on the keyboard height scale (dashed horizontal line) corresponds to 'at about the right height'. All the means are significantly different

FIGURE 5. *The relationship between keyboard height and mean keyboard height rating (upper curve, left vertical scale). The vertical line segment passing through each mean represents a range of two standard errors. The horizontal dotted line represents a rating of 4·0 ('at about the right height'). The lower curve was derived by calculating the absolute difference between each mean and 4·0.*

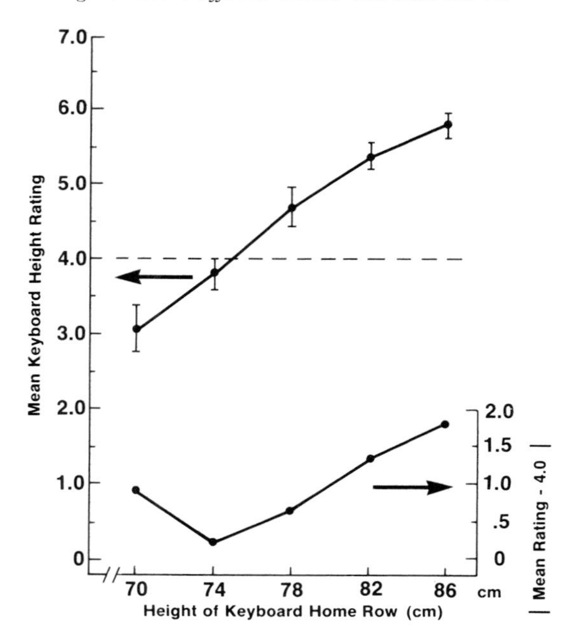

FIGURE 6. *Mean posture discomfort rating as a function of keyboard height. The vertical line segment passing through each mean represents a range of two standard errors.*

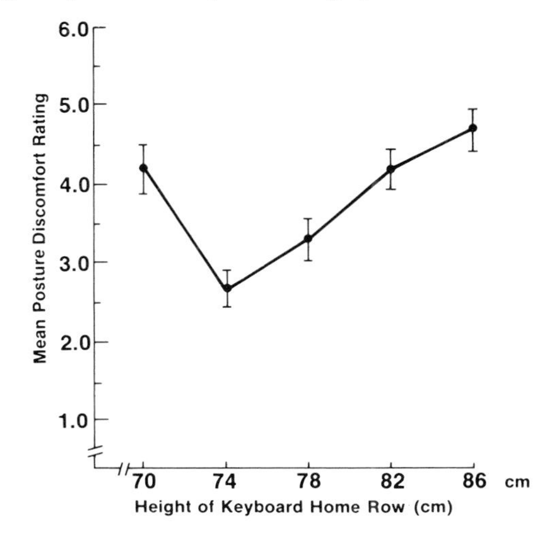

FIGURE 7. *The relationship between mean errors and keyboard height rating. The vertical line segment passing through each mean represents a range of two standard errors. A value of 4·0 corresponds to 'at about the right height'.*

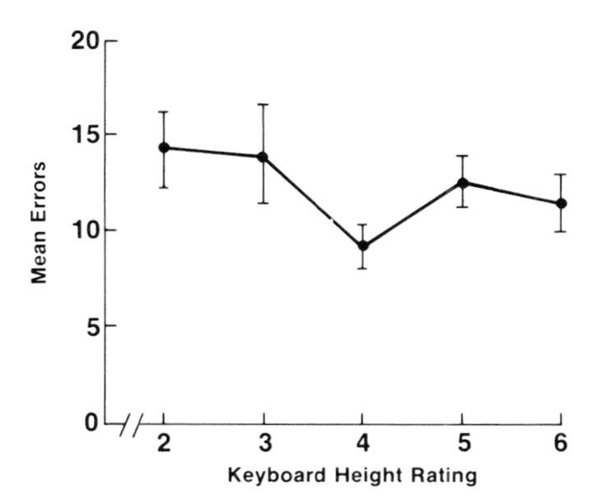

(Duncan Multiple Range Test, $p < 0{\cdot}05$). Apparently the subjects could discriminate among the five keyboard heights without difficulty. The lower curve in Figure 5 was derived by calculating the absolute difference between each mean and 4·0. Note the similarity between this curve and the mean error curve in Figure 1.

Figure 6 shows how that mean posture discomfort ratings varied with keyboard height. This relationship is also very similar to the relationship between keyboard height and mean

FIGURE 8. *The relationship between mean errors and posture discomfort rating. The vertical line segment passing through each mean represents a range of two standard errors. (3) means neither comfortable nor uncomfortable.*

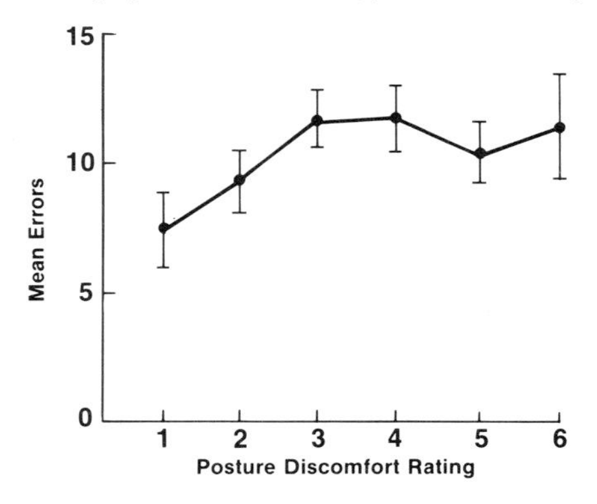

errors in Figure 1. The mean discomfort ratings for keyboard heights of 74 cm and 78 cm are significantly lower than the others. In other words, the operators reported that their working postures were more comfortable when the keyboard height was either 74 cm or 78 cm.

Figure 7 summarizes the relationship between keyboard height rating and mean errors. Subjects made the fewest errors when they reported that the keyboard was 'at about the right height' (category 4). Similar data for the posture discomfort ratings are given in Figure 8. Again, subjects made fewer errors when they reported that they were comfortable. The differences appear to be reliable; however, no test of statistical significance was performed because there were considerably fewer observations for categories near the ends of the scales and because there was frequently more than one observation per subject for the categories near the middle of the scales (i.e. categories 3, 4, and 5).

A workplace configuration that received a keyboard height rating of 4 (keyboard 'at about the right height') and a posture discomfort rating of 1, 2, or 3 ('very comfortable', 'comfortable' or 'neither comfortable nor uncomfortable') should be satisfactory for prolonged VDT use by the individual making the ratings. Similarly, a workplace that received a keyboard height rating of 3, 4, or 5 and a posture discomfort rating of 1, 2, 3, or 4 should be satisfactory for occasional VDT use.

Figure 9 shows the percentage of subjects that gave ratings in each of the four category groups described in the preceding paragraph as a function of keyboard height. A keyboard height of 74 cm appears to be satisfactory for more subjects than any other single height. Nevertheless, only 60% of our experienced female VDT operators reported that the 74 cm keyboard height was near optimum (KB 4 curve, open circles), while only 70% indicated that the working posture for that height was satisfactory (P 1–3 curve, open triangles). The corresponding percentages for the 78 cm keyboard height—the second best height—were much smaller: 30% and 45%, respectively. Note that the 82 cm keyboard height (height of the keyboard when placed on a standard desk) was rated 'at about the right height' by only 5% of the subjects (KB 4 curve).

FIGURE 9. *The percentage of subjects that rated each keyboard as 'at about the right height' (KB 4 curve) or 'a little too low' or 'a little too high' (KB 3–5 curve). Also, the percentage of subjects that rated the working posture for each keyboard height as 'neither comfortable nor uncomfortable' or better (P 1–3 curve) and 'slightly uncomfortable' or better (P 1–4 curve).*

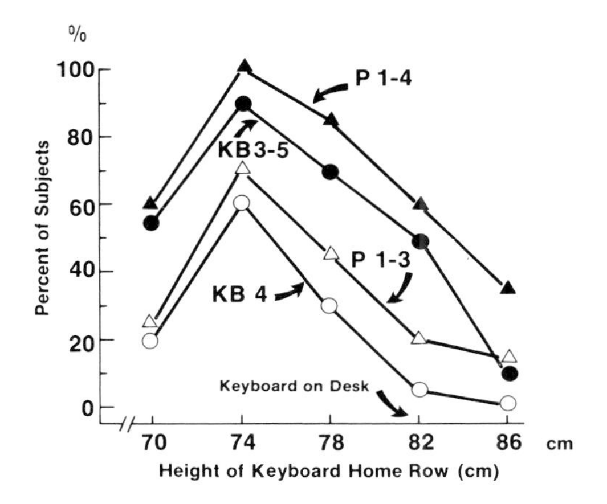

4. *Discussion*

The 76 cm mean preferred keyboard height for female VDT operators in this study is very similar to the 77 cm mean preferred keyboard height for females in the Grandjean *et al.* (1982) laboratory study. Two other laboratory studies involving keyboard height preferences reported slightly lower mean preferred keyboard heights of 74 cm (Brown and Schaum 1980) and 70 cm (Miller and Suther 1981), however. This is somewhat surprising because about 60% of the subjects in both studies were males. In contrast, Grandjean *et al.* (1983) have reported a mean preferred keyboard height of 80 cm for a group of mostly female VDT users in an actual working environment. While the origin of the differences among these findings is not clear, many factors could be responsible including experimental methods, chair design (e.g., whether the chair has a high backrest with a tilt adjustment), configuration of each subject's previous VDT workplace (see Table 7 in Grandjean *et al.* 1983), keyboard inclination (Arndt 1983), and possibly the presence or absence of palm rests and document holders (Hünting *et al.* 1981, Grandjean *et al.* 1982).

The maximum keyboard height of 75 cm recommended in *The VDT Manual* (Cakir *et al.* 1979) and ZH 1/618 (TCA 1980) is lower than that preferred by many VDT operators. In our study, over 50% of the subjects preferred a keyboard height that was greater than 75 cm. The reason for the difference between recommended and preferred heights is that VDT users prefer to have the keyboard home row about 5–10 cm higher than the elbow (see position of arrow in Figure 3 or Figure 4). The authors of *The VDT Manual*, however, have assumed that VDT uses will be most comfortable and most productive when the keyboard home row is approximately at elbow height so that the lower arms are parallel to the floor. They have used this assumption—which probably is not valid (e.g., see Arndt 1983,

Grandjean *et al.* 1982, and Figures 3 and 4 above)—along with anthropometric data, to derive their recommendation for a maximum keyboard height of 75 cm. In view of the above and the additional finding reported here that VDT users may make fewer errors when the keyboard is at or near the preferred height (Figure 7), the validity of the recommendation for a maximum keyboard height of 75 cm is questionable.

Acknowledgements

I wish to thank Bob Crawford, Margaret Krier, Carol McCreary, John Stevens and Jackie Wernle for technical assistance. I also wish to thank Harry Davis for granting permission to conduct the study.

References

Alden, D.G., Daniels, R.W. and Kanarick, A.F., 1972, Keyboard design and operation: a review of the major issues. *Human Factors*, 14, 275–293.

Arndt, R., 1983, Working posture and musculoskeletal problems of video display terminal operators —review and reappraisal. *American Industrial Hygiene Association Journal*, 44, 437–446.

Brown, C.R. and Schaum, D.L., 1980, User-adjusted VDU parameters. In *Ergonomic Aspects of Visual Display Terminals*, edited by E. Grandjean and E. Vigliani (London: Taylor & Francis) pp.195–200.

Cakir, A., Hart, D.J. and Stewart, T.F.M., 1979, *The VDT Manual* (Darmstadt: Inca-Fiej Research Association).

Grandjean, E., Nishiyama, W., Hünting, W. and Piderman, M., 1982, A laboratory study on preferred and imposed settings of a VDT workstation. *Behaviour and Information Technology*, 1, 289–304.

Grandjean, E., Hünting, W. and Pidermann, M., 1983, VDT workstation design: preferred settings and their effects. *Human Factors*, 25, 161–175.

Helander, M., 1981, A critical review of human factors standards for visual display units. Paper presented at the AIHA seminar on VDUs, Arlington Virginia, 13–14 April.

Hünting, W., Läubli, T. and Grandjean, E., 1981, Postural and visual loads at VDT workplaces. I. Constrained postures. *Ergonomics*, 24, 917–931.

Miller, I. and Suther, T.W., 1981, Preferred height and angle settings of CRT and keyboard for a display station input task. *Proceedings of the Human Factors Society's 25th Annual Meeting*, pp.492–496.

Stammerjohn, L.W., Smith, M.J. and Cohen, B.G.F., 1981, Evaluation of work station design factors in VDT operations. *Human Factors*, 23, 401–412.

Trade Cooperative Association (TCA), 1980, Safety Regulations for Display Workplaces in the Office Sector, No. ZH 1/618, 10/80 edn (Hamburg: TCA).

Index